The Atomic World Spooky? It Ain't Necessarily So!

Theo van Holten

The Atomic World Spooky? It Ain't Necessarily So!

Emergent Quantum Mechanics, How the Classical Laws of Nature Can Conspire to Cause Quantum-Like Behaviour

ATLANTIS PRESS

Theo van Holten
Hilversum
The Netherlands

ISBN 978-94-6239-233-5 ISBN 978-94-6239-234-2 (eBook)
DOI 10.2991/978-94-6239-234-2

Library of Congress Control Number: 2016956171

Printed on acid-free paper

For Max, Daan and Hugo

It ain't necessarily so
It ain't necessarily so
The t'ngs dat yo' li'ble
To read in de Bible
It ain't necessarily so.

Wadoo, zim bam, boodle-oo
Hoodle ah da wa da
Scatty wah!
Oh yeah!

(Sportin'Life in Porgy and Bess)

Flaptext

The world of atoms is strange. It is a world where—to paraphrase Einstein—"der Alten würfelt", where particles at the same time can be in different places, where particles communicate faster than light by telepathic ("spooky") contact, and where "Schrödinger's cat" can be dead and alive at the same time. Quantum mechanics, the theory of this world, is magic.

Or is it? Recent discoveries showed that quantum-like behaviour is not solely reserved to atomic particles. Tiny droplets of oil bouncing round on a fluid surface can also mimic the world of quantum mechanics. Macroscopic objects, described by "normal" laws of nature, behave in a quantum-like way! For the layman—for whom the main part of this book is written—this is good news. If the everyday laws of nature can conspire to show up quantum-like phenomena, there is hope to form mental pictures how the atomic world works.

This present book takes this discovery one step further. If electrons—by way of a tentative working hypothesis—are modelled as vibrating droplets instead of the usually assumed point objects, and if the *classical* laws of nature are applied, then *exactly* the same behaviour as in quantum theory is found, *quantitatively correct*! A reader, armed with just some basic knowledge of school physics, will thus be able to see through the tricks of the great illusionist that nature is. No specialised quantum theory nor magic is involved in the visualisations that may be obtained.

This book is almost formula-free and explains everything by using many sketches and diagrams. The mathematical derivations underlying the main text are kept separate in an appendix. The mathematics may serve to convince the experts but can be ignored by the general reader.

The author, a retired professor of Flight Mechanics and Propulsion at the Delft University of Technology, by accident (serendipity?) stumbled on these exciting conclusions. He chose to publish his findings in this mixed popular and scientific form, because he found that interested laymen more often than professional physicists feel the need to form visualisations of quantum phenomena and thereby to better "understand" them. If professional physicists are interested, they are invited to read the—peer-reviewed—formula appendix.

Theo van Holten

Foreword

Every physicist knows that physics is broken. A hundred years on from Einstein's theory of general relativity in 1915, you just need to open any school textbook to learn that nobody has been able to make his theory compatible with the other great theories of physics and quantum mechanics.

A small band of rebels believe that the answer lies not in retreating to ever more complex mathematics or ever more complicated models—such as assuming multiple universes which can never be actually observed—but in revisiting the foundations.

Progress has been made recently in emergent quantum mechanics.

Droplets bouncing on the surface of a vibrating tray of oil exhibit double-slit diffraction, quantised levels and tunnelling. They show that the nonlocal behaviour thought to be unique to quantum mechanics arises in completely classical systems.

More progress has been made in the field of analogue gravity, which finds some of the key equations of general relativity emerging from the so-called acoustic metric of a compressible fluid.

Experimentalists create analogues of black holes in extremely cold quantum liquids.

Nobody has a full answer but there is growing confidence that we might at least be starting to ask the right questions.

In this book, Theo van Holten tells of his work as one of a band of experts working on these exciting new theories. He offers a further step towards one of the most important goals of science: understanding the deepest physical nature of the world in which we live.

Prof. Ross Anderson FRS
Dr. Robert Brady
University of Cambridge Computer Laboratory
Cambridge, UK

Preface

To begin with a warning: the contents of this book may be controversial.

The readers the author had in mind when writing this book are interested laymen, typically the kind of reader who searches bookshops for the latest popular-scientific books on developments in cosmology, on recently found fundamental particles, or on the ever more magical findings of quantum physics. These readers presumably have some background of classical school physics (although most of it may have been forgotten). It is the kind of reader who does not like to be bothered with formulae or is even allergic to them, but who has the interest and tenacity to read sentences twice if necessary. But complete novices in the matters of the atomic world should be warned: the stories told in this book are *not* the same as usually found in books about quantum phenomena. This book does *not* give the conventional explanations. In order to read the usual stories, it is better to start in one of the many other popular-scientific books.

What then is this book about? This book certainly does not pretend to contain a new theory of quantum mechanics, nor does it have the intention. Quantum theory in its present form is an almost perfect tool to calculate the behaviour of elementary particles. But the theory is "strange", it is not something that intuitively can be understood. What this book tries to add are visualisations or mental pictures, closer to the intuition, because they are based on classical physics. However, the mental pictures in this book are not just half-baked analogies or metaphores, they are solidly founded on a large body of mathematical theory (for the diehards: the theory can be found in the appendix). This aspect makes this book different from other popular-scientific books.

Readers who just want to get a "quick" (60 odd pages long) impression are advised to read the first chapter and may stop there. This would be sufficient to get an acquaintance with the main ideas. If you still have some patience left, you might want to browse through Chap. 2 and the later Chaps. 14 and 15 too. Chapter 2 gives a more complete list of strange quantum phenomena than Chap. 1. Chapter 14 then goes through the entire list of quantum oddities again, together with the interpretations suggested by the theory developed in this book. They are interpretations in terms of everyday physics, intended to free quantum mechanics from its magical image and weirdness, and making everything more intuitively understandable. Finally, the more speculative Chap. 15 is about the really big magic: telepathic ("spooky") contact between particles, particles that at the same time can be in different places, "Schrödinger's cat" and more.

Even more efficient: people who do not need a refresher about the contributions to quantum theory by Planck, Einstein, Bohr, De Broglie and Schrödinger can skip the first part of Chap. 1 until Sect. 1.9. It will save them some 25 pages of reading, because the new stuff begins in Sect. 1.9. The remainder of Chap. 1 gives a global outline of the ideas developed by the author. It is a global outline, many more details and the dotting of the i's will be found in later chapters.

It is not strictly forbidden that professional physicists also read this book, but they are less likely to feel a need for an understanding of quantum phenomena in terms of classical laws of nature, since they are used to the strange aspects of quantum mechanics and probably do not find them "strange" anymore. Many professionals will as a first reaction even deny that it is possible at all to "explain" quantum theory in terms of classical physics. They will probably consider this book as heresy. However, if they want to assess whether the stories told in this book are utter nonsense or whether they contain more truth than just fairy tales, they are referred to the—peer-reviewed—mathematical appendix.

Let it serve as a warning for the general reader: this book will be controversial. But form your own opinion. Hopefully you find it enjoyable reading anyway.

Hilversum, The Netherlands Theo van Holten

Acknowledgments

During the many years of theory development and writing the results in the form of readable formula-free text, there was one person who all this time encouraged me to continue the work. This was my friend and ex-colleague Herman Schöyer, who painstakingly went through all the formulae (now collected in the appendix) and who was such an attentive reader of the main text that it has several times caused—useful—setbacks of many months, when rethinking was required. He was—sometimes in contrast to myself—convinced that the general approach of the research was right, and he thereby was to a large extent instrumental in the completion of this present book. Alas, he died by heart failure, too early to see the result in print. I dedicate this book to him.

Also many thanks go to a number of proof readers of (parts) of the main text. They came from my own circle of family relations and other friends and acquaintances. The intention was, to collect the opinions of potential readers, widely ranging from people without any background in physics to educated laymen who are well read in the popular-scientific literature, as well as professionals not easily frightened by mathematics. I may mention in particular brothers and sisters of mine (Jan van Holten, Saskia van Holten, Jan van Wijk), my wife Pia, my friend Kees Kleinjan and my former Ph.D. student Dr. Jessica Holierhoek. Their comments were very valuable and resulted in many, sometimes extensive, adaptations of the original concepts.

Finally, thanks are due to the editor of this book Mark Eligh, who brought me in contact with some real experts in the field of emergent quantum physics, Ross Anderson and Robert Brady (both working at the University of Cambridge, England). When starting the research, I was not aware of the beautiful experiments by Couder et al. on bouncing droplets mimicking quantum phenomena. In the original manuscript only brief mention was made of these experiments, after Herman Schöyer had brought an overview article to my attention. The comments by Anderson and Brady resulted in the addition of two new chapters discussing this subject in more detail.

To all these people I express my sincere thanks. Nonetheless, I am aware that the contents of this book will probably be controversial and that I myself am fully responsible for the text, including any remaining errors in the theory or misinterpretations thereof. This book must be considered as a first, tentative attempt to form mental pictures about quantum phenomena, and the—undoubtedly many—criticisms should be solely directed to the author.

Contents

Chapter 1
Introduction and Outline

1.1 Introduction

The stimulus to write this book was irritation. Do you recognise it? When I for the first time was reading about quantum mechanics, about the science-fiction like, almost magical behaviour of particles on an atomic scale, I was thrilled and read on in a hurry to learn how all this could be explained. And then came the grave disappointment, bordering on annoyance, when I was told that physicists cannot offer any explanation.

None of these weird phenomena can be understood on the basis of the physical laws we are used to and have been familiar with since our school days. I was told to forget all the mental pictures I had about the functioning of nature, because they are all pictures applicable to the human, macroscopic scale. The physics on an atomic scale is entirely different, and any attempt to get some intuitive feeling for it would unavoidably be based on everyday physics, and would fail. In fact, the behaviour of elementary particles often seems to directly violate the physical laws that we are used to.

Recently, breaches have been made in this dogma, since it was experimentally found that tiny droplets of oil bouncing along the surface of liquids can show quantum-like behaviour as well. A separate textbox tells you more about it, shows some pictures of the experiments, and outlines an analysis why this is so. The textbox is found at the end of the present Chap. 1, because the explanation of the phenomenon requires some familiarity with the theories by Maxwell, Planck, De Broglie, etc. "Have we been interpreting quantum mechanics wrong this whole time?" is the provocative title of a review article on the bouncing oil droplets (Wolchover in Quanta Magazine, 06.30.14). Many physicists, however, do not believe that this really could lead to a reconciliation of quantum mechanics and "normal" physics.

The story in the textbooks went on, and the mild annoyance became downright irritation. I was subsequently told that the only way to get some real insight into the

T. van Holten, *The Atomic World Spooky? It Ain't Necessarily So!*
DOI 10.2991/978-94-6239-234-2_1

magical world of quantum mechanics is to use the rather obscure mathematical language "spoken" by professional physicists. Extremely irritating! After all, the whole world is built up from elementary particles, and in effect I was told that the understanding of the world should be left to the experts who have been trained in it during long years of study.

This book tells a different story. The story here narrated initially started with an investigation of the "no man's land" between the scale governed by conventional laws of nature, and on the other hand the realm of quantum mechanics. There is a gradual transition between these two territories, so that it can be hoped that glimpses of quantum-like behaviour may be seen in this no-man's land, even if we still describe the physics by the laws of the macroscopic world. If this would appear to be true, then one would get a clue how even the classical laws of nature can conspire in such a way that quantum-like behaviour is described by them. At the very least, it could provide a few mental pictures, pictures based on all the laws we know from everyday life, and without any magic at all.

In fact, this hope was fulfilled. If one takes in a theoretical model "blobs of electrical charges" and does not treat them as rigid "marbles" but instead as deformable "droplets", then their behaviour is strange, and it does remind one of the behaviour of elementary particles in quantum theory. This is perhaps not quite unexpected, since we now have something like the above mentioned oil droplets, with the difference that in our case electromagnetic effects are involved instead of a liquid surface influencing the motion. Electromagnetism introduces its own problems of course, for instance: how to keep together a "blob of charges" which has the tendency to fly apart under the action of internal repulsion forces? Obviously, to make a consistent and credible model required something more than just assuming a "blob" without a further specification how it could be kept in existence. Nevertheless, a satisfactory model proved to be feasible (in the opinion of the author that is; if you are not convinced yet, form your own opinion by reading on).

What was discovered surpassed the expectations, certainly mine. By further extrapolating the model and the results found in the "no man's land" right down to atomic scale, not only interesting mental pictures resulted. Even *quantitatively* correct quantum behaviour was found! The—admittedly—unorthodox model of electrons so obtained yielded the correct quantisation of energy, the same as found in usual quantum mechanics. And most amazing: a fundamental constant of nature, "Planck's constant" governing the realm of atomic particles, came out expressed in terms of classical constants, with the correct value! At the very least, these findings may provide us with "pictures in the back of one's mind" to visualise in a more intuitive way the strange behaviour of the fundamental particles of nature ("intuitive": because of the use of everyday physics).

Ironically—remembering the cause of the author's irritation—the book is based on a piece of mathematical research by the author. Before the reader exclaims "not again!", the author must immediately add that this particular piece of mathematical research had the explicit aim to construct *mental pictures* about quantum mechanical phenomena that can be understood *without* mathematics. So: mathematics used to make mathematics superfluous.

In consequence, the book is divided in two parts. The first and main part is almost entirely free from formulae, except for a few very simple ones which in the opinion of the author can be understood by anyone and make things easier to explain than masses of words. The second part (to be found in the appendix) is *non*-essential reading for most readers; it is full of mathematical formulae, and for readers who only want to have a global understanding, the mathematics is not necessary and certainly not very entertaining.

The real understanding will not be found in the formulae, and for 99 % of the readers (optimistically assuming that there will be at least 100 readers) the appendix will therefore just serve the purpose to convey the comfortable feeling that in this book not all can be nonsense because so many—never to be read—formulae seem to underpin it.

The mathematics does allow, however, to convince the doubting Thomas. The remaining 1 % of the readers that perhaps likes mathematics—which is not necessarily a despicable hobby and can for some be a very nice way to kill the time—is invited to carefully study the mathematical appendix. They will—I hope—be able to confirm that the mental pictures in the first part of the book are not pure fairy tales but reflect a sound reasoning. I do hope that this second category readers will help me to correct any errors.

1.2 Why "Magic and Anti-intuitive"?

I have several times used words such as "magic", "weird", etc. You may wonder what it exactly is that makes quantum mechanics so strange.

It all started with publications in 1900 and 1905 by Planck and Einstein respectively. Leaving aside early discussions about the nature of "light", until the discoveries by Planck and Einstein, physicists had looked upon light as a wave phenomenon. This had sound reasons, based on experiments as well as on theory. Building on mostly experimental work by Faraday, Maxwell had finalised the theory of electricity and magnetism and published in 1873 a book containing what we now call "Maxwell's laws of electromagnetism". One of the conclusions in this book was that oscillating electrical charges are surrounded by a field of waves, so-called electromagnetic waves. Nowadays everybody is familiar with radio waves, one of the manifestations of the electromagnetic waves predicted by Maxwell's theory. The predicted radio waves were indeed discovered somewhat later by Hertz in 1886. What Maxwell also concluded was, that light is just another form of this type of waves. Light is only special because it has a frequency for which the human eye is sensitive.

And now, in the beginning of the 20th century, Planck and Einstein came to a different conclusion. Planck was rather annoyed by his own theory, but did not know how to avoid it. The work of Planck and Einstein, taken in combination, entailed that a ray of light sometimes indeed behaves as a bundle of waves (the "electromagnetic waves" described by the classical laws of Maxwell), but in other cases as a shot of particles ("photons" or light particles). It depends on the type of experiment whether the view as waves or the view as particle is most useful. In "reality" (whatever we mean by that) light is apparently both wave and particle, how odd and incomprehensible this may seem. By the way, an implication of "particles of light" is that the energy that is present in light radiation also comes in the form of a stream of separate "packets of energy", each packet bound up with a photon.

This was the prelude to the further development of quantum theory. One had encountered the fact that on the smallest scales of physics things can become "granular" or "lumpy" instead of smooth and continuous. Of course, it was already known that something similar is the case with matter: the smallest lumps of matter that have the recognisable properties of chemical elements are atoms. During the course of the 19th century this idea had been accepted as an established fact. Another, more recent discovery that in 1897 had come out of the laboratory of Thomson, was the "electron". It was found to be a particle carrying negative electrical charge, and it was thought to be one of the components of atoms. Electrical charge thus appeared to be lumpy too: no smaller amounts of charge than the charge carried by electrons have ever been observed. This smallest amount is called the *unit charge*.

For a time the idea existed that the negatively charged electrons inside an atom were embedded in a positively charged blob of material, like the raisins in a pudding. Hence the name "plum pudding model" of an atom. This idea had to be thrown overboard when Rutherford in 1911 came to the conclusion that all the positive charge inside an atom is concentrated in a small core. This core or "nucleus" is very much smaller than the atom itself. It has been compared to a fly in a cathedral, but a peculiar fly since it has the weight of practically the whole cathedral. The positive charges in atoms that have to balance the negative charges of the electrons, are apparently also concentrated in the form of small "grains of charge".

In hindsight, the time was ripe to develop a more comprehensive theory about this apparent lumpiness of matter, light, energy, charge, etc.: quantum theory.

One of the first things quantum theory concerned itself with, was the structure of atoms.

A revolutionary step was taken by Niels Bohr in 1913 when he proposed the model of atoms sketched in Fig. 1.1. Nowadays, some purist physicists would object to a drawing of an atom like this for reasons I will touch upon later. I use Euratom as an excuse to reproduce the picture here, and anyway Bohr did not know any better at the time.

What the picture shows is the idea that an atom is built similar to the solar system, with the atomic nucleus in the place of the sun, and electrons orbiting around it like planets. In the case of the solar system the planets are attracted to the sun by gravity, whereas in the atom electrical attraction forces between the positively charged nucleus and the negatively charged electrons do the job.

So far, so good. Why was this such a bold, revolutionary, thought of Bohr? At first sight it does seem an idea all of us could have had. Actually, the planetary

Fig. 1.1 An early logo of Euratom

model of an atom had been proposed before Bohr. The point is, that Bohr tried to explain with this model some characteristics of atoms that had been observed experimentally. He investigated the light emitted by the atoms of heated gases, such as the glowing gas in a candle, or the white hot gases on the sun. The light coming from such gases can be split in separate colours (the so-called light spectrum), like the sun's light is unravelled in the rainbow. It was known that every kind of atom has its own characteristic rainbow "signature". One of the saga's of physics is, that the element Helium was discovered by studying the solar spectrum, before the element was ever found on earth. Bohr wanted to find out why these spectra are so characteristic for each element, like a sort of finger prints. For this purpose he was forced to add to his planetary model a few things that would be alien to the real planetary system, or else he could not make any progress. His additions were truly courageous, and not many other physicists would have dared to propose them.

Have a look at another, well-known astronomical picture, the rings around the planet Saturn (Fig. 1.2).

The rings consist of smaller and larger rocks, orbiting Saturn just like the planets orbit the sun. What is clear is, that these orbiting pieces of rock may have any arbitrary distance to Saturn. Newton's laws, governing their motion, do not form any restriction to the radius of their orbits. It just depends on the energy these rocks happen to possess. The rocks further away from the planet have more energy than rocks closer to the surface of Saturn, and therefore they fly higher. From the almost continuous spread of the orbits in the rings it is clear that any arbitrary value of the energy can be present.

Fig. 1.2 Saturn's rings

This continuous spread of energy was something Bohr could not have in his atom model. In order to explain the measured light spectra, he needed something else. He therefore postulated that electrons cannot have any arbitrary energy, and that they cannot circle the nucleus at any distance they like. Only certain orbital radii are possible and nothing else is permitted, an electron cannot circle at a distance in between any of the allowed radii. This postulate is called the "quantisation of energy" of the electrons in an atom. If such a harsh rule would apply to the rocks flying around Saturn, the beautiful, broad ring could not exist. Why electrons within an atom would have to obey this extra rule thought up by Bohr, was not clear to Bohr himself, nor to anybody else. And what Bohr essentially said was, that Newton's laws had to be forgotten in the case of tiny objects like atoms. Quite a step! Personally, I would never have dared to publish an idea like that, out of fear to be interned in an asylum. It illustrates what a special kind of person Bohr was, and fortunately for his reputation: he proved to be right!

But there was more. Bohr not only shoved Newton aside but—as if it was not enough heresy already—also Maxwell. Just as fundamental as Newton's laws are for the dynamics of bodies, are Maxwell's laws for electrical and magnetic phenomena. Maxwell's laws and Newton's laws together almost form the whole of classical physics, or rather: the whole fundament. This complete fundament was damaged by Bohr, because he also had something to say about the applicability of Maxwell's classical electrodynamic laws in the realm of the atomic world.

To understand what is the issue, consider again an example taken from astronomy. Artificial earth satellites in a low orbit cannot escape entirely the influence of the atmosphere. However small this influence may be, they meet molecules of air from time to time, which causes some air resistance. Consequently, they gradually lose energy, even though it may take a long time before this effect becomes noticeable. Lower energy means that they will descend to a lower orbit, where the influence of the atmosphere is somewhat larger. The end result will be that these satellites spiral to the earth, to find their death by burning in the more dense layers of air.

Electrons, according to electrodynamic theory based on Maxwell's laws, would await a similar fate. Naturally, there is no air resistance in this case, but there is the so-called "radiation resistance". Every electron which experiences an acceleration (or deceleration) will emit radio waves. This is a well-known fact, and nearly everybody nowadays has a demonstration of this phenomenon in his/her pocket or handbag. If we pump masses of electrons up and down or back and forth in a copper wire, this device will cause electromagnetic radiation (radio waves), and the waves can bring in motion other electrons in another copper wire. We are all used to it and call it an antenna, a component of every mobile telephone. Now, these radio waves represent energy, so that the electrons in the antenna would lose energy, were it not that this lost energy is replenished by the battery in your mobile.

An electron circling an atomic nucleus all the time is in an accelerated motion. Without the attraction by the nucleus it would fly away in a straight path, but the attraction towards the centre causes the path of the electron to be curved so that it moves in a circle or follows an elliptical trajectory like a planet around the sun. This deviation from the straight path is called the "centripetal acceleration" the electron experiences, and classical theory says that the electron accordingly must emit radiation. Classical physics thus predicts that an electron is bound to gradually lose its energy and would spiral into the nucleus, to be swallowed there as is shown in the gruesome picture of Fig. 1.3. "Gruesome" because it would mean that all the atoms in the universe would long have gone today, and there would be no world as we know it.

Obviously, Bohr could not have that. He declared the classical electromagnetic laws to be invalid in the case of his atom model. He stipulated that electrons, especially when they are circling in the lowest permitted orbit, should maintain "radio silence" to escape the fate to be devoured by the nucleus, even though

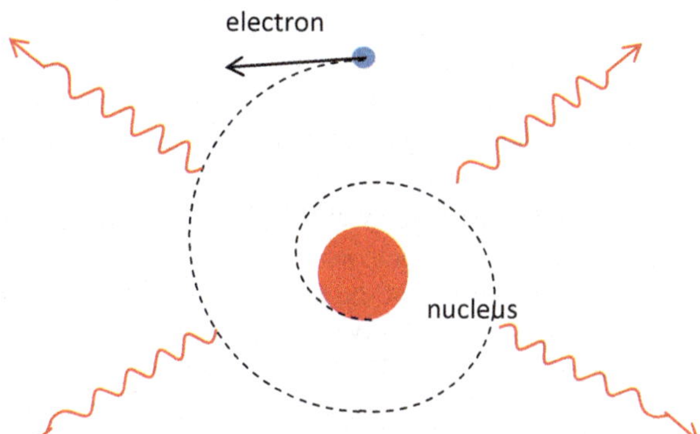

Fig. 1.3 Electrons circling a nucleus must according to Maxwell's laws lose energy by radiation, and will spiral into the nucleus

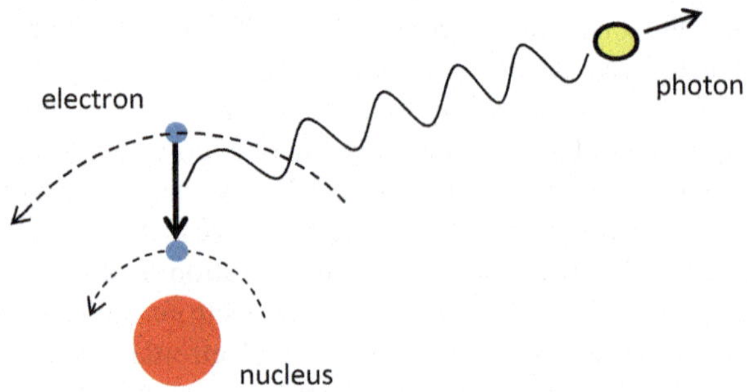

Fig. 1.4 Jump of electron from higher to lower orbit, releasing a photon

Maxwell's laws say such a radio silence cannot happen. Once again: a bold step! Imagine this in every day life: any time we do not like a law, we say that it is not applicable to us. The police in turn would not like us.

Bohr's "legislative revisions" did not stop there. If an electron is in a higher orbit than the lowest one, he said that exceptions to the rule to maintain "radio silence" are allowed, but these exceptions are rare and take a peculiar form. It is sketched in Fig. 1.4.

An electron, after say one million orbits, may become "bored" with its endless circling and may then "decide" to jump down to a lower orbit. It thus comes in a lower energy orbit, and sheds its surplus energy by sending out a light flash (known as a photon, a "particle of light").

At what time exactly such an event will happen is unpredictable. It is a matter of pure chance, there is no special cause which triggers the event. Sometimes the electron becomes "bored" after a thousand circles, at other times after a million or a "zillion" orbits, the time it happens is quite arbitrary.

You may realise that this idea is perhaps the most revolutionary of all Bohr's ideas. In "normal" physics every event is the result of a definite cause. We say that classical physics is causal. The event shown in Fig. 1.4 is, according to Bohr, a matter of pure chance, it happens without a clear cause, it just happens. The best we can do is to determine a *probability* when the light will be emitted by the atom: there is not much chance that it will happen after just one orbit of the electron, and neither is it very probable that the electron will stay forever in its original orbit, although neither chance is completely zero. The exception is of course the lowest orbit, where an electron is destined to stay forever unless by some action from outside, for instance the meeting with a photon coming from elsewhere, the electron is launched into a higher orbit.

At any rate, Bohr had done away with the sacred concept of causality. It was a reckless proposal, but he was right. Or at least he had found the right track. Nature itself behaves "unnatural" on the tiny scale of atoms. Nature itself ridicules its own laws in this sub-microscopic world. And there is no policing force to call nature to the order.

1.3 The Modern Quantum Theory

Now, you may have read in some popular-science books that everything was explained by the later developed "modern quantum theory".

Bohr's atom model did good service during a decade or so, but finally it broke down, refinements were clearly necessary. Around 1925 an entirely new description of the phenomena going on in atoms was found. Bohr's picture of a planetary system was changed, as well as the mathematical description of it, and from that time on one spoke of the "old quantum theory" of Bohr, and the "modern quantum theory" invented by people like De Broglie, Schrödinger, Heisenberg and many others.

However, to say that everything could be explained by the new quantum theory is nonsense, in my opinion. It depends of course on your definition of what you mean by the word "explanation". But if you expect that a further scrutiny of Newton's laws had by that time revealed that quantisation of energy can be in agreement with Newton's laws, then you will be disappointed. Quantisation of the energy levels in an atom still does not have a place in the Newtonian world. And, if you had expected that after scrutinising Maxwell's laws it had appeared that under certain circumstances "radio silence" is indeed a possibility even if the electron is subject to continuous accelerations, you will be disappointed once again.

What had happened was rather, that physicists had become accustomed to the idea that there are two separate worlds, the macroscopic world in which we humans live, and the world of the atom. Both these worlds have their own set of laws and rules, and the "legislation" is quite different in the two worlds. Indeed, these separate sets of laws are often in direct conflict with each other. After having taken the mental step to accept that there are two different physics, the question was: which law comes in the atomic world in the place of Newton's law for the motion of bodies? Attempts to reconcile the two worlds had been given up.

It was the great achievement of Schrödinger in 1926 that he found the law that is able to describe and predict the strange behaviour of atomic particles: Schrödinger's equation, or "quantum mechanics". Schrödinger's equation is entirely different from the classical equations of Newton. It is not just an extra correction factor to be put into Newton's "$F = m \cdot a$" such as Einstein's correction that says that the mass m is not constant as thought by Newton, but is a velocity dependent quantity. No, Schrödinger's equation would not be recognised by Newton, it is completely different. It has to be, because it must describe completely different behaviour.

How ingenious this equation may be, it does not provide an *explanation* for the phenomena observed in the atomic world. It is more a matter of "strange phenomena have been observed? OK, let us try to think up a strange formula out of the blue that can correctly describe these phenomena, *without having any recourse to other physical principles*".

This is not meant to be a denigration of Schrödinger's extremely clever equations, because we can just as well ask "what are Newton's equations else than empirical formulae"? They came falling out of the blue as well (apart from the famous but unhistorical apple that must have come falling from an apple tree).

Be this as it may, the modern quantum theory actually made the list of anti-intuitive phenomena longer, rather than explaining them away. "Anti-intuitive", because human intuition is unavoidably based on the laws in the other world, viz. our own macroscopic world. Let us therefore consider a few more "magical tricks" of nature, as revealed by the modern quantum theory. For this purpose it will be necessary to say a few words about the new theory.

1.4 Matter Waves

One of the foundations of the modern quantum theory is the so-called relation of De Broglie. It was one of the essential stepping-stones which helped Schrödinger to formulate his equations. The contribution by De Broglie was first published in his dissertation (1923) at the Paris university, and it contained two central ideas. By the way, his ideas were considered so unorthodox, that the help of a "habitually out-of-the-box thinker" like Einstein was asked, before anyone dared to admit and approve De Broglie's work as a doctor's thesis.

As already briefly described in Sect. 1.2, at the time De Broglie made his unorthodox proposals the "wave/particle duality" of light was considered an established fact. It means that a ray of light sometimes behaves as a bundle of waves, and in other cases as a stream of particles, depending on the type of experiment that is being performed. Despite the fact that our intuition protests against the idea that light can be both particle and wave at the same time, the experimental results were convincing and had to be accepted. Saying that these results were understood goes too far, understanding and accepting the facts are different things.

De Broglie now proposed that the same thing could be true in the case of elementary particles. It was called the *particle/wave duality of material bodies*. According to his ideas a material particle is always accompanied by a wave of "something" (though he could not tell what that something was), which he described as a "pilot wave", because the wave would, he thought, guide the movements of a particle. Nobody, including himself, could tell what sort of wave this would be, and what would be the medium through which it would propagate. So much was clear, that it would *not* be an electromagnetic wave like radio waves, light, X- (or: Röntgen-) rays or even Gamma-rays, all of them belonging to the category of electromagnetic waves described by Maxwell's laws. Other people therefore called it a "ghost wave" to stress that it was all rather mysterious, but finally these waves were baptised with the neutral name "matter waves", which is still in use.

Indeed, not very much later experiments were performed with electrons that could only be explained by describing the electrons as "wave-like things". A famous experiment that can only be explained by assuming electrons to possess a particle/wave duality will be described below. It is called the "double-slit" experiment, and confirms beyond any doubt the ideas launched by De Broglie.

The second essential contribution made by De Broglie was also an analogy of well-known electromagnetic wave phenomena. We are all well aware that radiation (i.e. electromagnetic waves) is the more "aggressive" the shorter the wave length it has. Think for instance of UV (ultra-violet) radiation, a part of the sun's spectrum that is invisible for the eye, because it has a wave length slightly *smaller* than what the human eye can detect. UV-radiation is quite harmful if you are exposed to it during a long time (sun burn, skin cancer). On the other side of the sun's spectrum is IR (infra-red) light, also invisible because it has a wave length which is too *long* to be detected by the eye. IR radiation represents heat radiation, and it is so harmless that it is being used for instance in the remote control of your TV-set. Of course, we here compare UV- and IR-radiation on the basis of equal intensities: very high intensities of heat radiation will cause burns as well, as every child has to learn empirically!

X- (or: Röntgen-)rays consist of electromagnetic waves with an even shorter wave length, and they are much more aggressive and harmful in the short term than UV-radiation, not to mention gamma-rays found in radio-activity (having the shortest wave length of all the examples given) that can cause almost immediate damage to the body tissues.

Now, De Broglie reasoned that if particles are accompanied by waves, there would probably be a similar relation between the "aggressiveness" of the particle, and the wave length of the accompanying "ghost wave". He assumed that there should be some mathematical relation expressing that the shorter the wave length, the more aggressive the particle would be.

How to express the aggressiveness of a particle? Well, classical mechanics already has defined such a characteristic, it is the "momentum". Momentum says something about how much impact a particle (or more generally: a body) can have, for instance during collisions with other bodies. Momentum is the product defined as (mass of the body) × (velocity), which is understandable because both mass and velocity will have their influence on the damage a body can do (again thinking about the scenario of a collision). In mathematical language: if we symbolise momentum by p, mass by the letter m and velocity by the notation v, then we have the relation $p = m \times v$, where now—very scientifically—the "aggressiveness" has been caught in terms of a formula, instead of just by semantics.

To remain "scientific", the wave length of the mysterious "ghost wave" will be denoted by the Greek symbol λ (lambda). Do you notice that the symbol λ, because it smells like mathematics, suddenly lends a bit of respect to this phenomenon of a ghost wave that is understood by nobody?

As a final step I introduce the notation " $\div$", the short for "proportional to". If we see the "impressive" formula $a \div b$ this means: if quantity b is made twice as large than it was before, then also quantity a becomes twice as large. What we in the case of De Broglie's theory need is "inverse proportionality": $a \div 1/b$, meaning that increasing b reduces a, for instance doubling b will mean halving a.

We need the inverse proportionality because the earlier conclusion was: the *smaller* the wave length λ, the *larger* the aggressiveness p (momentum). Expressed in math form:

$$momentum \div \frac{1}{wavelength\ of\ matterwave}, \quad or: \quad p \div \frac{1}{\lambda}$$

And this is all! Here you see the fundament on which the modern quantum theory is based. Essentially this is the famous De Broglie's formula which directly gave rise to Schrödinger's equations.

A slight refinement is still possible, but it does not really change anything, certainly not the essence. The point is, that such a proportionality can always be written as an equality:

$$momentum = \frac{h}{wavelength\ of\ matterwave}, \quad or: \quad p = \frac{h}{\lambda}$$

using a proportionality constant, in this case denoted as h. In itself this does not add any new information to the fact that momentum of a particle is proportional to the inverse wave length. But the interesting thing is that the same proportionality constant h happens to crop up everywhere in quantum theory, in all sorts of relations. It appears to be a universal constant, one of the fundamental physical constants like e.g. the gravity constant g in Newton's laws. The constant h was already present in the formulae of Planck and Einstein, mentioned in Sect. 1.2., concerning the wave/particle duality of light. To honour Planck, who first introduced the constant, and to give him credit for truly being one of the founding fathers of quantum theory, the universal constant h is called "Planck's constant".

De Broglie's formula was proven correct by numerous experiments. One of the most striking experiments is worth mentioning in somewhat more detail. It is the so-called "double-slit" experiment, and it is a good illustration of the "new magic" that is implied by the "new quantum theory".

The description of the double-slit experiment is a rather long diversion, and there is a danger that you will lose sight on the main story line. Later, more of such diversions are to come, and for this reason I have inserted at several places text boxes with a "road map", like the following one.

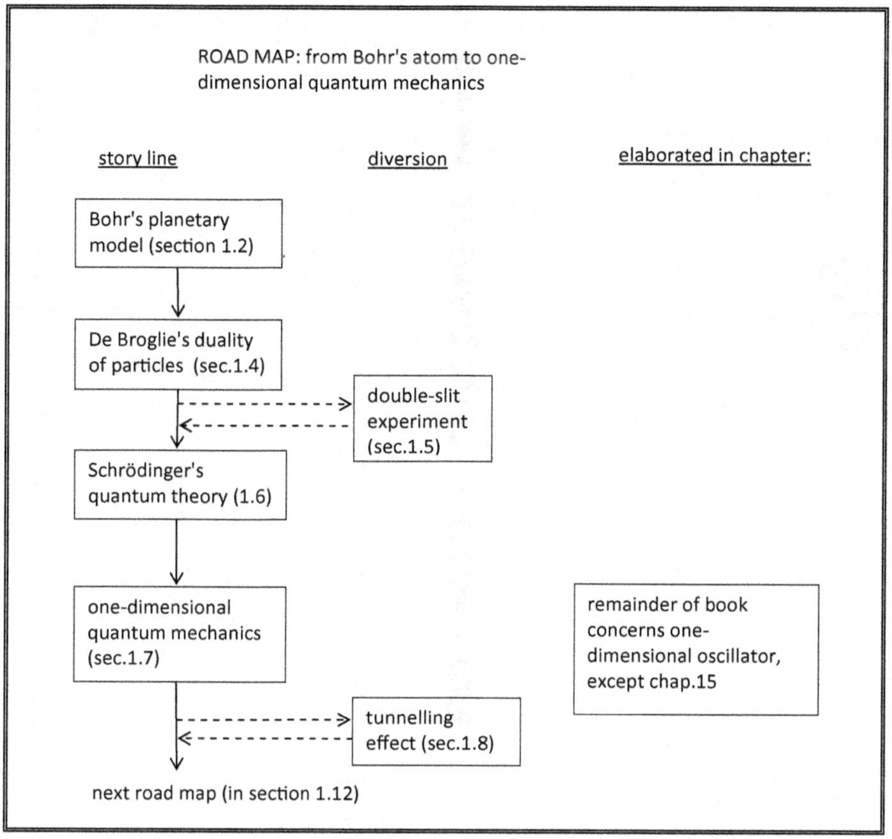

1.5 The Double Slit Experiment

As a preliminary step, imagine we have a pond where waves are produced by some device, e.g. by periodically dimpling a body in the water and retracting it again. Next to this source of waves there is a barrier, with two gaps in it (Fig. 1.5).

Each gap receives waves and sends them away again on the other side of the barrier (you may remember from your school days that this is called "the principle of Huygens"). The gaps have thus become new sources of waves, but the waves going through the gaps have the property that they are in step with each other (in jargon: they are in phase). The water level is going up and down in each gap in the same way. The waves that have passed the openings spread in all directions and reach the edge of the pond. It will be evident that the waves from one gap will interfere with the waves coming from the other one. The interference means that at

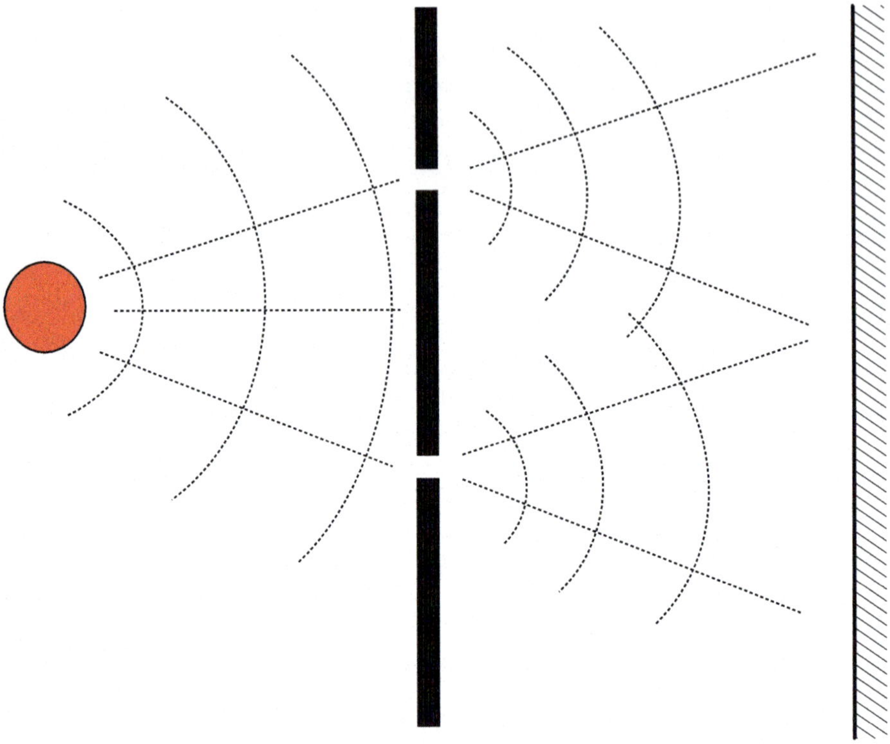

Fig. 1.5 The double slit experiment in water

some places the crests of the waves coincide and magnify each other: so-called constructive interference, resulting in an increased up and down movement of the water. In other places a crest and a trough meet each other, and would result in cancellation of the waves: here we have destructive interference. Walking along the shore on the right one sees places with high waves and between them places with no water movement at all. The photograph below shows very clearly the interference pattern in a pond due to two sources of waves (Fig. 1.6).

In the case of light the same phenomenon can be observed. We then have a light source, throwing its light on a screen in which two slits have been cut. Each slit passes the light it receives, and each slit by itself becomes a new source of light for the space to the right of the screen. The light from each slit finally arrives at a screen where the brightness is observed. What will be observed is that there will be bright and dark bands on the illuminated screen, due to interference between the light sources (Fig. 1.7).

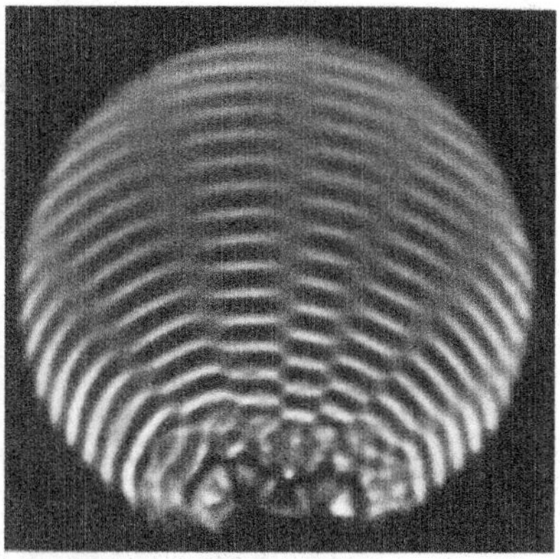

Fig. 1.6 Photograph of the double-slit experiment in water

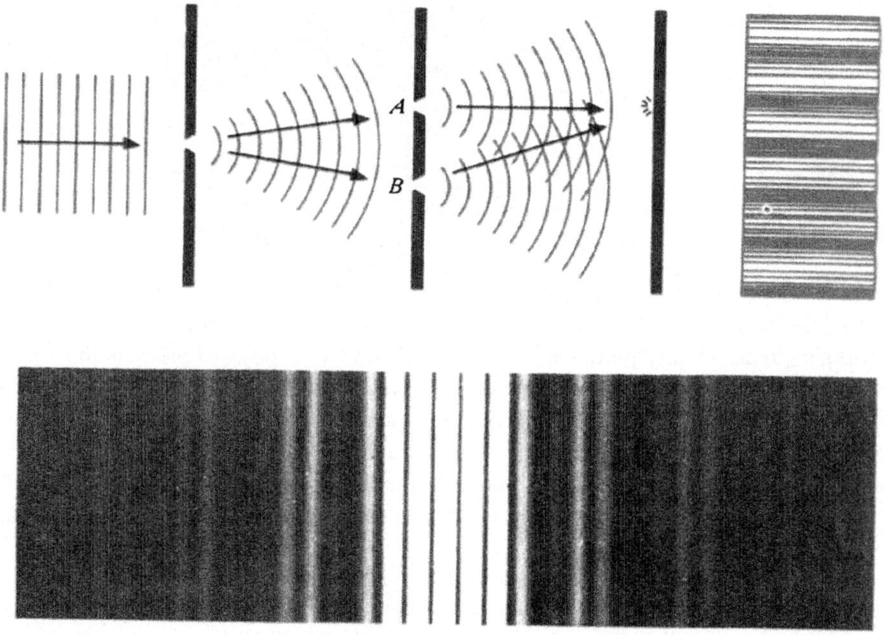

Fig. 1.7 The double-slit experiment using light. The *lower part* is a photograph of the resulting interference pattern

Closing one of the slits destroys this so-called interference pattern. Light coming through just one slit will illuminate the screen on the right evenly, the brightness slowly fading away the further we are from the centre. No alternating bands of darkness and brightness will be observed, because there cannot be interference.

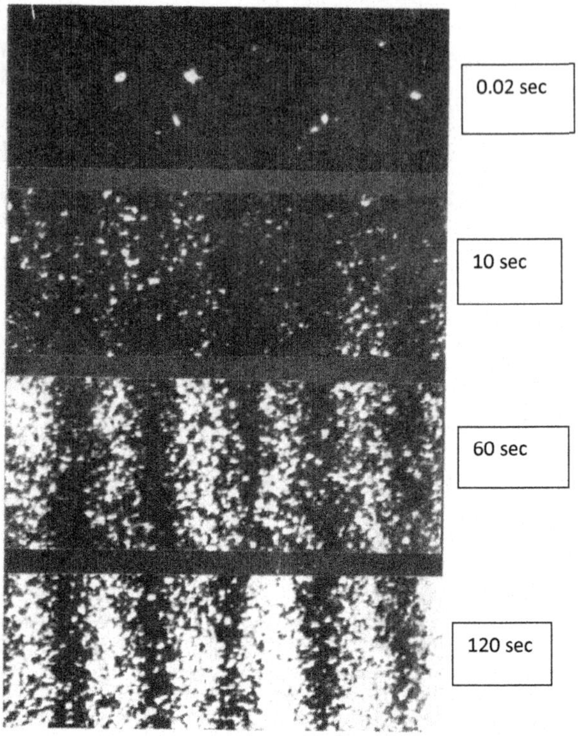

Fig. 1.8 Electrons are shot at a photographic plate through a double-slit arrangement. The plate is developed after different exposure times

Now remember that in quantum physics, according to De Broglie's hypothesis, particles can also behave like waves. If we replace the light source by a source that shoots a stream of tiny particles like electrons, quantum physics predicts that again we would see an interference pattern on a photographic plate which makes the impacting electrons visible. The prediction is correct: surprising as it may be, small particles of an atomic scale do show this behaviour! Figure 1.8 shows it. To begin with, the electrons arrive as individual particles, one at a time. However, the more electrons have hit the photographic plate, the more clearly a wave-like pattern can be observed.

Closing one of the slits has the same effect as described in the case of waves, the particles then no longer concentrate at certain points on the receiving screen and avoid other points, they spread evenly after having passed through the slit that is still open.

What is so strange and incomprehensible about this phenomenon is, that we imagine particles to pass either through the lower slit or through the upper one, a particle cannot pass through both at the same time. Nevertheless, a particle passing e.g. the lower slit seems to "know" whether the other slit is open or not. If the other slit is open the particle avoids to fly in certain directions, so that a clear interference pattern is formed, visible after many particles have passed the set-up! This kind of behaviour cannot be understood if we would stick to the view that these electrons are nothing more than a sort of bullets, one by one shot through the arrangement.

The double-slit experiment is a fine (but puzzling) illustration that De Broglie's ideas about these mysterious "matter waves" or—to use the terminology of that time—"ghost waves" are not pure fiction. De Broglie himself called them "pilot waves" to suggest that these waves in some way determine how the particle is going to move. We must thus accept that on an atomic scale things can be both particle and wave at the same time (so-called *duality*). Who said that the modern quantum theory "explains everything"? In my opinion all this only adds to the mysteries.

1.6 Schrödinger's Equations and the Model of Atoms

The new equations of the "modern quantum theory", i.e. Schrödinger's equations, are an extension of De Broglie's ideas. They consist of equations describing how the matter waves are spread out in three-dimensional space, and how such a complete "field" of matter waves develops in time. An example is given below in Fig. 1.9, which will now be explained.

First, it should be mentioned that the physicist Max Born had found that these mysterious "matter waves" do have some relation with more understandable properties of particles. There is a relation between the field of matter waves and "the chance to find the particle at a certain place". If the matter waves are more intense in certain places within an atom, the probability to find there an electron is larger than in places where the matter waves are weak. For this reason matter waves are also called "waves of probability". Of course, this still does not explain what these matter waves really are, but at least their function as "pilot waves" governing the whereabouts of electrons is made more concrete.

In the modern quantum mechanics we still have the quantisation of energy like it was postulated by Bohr. Remember that Bohr had added the rule that in an atom the electrons can only have discrete values of the energy, their energy cannot have any arbitrary value. In fact the quantisation can be calculated more accurately by the new quantum theory than by Bohr's old theory, which for certain kinds of atom made wrong predictions.

Figure 1.9 depicts the layout of a hydrogen atom. Each square of the figure concerns a particular value of the allowed energy (or rather: a particular "state" of the electron, to use the language of physicists). The brightness in the figures indicates how large the probability is to find the electron in specific areas around the nucleus. The patterns are beautiful, but one cannot say that these pictures are less mysterious than Bohr's clear-cut circles and ellipses. A bit of the "vagueness" of Saturn's rings has returned, although the idea of "regions which are practically forbidden for the electrons" (the black areas) is clearly still there.

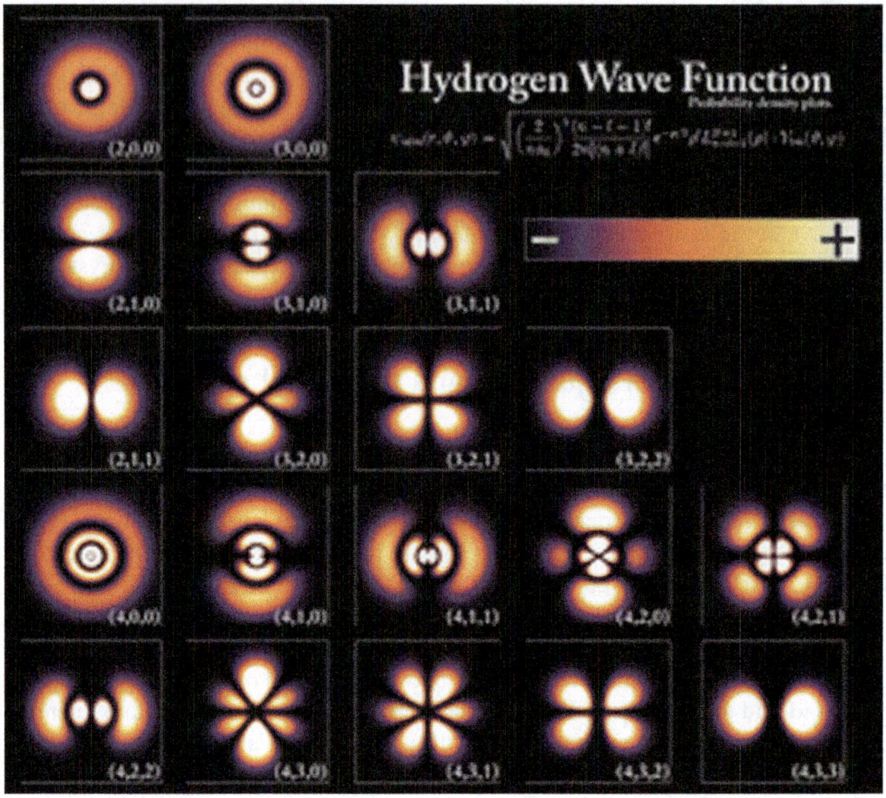

Fig. 1.9 The probability to find an electron in regions within a hydrogen atom (the lighter the colour, the higher the probability). From Wikipedia: electron

I will not go into more details, that is a matter of later chapters. Let it be an illustration that the "modern quantum physics" does not answer the question where an electron exactly is at a certain moment, it only indicates where the chance to see an electron is large or small. Even if at some instant of time we would know, by some kind of measurement, the position of an electron then at the next instant there would again be uncertainty about its exact place, and the "vagueness" sets in again. However, Schrödinger's equations do tell us where we have to look for the largest *chance* to find it again. But this is just a chance, it may be at another place as well. Such are the laws of probability.

All this is in complete contrast to the 100 % causal (therefore: deterministic) theory by Newton about the trajectories of heavenly bodies. The laws of Newton allow us to predict the future over periods of hundreds of years or longer, e.g. we can with certainty predict where the Moon will be at 23:00 h on May 3rd in the year 2042 (when there is a chance, although remote, that I will be a 100 years old).

1.7 One-Dimensional Quantum Theory

This book deals with the "new quantum theory". Do not worry, the book does *not* have the intention to change anything in this almost perfect theory. It just attempts to get rid of some of the magic surrounding the theory, by adding interpretations based on the classical laws of nature. How this has been achieved is the subject of the remainder of the book.

But, as conveyed by the above shown picture, the atom itself is pretty complicated. The pictures of Fig. 1.9 concern the simplest kind of atom, viz. an atom of the first element in the Periodic System: Hydrogen. You can imagine how complicated everything may look in more general cases.

In order to develop the theory of this book, and in order to explain it, I concentrated on another quantum situation, which is called the *one-dimensional oscillator*. In a minute I will tell you what it is, but before you get the feeling to have been cheated: in the one-dimensional case we meet *exactly the same* "magical" quantum phenomena as have been described for the atom. The same Schrödinger equations apply, we again have to disregard all the classical laws of nature, and an electron behaves just as weird. Only the sums are simpler, as well as the explanatory sketches (because they need not be pictures drawn in perspective).

What is a one-dimensional oscillator?

First imagine in our own macroscopic world a marble that has been charged electrically. Something like a plastic comb which can be charged by a few strokes through one's hair, but then in a spherical shape. The marble is fixed on a toy railway wagon so that it can move back and forth between two larger spheres carrying electrical charges of the same sign. The situation is sketched in Fig. 1.10.

As you notice, I have sketched all the charges in a red colour, with a plus-sign, and I will continue this convention throughout this book. You may wonder why, because electrons are negatively charged particles. It is just a matter of convenience: in the theory the direction of an electrical current is opposite to the flow of electrons, so that one may save one self a lot of minus-signs in the formulae by taking the charges positive. It does not affect the outcome, of course.

If the moving sphere is exactly in the middle between the fixed charges, it is repelled by both the fixed charges by the same amount, and it will stay in the middle. The middle is apparently an equilibrium position. If we now by hand displace the railway carriage away from the centre, for instance to the left, the moving charge will feel a larger repulsion from the left than from the right. Upon releasing the carriage the charge will move to the right, and will try to go back to its

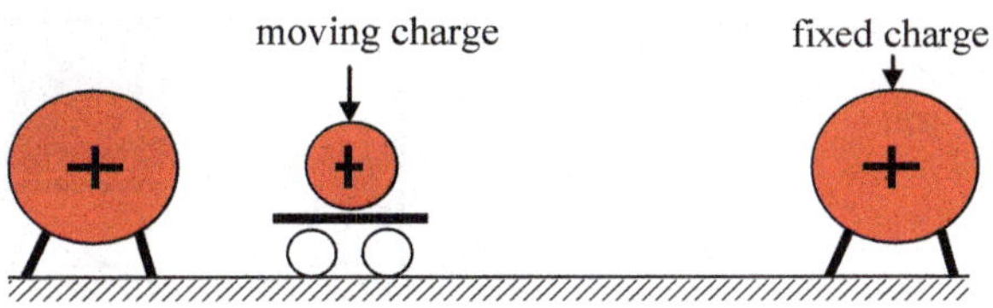

Fig. 1.10 Electrical oscillator formed by a charge in a potential well

equilibrium position in the middle. However, at the time it has arrived in the middle, it has obtained some velocity and by its inertia will overshoot that equilibrium point. Finally it will come to rest somewhere to the right of the middle, and is then driven in the opposite direction to the left, so that everything repeats itself. The carriage is going to perform a regular motion back and forth. In jargon this is called a smooth, so-called *sinusoidal* motion.

Of course, this behaviour of our charged sphere is exactly the same kind of motion as we observe in the case of a clock pendulum, a children's swing or a marble rolling back and forth inside a bowl, though in these latter cases the motion is driven by gravity instead of by electrical forces. It is the behaviour everybody is used to. Classical books on mechanics devote many chapters to the mathematical analysis of these so-called *oscillators*, because the analysis of these simple "swing systems" forms the basis to also understand much more complicated things like the stability and control of airplanes and ships, the vibration of machine components, musical instruments, etc., etc.

Using jargon, an oscillator such as shown in the above figure is often called a "mass-spring system", since it is equivalent to the kind of set-up sketched in Fig. 1.11.

In this case the force driving the mass towards the middle is derived from mechanical springs. For several reasons the mechanical spring model is convenient to analyse, therefore it is a common model for oscillators in general. Anyway, the nature of the driving forces does not matter. Its mathematical description is the same, no matter whether the driving forces are of a gravitational, mechanical or electrical nature, the model always leads to a smooth sinusoidal motion.

So far you will not have read anything new. But wait! Before you stop reading because already boredom threatens to set in, let us now have a look at an oscillator at atomic scale.

At atomic scale we are most of the time dealing with things like tiny charged particles such as electrons (negatively charged) moving under the influence of other charges so that again the model of Fig. 1.10 is applicable. Because the fixed charges need not have exactly the same form as the spheres depicted in the figure, often the situation is more generally called "an electron moving within a *potential well*", where the word "potential well" refers to an electrical field which tries to drive the electron back towards an equilibrium position whenever it is disturbed. Let us not complicate matters unnecessarily: Fig. 1.10 is a sufficiently good illustration of the model which is analysed in the first few chapters of textbooks on quantum mechanics. Almost every book on quantum mechanics starts at this point, because it

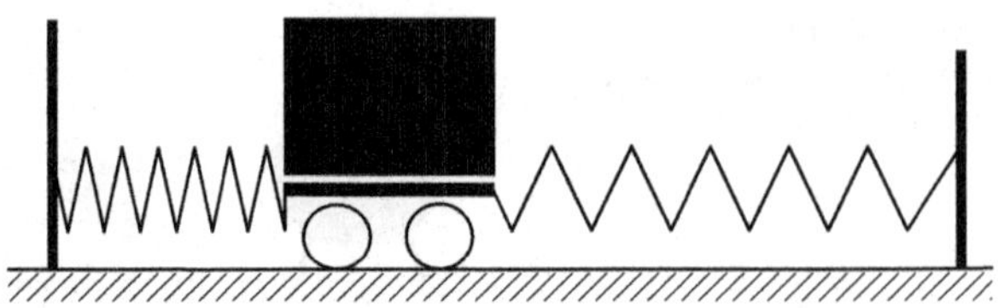

Fig. 1.11 Mechanical oscillator; the mass-spring system

illustrates practically all the weird phenomena encountered in the strange world of elementary particles. Like the books on classical mechanics, books on quantum mechanics take the simple oscillator as a basis to understand more complicated things, in quantum mechanics for instance the clouds of electrons in the vicinity of an atomic nucleus, together forming a complete atom.

Having concluded that in both the macroscopic as well as in the atomic world the same model for a simple oscillator can be used, one would of course expect that the behaviour of the moving charge is also comparable in both cases. Forget it! Experiments show that an electron does *not* move so smoothly in the regular sinusoidal way we would expect. Rather, it jumps irregularly from place to place, and shows in no way any similarity in its behaviour to the well behaved macroscopic swing or pendulum. Chapter 2 of this book sums up all the differences between macroscopic behaviour and behaviour at atomic scales. It amounts to a long list of pathological behaviour of electrons or other elementary particles. Let me give one example about a phenomenon that is present in the three-dimensional atom as well, but in a form that is much less clear and surprising than it is in the one-dimensional oscillator.

1.8 Tunnelling of Electrons

One of the most striking observations is, that an electron sometimes jumps completely *out* of the potential well. Imagine that you would look at a marble (Fig. 1.12), peacefully rolling back and forth inside a bowl, and would see it suddenly jump over the rim to escape completely! You would be surprised, to say the least. Physically minded people would also be puzzled for another reason, because they would say: my marble had a certain amount of energy, and that energy was in no way sufficient to get up so high on the wall of the bowl as to reach the rim. How could it ever climb over the rim? It did not have the energy to do that! Is the law of energy conservation suddenly violated, or what?

Fortunately, you will never see things like this happening in our human world. Nevertheless, marbles on an atomic scale do it all the time! This behaviour is called "tunnelling" (the reason for this strange name will be explained in a later chapter), and the effect is used in practical devices like "scanning tunnelling microscopes". It would seem as if the law of conservation of energy indeed is not always observed by these tiny particles. And it is not only this one law that does not apply.

This is the kind of set-up that will be dealt with in this book, instead of the three-dimensional atom. However, from Chap. 2 where a more detailed description of the quantum behaviour of such an oscillator is given, you will see that essentially all the atomic mysteries mentioned earlier are present in the one-dimensional oscillator.

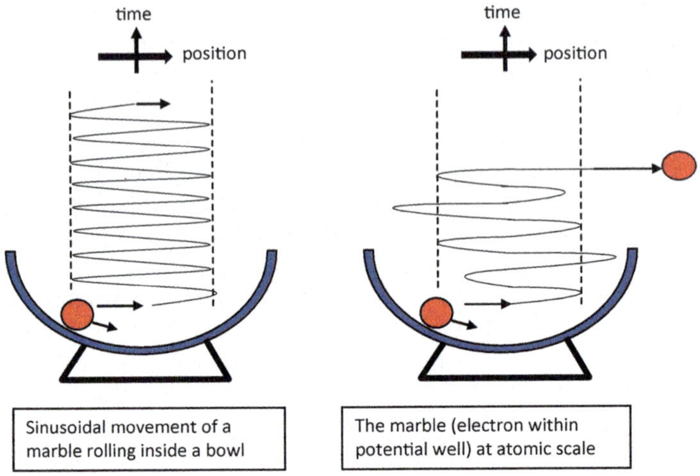

Fig. 1.12 Comparison of macroscopic and atomic oscillator. On the *right* the tunnelling effect is shown

1.9 Why "It Ain't Necessarily so!" in the Title?

This song text from Gershwin's opera Porgy and Bess expresses the doubts Sportin'Life has about all sorts of stories he should, according to his more religious fellow villagers, hold for "the truth". He doubts if Jonah could really have survived in a whale, for instance. And who can blame him (even though Sportin'Life is a rather despicable person whilst his fellow villagers are undoubtedly very honest people)?.

Likewise, I am so bold to ask if we really have to accept the story about the two separate worlds in physics with their two different sets of laws and rules. After all, one can imagine that there is an "intermediate world", the scale somewhere in between the scale of the quantum world and the human scale, where the two worlds overlap. Which of the two sets of rules would one have to apply on this scale? It has always nagged me: when must I use Newton's equations and when do I have to apply Schrödinger's equation? At some intermediate scale we will have to step over from one type of equation to the other. And more serious than choosing the applicable theory: at some scale we expect a change of the *actual* behaviour of objects, from "normal" to "weird". What is that scale? And even more intriguing: is it an abrupt change or rather a gradual process? A sharp boundary where we would see a sudden change of behaviour is not really credible. When scaling down an electrical oscillator, either during an actual experiment or by a thought experiment imagining that the total charge is gradually reduced and the dimensions are made smaller, one would not expect that at some point, by taking away a single electron, suddenly the boundary has been crossed and the oscillator shows different behaviour.

In fact, quantum theory indeed predicts a *gradual* transition between the two worlds of classical and quantum mechanics, as we would expect. Most textbooks on quantum mechanics devote a chapter to what happens if the energy of a system is increased bit by bit. The conclusion is, that the behaviour of for instance an oscillator gradually loses its "weird" characteristics when scaling up from the atomic scale, and it is going to approach "normal" behaviour. Because the existence of such a gradual transition region is so important—especially for the sake of this book's argument—Chap. 3 sums up what quantum mechanics can tell about it. Note that all that is mentioned in that chapter concerns the transition *coming from the small scales* where quantum mechanics is applicable, and increasing the dimensions from there.

The next, obvious question is of course: what does *classical* mechanics tell us about a possible transition region? In other words: can the transition region also be approached from the side of classical mechanics by decreasing the dimensions? If we assume a large scale oscillator, described by Newton's and Maxwell's laws, and scale it down in size, do we see a transition to quantum behaviour? Or do we discover at least "glimpses of quantum-like" behaviour?

The answer is obviously: no! If the possibility of such a transition had been implied by the classical laws, the discovery of quantum behaviour in the beginning of the 20th century would not have been the great surprise it in reality was. The reason is that there are no obvious scale factors to be found in either Newton's or Maxwell's laws. Of course, rather trivial scale effects do exist and are predicted by these laws. For instance, mounting a lighter object in the mass-spring system depicted in Fig. 1.11 would lead to a higher frequency of the oscillation. Everybody who ever had the task to fine tune an old fashioned pendulum clock knows that shortening the rod of the bob weight makes the clock go faster. If we scale down an electrical oscillator such as depicted in Fig. 1.10 by taking away mass and charges the frequency of the oscillation will become higher. It explains why a *large* amount of electrons pushed up and down in an antenna rod emits radio waves with typical wavelengths from centimetres to hundreds of meters, whereas a single electron jumping much faster up and down inside an atom will emit electromagnetic waves in the frequency band of light.

But that is not what we mean by "transition region". A higher or lower frequency of the oscillation does not mean that the *character* of the nice, regular sinusoidal motion is lost. The classical laws always keep predicting a smooth sinusoidal motion, they can never predict the radical change of behaviour such as the spasmodic jerks of real small objects.

Or can they? The above written statement is not entirely true in fact. The statement should be: "the classical laws never predict a radical change of behaviour for a *completely rigid* object". My own investigation shows something remarkable:

if we take a suitable model of an electron (*not* the rigid marble-like thing that mostly is considered in the textbooks) and do the downscaling, then glimpses of quantum-like behaviour are seen in the formulae, even when the classical physical laws are applied! *The classical laws of nature indeed can conspire in such a way that hints of quantum-behaviour are found.* This being so, one then can build up mental pictures about these "weird" quantum phenomena that are a lot less weird, because these phenomena then can be interpreted on the basis of "normal" physics. This is precisely the purpose of this book.

1.10 From Marbles to Droplets

To make a long story short: when we drop the assumption of an *infinitely* rigid "thing" inside an oscillator, the classical laws of nature do predict scale effects, in such a way that very small objects are going to display really drastic changes in behaviour.

In fact, what we have done in this book is to assume that the charge inside the potential well is not a rigid "marble-like" object, but instead something like a *droplet of charges*. Such a droplet is assumed to be coherent, but it has the freedom to deform and change its shape. The kind of deformations that might occur are somewhat similar to what we see in the case of water drops and soap bubbles: flattening or bulging in different directions, vibrating and sometimes even "shivering". See Fig. 1.13.

If we now perform the thought experiment to gradually reduce the total amount of charge in the droplet and reduce the dimensions accordingly, a transition is found where we do see that the *character* of the motion changes. It will be shown in Chaps. 8, 9 and 10 that the change is such that it reminds one of what is being observed in the quantum world. The reduction of scale gives a transition to quantum-like behaviour—at least we get definite hints of it—even if we still describe the oscillator by our "normal", classical laws of nature!

The most surprising (although a bit puzzling) is, that if we go on with the downscaling so far that the total amount of charge in the "droplet" becomes the unit

Fig. 1.13 From rigid marble to vibrating droplet

charge (that is the charge of one electron) *we find the same quantum behaviour as predicted by Schrödinger's equations*. And when I say "the same" I mean literally "the same", i.e. there is not only agreement in a qualitative sense, no: the agreement is even *quantitatively* correct. For instance, we find that the droplet complies with De Broglie's relation, and we find the correct energy quantisation if the droplet is moving within a potential well (in a well only discrete values of the energy are allowed, just as in an atom). And this correct quantisation is found despite the fact that we are still using the classical laws of nature!

Now, a lot more has to be said about such a model of the electron. You will immediately have thought already that this *never* can be a consistent model of an electron. And you are right, it cannot. Obvious problems are of course, that if we assume the electron to have finite dimensions (in contrast to the usual model of electrons), then the charge has been "smeared out" in a finite volume and the parts of this volume would have a smaller charge than the unit charge. Smaller amounts of charge than the unit value have never been observed in any sort of experiment, and are thus a dubious concept. Furthermore, still a graver objection is that all the parts of the finite volume would repel each other. The electron, if it were indeed a smeared out thing, would have the tendency to explode. You will see in a minute that nevertheless the model of a "droplet of charge" can be made consistent, and that—surprisingly—it is not even a very radical departure from existing ideas in physics. To explain this, I need to tell you how this—seemingly—unorthodox model originated, and how it can be reconciled and made compatible with the usual physical ideas.

1.11 A Credo Concerning Infinities and Singularities

The usual idea about electrons is, that they do *not* have finite dimensions, and that they neither have any internal structure. To formulate it differently, they are just "singularities" of the electromagnetic field, points where field lines come together (and where things get out of hand, because all the physical descriptors become infinite).

I am just a simple engineer (originally in aeronautics and aerodynamics), and let me state a "credo": infinities or singularities, i.e. points where things become infinite, are—often useful—theoretical concepts, but are not physical realities. In engineering we often enough use concepts like "infinitely stiff bodies", "infinitely small viscosity of fluids", "point objects" (i.e. infinitely small things), and theoretical aerodynamics abounds with "singularities of the flow". But these concepts are just artefacts, often leading to very useful, very practicable theories but nevertheless artefacts. Such theories, although they have useful applications, always have their limitations: an engineer would never be able to determine the strength of structures by making the assumption that the structure is infinitely stiff. Every object has some flexibility, even the most rigid looking things like massive concrete dams of storage lakes. In fluid dynamics, calculations based on "infinitely small viscosity" are very useful, and are in frequent use, e.g. for doing calculations of the pressure distribution on aircraft, but nevertheless, if one wants to calculate the drag of an aircraft, we must take into account that air has some viscosity, however small it is. "Singularities", i.e. points where things become infinite, are likewise extremely useful in lots of theories, but they are the product of simplifying model assumptions. If singularities sometimes are no longer found useful but instead become bothersome, then it is time to look more carefully, to "zoom in" on the singularity so to say, so that we again see internal structure.

During the 60s and 70s of the last century special mathematical methods were rather popular in aerodynamics, which precisely had this purpose to "zoom in" on singularities, and "open them up" so that infinities were removed, with the purpose to make the internal structure visible that was hitherto masked by the infinities. One could call these methods a sort of "mathematical microscopes" so that it was possible to study what goes on in detail in points of the flow that at first sight had the character of a singularity. One of these methods had the ugly name "matched asymptotic expansion approximation for singular perturbation problems" or briefly: max-technique. I advise the reader to forget this horrible name immediately, but I did use the max-technique in my investigations. The principle is, to "stretch the coordinates" near the point of interest, but in a very careful way such that the correct connection with the other—"unstretched"—parts of the flow field is retained. An analogy is, to use detail maps for hiking, choosing maps of course that are up to date so that they are properly embedded in the larger scale maps used for driving by car to the starting point of the walk. In a later chapter (Chap. 5) I will tell more about it, as far as is necessary for the understanding of the story told in this book. Unfortunately, this branch of mathematics was almost forgotten completely when the electronic computer came to take over and number crunching seemed the solution to any problem.

1.12 The Cloud as an Initial Model

To return to the modelling of electrons: from my "credo" it has become understandable why, for an exploration of the transition from the macroscopic world to the quantum world, I started macroscopically by taking a sort of "cloud" of many electrons in a *finite* volume that during the whole process of downscaling would remain *deformable*. One should imagine something like a shoal of fishes, fastly changing in shape, but nevertheless keeping some coherence (Fig. 1.14). Clouds of

Fig. 1.14 A shoal of fishes, like the assumed cloud of electrons

electrons also existed in old fashioned radio vacuum tubes, the electrons being emitted by a heated "cathode" and locked up by the electrostatic fields within the tube. This is the kind of picture I had in mind at the start, just as a tentative model to see what it would lead to.

During the first stage of the investigation I did not leave the macroscopic world yet. I started to study the behaviour of such a cloud. How? In a way that would be considered idiotic by modern scientists who immediately would step to a big computer, would do numerical calculations and would then make beautiful pictures and animations of their numerically simulated cloud. My approach was different, and really very old fashioned: I tried to simplify the situation so much (reported in Chap. 4) that analytical equations could be obtained, i.e. just a set of formulae that, if wished, could be used for animations, but can also be inspected in a qualitative way. I obtained what in jargon is known as "equations of motion". Contrary to what the name suggests, these are *not* equations describing explicitly the motion of the cloud. It is more extensively explained in Chap. 7, but in short: these equations of motion are a sort of "semi-manufactured article", it is only upon solving them that one finds the motion in explicit terms. But often it is not possible to find explicit solutions, and then the equations of motion can be inspected to derive some general characteristics of the motion (even though the motion itself in such a case cannot be determined precisely).

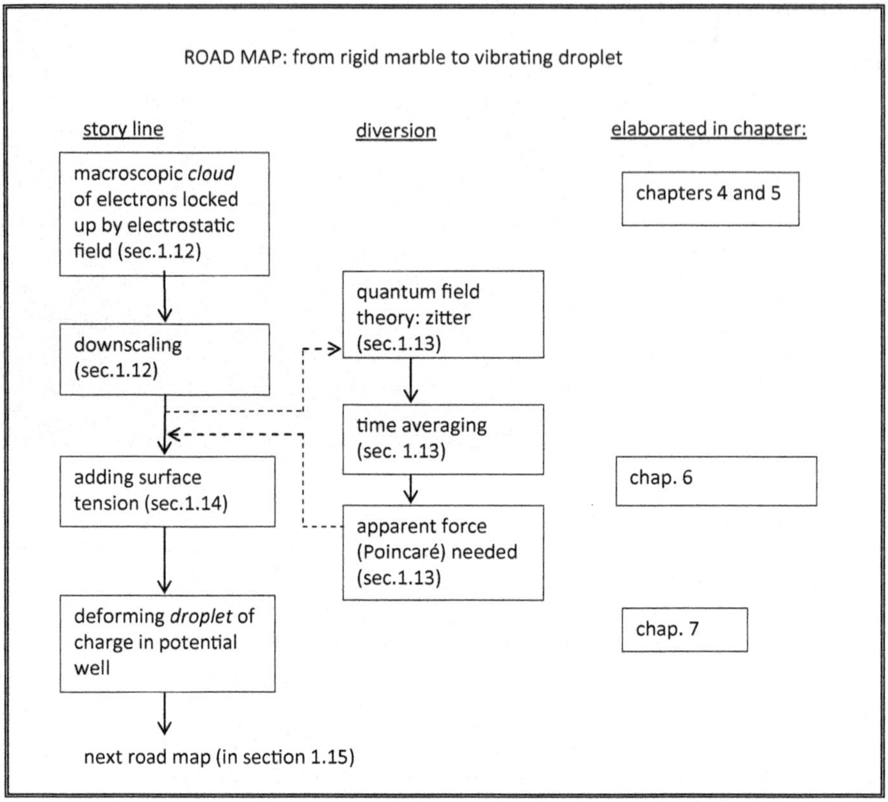

In casu: it is well known how equations of motion would look when they—upon "solving" them—would lead to smooth sinusoidal back and forth motions of the centre of the cloud, if it is locked up inside a potential well. Now, having set up the equations for a *deformable* cloud, I could inspect whether there are extra terms, next to the terms belonging to sinusoidal motion. And, due to the fact that my cloud could deform in its overall shape, there indeed were such terms. Next questions about these "perturbation terms" were: How large are the perturbations? Are the perturbations dependent on scale? What is the character of the expected deviations from pure sinusoidal motion? etc.

Then the next step: downscaling. If it is assumed that there is less and less charge in the cloud which accordingly is made smaller and smaller, one can try to determine at which point the extra perturbations in the equations of motion would become so large that they are no longer insignificant, but begin to dominate the character of the motion.

And the final step: assuming that the total charge of the cloud would be the unit charge, and that the total mass would be the same as that of an electron, what would then be found?

Now, concerning that final step, I had been warned by no less a person than Richard Feynman to expect that everything would go haywire during that last step, and that infinities would crop up everywhere in my formulae. I do not mean to say that I ever met Feynman in person, alas no! You will have heard about Feynman, the famous, humorous physicist who had the gift to explain difficult physics in plain, every day language, and who wrote his books with a lot of winks. And I almost forgot: he also won a Nobel Prize. Now, there exists a three volume physics textbook, "The Feynman lectures on physics" which, written for physics students, explains everything in a very original, unorthodox way. The book is famous, and it is still on the shelves of bookshops, even half a century after its first publication. In this book (volume II, Chap. 28) Feynman tells about early (now abandoned) models of the electron (amongst them Lorentz's model, that we will meet later again), and about their successes and failures. Feynman himself also made an attempt to improve such models (in vain), and he ridicules his own work. From this chapter came the warning about infinities I mentioned above. Though these models of the electron are now abandoned, infinities still pester even the modern theories. In the book by Abraham Pais: Inward Bound, a history of atom and nuclear physics written for experts (it includes a lot of formulae) there is even a separate chapter (Chap. 16) titled "Battling the infinite".

So, this was sufficient warning for me to use the above mentioned "max-technique" to set up my equations governing the motion of a droplet of charge, because this mathematical technique is better able to deal with singularities than more conventional mathematical approaches. The result of the final down-scaling to sub-atomic sizes was already revealed above: it led to surprising, and encouraging results.

As a pure mathematical exercise this is fine. But consider the model of the electron that we arrived at during that last step, now considered from the physical point of view: it is an impossible model. We now still have a cloud of charge elements, each element apart carrying a small electric charge, much less than the smallest possible charge in nature. Furthermore, a shoal of fishes would have some coherence because of an innate instinct. But where would coherence come from in the case of a cloud of charge elements, that has on the contrary the "innate" tendency to explode due to the mutual repulsion between its elements?

1.13 The Cloud as a Time-Averaged View of a Trembling ("Zittering") Electron

Nevertheless, I am going to show you that the electron model I arrived at is not so impossible at all, provided we add a correction. To show that this model can even be derived from "ordinary" quantum mechanics, I will have to make a small diversion to the so-called "quantum field theory". This is the part of quantum mechanics that deals with the interactions between photons and elementary particles.

One of the results from quantum field theory is rather magical. So much indeed, that some people doubt its reality. As I do not intend to criticise quantum theory as it is, I take it just for granted. The particular result I want to discuss here says that in a vacuum, i.e. in an environment which is completely void of anything, particles can come into existence, quite spontaneously, out of nothing. It happens all the time, everywhere, and in unimaginably large numbers. Always in pairs, a particle and its anti-particle, so that the total charge and the total momentum of the pair is zero. In fact, all the characteristic quantities you can think of, like spin for instance, add up to zero if the pair is taken together, except their energy (equivalent with their mass). These "ghost" or "virtual" particles stay in existence during an extremely short time, after which the pair disappears again by cannibalising each other. Sometimes vacuum is described as a seething, boiling ocean, in contrast to the idea that vacuum indicates a dull emptiness where nothing is happening.

Actually, amongst these so-called "virtual" particles also virtual electrons are found, together with their anti-particles (so-called *positrons*, i.e. electron-like particles with a positive charge). Now, if somewhere in this seething ocean a "real" electron is floating around, something special can occur. In case a virtual pair is coming close to the real particle, the possibility exists that the "virtual" positron melts together with the "real" electron. The result is that both are annihilated. Fortunately charge and all the other relevant properties are preserved, thanks to the "virtual" electron which is now left on its own. This leftover particle now assumes the role of the "real" electron.

What we see as a result of this role swapping is, that the original real electron seems to have made a jump towards a slightly different position. This happens so often that we "see" an electron always jittering around a bit, it does not stay in a fixed position (see Fig. 1.15). The phenomenon is called the "zitter" of electrons, the German word for "shudder" or "tremble". All this is going on at an extremely small time- and length scale. The jumps of the electron take a time in the order of 10^{-21} s, an exceedingly small time. The relevant time scales in the human world are

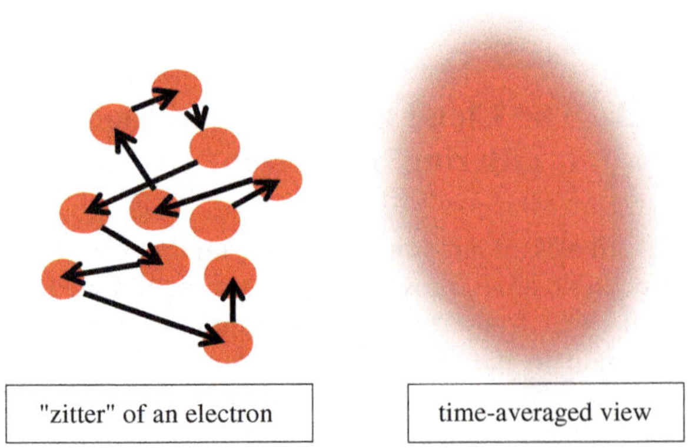

Fig. 1.15 Electron zitter, and the time-averaged view of it

far larger, even using sophisticated laboratory equipment. Depending on the time scale considered relevant, we will mostly see the "zitter" of an electron just as a blur. It is the same effect as when we try to look at the spokes of a fast turning bicycle wheel: the only thing that can be seen by the naked eye is a grey disc. We obtain a so-called "time-averaged" view of the wheel with spokes. Likewise, a time-averaged view of the "zittering" electron will appear as a "smeared out" region of electrical charge. One could call it a "cloud of charge", and this cloud has dimensions far larger than the actual electron (which anyway had zero dimensions according to the present theory of electrons).

This picture of a smeared out blob of charge can also justify the model we have adopted, consisting of a "cloud of charge" with finite dimensions, where the unit charge has been distributed continuously within a small volume. It is important to realise that such a model has come about by the process of "time averaged observation" of the "zittering" point-like electron, and we must realise that this averaging is a wholly artificial process we as observer have introduced. It is just a model simplification, and one should be on the watch that no inconsistencies are introduced by this artificial simplification.

Now, the actual electron, jumping around, at each instant of time is the only charge present, and it is not pushed away by any other charge. The "drunkard's walk" sketched in the left part of Fig. 1.15 will not show any tendency to explode. At most the region where the electron is jumping around can, after some time, have drifted away a bit. But now compare the smeared out cloud on the right of Fig. 1.15, this does have the tendency to explode. The culprit is the fact that, inherent in the act of time-averaging, all the elements of the charge are there at the same time, and thus repel each other. The explosion tendency is clearly an artificially introduced phenomenon, only caused by our model simplification. To obtain a *consistent* simplified model which reflects the real phenomena, we have to add a correction.

The procedure of making simplified models consistent by adding suitable corrections is not uncommon in physics. Think of the rotating earth. Very often we happily apply Newton's laws to the motion of objects moving over the earth's surface and we are often justified in doing so, but it depends on what we want to calculate and which accuracy is required. Actually, what we do is wrong because Newton stipulated that his laws are applicable to frames of reference that are standing still (whatever that is) or at most have a constant speed (both in magnitude and direction). The rotating earth surface certainly does not comply with these conditions, and we are confronted with our error when doing more precise calculations and comparing them with experiments. What should, correctly, be done is the following: we are *allowed* to use Newton's laws as if the earth were not rotating, but *only after having added some correction forces*: an extra centrifugal force and the so-called Coriolis-force, which affects moving objects. Only then we find e.g. the measured difference in gravity when comparing the poles with the equator, and only then can we explain why flowing masses of air (wind) tend to form "whirlpools" on the meteo maps. Correction terms like "centrifugal- and Coriolis-forces" are often called "apparent" forces, one could also call them "correction forces" to make up for the errors we have introduced ourselves.

1.14 Surface Tension in the Role of "Apparent Force"

Now back to the time-averaged model of an electron: we have to add "something" to the model to prevent the model from exploding, because such an explosion tendency is not real. We need a correction force or, in the above used terminology: an extra "apparent force" to contain the smeared out charge. In the century old, now abandoned electron models discussed in the book by Feynman, a similar correction was needed so that it henceforth would be allowed to use Maxwell's equations again without restrictions. Feynman speaks in his characteristic style about "the rubber bands to keep the thing together". A more posh word for these rubber bands is: Poincaré forces. It cannot be emphasised too much that, however impressive this word may be, a Poincaré force is *not* a real force of nature, it is just an "apparent" force, a correction needed to repair an oversimplification when we modelled the reality.

Naturally, when I was setting up the equations of motion (as described earlier in Sect. 1.12) I had to become concrete, and had to think up a mathematical form to express this Poincaré force and to include it in my formulae. This was almost a show stopper, until I realised that the—fictitious—explosion tendency is mostly a surface effect. Elements of charge near the middle of the cloud "feel" repulsion forces from all sides, with a net effect of zero or at least very little. On the other hand, the elements near the edges of the cloud are driven outwards because the repulsion forces all come from within the cloud. A suitable repair would thus be to add a sort of "surface tension" effect, as apparent force. Surface tension is well known in the case of drops of water. Water is so common in our lives, that nobody is surprised by its behaviour, even if it is somewhat puzzling when thinking deeper about it. Water, as well as most other liquids, has a strong tendency to form drops. Water drops are not even very easily destroyed, sometimes they can collide without merging into each other. Or they can roll over a water surface, like balls over the ground. The latter can also be seen easily when one drips coffee from a spoon on the coffee in a cup. Under many conditions (coffee strong enough, not too cool or too hot), one sees small "balls of coffee" rolling over the surface (Fig. 1.16).

All this is familiar stuff, and almost everybody knows the name for it: it is the consequence of *surface tension*. Cohesion forces exist between the water molecules, but the molecules in the surface are attracted only in one direction: towards the inside of the drop. In this way the molecules on the outside of the drop form a sort of skin, and exert an inward pressure on the inside of the drop. They keep the drop together and put it under pressure. Surface tension even gives such a strong "film" on a water surface that insects can walk on it (Fig. 1.17).

Classical physics tells in which mathematical form the surface tension can be formulated, and I could include this correction term in my equations. From that time

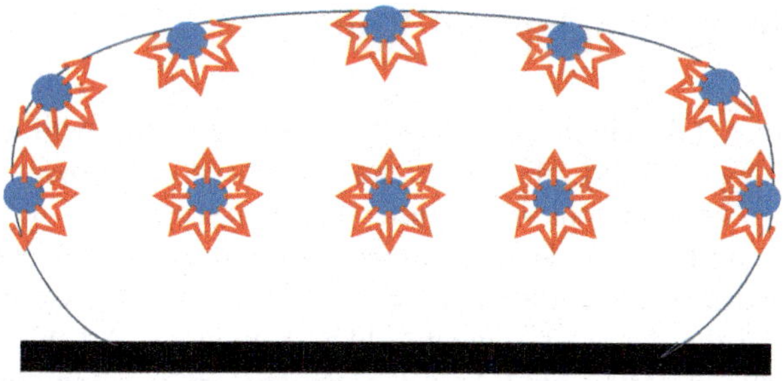

Fig. 1.16 Mutual attraction of the molecules inside a water droplet gives rise to surface tension

Fig. 1.17 Surface tension makes walking on water possible

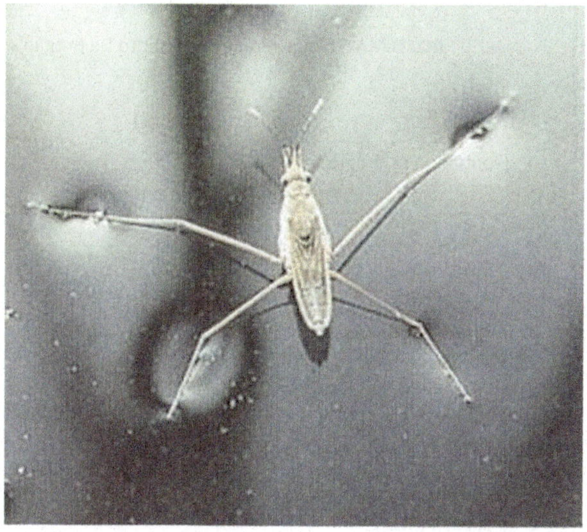

on, it was better to speak about a "droplet of charge" instead of "a cloud", and this is the word used in the rest of this book.

To summarise: the "droplet" model thus corresponds with a time-averaged view of the real "zittering" electron, with an added "apparent force" in the form of surface tension, so that without restrictions Maxwell's equations may be applied.

There was still a problem, though. Although classical physics books could give me the *form* of equations expressing the surface tension, they did not give me the absolute *magnitude* of the apparent force to be added. Of course not: my surface effect, intended to correct false electrostatic phenomena, is physically quite a different thing than the cohesion in a water drop. Later, I was able to determine the actual magnitude, but I leave that as a story to be told in later sections of the present chapter. For now, let me tell you what could be concluded even without the full quantification.

1.15 Pulsating, Hopping and Beating

I told you in Sect. 1.12 about the games I had been playing with the equations of motion describing a cloud of charge, now baptised a "droplet of charge". It involved setting up equations of motion, inspecting them for terms not usually found when a marble-like object is moving inside an oscillator, and investigating whether these extra terms would gain significance when downscaling and whether they so could be the cause of "quantum-like" behaviour. All this will be told in more detail in the present section, and the whole story is summarised in the road map below.

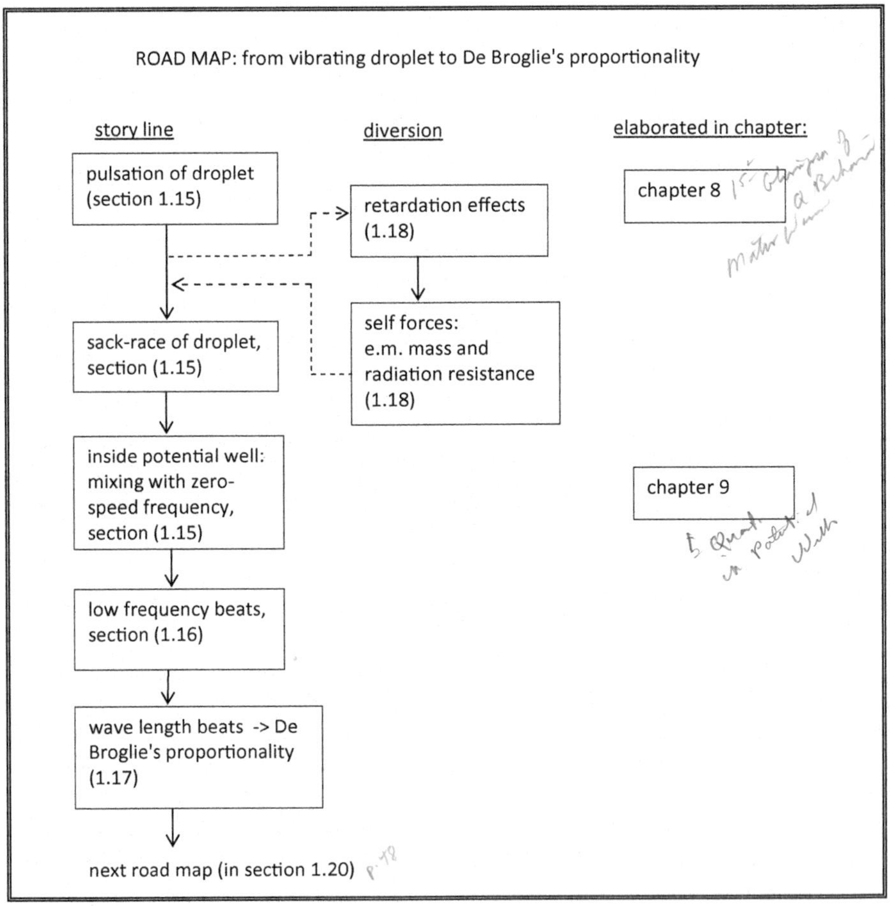

Let me start at the end: I will first explain what sort of quantum-like phenomena were found during the first step in the process of scaling down from macroscopic scales (not yet assuming the droplet to have a charge as small as the unit charge).

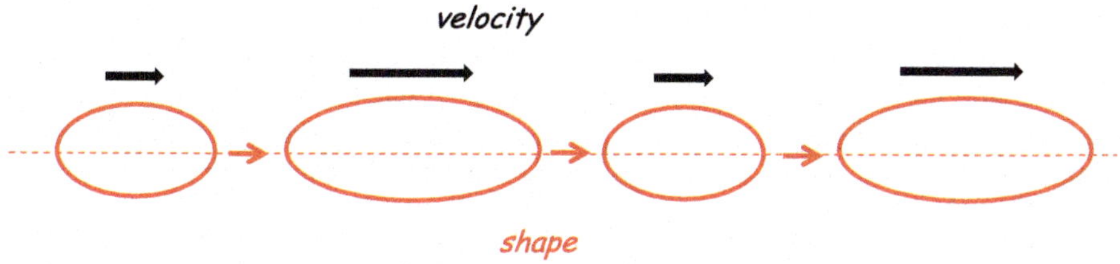

Fig. 1.18 Pulsation and velocity fluctuations of the droplet

What immediately attracted attention was, that the relation of De Broglie appeared. Not yet in the final quantified form, but definitely in the form of the proportionality $p \div \frac{1}{\lambda}$. As you perhaps remember, this relation says that the momentum p of a particle (momentum is mass times velocity: $p = m \cdot v$) is inversely proportional to the wavelength λ of these mysterious "matter waves", waves no one knows what they really are except that they are functioning as "pilot waves".

Whatever in the case of a droplet takes on the role of "matter waves"? Well, it is something more down-to-earth than "ghost waves", but it needs a bit of explaining.

According to the equations of motion the droplet is going to oscillate in *size* and *shape* next to its back and forth translation motion in the potential well. It periodically swells and shrinks, a motion soon baptised in my investigation as the *pulsation* of the droplet. Because, as I have explained, the theory was simplified from three dimensions to one, the pulsation then looks more like a periodic stretching and shrinking of the droplet, as sketched in Fig. 1.18. This pulsation has a very high frequency, and is accompanied by variations of the speed of the droplet (the coupling between the two kinds of motion may not immediately be obvious, and I have to come back to this matter later). The translation motion of the droplet thereby becomes "jerky", the droplet traverses its trajectory in a series of jumps, so to say. I sometimes compare it with the "sack race" poor parents have to take part in at the last school day of their children.

The crucial point is that the frequency of the pulsation depends on the average velocity of the droplet. If the droplet is standing still we find a certain—already very high—value of the frequency, which will be called the "zero-speed frequency" f_0.

When the droplet acquires some velocity, the frequency of pulsation f becomes even higher: $f > f_0$.

We can define a *shift* of the pulsation frequency due to velocity:

$$(\textit{shift of pulsation frequency}) = (\textit{pulsation frequency at speed}) \\ - (\textit{zero speed frequency})$$

or, in the short notation of mathematics:

$$\Delta f = f - f_0$$

This is just a definition, explaining the word "frequency shift", and it is not yet very interesting. But now we come to something that will appear crucially important in the remainder of this chapter, and which is really the key to understanding how "De Broglie" is found. It is the following property of the frequency shift, as determined by the detailed mathematical analysis in the appendix:

$$(\textit{shift of pulsation frequency}) \div (\textit{average speed of droplet})^2$$

The shift of frequency due to velocity thus increases progressively, in fact it increases very fast in a quadratic way. In mathematical short notation:

$$\Delta f \div \left(v_{average}\right)^2.$$

I have put this proportionality in a grey box, because everything in the following, including the noted quantum-like behaviour of the droplet hinges on this relation.

If the droplet is moving back and forth within a potential well, its velocity is varying from zero (at the instant it reverses its direction of motion upon reaching one of the walls) to a maximum value somewhere in the middle of the well. The frequency of the pulsation therefore varies between the "zero-speed frequency" and some higher value.

Let us first consider a special case for the sake of simplicity, the so-called potential *box*. This may in the macroscopic world be compared with a flat bowl having high vertical walls (Fig. 1.19).

In such a potential box the droplet has a constant speed, until it hits one of the walls and is bounced back. After the bounce it will have the same speed in the other direction. At first sight one would presume that the droplet would have the same frequency of pulsation all the time. However, this is incorrect because there are brief moments interrupting the constant velocity, when momentarily the speed is

Fig. 1.19 The macroscopic analogue of a potential box

zero during the encounters with the walls. In the mathematical appendix it is shown that these interruptions, no matter how short they are, have a profound influence on the pulsation. The pulsation will now have the character of a superposition of two frequencies, viz. on the one hand the frequency which corresponds with the velocity in between the walls, and on the other hand the zero-speed frequency.

Perhaps the idea that an object can vibrate in two different ways at the same time is something you find a bit strange. But think about the strings in a musical instrument like a violin or a piano, or the column of air in a trumpet or an organ pipe: their vibrations are also a superposition of the ground tone and "overtones", these overtones giving the special timbre to the instrument.

The mix of the two frequencies in the case of the pulsation of a droplet of charge takes a peculiar form. The two frequencies are very close to each other, and we will observe a special kind of interference, which is called a "beat phenomenon", explained in more detail in the next section.

It is not only the pulsation that becomes the superposition of more than one frequency if the droplet is trapped in a potential well. The same applies to the fluctuations of the velocity around its average value. Once again, I will have to come back to the matter, but it was already mentioned that the velocity is "dragged along" with the pulsation. The "sack race" of the droplet will therefore be a mix of more frequencies too.

1.16 Interference and Beats

Let us return again to the simplest situation, viz. the potential *box*, where we are dealing with the mixture of two different frequencies. It was announced that this leads to a special kind of interference, the so-called beat phenomenon.

In the days of large multi-engined propeller airliners, one could easily hear—and feel!—whether the engines were properly synchronised with each other. The sound of a large piston engine with a propeller is a sort of low rumbling noise, in the frequency range of several hundred cycles per second. The pressure variations in the air caused by each engine separately are perceived as a low tone. But if any one of the engines is slightly "out of step", running a little bit faster than the other ones, a periodic variation of the loudness of this rumble is heard. The loudness of the engine rumble then slowly swells and drops again, which cycle will repeat itself continuously. We call these slow variations of the noise level "beats", caused by the interference of two noise sources. The word "beat" describes almost literally what is noticed within the airliner: successive shakings occur in the aircraft structure, slowly rising and falling in intensity, and felt as "shivers" going through the airframe.

The frequency of the train of beats (i.e. the number of beats we hear per second) is low, it is equal to the *difference* in frequency of the two sound sources.

The cause of the beat phenomenon is easily explained by drawing some lines on paper, as has been done below (Fig. 1.20). Patterns of white and black lines can be

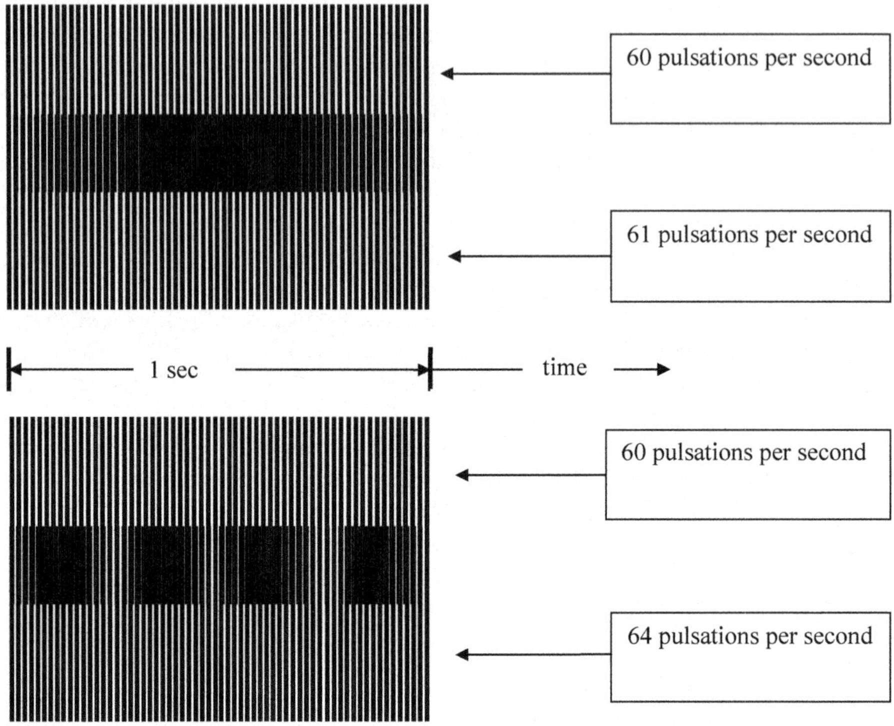

Fig. 1.20 Beats formed by overlapping line patterns

imagined as ripples or waves on a surface. The uppermost part of the line diagram shows 60 black and white lines in a length that is taken to correspond to 1 s on the time scale. This part thus symbolises waves having a frequency of 60 Hz (Hertz) or, which is the same: 60 cps (cycles per second), meaning that there are in every second 60 ripples.

Lower down, a pattern of 61 lines is drawn in the length that symbolises 1 s on the time scale. The two sets of lines overlap in the middle, and we can clearly see the so-called beat-pattern, caused by the interference. In this example we have obtained 1 beat per second. In other words, the beat frequency is 1 cycle per second, or 1 Hz.

An illustration that the number of beats in a second is always equal to the difference between the two interfering frequencies is given by the second line diagram. Here we have one set of ripples, again with a frequency of 60 Hz, and another with a frequency of 64 Hz. In the overlapping region we now see 4 beats. The number of beats in the length symbolising 1 s is thus 4, in other words the beat frequency is 4 Hz. And this is precisely the difference between the two frequencies of 64 and 60 Hz.

Piano tuners use the phenomenon of beats to discover if there is still a tiny difference between the tone of a tuning fork and that of the instrument. They will hear beats if the piano is slightly out of tune. The standard a-tone on the piano is 440 Hz, i.e. the a-string should correctly vibrate 440 times every second, and this is the tone given off by the tuning fork. If the string is actually vibrating with 442 cycles per second, we will hear the intensity of the sound go up and down twice per second when the tuning fork and one of the strings have been struck simultaneously. The beat which is being heard has a frequency of 442 − 440 = 2 Hz. Such a slow variation of the sound intensity can be heard easily, certainly by a trained ear.

Let us now return to our droplet of charge moving inside a potential box. The pulsation is in this case a superposition of two frequencies, viz. the zero-speed frequency and the slightly higher frequency the pulsation acquires when it is moving at a certain speed from wall to wall. We will thus obtain a beat phenomenon, meaning that the intensity of the pulsation periodically increases and decreases again. The number of beats per second is given by:

$$(\textit{beat frequency}) = (\textit{pulsation frequency at speed}) - (\textit{zero speed frequency}).$$

One could also say:

$$(\textit{beat frequency}) = (\textit{shift of the pulsation frequency due to velocity})$$

But now remember that this shift was proportional to the square of the velocity (look back at the relations in the grey box above):

$$(\textit{shift of pulsation frequency}) \propto (\textit{average velocity of droplet})^2$$

Taking everything together, the final result is:

$$(\textit{number of beats per second}) \propto (\textit{average velocity of droplet})^2$$

If some readers would prefer the short notations used in mathematics, this relation reads simply

$$f_{beats} \propto (v_{average})^2$$

The droplet that is moving in the potential well will now show a pulsation with a slowly rising and falling intensity (in jargon: the *amplitude* of the pulsation is varying).

Similarly the fluctuations of the droplet's velocity periodically become more intense and then subside again until the next beat comes on. To illustrate it, Fig. 1.21 shows—schematically—how the "sack race" of the droplet looks after it has been modified by the interference effects.

To read Fig. 1.21, imagine that the droplet is passing you on its way from left to right in the potential well. At each instant of time you will see the droplet in another

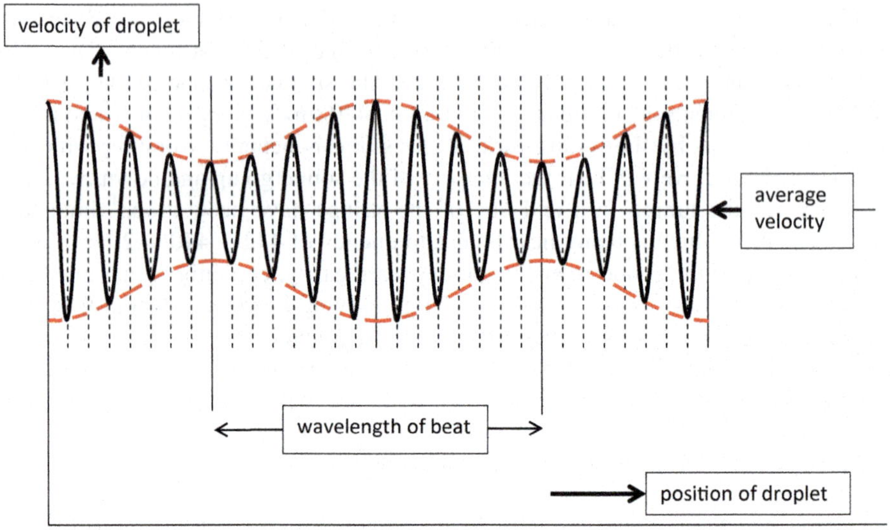

Fig. 1.21 The beats in the velocity fluctuations

position. The position of the droplet is given along the horizontal axis in the graph. Vertically, the velocity of the droplet is shown. It is a fast fluctuation around an average value. The graph shows that beats are the phenomenon that this fluctuation becomes periodically larger and smaller. In the figure it is also indicated what we call the *wavelength of the beats*: it is the distance travelled by the droplet between two consecutive beats.

1.17 The Wavelength of the Beats Depends on Velocity as in De Broglie's Relation

As you will suspect from all that has been said above, the wavelength of a beat is bound to depend on the average velocity of the moving droplet. Algebraically, the relation between this wavelength and speed of the droplet requires just a few lines to derive. However, I promised that you would not be bothered by formulae, so I will explain it by giving an example from everyday life. Readers who want to see the algebra are referred to the separate textbox, and others can disregard the textbox.

In Fig. 1.22 a truck is shown unloading pylons, for instance to mark a piece of the road where repairs are necessary. The operator on the truck heaves pylons from the truck and puts them down on the road with a certain frequency, where "frequency" in this case means: so and so many pylons per second are handled. This is sketched in the upper schematic. The distance travelled by the truck between two consecutive pylon droppings may be called the wavelength (also shown in Fig. 1.22).

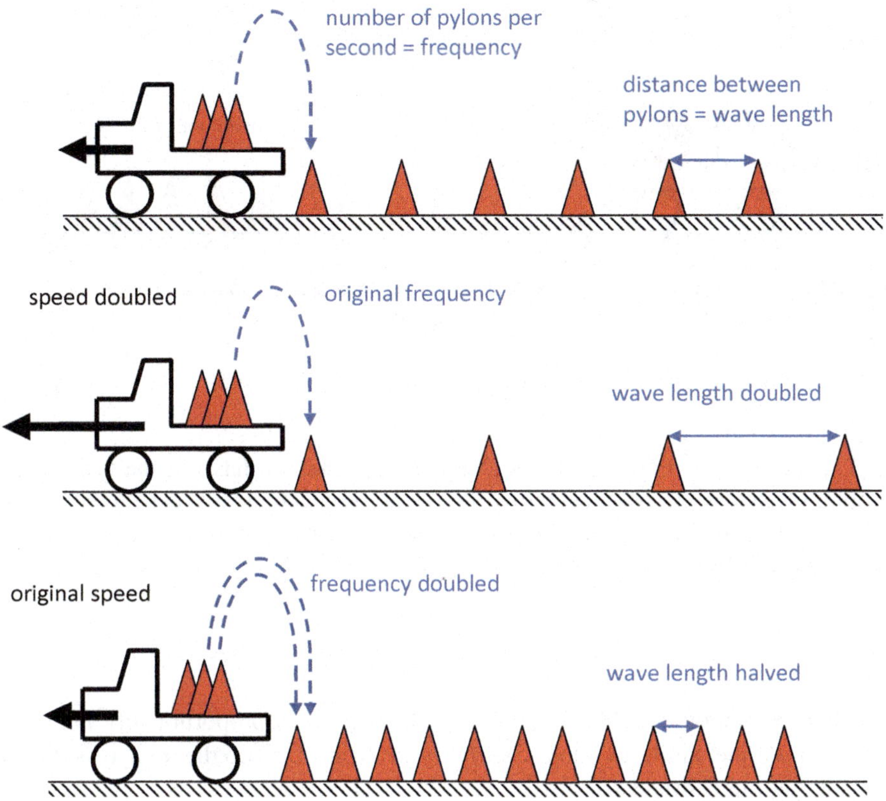

Fig. 1.22 (*First part*): analogy showing relation between speed, frequency and wavelength

On the second line is sketched what happens if the truck has doubled its speed, whereas the operator maintains his original handling frequency. The result is that the distance between the pylons on the road (the "wavelength") has become twice as large.

In the third line another situation is sketched, where the truck has maintained its original speed, but the operator works twice as fast (frequency doubled). The "wave length" has now halved.

But now the trick. We have a special kind of operator, who becomes more active the more the truck's speed increases. An increase of speed now has two counteracting effects, one directly due to the truck's speed, the other due to the fact that the operator works faster the higher the speed of the truck is.

In the special case that the frequency of putting down the pylons is proportional to the square of the speed, i.e. if $f_{pylons} \div (v_{truck})^2$, the effect of the frequency dominates over the direct effect of the truck's speed. We see from the schematic below that doubling the truck's speed now leads to halving the distance between the pylons on the road, or:

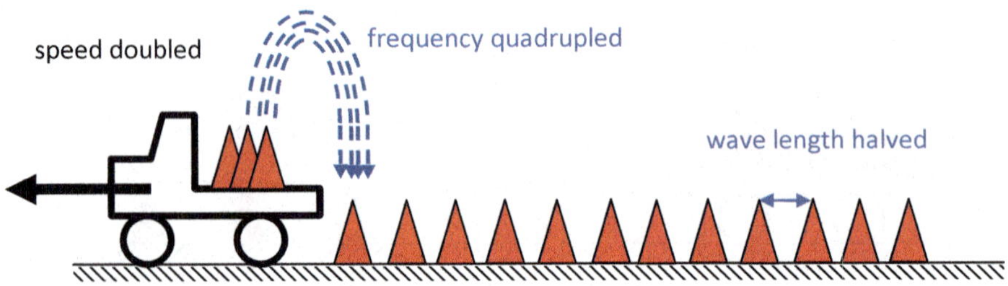

Fig. 1.22 *(continued)*: the wavelength if the frequency depends quadratically on the speed

$$\lambda_{pylons} \div \frac{1}{v_{truck}}$$

The situation of Fig. 1.22 (cont'd) is exactly what we have in the case of the droplet of charge. The frequency (of the beats) is quadratically dependent on the speed (of the droplet), so that the wavelength of the beats shows the same relation as the wavelength of the pylons:

$$\lambda_{beat} \div \frac{1}{v_{average}}$$

which is a relation remarkably similar to De Broglie's proportionality $m \cdot v \div \frac{1}{\lambda}$. You can see the similarity more clearly by inverting De Broglie's expression and leaving out the constant factors:

$$\lambda_{matter\ wave} \div \frac{1}{v_{particle}}$$

The conclusion in words is:

> These results suggest that "matter waves" could perhaps be compared with the "trains of beats" found in the pulsation and velocity of a droplet of charge.

I am somewhat cautious in this formulation, because it is not allowed to definitely *identify* matter waves with trains of beats, as long as everything is still so vague as a proportionality relation. One has to quantify the expressions, i.e. we have to find formulae where the proportionality sign "$\div$" is replaced by an equal sign " = ". As you have seen, it all hinges on the proportionality $\Delta f \div (v_{average})^2$ which earlier was put inside a grey box because of its importance, and which said that the shift of the pulsation frequency due to the droplet's velocity is quadratic.

If the proportionality constant in $\Delta f \div (v_{average})^2$ can be quantified, then we will have a real equation, and only then are we allowed to make a more definite identification. Do not worry, it will be done.

Text Box: wavelength of beats
The wavelength λ of a beat is defined as the distance travelled by the droplet beween two beats. Hence, it is the velocity of the droplet multiplied by the time one single beat takes:

$$\lambda = v_{average} \cdot T_{beat}$$

If it is said that the beat frequency is f_{beat} we mean that in 1 s there are a number of cycles equal to f_{beat}. The time per beat cycle is thus $T_{beat} = \frac{1}{f_{beat}}$. According to the first formula we thus find

$$\lambda = v_{average} \cdot T_{beat} = \frac{v_{average}}{f_{beat}}.$$

But we also have the relation $f_{beats} \div (v_{average})^2$, so that the final result is

$$\lambda = \frac{v_{average}}{f_{beat}} \div \frac{1}{v_{average}}$$

and this is precisely the same as De Broglie's proportionality $m \cdot v_{average} \div \frac{1}{\lambda}$.

1.18 Retardation and Self Forces à La Lorentz

Before proceeding as promised in the last sentence, there is still another job to be done. In the road map in Sect. 1.15 this job is indicated as the diversion from the straightforward story line. In the text accompanying Fig. 1.18 it was stated that the pulsation of a droplet also causes fluctuations of the droplet's velocity. The pulsation "drags along" the velocity, so to speak. Now, this is not at all evident. On the contrary, it is one of the famous results of classical mechanics that deformations of a body do *not* influence the motion of the centre of gravity. The most extreme example, always mentioned in school books, is the exploding cannon shell. The shell falls apart in a thousand small pieces, but the centre of gravity continues as if nothing had happened, the centre of gravity moves on and its trajectory is curved by gravity like before the explosion.

This seems to be in conflict with the coupling which was stated to exist between the pulsation and the translation of a droplet of charge. An explanation is certainly required, the more so since such an explanation will highlight a few peculiarities of

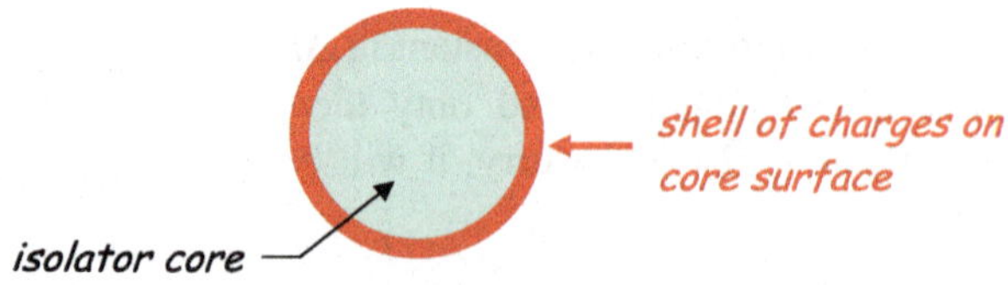

Fig. 1.23 Lorentz's electron model

a droplet of charge that will be needed later too, when we come to quantifying De Broglie's relation.

Now back to the question why the pulsation of a droplet influences its translation and makes it a "sack race". Thanks to the road map, I no longer hesitate to make further diversions, and go first to a well known phenomenon in astronomics.

The light from even neighbouring stars reaches us only after many years, so that if we see such a star it might be possible that it no longer exists at that very moment. Anyway, it is certain that the observed star is at the moment of observation no longer at the same place where we see it. The signal delay is, with a posh name, called *retardation*. It is present in all electromagnetic phenomena (after all, light is an electromagnetic wave), even on the scale of sub-atomic particles such as electrons. Considering how small an electron is and how fast electromagnetic waves propagate, one can hardly believe that the retardation effect could be of any importance on the atomic scale. Nevertheless, it is. This was already shown by Lorentz, who studied one of the early, now abandoned, models of an electron shown in Fig. 1.23.

His model was a rigid one, in contrast to our droplet model. It also is in contrast to the presently in physics adopted view of electrons as point-like objects (singularities), Lorentz assumed that the charge was distributed in the form of a shell covering a sphere of isolator material. The isolator core was needed to prevent an explosion due to repulsion between all the parts of the charge, the binding forces between the charge and the isolator would prevent such a calamity. The spherical shell of charges is "glued" on the surface of the core, to use the pictorial language of Feynman.

Now consider two different elements of the charge distribution, and let us assume for now that the shell as a whole is at rest. The two elements considered will mutually repel each other, the repulsion force on each is equal in magnitude but opposite in direction. Therefore, the *total* force on the pair taken together is zero, and the same conclusion applies to all pairs of elements, and indeed to the entire shell of charges. The electrostatic repulsion does not cause a force on the shell as a whole.

Let us now assume that the shell is accelerating, which means that any element has a velocity increasing in time. As an outside observer we see that all the elements at any instant of time have the same velocity, although it is increasing. But that is not what is being "seen" by the elements themselves. Due to the retardation effect an element will "see" other elements at a place where they were some time ago, and what is more: it will "observe" that the other elements have the velocity of some time ago, which is smaller than its own velocity at the present time. The situation

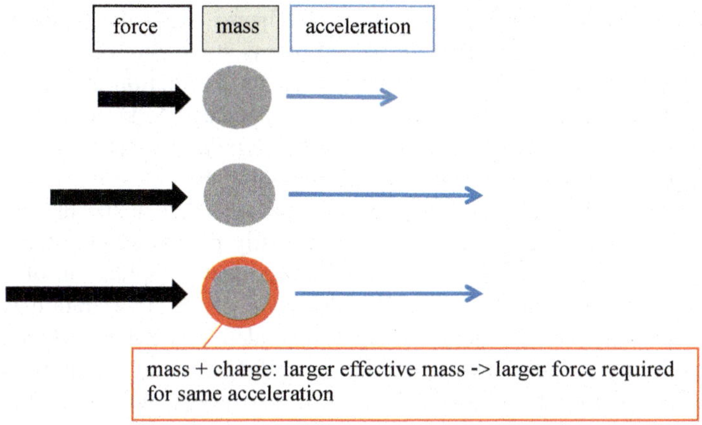

Fig. 1.24 Electromagnetic mass

becomes extremely complicated by all the different transmission delays between all the different elements of the total charge. One can easily believe that the nice balance of forces existing when the thing was at rest is completely lost during an acceleration.

This is what Lorentz worked out. He wanted to know how large the forces due to the unbalance—baptised *self forces*—would be.

First he found that electrically charging a body adds to its inertia. This appeared to be one of the consequences of the retardation effects. The effect is pictorially illustrated in Fig. 1.24.

An uncharged body needs a certain push to give it an acceleration, and a proportionately larger push if one wants to give it a larger acceleration. If the same body is given an electrical charge, then suddenly for the *same* acceleration a still *larger* force is needed. The inertia has been increased by the electrical charge. Formulated differently: it is as if the mass of the body has become larger. This mass associated with charge was baptised *electromagnetic mass*. It is a consequence of the retardation effects, in other words it is caused by the fact that light (or in general: electromagnetic waves) has a high velocity but not infinitely high.

Once again: considering how small an electron is and how fast electromagnetic waves propagate, one can hardly believe that the retardation effects could be of any importance on the atomic scale. But they are!

The electromagnetic mass, according to Lorentz's sums, would be inversely proportional to the radius r of his spherical model: $m_{em} \div \frac{1}{r}$. It thus becomes larger the smaller the sphere. This is one of the reasons for the infinities in the theory about electrons, for if it is assumed (as in present quantum theory) that electrons do not have dimensions, the mass would become infinite, which is clearly not true. Feynman in his book tells that numerous attempts have been done to revise the theory, including an attempt by himself. One would think that in the modern quantum theory such problems have been solved. Alas, no success as yet. One has found ways to work around it (you may have heard about the method of renormalisation) but this is rather a cure of the symptoms, not a cure of the basic disease.

Lorentz also found another self force caused by retardation effects, which was called *radiation resistance*. Radiation resistance appeared to be proportional to the "jerk" of the charge, i.e. it is proportional to the *change* of acceleration of the shell. The word "jerk" is not very well known although the phenomenon itself is certainly well known by bus passengers. A steady deceleration when the driver smoothly applies the brakes can be overcome relatively easily by passengers standing in the aisle, they just tense their muscles. But a sudden heavy application of the brakes will launch them all over the place. The *increase* of the deceleration is here the disastrous factor. It is called "jerk". Electrons are also sensitive to jerk, and resist it. They respond to the jerk by generating a self force, the radiation resistance. Whenever an electron is subjected to jerk, it also emits radiation. The radiation resistance may be viewed as a recoil force, similar to the kick to one's shoulder if a gun is fired: "shooting" radiation away gives radiation resistance as a reaction. The electron loses hereby energy, and the surplus energy is found in the energy of radio waves. We might say that radio waves are analogous to the cries of complaint by maltreated bus passengers. The radiation resistance force, and the many riddles surrounding it, will later be considered in more detail, so do not despair if all is not too clear yet. For now, the most important thing is to remember the term "jerk", because that will occur more often in later sections.

My own sums (written out in full detail in the formula appendix), yield essentially the same results as Lorentz's computations. Of course, my model is quite different, and the mathematical method I apply is entirely different as well. And naturally, since my model is deformable, I find more self forces than only electromagnetic mass and radiation resistance.

What is relevant for the present discussion is the electromagnetic mass. From the mathematical analysis in the appendix it is found that this mass depends on the elongation of the droplet of charge. A proportionality is derived, with the following form:

$$(electromagnetic\,mass\,of\,droplet) \div \frac{1}{(length\,of\,droplet)}$$

Evidently, this expresssion is very similar to what Lorentz had found for his—entirely different—model. The crucial difference with Lorentz's result is, that in the case of the droplet of charge the length is variable. The continual stretching and shrinking of the droplet was called the pulsation of the droplet. We see from the result shown above that there is a corresponding variation of the electromagnetic mass. Now having another look at Fig. 1.18 this explains why the pulsation also affects the translation velocity of the droplet: together with the pulsation, the *mass* of the droplet will be *varying*. When the elongation of the droplet is large, its mass will be small, and vice versa. This is bound to have an effect on the velocity of translation, of course. Here is also found what is the difference with uncharged bodies, where deformations do not affect the total mass. Much more could be said about it (see Chap. 5), but for now the explanation of Fig. 1.18 will here be concluded.

1.19 Quantifying "De Broglie's Equation" of the Droplet Theory

An identification of "matter waves" with "trains of beats" would be more convincing of course, if one could quantify the proportionality constant in the expression

$$\lambda_{beat} \div \frac{1}{v_{average}}$$

so that an expression would be obtained like

$$\lambda_{beat} = \frac{K}{m \cdot v_{average}}$$

If it would be found that the value of K, as predicted by the droplet theory, is practically the same as Planck's constant h in De Broglie's formula

$$\lambda_{matterwave} = \frac{h}{m \cdot v_{particle}}$$

then the identification would be complete. One can hardly expect that K would ever be found to be *exactly* equal to h, because of the many approximations in the droplet theory which have been introduced to make the mathematics manageable. But a correspondence of the orders of magnitude of the two constants would, in my opinion, be great. It would make people wonder if such a similarity between the two results can really be pure coincidence.

But how would I ever be able to determine the value of "my" constant K? Recall that in the droplet model of an electron all the electromagnetic self forces could be calculated with some precision (apart from the approximations just mentioned). The problem is in quantifying the surface tension. This was just an "apparent" force, something added to correct an oversimplification in the droplet model. So, it is not even a real physical force, and one cannot look up its properties in physical handbooks.

The solution was eventually found, after a long time. As so often, *after* having found how the problem could be tackled, the solution appeared simple. And it is really not very difficult to explain. Again, a road map may help to see how we navigate, including some necessary diversions, to De Broglie's equation. The road map is shown below, on the next page.

1.20 The Equilibrium Length of a Droplet

Figure 1.25 is essentially the same as the earlier Fig. 1.18, but sketched from a different viewpoint. We are now not concerned with the translational speed, and want to focus on the pulsation. What is shown in Fig. 1.25 is what we see when sitting on the droplet: pulsation, in the form of periodically stretching and shrinking of the droplet. The red contour depicts the so-called equilibrium shape, where the squeezing due to surface tension and the expansion due to electric repulsion just balance each other. The pulsation comes about in the same way as with any other oscillator: if the droplet at any instant happens to be longer than the equilibrium shape, it tends to shorten itself, in the direction of the equilibrium. However, due to inertia it will overshoot the equilibrium point, and thus continues its oscillation.

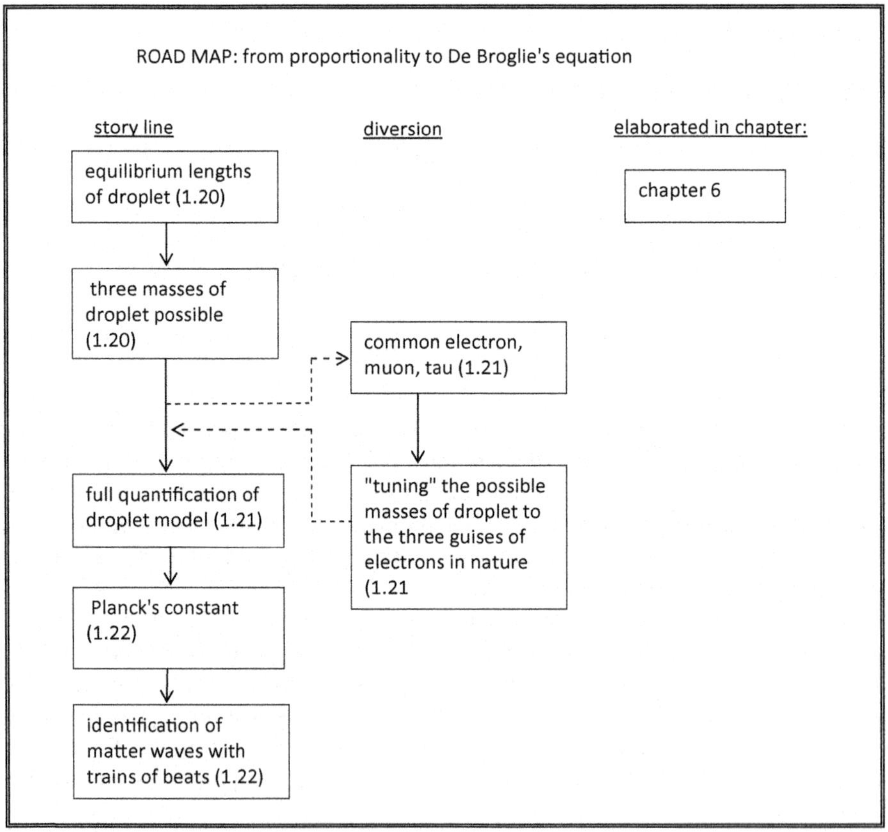

It is interesting to have a closer look at the equilibrium configuration. How would we be able to determine the length of the droplet where the explosion tendency (due to repulsion forces) and the squeezing tendency (due to the

Fig. 1.25 Pulsation around an equilibrium length

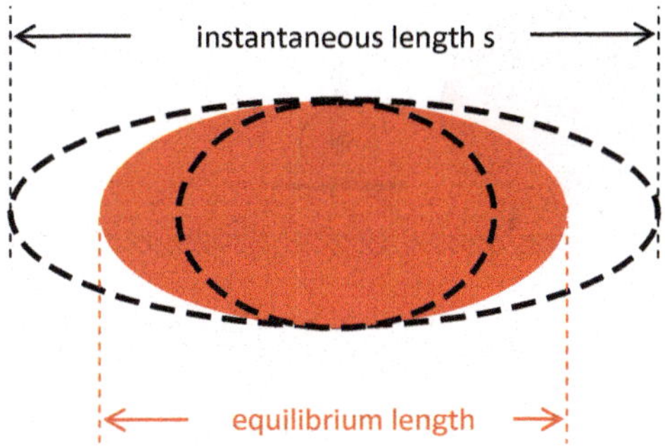

surface-tension-like forces) are exactly equal to each other? Let us go back to the simpler oscillator drawn in Fig. 1.10. Here we had a mobile charge in between two fixed charges, and it was immediately obvious that the middle would be an equilibrium position for the charge mounted on the toy railway wagon, if only because of the symmetry of the set up. Now look at the slightly more complicated set up drawn in Fig. 1.26. What the drawing symbolises is that on the left side there is more fixed charge than on the right side. In this case we can find the equilibrium position with the help of a force diagram, as is also shown in the lower part of Fig. 1.26.

The red line shows the repulsion by the large fixed charge and how it depends on the position of the mobile charge. It is a force on the wagon from left to right, large when the wagon comes close to the charge and decreasing the further the wagon's position is away from the fixed charge. Similarly, the blue line shows how large the repulsion is by the smaller fixed charge on the right. If we have calculated this diagram, we can immediately pinpoint the equilibrium position of the wagon: it is the place where the red and blue lines cross each other, because there the force from the left is equal to the force from the right.

The mobile charge can perform oscillations around this equilibrium point. Small oscillations are symbolised by the two dashed lines, giving the positions between which the wagon will roll back and forth after it has been disturbed.

In the same way we can determine what will be the equilibrium length of a droplet. It is slightly more complicated than the above shown simple oscillator, but the principle is the same. Have a look at the blue line in Fig. 1.27. This line gives how large the expansion tendency is, depending on the elongation of the droplet. The general shape of the blue curve is quite understandable: on the left side the droplet is assumed to be short and compact. The separate elements of the charge are here closely packed together, and the tendency to expand will be large. On the other hand, on the right side of the figure the length of the droplet is assumed to be relatively large, the charge elements here have more living space and are not too frantically striving towards more room. Nevertheless, the expansion tendency never vanishes completely, however far the droplet has been stretched. The blue curve can be calculated accurately (within the limitations set by the modeling approximations), since everything is known and quantifiable in electromagnetic theory.

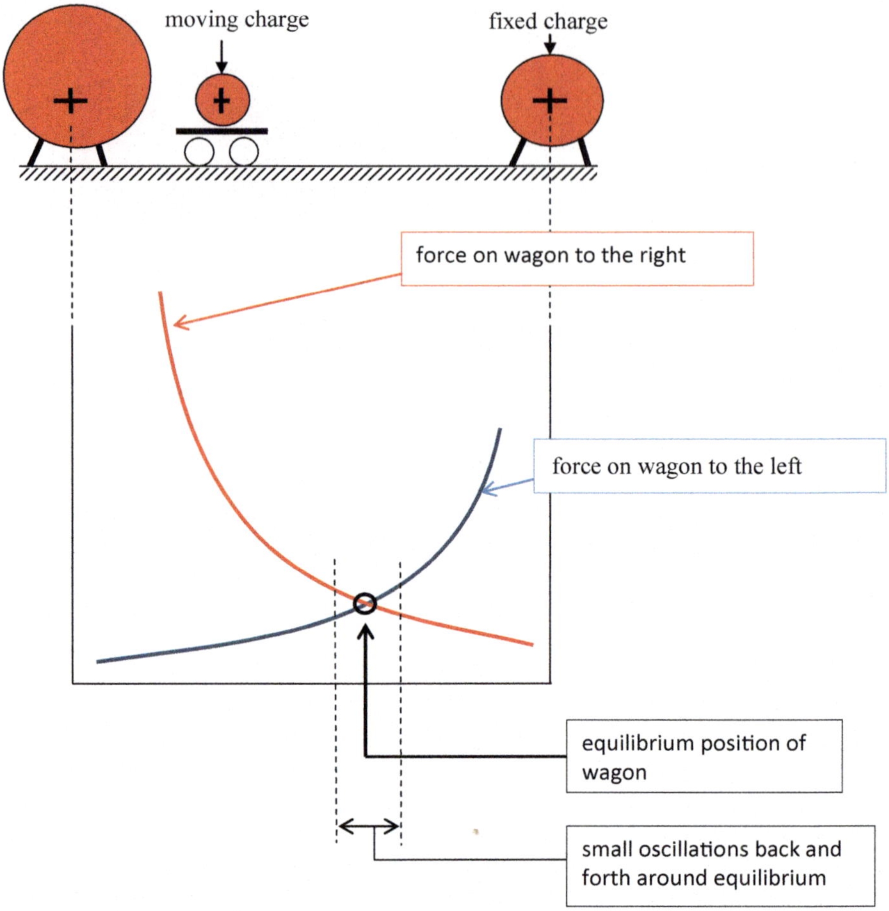

Fig. 1.26 Asymmetric electrical oscillator formed by a charge in a potential well

The red line is a different kettle of fish. The red line depicts the tendency to squeeze the droplet by the surface-tension-like apparent forces. It is (not yet) quantifiable, but it is neither completely arbitrary in shape. Surface tension depends both on the *size* as well as on the *shape* of a droplet. A small size (left hand side of the figure) gives a large squeezing tendency, since there is relatively more surface (compared with the volume) than in large droplets. Compare it with water drops: small droplets are very stable, but large drops can easily fall apart into smaller ones, so much so that very large drops cannot even exist. The left side of the red curve therefore bends upwards, but scaling laws show that the red curve must be less steep than the blue curve. Apart from size, shape also plays a role: the more the shape of the drop differs from a sphere (right hand side of the figure), the larger the tendency to contract back into a spherical shape. It causes that the red curve bends

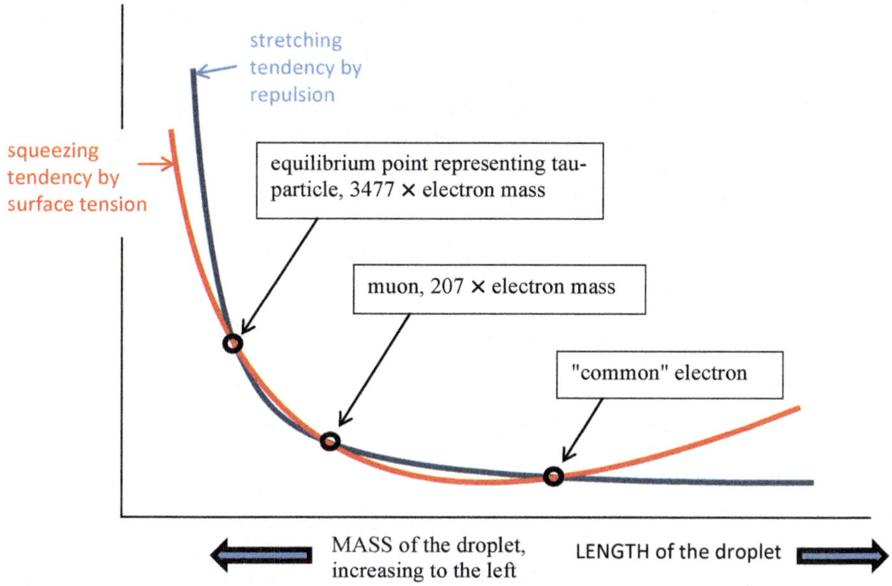

Fig. 1.27 Stretching and squeezing tendency as a function of the droplet's length. The three equilibrium points are identified with the common electron, the muon, and the tau particle

upwards again on the right side (i.e. it has an increasing tendency to squeeze back to the sphere shape, the longer it is stretched out).

One suspects (and this is supported by general algebraic rules) that there will *always* be *three* points where the blue and red lines intersect. An intersection point physically means that here the squeezing tendency and the expansion tendency are equal, so that such a point represents equilibrium. We thus see that there are three possibilities for having equilibrium, at three different values of the elongation of the droplet.

As in the earlier Fig. 1.26, in general there will be oscillations around an equilibrium point. In the case of Fig. 1.27 around *each* of the three equilibrium points we can have small oscillations. These small oscillations in this case are what we called the pulsation of the droplet around its equilibrium length, as depicted in Fig. 1.25.

The exact position of these three equilibrium points is not known, because the mathematical expression for the surface effect contains several unknown coëfficients. One could say that by properly choosing the value of these unknown coëfficents we have "knobs" that can be turned to adjust the shape of the red curve. By turning one of the knobs we can for instance shift the curve up or down. Turning another knob can lift the left part of the curve, without affecting too much the part on the right hand side. And there is a third knob that can do the same for the right hand side.

You may remember the phenomenon of electromagnetic mass, and the fact that its value depends on how large the droplet has been stretched (Sect. 1.18). Therefore, the three possible equilibrium points also represent three different possible values of the *mass* of the droplet. We come to the conclusion that our model of a droplet of charge implies that there are three different electron masses permitted, if we assume that the droplet is a valid model for an electron. At first sight this may seem odd. But now the "pièce the résistance": in nature the electron comes in three different guises!

1.21 Electrons in Three Guises

The fact is, that in nature there are *indeed* three different kinds of electrons! The "normal" electron is the most abundant one and was known for a long time (discovered by J.J. Thomson in 1897). Later, in 1936 the so-called "muon" was found by Anderson in cosmic rays, showers of particles raining down on the earth surface. The muon particle in all aspects is similar to the electron, except that it is approx. 207 times heavier. And still later, in 1975 during tests by Perl in artificial particle accelerators, amidst the showers of debris of broken nuclei a third electron was found, the so-called "tau particle", which is approx. 3477 times as heavy! At the time of their discovery physicists were annoyed, because these things were not really anticipated either by theory or by extrapolations from experimental results. A well known story goes that the famous physicist Rabi said a bit dismayed when he heard about the muon: "who ordered this?".

It is obvious what was my next step, and you probably already expected it: I tentatively "turned the knobs" in the theoretical model of my droplet of charge, until the intersections of the blue and red line in Fig. 1.27 exactly gave me the masses of the three species of electrons. Doing so, this provided me with a complete quantification of my model: there were no longer any undetermined constants in the formulae. Everything could now be calculated precisely (well, "precisely", as far as permitted by the approximations introduced in the maths).

Let us now return to the beginning of this section, where I expressed the desire to quantify the constant K, so that a proper comparison would be possible between the results from the droplet theory and De Broglie's formula. At this stage such a quantification became possible, and lo and behold: my factor K and Planck's constant were almost the same!

1.22 Planck's Constant

As I said before, in my opinion one would be glad if the agreement between K and Planck's constant h would be an agreement of orders of magnitude. Actually, it appears that K and h are much closer together than this, they differ just 1.5 % from each other.

I can imagine that lots of readers want to see the expression from which the constant K was calculated. The puzzling thing about it is, that the droplet theory is wholly based on classical laws of nature, so that K is also expressed solely in terms of classical constants, such as the value of the unit charge, the speed of light, etc. On the other hand, Planck's constant is a fundamental constant of nature, occurring only in the realm of quantum theory. One would not expect that it is related to the constants of nature occurring in classical physics. Remember the two separate worlds (Sect. 1.3)? For the curious readers who are not easily frightened by a somewhat complex formula, I have prepared a text box that can be ignored by other readers, and that shows you the formula for K, as well as the numerical values I used to evaluate K. The bridge that connects the two worlds in my formula is the fact that two empirical constants occur in the expression, viz. two mass ratio's that are defined as

$$\mu = \frac{(mass_{muon})}{(mass_{common\ electron})} \approx 207 \text{ and } \tau = \frac{(mass_{tau\ particle})}{(mass_{common\ electron})} \approx 3477.$$

How these parameters did creep into my theory has been explained above, they were the result of fitting the surface tension effect so that the three guises of the electron were reproduced.

All of the above may be summarised in the form of two conclusions which in my opinion have been made sufficiently plausible by now. The first is:

> The fundamental constant of Planck, governing the quantum world, can be related to a combination of fundamental constants belonging to the classical world. These two worlds are not so separate after all.

It may be that this way to formulate the results is too bold. Let me be more cautious, and just say that using the model of a droplet of charge gives the possibility to build up "mental pictures" about the relation between the two separate physics. Maybe the model of a deformable droplet of charge does not represent reality (whatever that is), but the results might serve to have "pictures in the back of one's mind" so that quantum phenomena will lose a bit of their strangeness. Just the feeling that the classical laws of nature might conspire to show up quantum-like behaviour is food for thought.

The second conclusion, to be treated with the same caution, is:

> The matter waves of quantum mechanics may be identified with trains of beats occurring in the dynamics of a particle, caused by interference effects.

Textbox: Planck's Constant

De Broglie's relation is $p = \frac{h}{\lambda}$, where p is the momentum of a particle, λ the wavelength of the associated "De Broglie wave", and h the constant of Planck.

In the theory of the droplet of charge a similar expression is found: $p = \frac{K}{\lambda}$ where

$$K = \frac{1}{c}\frac{q^2}{4\pi\varepsilon_0}\sqrt{\mu \cdot \tau} \cdot \beta \, \text{with}$$

c =velocity of light, $2.9979 \times 10^8\, m/s$

q = charge of droplet, for electron the unit charge $e = 1.6021 \times 10^{-19} C$

ε_0 =vacuum permittivity, $8.8544 \times 10^{-12} N^{-1} m^{-2} C^2$

$\mu = \frac{m_{muon}}{m_{electron}}$, the ratio of masses of the muon and electron, $\mu = \frac{105.7}{0.511} = 206.85$

$\tau = \frac{m_{tau}}{m_{electron}}$, the ratio of masses of the tau particle and the electron, $\tau = \frac{1776.8}{0.511} = 3477.1$

β =factor depending on the non-dimensional shape of the charge distribution, of order O(1).

If the values are substituted, we find

$$K = 6.52 \times 10^{-34} \cdot \beta \, \text{J} \cdot \text{s}$$

The factor β depends on the—unknown—non-dimensional shape of the charge distribution within an electron. The only thing known about it is, that its numerical value is of order $O(1)$.

The best estimate that presently can be made is, to take $\beta = 1$.

We must compare this with the value of Planck's constant:

$$h = 6.6256 \times 10^{-34} \, \text{J} \cdot \text{s}$$

The difference between K and h is thus 1.5 % for $\beta = 1$.

1.23 To Complicate the Picture: Chaos

In all the text above it was suggested (on purpose, for the sake of explanation), that the pulsation would be a nice, smooth, sinusoidal phenomenon. It sometimes is. But more often, the pulsation is more whimsical. It then has a *chaotic* character, as is sketched in Fig. 1.28. Chaotic motion is irregular, it never repeats itself, but it is not completely random. As may be seen in the figure, one can discern a certain pattern

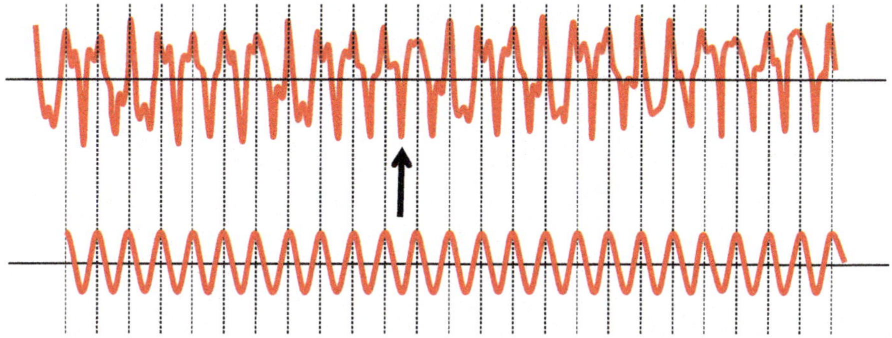

Fig. 1.28 Chaotic instead of sinusoidal pulsation

in it, not dissimilar to the pattern of the sinusoidal motion shown on the lower line. We see the same kind of rhythm: in the figure very often the peaks and troughs of the chaotic signal coincide with the maxima and minima of the sine wave. This is not surprising, because both signals are solutions of the same equation of motion.

According to the equations of motion the occurrence of chaos will depend on how the total energy of the droplet is divided between the energy in the translation and the energy in the pulsation. Do realise that at the time we adopted the model of a droplet that is able to deform, from that point on the total energy of the droplet is found in two forms: translational energy and energy bound up with the variable amount of stretching of the droplet. There are, as it were, two reservoirs of energy, and they can communicate with each other. On average the energy in both reservoirs is the same, but that is only a long-term average. During certain short time intervals the energy in the translation may be larger than average, at the cost of the energy in the stretching. At other times the reverse can happen. The equations of motion now tell us, that the smooth sinusoidal pulsation will happen at the moments that there is relatively little energy present in the translation fluctuations. Most of the time, however, when the energy is more fairly divided or when the translation even takes an unfairly large part of the total energy, we will find the chaotic character of the pulsation.

The variations of the velocity are dragged along with the pulsations. If the pulsation is chaotic—as most of the time is the case—the velocity of the droplet also will show a chaotic variation around its mean velocity. And in reverse: if the pulsation is sinusoidal then the fluctuations of the velocity are also smooth.

Figure 1.21 showed the beats in the velocity caused by mixing with the zero-speed frequency. This figure evidently applies to the situation of nicely ordered, smooth motion. Often the picture of the beats will be somewhat more whimsical due to chaos, but that concerns mainly the form of the short ripples (the parts of the red curve in Fig. 1.28 in between the vertical dashed lines). The overall picture of a beat, i.e. the process of slowly rising and falling of the amplitude, will not be affected too much.

Therefore, all that has been said about the De Broglie wavelength can still be taken seriously, despite the often existing chaos. At most there is a slight "fuzziness" in the De Broglie wavelength, it does not have a sharply defined, distinct value but instead will show a small spread around a "central" wavelength.

1.24 The Remainder of the Story Can Be Brief

So far, I have spun out the story in rather much detail, despite the fact that the title of the present chapter is "Introduction and *outline*" which suggests a short story. The reason for this choice is, that in quantum mechanics De Broglie's equation is very fundamental and is at the basis of the whole of the modern quantum theory. In Chap. 11 I will give a heuristic derivation of Schrödinger's equation, which makes it plausible that for any physical system where De Broglie's relation is applicable, the Schrödinger equations automatically follow from this, and are valid too. These equations, which describe the whole of the quantum world, are a consequence of De Broglie's first step.

Now that I have made it plausible that the classical laws of nature can conspire in such a way that De Broglie's relation is found, it cannot surprise the reader that all the other quantum phenomena can also be described by the classical laws of nature. It is the subject of the later chapters of this book to make this plausible in more detail. Since it was left to Chap. 2 to give a full list of quantum phenomena occurring in one-dimensional oscillators, we cannot delve more deeply into all these phenomena here in this introductory chapter. Most exhibitions of the strange quantum behaviour simply were not mentioned yet. However, a few phenomena have already been given as examples, so that it is tempting to tell you more about the possible interpretation of these few phenomena, if only to whet your appetite.

1.25 The Quantisation of Energy in a Potential Well

For instance, what about the quantisation of energy in atoms as well as in one-dimensional oscillators? Where does it come from, and how can classical laws conspire so that they require such restrictions which at first sight are alien to the classical world?

The explanation can only be superficial in this chapter. The later Chap. 9 will give a more extensive discussion, and the full mathematical details are found in the formula appendix. The later chapters also explain why this quantisation will be noticeable only when we descend to very tiny scales, such as the atomic dimensions. For now, let me try to give you an inkling about the fate of a pulsating droplet of charge that is locked up inside a potential well, by using an analogy.

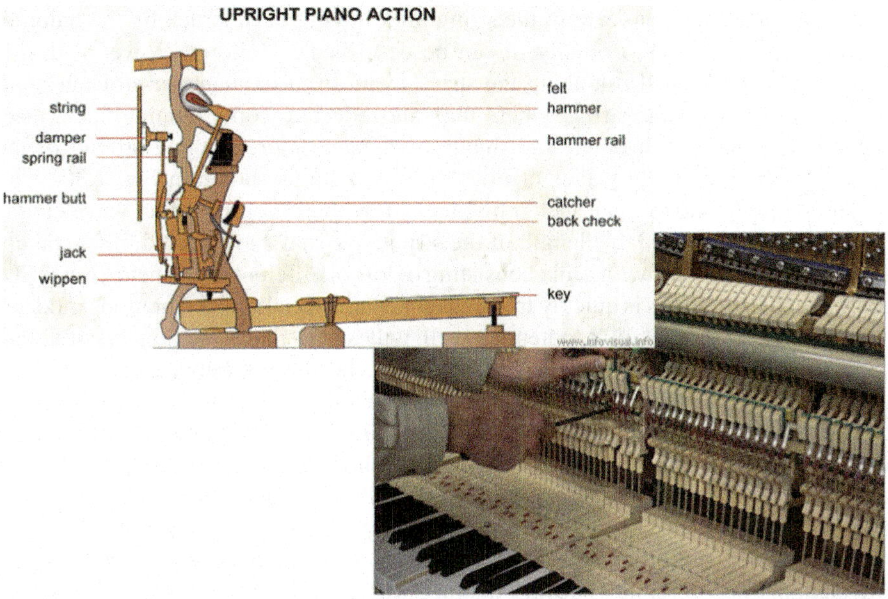

Fig. 1.29 The ingeneous mechanism in pianos to strike the strings

You may have seen the ingenious mechanism inside home or concert piano's, Fig. 1.29. If a key is pushed, the damper on the corresponding string is redrawn, and immediately following this a hammer strikes a string or a multiplet of strings and is also immediately drawn back again. As long as one keeps the key pressed down, the damper remains lifted so that the string is free to vibrate and—through the mediation of the sounding board—can make a musical tone audible. It is really a wonder of technique, the mechanism is extremely accurate and reliable, and has to be replicated as many times as there are notes on the piano (something like 87 of these precision mechanisms are present in an "ordinary" home piano). The mechanism even gives "feel" to the keys, and in reverse the "touch" of the player influences the timbre of the tone.

Fig. 1.30 Standing waves in a string: ground tone (*first line*) and overtones (*second* and *third line*)

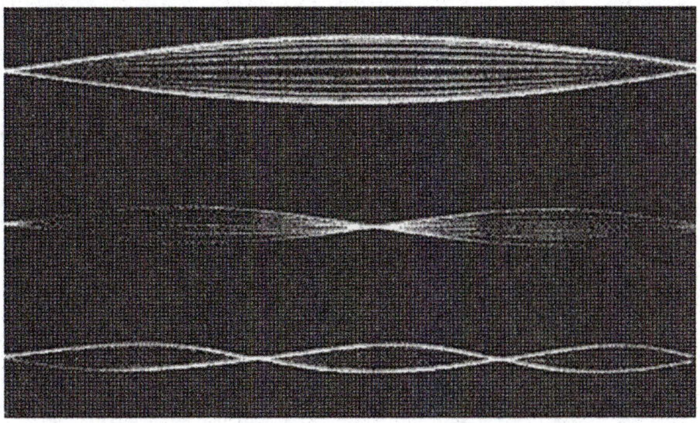

What exactly happens within the string after having been struck by the hammer is complicated physics. Roughly, it can be described as follows. Waves with different wavelengths will run along the string away from the hammer-stricken point towards the ends of the string, where they are reflected, come running back, meet and interfere with each other, and finally after some more running around of the waves result in *standing* waves of the string, looking as shown in Fig. 1.30.

The standing waves have such a frequency that their corresponding wavelengths exactly "fit" in the available length of the string. We may have started, immediately after the hammer blow, with a mix consisting of lots of different frequencies, but all the energy in the vibrations is quickly transferred towards the allowed vibration "modes".

Similarly, a droplet in a potential well will settle into a stationary situation. "Stationary" meaning that during each pass of the droplet between the potential walls the beats are repeated in the same way. This final situation is only possible if the wavelengths of the beats can exactly be "fitted" in the width of the potential well. One could, during a start-up phase, have short intervals of time where things are not so neatly ordered and are not repeating themselves during each pass of the droplet, but in the end we will have a stationary motion of the droplet.

Since under these stationary circumstances only distinct values of the beat wavelengths are possible (like the distinct wavelengths possible in the string of Fig. 1.30) this has consequences for the energy. According to De Broglie's formula, the wavelength of beats is coupled to the average velocity of the droplet. If only certain discrete wavelengths are allowed, this means that only certain velocities are allowed. The droplet cannot have any arbitrary velocity, because the accompanying beats most of the time would not fit in the available width of the potential well. One could call it a quantisation of the average speed of the droplet. The energy of the droplet, closely related to the velocity, must therefore be quantised too. I realise that this brief explanation is somewhat "sloppy" and may be found unsatisfactory. The intention here is no more than to roughly indicate why quantisation of energy will be a consequence of the pulsation of the droplet. The full story can be found in a later chapter, and in the mathematical appendix.

1.26 Tunnelling

Another phenomenon which was given as an example of weird quantum behaviour was the so-called tunnelling of atomic particles. It meant that such particles can sometimes come in places where they energy-wise could never be, according to the classical laws of nature. An indication how this may be explained by the droplet theory will now be given. The tunnelling phenomenon perhaps looks more magical than the quantisation of energy, but its explanation is easier.

It was mentioned that the mutual influence of the two types of motion, translation and stretching, also implies an exchange of energy between these so-called "dynamic modes" . The total energy is conserved, but at some time the translation may have taken some energy from the stretching, and at other times vice versa. Focusing on the translation, we see the energy in this "mode" fluctuating, sometimes it is smaller than the average value, and sometimes it is larger again.

It is easy to imagine what might be the consequence of all this, especially when we remember that the motion is chaotic. Situations may occur that the droplet, after having completed a few back and forth cycles in the well, approaches the point from which it was released when the motion started, but overshoots this point because it happens to have some translational energy left over. In extreme cases, when very much energy has accumulated in the translation motion (at the cost of the energy in the stretching) the particle "flies out of the bowl". This is just the kind of behaviour that is indeed observed in reality. As is clear from this visualisation we do not have to assume a violation of the law of conservation of energy to explain the behaviour.

In usual quantum mechanics only the translation is considered, the fact that there may exist a second "reservoir of energy" (the stretching) from which energy can be borrowed is "invisible". In this view, the surplus of translational energy that makes tunnelling possible, can only be explained by assuming that the laws of conservation of energy *can* be violated during short time intervals. In usual quantum mechanics such short-time violations are said to be due to a fundamental "uncertainty" in the energy (Heisenberg's uncertainty principle).

1.27 How Can Bohr's Radio Silence Come About?

In the main text, following the chapters where all the above is considered in more detail, you will find a few chapters (Chaps. 12 and 13) dealing with radiation phenomena. Radiation was briefly touched upon earlier in the present chapter (Sect. 1.18). Perhaps you remember that I somewhat frivolously compared radiation with the cries of complaint by bus passengers, suffering the rough braking actions by a not too attentive driver.

Because the self force of radiation resistance is rather complicated, whereas many arguments in the present section hinge on it, it may be well to elucidate the matter somewhat more than in Sect. 1.18. Imagine driving your car on a rather narrow road. You are closing in on a slow moving farmer's tractor, and you accelerate smoothly in order to have after a few seconds a higher speed at the moment that you will actually overtake the "obstacle". Then suddenly, you see in the far distance another car coming from the opposite direction. The distance is still so large that you consider it safe to continue the overtaking manoeuvre, but nevertheless you feel it wise to step more fiercely on the accelerator and accelerate the car somewhat faster. So far so good, and you can just as safe overtake the tractor, thanks to the increased acceleration and the resulting quicker completion of the whole manoeuvre.

But now compare this with the situation that your car carries electrical charges on its outside, it has picked up some static electricity (sometimes you can feel a spark when touching the metal of the doorknob). During the attempt to increase the acceleration it would feel as if there is at the same time some extra drag. In order to achieve the desired extra acceleration you would notice that the accelerator pedal has to be pushed more than you anticipated. It is as if during the overtaking manoeuvre the car splashes through a pool of rainwater, so that it needs a larger effort to achieve the desired larger acceleration. In fact, the extra drag you would feel is not due to an unexpected wet road surface, it is due to the fact that your car is functioning like a radio station. As long as there is a change of the acceleration (this was called "jerk" as you may remember), the electrical charges start to emit electromagnetic radiation. There is a recoil force associated with this emission of radio waves, and that is what you felt as the extra drag. Equivalently, one could also say: the engine has to power this radio station, and no longer all the power is available for its prime purpose, viz. to accelerate the car.

Naturally, in an actual car these effects are completely negligible, and cannot be noticed. Do not fear that overtaking manoeuvres will be hampered by some static electricity which has collected on your car. But things work out in different proportions for small charged particles. If radio waves are "shooting away" from a moving charge because it is subjected to jerk, a "thump back" is felt by the transmitting particle, similar to what happens when a gun is fired. The additional force accompanying the "jerk" is the "radiation resistance".

There are weird consequences of a self force which is proportional to jerk (to be more precise: jerk is officially defined as the increase of acceleration *per second*). The above described relation between radiation resistance and jerk was found and quantified by Lorentz when he analysed his model of an electron (shown in Fig. 1.23). It caused almost immediate consternation among physicists. Why consternation about a beautiful theory?

Let us do a few thought experiments. In Fig. 1.31 we imagine that we pinch an electron between thumb and forefinger, and smoothly shake it horizontally. One must also assume that we have been doing this already for a long time and continue to do so. It is simple mechanics to work out what will happen. For a change we have an equation of motion which has an easy solution in closed form, and what we

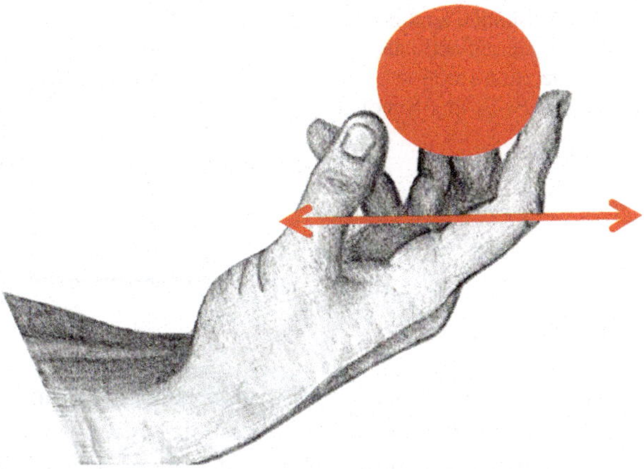

Fig. 1.31 Moving a charge sinusoidally

find is business as usual: the electron is subjected to a regular jerk, and emits radio waves. It thus loses energy, but the accompanying radiation resistance is counteracted by our hand, so that the lost energy is replenished by our hand. One can go on for ever without problems. Nothing to worry about!

But the radiation resistance is a wolf in sheep's clothing. Next do another thought experiment, as sketched in Fig. 1.32. Whilst the electron is at rest, we touch it gently, so that it is slightly disturbed. Retracting our hand immediately again will not help to prevent a catastrophe: the electron will move away with ever increasing speed, until it disappears over the horizon out of sight. This will not happen in reality of course, but it is the prediction by an analysis in which no one could point out where the error was. The unrealistic behaviour was in Lorentz's time called "the runaway motion", and it was a strong warning that something—but what?—was wrong with the theory.

Still another experiment, sketched in Fig. 1.33.

Here it is imagined that one pushes against an electron with a force that suddenly comes into existence. "Suddenly" is here meant literally: there is no gradual application of the force. The force is there without any delay, with its full strength. Again it is relatively simple to calculate what will happen, but the meaning of the solution is hair raising! You will not believe this, but the solution of the equations of motion now tells us that the electron would start to move *before* the force was

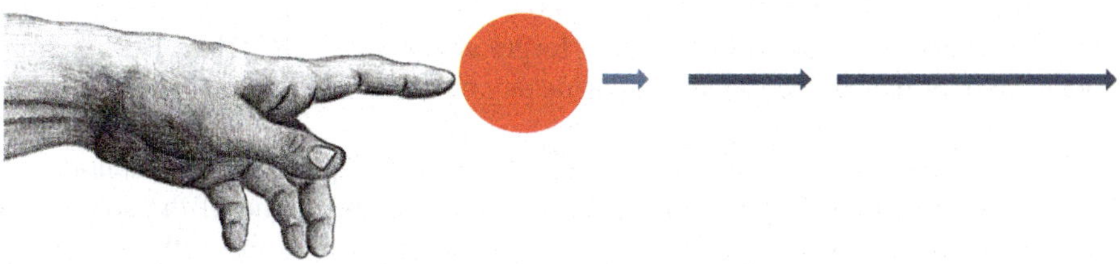

Fig. 1.32 Gently touching an electron at rest will cause a runaway motion

Fig. 1.33 Sudden application of a force causes anti-causal pre-acceleration (motion before there was any force)

applied. This is entirely anti-causal behaviour, and it is all due to the radiation resistance. By the physicists in Lorentz's time it was called "pre-acceleration".

The particular electron model of Lorentz was in the course of the 20th century forgotten, and one did not bother any more about "radiation resistance", "runaway motion", "pre-acceleration" and "anti-causal behaviour", except people like Feynman who could not stand the thought that nobody had been able even to find where the error in the theory was. This was a handicap for my own work on the deformable droplet of charge. Although mine is an entirely different model than Lorentz's, nevertheless from my own, entirely different type of analysis came again the same—cursed—radiation resistance term.

Fortunately, I came across a relatively recent book by Arthur D. Yaghjian: Relativistic Dynamics of a Charged Sphere (1992), that provided the beginning of a solution. Probably triggered by the phenomenon of pre-acceleration Yaghjian's reasoning was: what if we *assume* that radiation does not start immediately when a charge begins to move, but that there is a delay? And of course a similar delay in the start of radiation resistance? This delay may be small, something in the order of the timespan that light needs to travel from one side of an electron to the other. Yaghjian showed that even such a tiny delay would solve all the problems: no pre-acceleration, no runaway motions, no crazily behaving electrons any more. Yaghjian unfortunately did not philosophise further about it, he did a careful job to prove that everything can be made OK under this assumption, but what was not mentioned by him was: are Maxwell's laws not violated by assuming that there is such a start-up delay of radiation? How can it be explained physically?

I did some further analyses, to find out whether the start-up delay of radiation can be reconciled with Maxwell's laws. Of course I was rather anxious to know the answer, because I myself had once again been saddled up with Lorentz's radiation resistance term. Writing a book about how the classical laws may conjure up quantum behaviour, I could hardly say "forget about Maxwell". Here Bohr had an advantage over me. He boldly shoved aside any classical law that he did not like,but I cannot do that in this book.

The start-up delay of radiation is, fortunately, easily explained and it is not at all in conflict with Maxwell's equations. To briefly give an outline of the solution, I can use an analogy again. Look at Fig. 1.34, which shows a rope with two people

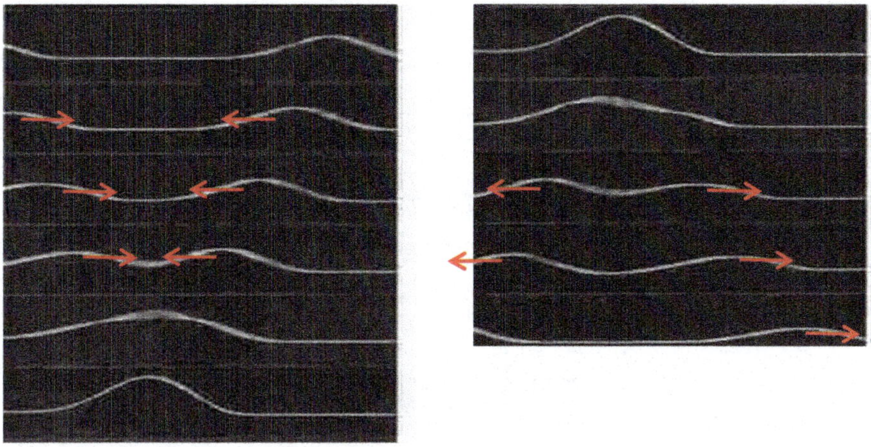

Fig. 1.34 Waves running through a rope. *From* PhysWiki, http://physwiki.ucdavis.edu

(out of sight) who have just given a single sweep to the ends of the rope. What you see is the result: two waves running through the rope from the ends towards the middle, where the two waves will meet, melt together and thus magnify each other. It is a nice example of what earlier was called constructive interference.

The next pictures (on the right hand side of Fig. 1.34) show that the two waves after this intimate encounter disentangle themselves again, and run along in the same direction as they had before.

In order to see more clearly what conclusions may be drawn, Fig. 1.35 pictures in a more stylised form the same event again. The waves have in this figure been given the form of blocks. A rope would have difficulty to take on this shape, but one could imagine that they are waves on a water surface (the block shape would be possible in a theoretical idealisation, but of course it is from a realistic point of view highly schematic). With respect to the vertical dot-dash line drawn through the middle, all the events are symmetrical. One may just as well think this middle line to be replaced by a reflecting wall. What happens, for instance on the right side of this wall, will exactly be the same whether there is such a wall or not. The only difference is, that with the wall present we will interpret the wave pattern as "a reflection" by the wall of the inward coming wave, after which the wave returns to where it came from. The latter part of the event, i.e. what follows the reflection is shown separately in Fig. 1.36.

We see that such a reflection against the wall entails first a gradual accumulation of water (as well as of potential energy): the water heaps up against the wall, as shown by the dotted contours. After this, the accumulated "blob" of water and potential energy releases itself in the form of a wave to the right.

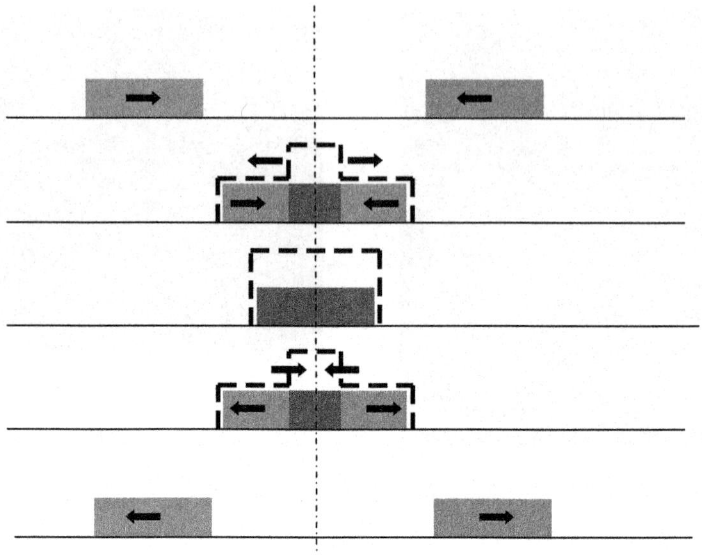

Fig. 1.35 Stylised waves on a water surface, running towards each other

Fig. 1.36 Start-up of a wave to the right, from "frozen" potential energy

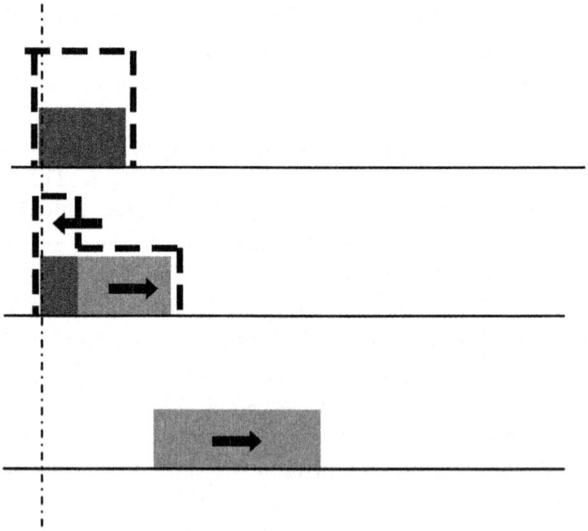

One might even go so far in the thought experiment, that at the moment when the accumulation is largest, we "freeze" the situation. What we then have, is accumulated energy which upon "unfreezing" starts up a wave which is going to carry off the energy to the right. Now, inspecting the situation during the start up of the wave carefully, and looking back at the earlier pictures, it is discovered that during the first phase of the starting up process water will move both to the right (the front end of the final wave) as well as to the left (the tail of the wave that still has to be reflected). At the *very first* moment these opposite motions are equally important. Almost immediately after "time zero", however, the amount of water in the outward moving wave ("outward" means to the right in this case) becomes larger, at the cost of the still incoming wave, and when the starting up process is finished, we only have an outward moving wave to the right. The crux is, that starting up a wave from an accumulation of potential energy (in jargon called a "transient") involves a rather complicated history of in- and outgoing waves.

What I did next was, to cast such a start up process in the terminology and the mathematics of electromagnetic waves, using Maxwell's equations. I then found that at "time zero" the radiation resistance is indeed zero! After this, the radiation resistance soon rises and will have attained its full strength at the moment the wave becomes fully detached from the wall. This behaviour of the radiation resistance is really not very surprising, because an ingoing and outgoing wave together do not carry away net energy, as long as they are of equal strength. I think that we have here found the physical explanation for the delay Yaghjian proposed in order to bring back causality to the experiment shown in Fig. 1.33.

We can now combine this conclusion with what was earlier told about the behaviour of droplets of charge and their pulsation. It was mentioned that most of the time the pulsation has a chaotic character, instead of the smooth sinusoidal character the motion sometimes, in more exceptional cases, assumes. The corresponding chaotic variations of the velocity of the droplet were schematically depicted in Fig. 1.28. The chaos entails that the motion of the droplet is very irregular, very jerky. At each instant of time the motion seems hardly correlated to what the droplet was doing at an earlier instant, the velocity makes almost discontinuous jumps. It is as if the motion at each instant of time undergoes a new start-up (for mathematically inclined readers who want a more exact description: the auto-correlation function approaches a delta-function).

What is the consequence for the radiation given off by the chaotically moving droplet? Well, there is obviously much more "jerk" in the motion than if we had smooth sinusoidal motion, one would therefore at first sight expect that there is more intense radiation compared with smooth motion. But now recall the story told above about the delays. If there is a sharp jump of the velocity it will take some time before radiation starts up. And before the radiation has started, already another irregular jump will have occurred, leading to the next delay. All the time radiation wants to start up, but it only leads to a "rejected take-off". We may conclude that *chaotic motion* of the droplet *suppresses the possibility to radiate.*

This might be the explanation of the occurrence of "Bohr's radio silence". At the very least, this can be the mental picture we can have about the phenomenon. As chaotic motion of the droplet is the kind of motion that exists most of the time, most of the time there cannot be radiation. But remember that the chaotic state is sometimes, in more exceptional cases, interspersed with intervals of more smooth motion. During these sparse intervals of time, the blockage of radiation is lifted, and the radio silence ceases momentarily. This deserves a separate section in this chapter, although the message will be clear already.

1.28 "Pure Chance" or "Causal but Unpredictable"?

In Sect. 1.2 the story was told of Bohr's revolutionary ideas, of which one was that the "other" world, the world of the atom, is not always causal. An example was, that at unpredictable times an atom may radiate electromagnetic waves, because sometimes an electron "decides" to jump down to an orbit with a lower energy. No special *cause* for such a "decision" can be pointed at in the quantum theory, only the probability that it happens can be calculated. By the way, there are more examples to be given where the quantum theory is non-causal.

In the particular case of radio silence we can now have another picture in the back of our mind, which is no longer non-causal. To sum up in schematic form the "mental pictures" afforded by the droplet theory:

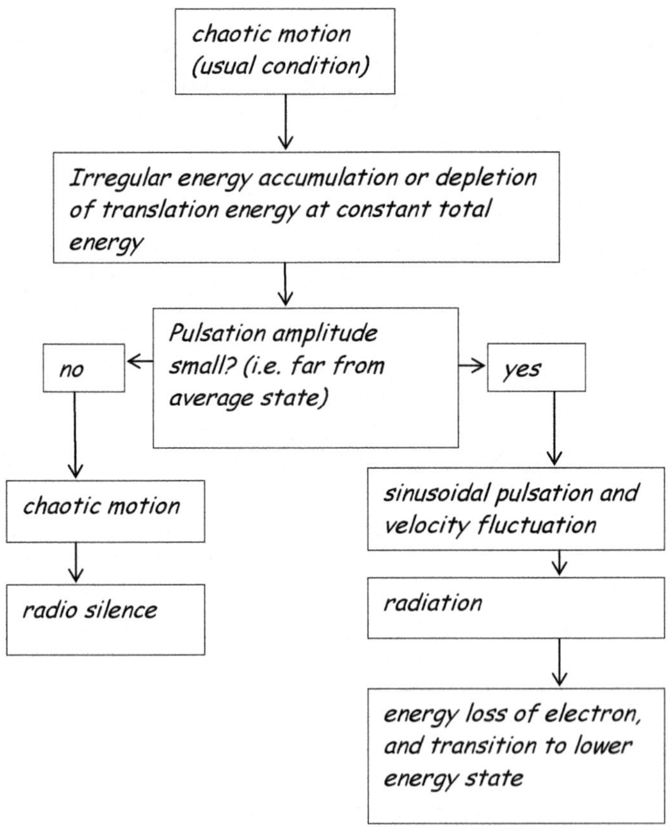

The diagram once again stresses the fact that the determining factor whether there will be radiation or not is the irregular energy division between translation and stretching. The changes in the division of energy are determined (at least for a large part) by the chaotic motion. Now remember that chaos is not the same as randomness. The chaotic motion is just as much a solution of the equations of motion as the sinusoidal motion. It just depends on the value of some parameters in the equation which one of the solutions will be realised.

The verdict about the question "pure chance" or "causal but unpredictable" is clearly the latter. Chaos determines everything, and chaos is the result of a causal process although the outcome is unpredictable.

1.29 The Most Prominent Hiatus in the Droplet Theory

Chapter 14 of the main text is titled "successes and failures" of the droplet theory. This chapter repeats briefly the weird quantum phenomena found in one-dimensional oscillators (Chap. 2), and then gives brief summaries of the interpretation according to the droplet theory, i.e. interpretations in terms of the classical laws of nature.

Apart from the phenomena already discussed in the present overview, you will see in Chap. 14 questions answered such as:

- why matter waves can give information about the whereabouts of particles,
- what the frequency will be of the radiation emitted by electrons jumping between two permitted energy levels,
- and what about the "bundle of incongruities": the photon? It is a particle of light, without a supporting electromagnetic field, but nevertheless with a certain colour (= frequency). Even in this case the droplet theory leads to a less problematic mental picture.
- etc.

This is not to say that all the mysteries of the quantum world have been solved, certainly not, there are many left. They will not be carelessly glossed over in Chap. 14, do not fear. But there is one overwhelming difficulty that has to be addressed here and now. The point is, that the droplet theory is wholly based on the electron model explained in earlier sections. An electron is a charged particle, and the theory was therefore developed by invoking Maxwell's equations. Now, quantum theory is not only about charged particles, it also encompasses neutral particles as well as neutral conglomerates of particles. What about them?

With regard to this problem one may have a sort of "déjà vu". Relativity theory existed before Einstein, he was not the inventor. In the theory of electromagnetism relativistic effects were long known, as is reflected in the name "Lorentz transformations" for the central set of equations of relativity theory. But the problem was, that these Lorentz transformations had been derived from and for electromagnetic phenomena, there was no clue why these relations would be applicable to other than charged things. The crucial step taken by Einstein was, that he showed from more general principles that Lorentz's transformations would be applicable to any sort of matter, and were not restricted to the world of electromagnetism.

Conclusion: another Einstein will be needed to supplement the here developed theory of a droplet of charge by a theory of a "droplet of *non*-charge", or something like that. Fortunately, the results found for the "droplet of charge" are intriguing enough in their own right to report them in this book. Once again, perhaps the most important message is that the classical laws of nature *can* conspire to give quantum behaviour.

1.30 The "Big Magic" of the Atomic World

Chapter 14 does not cover *all* of the magical phenomena occurring in the atomic world. The chapter has a limited scope: only those phenomena are treated for which in the formula appendix the mathematics have been worked out completely. We have adopted the droplet of charge as a model for the electron, instead of the usual point-like singularity. It was just a working hypothesis, to find out what would be the consequences of such a model. Determining these consequences was a matter of conscientiously working out the mathematics, and interpreting the results in physical terms. The results of this prosaic exercise were described in Chap. 14.

However, the mathematics also leads to *hints* about other quantum phenomena, even if the theory is not complete as yet. Chapter 15 describes these hints and the pictures that can be based on them, on condition that we accept some additional speculation. It is compared with a bridge that rests on firm piers and pillars, but with the main superstructure connecting the piers still somewhat vaguely seen in the mist.

The subjects treated in Chap. 15 include for instance:

- Superposition of states (the magic of Schrödinger's cat), or superposition of frequencies (a common phenomenon)?
- Telepathic ("spooky") contact between widely separated particles (or: superluminal signalling?)
- One particle, at the same time in different places.

Perhaps it is exactly this kind of magic that you have been waiting for in this book, because it illustrates most vividly why the atomic world is so strange. Chapter 15 sketches possible mental pictures about them, again based on everyday physics. But remember that some speculation is involved.

1.31 The Ultimate Test of the Theory

The ultimate test for any theory on particle dynamics is, whether it can explain the double-slit experiment (Sect. 1.5). If a theory can correctly predict the outcome of the "double-slit experiment" there is a good chance that the theory has caught the essentials of quantum mechanics. If it cannot, then apparently something is still missing in the theory, or it is not a valid theory at all.

Everybody will be extremely curious at this point about the performance of the "theory" developed in this book. The problem is of course that the set-up of the double-slit experiment is entirely different from the oscillator that we have been studying in this book. The main stumbling block is that we have been studying a one-dimensional motion, i.e. the translation is in one direction along a straight line, and the pulsation was also assumed in that same direction. In contrast, the essence of the double-slit experiment is that it is more-dimensional.

We must unfortunately conclude that the "theory" developed in this book cannot be subjected to the "ultimate test" and therefore cannot (yet?) be called something that aspires to be a full fledged theory. For the time being we must consider the whole effort as an attempt to build up a series of mental pictures, and not more than that. But who knows what the future will bring?

Actually, in the opinion of the author there may be a good chance that in future the double-slit experiment can be explained using the model of a deforming droplet of charge. What this optimistic expectation is based on will be further explained in Chap. 15.

1.32 How Useful Is the Whole Exercise?

What then is the use of the whole exercise described in this book? It depends on the expectations of the reader. Quantum theory as it stands is an almost perfect theory, able to predict particle phenomena with exceptional accuracy. There is really no need at all to argue about this. The only problem with the existing quantum theory is that "it is strange". It is well known that one of the founding fathers of quantum mechanics, Albert Einstein, himself did not like quantum theory. He always kept the feeling that there should be some level underneath quantum theory, by which he meant that quantum theory in its present form is not complete and that there is some kind of missing "hidden variable".

Sometimes one encounters people in the street who at first sight belong to the category of the "village idiot". They are shouting incomprehensible things and are without any reason cackling with laughter, whilst at the same time gesturing ferociously. Our first impulse is to shy away from such loco's, one never knows whether they could become aggressive. Until you see that this person has an ear-phone in his ear and a small microphone hanging from it. That suddenly alters our perception, because we know that the person in question may be perfectly normal, and his behaviour is not really alarming. The person on the other side of the telephone connection is in this case the "hidden variable", whose presence can explain everything. Likewise, Einstein had the feeling that the "strangeness" of quantum physics could be lifted once we had discovered what the missing hidden variable is. In particular the fact that quantum theory only predicts the probability of events irritated Einstein, since it would imply that the driving factor behind events is just "pure chance", whereas according to Einstein's often quoted exclamation "God does not throw dices" ("Der lieber Herrgott würfelt nicht").

The present approach to quantum mechanical phenomena would, if it could be proved to be not all nonsense, indeed make the connection between classical laws of nature and Schrödinger's equations. It could also provide the answer to the question what is the "hidden variable" in quantum theory. But let me emphasise that the state of it all is certainly not that far advanced that such a claim already can be made. In the opinion of the writer such a claim would only be legitimate if the "double-slit experiment" could satisfactorily be explained on the basis of a deformable droplet model. And even if that would prove possible, still the theory of the deformable droplet would be clumsy to use for practical problems. Existing quantum theory would be much more efficient.

What then is the use of everything in this book? Let us keep it low key, small (and hopefully beautiful) and conclude that for some people, like myself, it is nice to have some "mental pictures" that would seem to remove a little bit of the magic, weirdness and anti-intuitiveness of quantum phenomena.

Textbox: Bouncing Droplets Showing Quantum-Like Behaviour
Fascinating experiments have been done on droplets of silicone oil bouncing on a surface of the same type of oil. See e.g. one of the first publications about the subject in Nature (2005) by Y. Couder, S. Protière, E. Fort, A. Badouad (Paris University). Using a special laboratory set-up they discovered that these droplets can mimic quantum behaviour. Following this, more of such intriguing experiments were reported by—amongst others—Moláček and Bush (MIT). Full references are given in the chapter "References".

When hearing the description "bouncing droplets", most people will—mistakenly—think of the game they played in their youth (or—because it is so fascinating—still as adults) by skimming stones along the surface of a pond and let them skip as long as possible. Almost everyone will recognise pictures like Fig. 1.37. However, it is best to forget immediately this kind of

Fig. 1.37 A well-known picture, NOT to be confused with the wave patterns caused by bouncing oil droplets

picture when it is tried to imagine what happens in the mentioned experiments.

The essential difference between the two “experiments” is the bouncing objects’ speed. To be more precise: what is crucial is the speed in comparison with the velocity of surface waves. The stone in Fig. 1.37 moves much faster than the water waves. As a result the stone during each jump lands on an unperturbed, practically horizontal water surface. It therefore always bounces in the same direction as it had before hitting the surface, and the stone follows a straight path (seen from above) until it finally sinks. An aerodynamicist would say that the situation is “supersonic“, by which he/she would mean that the speed of the object is larger than the wave speed.

In contrast, the motion of oil droplets in the experiments by Couder and coworkers could be called “subsonic“. The droplet itself makes jumps—although small—like the stone in Fig. 1.37, but is now overtaken by the surface waves, and always lands on a place where the surface is already disturbed by the waves coming from earlier bounces. No longer it meets a horizontal surface all the time, and the slope of the fluid in the point where the droplet skims the surface will determine in which direction the droplet will continue its jumping. If there are perturbations of this wave pattern, e.g. by the presence of other objects, the droplet can thus deviate from a straight path, and in the extreme may even go round in circles. In publications about the droplet experiments the term “high memory regime” is mostly used instead of “subsonic”, but as an aerodynamicist the author should be forgiven to think in terms of sub- versus supersonic.

The experiments are performed with a tray of fluid which oscillates vertically (Fig. 1.38). This keeps the droplets aloft, so that they can continue bouncing for a long time. The vertical accelerations to which the fluid is subjected inevitably have some influence on the behaviour of the fluid as well. One effect is for instance, that surface waves are slowed down compared with a fixed tray. In the “subsonic” situation, the speed of the droplets is therefore really small, in the order of 10 mm/sec as can be seen on movies (beautiful movies can be found on YouTube, when Y. Couder or John Bush

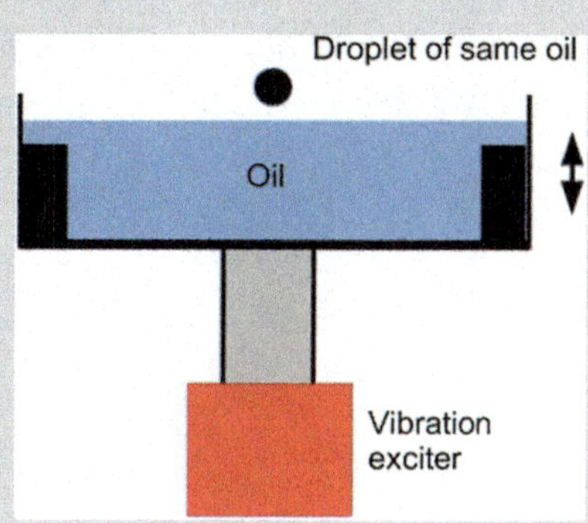

Fig. 1.38 Oscillating tray of fluid with bouncing oil droplet

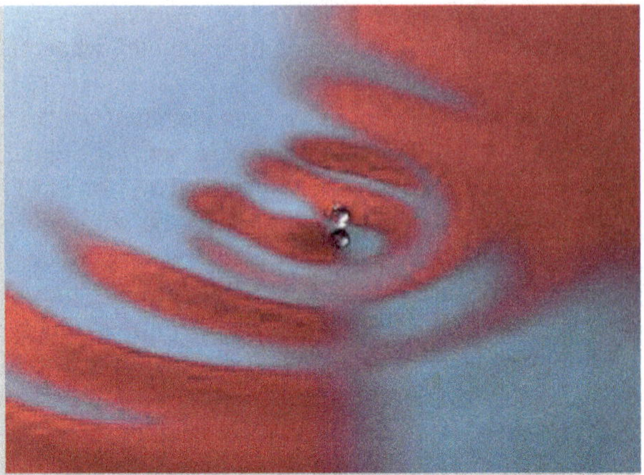

Fig. 1.39 Photo (by John Bush) of the wave pattern around a "walking" droplet

is googled). What also can be seen in these movies is the shape of the wave pattern that moves along with the droplet. Figure 1.39 is a photograph of it.

It can be concluded that there is a peculiar interplay between the waves and the droplet: the droplet causes and maintains the waves, but conversely its bounces are also influenced by the waves. The waves act, so to speak, as pilot waves. Here the first glimpse of similarity with quantum mechanics is found: De Broglie too spoke about "his" matter waves as pilot waves, which he thought would guide a particle and would determine its path.

Actually, it appears that the similarity between bouncing droplets and quantum mechanical effects goes much further than this, and it is quite striking. For instance, the set-up shown in Fig. 1.40 resembles the double-slit experiment of quantum mechanics. It is the experiment which proves the particle/wave duality of particles and which, according to Feynman, shows the essence of the strange behaviour of elementary particles. Surprisingly, the droplet in Fig. 1.40 is going to behave in the same way as an electron in the double-slit experiment: after having passed one of the passages it will avoid to walk further in certain directions. The fluid droplets, although being

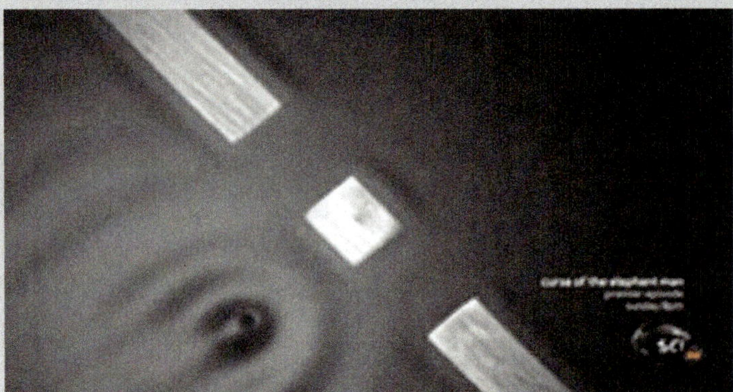

Fig. 1.40 Oil droplet analogy of the double-slit experiment

macroscopic objects obeying the "normal" laws of physics, are thus an analogy of atomic particles obeying the rules of quantum mechanics!

Many more manifestations of quantum-like behaviour are found in the case of these bouncing droplets, e.g. tunnelling when a submersed barrier is met. Also quantisation of the energy is observed when the droplets are made to circle around (rotation of the tray causes a concave shape of the surface which mimics a central force).

How can it be explained that the interaction between the bouncing droplets and their self generated wave patterns is so similar to the interaction between elementary particles and their field of matter waves? Two scientists of the Cambridge University, Ross Anderson and Robert Brady, solved the mystery (R. Anderson, R. Brady: Why bouncing droplets are a pretty good model of quantum mechanics, see the chapter "references"). The departure point of their thinking was a long known phenomenon in physics: the Bjerknes force between two vibrating objects immersed in a fluid. A picture taken from a historic article (Popular Science Monthly, 1882) is shown in Fig. 1.41. By properly actuating a pump, membranes in the two submersed devices can be made to oscillate. It then appears that two vibrating objects in a fluid will

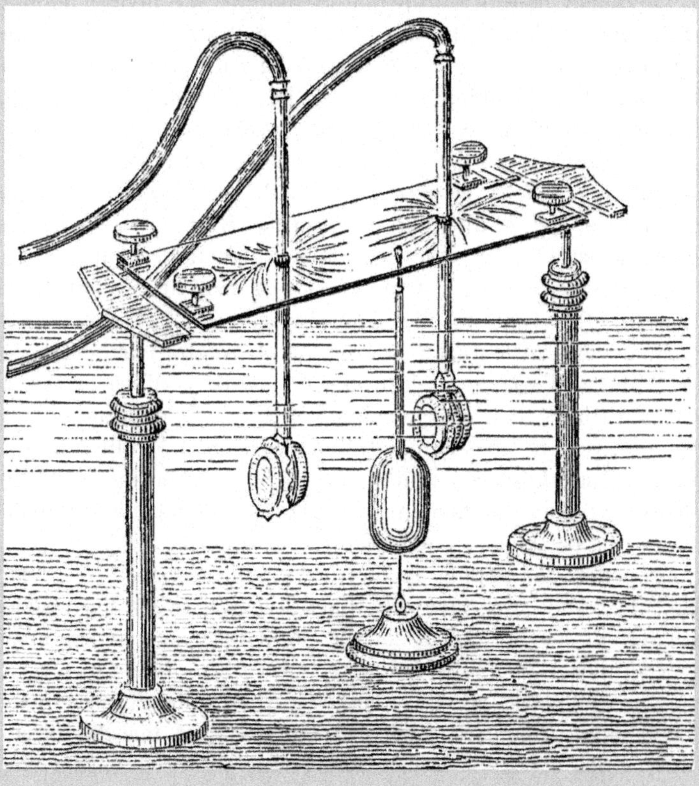

Fig. 1.41 Bjerknes' experiment showing Coulomb-like force between two vibrating objects in a fluid (Popular Science Monthly, 1882)

attract or repel each other, depending on whether they vibrate in step with each other, or move in so-called "anti-phase". This is due to the field of velocities around them and the associated pressure field. It works out in such a way that the mutual force is inversely proportional to the square of their distance, like the Coulomb attraction or repulsion between two electrical charges.

This phenomenon was combined with a piece of theory concerning the waves caused by bouncing droplets. Anderson and Brady proposed a mathematical expression describing the wave pattern as shown in Fig. 1.39. Although it is a simplification of the real pattern, it cleverly captures the essentials, and enabled them to calculate the slope of the fluid surface as experienced by a skimming droplet. A crucial property of this approximated wave pattern is that its shape depends on the speed of the droplet, morphing in the same way as relativistic objects do, i.e. objects obeying the rules of special relativity theory (for the insiders: the wave pattern is in agreement with the Lorentz transformation). The validity of this theoretical model was verified by a calculation how the "walking speed" of the droplet would depend on the frequency of the oscillating tray. It appeared that the model predictions were in good agreement with the experimental results obtained by Couder. Obviously, it was worthwhile to continue the investigations on the basis of this description of the wave pattern, with its relativistic properties.

The next thoughts went along the following lines. Consider two bouncing droplets without a "walking speed", to begin with. They experience a Coulomb-like mutual force (the Bjerknes force). Now set the droplets in motion (i.e. assume a non-zero walking speed), and remember that in that case everything changes in accordance with the laws of relativity. The "mutual forces at speed" then will be described by Maxwell-like laws. This last statement may require somewhat more explanation, which can be given by recalling the history of Einstein's theory of relativity. Relativistic effects were known before Einstein, they had already been found in the realm of electromagnetism (Einstein later showed that these effects are more general). Therefore, Maxwell's laws of electromagnetism do conform with relativity, and these same laws reduce to the Coulomb forces in case of zero speed. If we now—conversely—have found that there is a Coulomb-like force between stationary droplets, and if we know that changes due to the speed of the droplets are relativistic, then the conclusion is inescapable that the moving droplets may be described by Maxwell-like relations.

This last conclusion implies a rather striking phenomenon. We then expect that there must also exist "magnetic-like" forces between the walking droplets, on top of the "electrical forces". And once again, scrutinising the droplet experiments, this appeared to be true!

Having arrived at this point, we could here finish the discussion about the strange behaviour of bouncing droplets. The reason is, that the present book deals with vibrating droplets of *charge*, and shows that analysing their behaviour by Maxwell's equations gives the same results as found by quantum theory. No wonder then, that fluid droplets behaving according to Maxwell-like equations also show quantum-like behaviour.

But let us return to the route Anderson and Brady followed to show the connection with quantum theory, which is different (more mathematically formalistic) than the stories in the present book. Have a look again at Fig. 1.39 which gives an impression of the wave patterns when the droplet "walks" over the fluid surface (in this photograph the walking speed is roughly in the direction of the lower righthand corner). We see that the pattern is highly asymmetric, i.e. there is a large fore-aft difference w.r.t. the instanteous position of the droplet. In contrast, when the droplet would bounce on the surface without a forward speed, i.e. when it would be stationary, the wave pattern obviously would be completely symmetric. The amount of asymmetry is therefore a measure indicating how fast the droplet is moving over the surface. In the theoretical wave pattern devised by Anderson and Brady we see something similar: there is a basically symmetric pattern, multiplied by a factor that introduces the asymmetry when the speed of the droplet becomes non-zero. This factor grows in importance, the larger the walking speed is. Mathematically, the multiplication factor has the form of a wave, and the factor has *more* influence (i.e. it gives more asymmetry, which in turn indicates a larger walking speed), the *smaller* its wavelength is. This "smells" like De Broglie's relation in quantum mechanics, stating

$$(momentum\ of\ a\ particle) = \frac{Planck's\ constant}{(wavelength\ of\ associated\ matterwave)}$$

And indeed, working it out Anderson and Brady found that a similar relation is applicable to the walking droplet! Needless to say, the constant in this "De Broglie like" expression pertaining to oil droplets does not have the value of Planck's constant: an oil droplet just shows an *analogy* of quantum mechanical behaviour and is not the real thing.

Once more, we could stop the discussion here. In the present book it is made plausible that in any system obeying De Broglie's relation also Schrödinger's equation is applicable. In other words, Schrödinger's equation follows from De Broglie's relation. However, in the present book only a heuristic argument is used. Anderson and Brady prove it by mathematical reasoning: the behaviour of oil droplets is indeed governed by a Schrödinger-equation, where Planck's constant has been replaced by another appropriate constant. After having shown this, their conclusion is: "same equations, same solutions". It explains "why bouncing droplets are a pretty good model of quantum mechanics" (as one of their papers is entitled). It then is no longer a mystery why bouncing droplets show quantisation of energy, display tunnelling effects, mimic the double-slit experiment, etc. They even proceed to explain a phenomenon like the spin of electrons, which has an analogy as well in the droplet behaviour.

Comparing the theory of the oil droplets due to Anderson and Brady with the theory of droplets of charge in the present book, it may be concluded that the two theories are complementary. The behaviour of oil droplets cannot be anything else than an *analogy* of the behaviour of electrons, but the theory is in certain aspects more advanced because it covers "spin", whereas this book hardly touches on this subject except in a rather speculative way. On the other hand, in the present book an attempt is done to go further than just analogies: modelling electrons as vibrating droplets of charge leads to a practically correct value of Planck's constant. Consequently, the derived behaviour of a droplet of charge is even quantitatively the same as that of electrons in quantum theory, despite the fact that classical laws are used to derive this behaviour. A future effort is certainly worthwhile to completely merge the views offered by fluid droplets and by droplets of charge.

Chapter 2
Strange Behaviour at Quantum Scale: The Full List

2.1 Introduction

In the preceding chapter a few examples already have been given to illustrate the strange behaviour of particles at atomic scale. It was announced that the full list of incomprehensible behaviour is much longer, and it is the task of the present chapter to sum it up completely. The proviso is, that we are talking only about a simple *oscillator*. The free electron, the electrons in atoms, or the double slit experiment mentioned in the introduction fall outside the scope of this chapter. Although a first understanding of the jumping up and down and back and forth of real electrons inside atoms may be obtained by studying the simple oscillator, it must always be kept in mind that the model of a simple oscillator only provides a *first*—although important—understanding, and is not the complete story. Nevertheless, the simple oscillator is so enlightening that all study books, both on classical mechanics as well as on quantum physics, start by considering and analysing it.

In order to make it a systematic story, we go back to the classical oscillator (such as the children's swing), and consider it in somewhat more detail than in the preceding chapter.

2.2 The Story of the Oscillator as Told in Classical Textbooks

Imagine a bowl in which a marble is rolling back and forth. To keep it simple, invariably the situation is first assumed to be *one-dimensional*, i.e. looking on top of the bowl one sees the marble rolling in a straight line, going down from one side of the bowl through the lowest point in the centre, and up again to reach a position on the other wall diametrically opposed to its starting point. After this the motion is reversed and the marble rolls back along the same straight line (seen from above),

T. van Holten, *The Atomic World Spooky? It Ain't Necessarily So!*
DOI 10.2991/978-94-6239-234-2_2

etc. Often one assumes furthermore that the excursions of the marble are not too large (the so-called "linearised one-dimensional oscillator"). That however is an assumption which is purely intended to simplify the mathematical analysis, and it does not concern us in this qualitative chapter. One could ask why this is called "one-dimensional", because the position of the marble is given by *two* coordinates, its distance from the centre of the bowl and its height. The point here is, that the marble is not free in the choice of its height, it is constrained to move along the contour of the bowl. We therefore need only specify *one* coordinate, e.g. its off-centre position, and its height is then automatically known. Hence the terminology "one-dimensional" (Fig. 2.1).

Obviously, a perfectly analogous situation is a cyclist free-wheeling down the slope of a hill to cross a valley. After the lowest point the road curves upward again to climb a hill on the other side. It may seem trivial to point out that this is an analogous situation, but the funny thing is that exactly this cyclist has given a name to one of the magical phenomena occurring in quantum mechanics (so-called "tunnelling"). Curiously, we will meet the cyclist again in the part on atomic phenomena.

Another analogous situation is the pendulum of an old fashioned clock, as will be evident (*mathematically* analogous that is; a cyclist will be glad to hear that we do not confuse his/her actual appearance with a marble or a clock).

In these three examples the motion is driven by gravity. Gravity need not always be the driver of the oscillator. One might also imagine a body, having a certain mass, sliding over a horizontal surface, and performing its motion back and forth

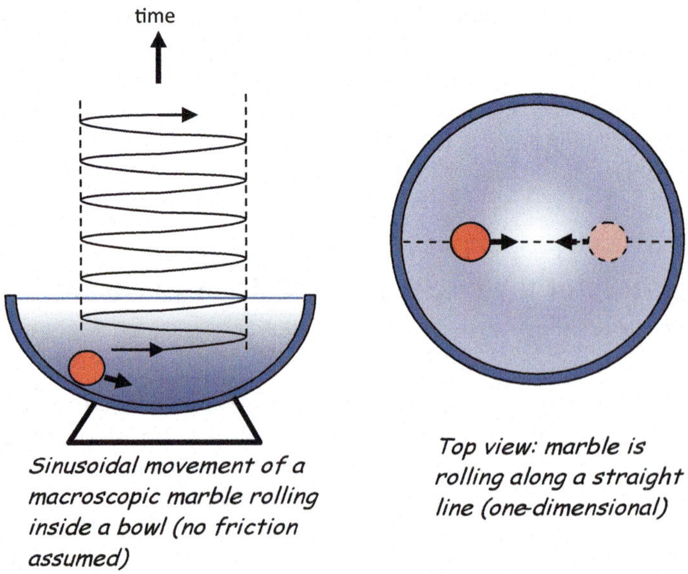

Fig. 2.1 One-dimensional oscillator driven by gravity: the marble and bowl

Fig. 2.2 One-dimensional oscillator driven by mechanical springs: the mass-spring system

under the influence of a system of springs. These springs must be arranged such that they tend to drive the mass back to a central point, like the effect of gravity which tries to move the marble back to the centre of the bowl. Such a so-called "mass-spring system" is also a perfect analogy—in the mathematical sense—of the marble-bowl system. We repeat here the schematic shown earlier of the mass-spring system (Fig. 2.2).

The driving force could also be of an electrical nature, in which case the body is assumed to have an electrical charge. Stationary charges in the vicinity cause an electrical field (a so-called "electrical potential well") such that the repulsive forces on the moving charge again have the same effect as gravity has on the rolling marble.

Note that we have drawn all the charges in the schematic below as positive ones, which seems to be illogical because an electron has a negative charge. This does not matter as far as the motion of the mobile charge is concerned, since like charges always repel each other, so that the oscillator is not affected by the sign of the charges. What is important to mention is, that everywhere in this book positive charges are assumed, because this choice has a few advantages. For instance, a flow of *positive* charges is called the electrical current, with the direction of flow and the current the same. It is an inconvenience dating back from the times that electrons were unknown yet, and one had to make a choice what to call a positive or negative charge. A choice that later appeared to be the "wrong"—or rather an inconvenient—one. Nowadays an electrical current is known to be most often be associated with a flow of electrons (negative charges), so we are stuck with the inconvenience that according to the historically chosen convention a current has a direction opposite to the velocity of electrons. Of course it is all a matter of definition and does not really make a difference (apart from the unexpected occurrence of minus signs in formulae), and it is reflected in schematics such as pictured below (Fig. 2.3).

The interesting thing is, that the actual form given to the experiment does not matter. The oscillator may be driven by gravity, mechanical springs or electrical

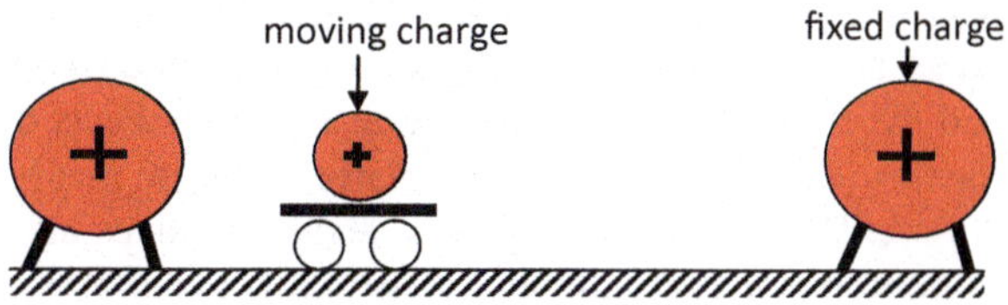

Fig. 2.3 One-dimensional oscillator driven by electrical forces: the charge inside a potential well

forces, it turns out that the mathematical formulation in all these cases is identical, and that the motion in principle displays the same characteristics.

What this characteristic motion entails is of course familiar to anybody, even to people who have never opened a physics textbook. The result of the experiment will be a smooth, periodic motion back and forth, slowly dying out. This more or less slow subsidence of the motion is due to additional forces on the moving mass, occurring as soon as the mass has obtained some velocity. Obviously a marble, and even more obviously the cyclist, will experience air resistance opposing their motion. But there are more so-called *damping forces* of another kind as well: think of the rolling resistance of a bike, due to friction in the bearings and due to the constant deforming of the tires.

People who have "internalised" their lessons at school will intuitively interpret the motion of a marble as a continuous exchange of different forms of energy. This mental model says that the marble at rest in a high position has *potential* energy and no *kinetic* energy. Then, when the marble starts to roll downwards it acquires kinetic energy at the cost of its potential energy. You could say that potential energy is transformed into kinetic energy, until in the lowest point of its trajectory all its energy is in the form of kinetic energy.

Were there no frictional effects such as air resistance and rolling resistance, the motion would go on forever. This means that the total energy in the marble itself, i.e. the sum of its potential and kinetic energy, would be constant and would not change in time. However, the unavoidable resistance forces (or: "damping" forces) all the time tap some energy from the marble and transform its mechanical energy into other forms of energy such as heat. The air resistance experienced by a cyclist means that his energy is partially lost in the form of turbulence in his wake, which turbulence slowly decays by the friction of the air and is finally transformed into heat too. The result is in the end that the marble is left without mechanical energy and finds itself at rest in the bottom of the bowl, all the initially present potential energy having been lost, mostly in the form of heat. Now heat is another form of energy, and it turns out that what the marble or cyclist have lost in mechanical energy is completely found back in the form of heat, so that again the total amount of energy (potential-, kinetic- and heat energy) is the same. This is called the *law of conservation of energy*: energy cannot be "lost", it is just transformed into a different form. Although it should be said that a cyclist will consider his loss of *mechanical* energy as a loss of *useful* energy, since the heating up of the atmosphere in his wake will not really interest him.

Apart from the production of heat, there are other interesting methods to lose energy. Think of the fact that the rolling motion of the marble is clearly accompanied by a distinct noise. An earthenware bowl will in this way produce even a rather annoying sound. The noise comes from the combination of the rolling marble and the bowl, the bowl vibrating as a consequence. The noise means that energy in the form of acoustic waves is radiated away, so that here is another source of energy "loss". We should add that these acoustic waves also slowly die out or "dissipate" in the form of heat, although that may happen at a considerable distance from the bowl.

In the case of the rolling marble the energy loss by noise radiation is tiny compared with other energy losses such as the loss associated with air resistance. It is another matter when we consider the electrical oscillator.

Whenever a charged particle is in motion, in particular if it accelerates or decelerates, it sends out electromagnetic waves. This too is something known to everybody. Everybody knows what happens when electrons in large numbers are pumped up and down through a vertical copper rod. This arrangement is more usually called a radio antenna, and it emits radio waves. In all conductor wires such as power cables the same happens, so that an alternating current in them can cause a steady "hum" in a nearby radio receiver.

Likewise, a charged body moving back and forth within a potential well, i.e. the electric oscillator, will emit radio waves. Like sound waves, radio waves—or using the more general terminology: electromagnetic waves—represent an energy loss. A charged body moving back and forth will by this effect gradually lose its energy, and if we want to keep it moving, some external force has to be applied. The external driving force has to oppose the so-called "radiation resistance", i.e. the recoil force on the electrons due to the emission of radio waves. The charges in a radio antenna have to be kept in motion by applying a varying electric potential to its ends, and by feeding energy into it in this way.

There is a close analogy here with the pendulum of a clock, which loses energy all the time by frictional effects, and must be kept in motion by the clock mechanism which gives it a kick during each swing, and replenishes the mechanical energy in the pendulum.

2.3 What Textbooks on Quantum Mechanics Tell About the Oscillator

And now the weird behaviour of a quantum oscillator! The oscillator may in principle have all of the above mentioned forms, be it on an extremely small scale, comparable with the size of atoms. Because in reality we are now probably dealing with charged particles such as electrons or protons instead of cyclists, the electrical model is the one closest to the physical reality. A single electron moving within a potential well is obviously not subjected to air resistance or other kinds of friction, so that radiation is the only mechanism by which it can lose energy.

What is observed is first of all that such a quantum oscillator does not always act as an antenna! In fact, most of the time it does not send out any electromagnetic waves, or so to speak: there is "radio silence" most of the time! This is a phenomenon that has no explanation in classical physical theory. Indeed, we would be astonished if an antenna in our human world would stay "silent" if we knew for certain that the masses of electrons within the antenna rod are moving in the proper way.

Less surprising is, that during these periods of "radio silence" the total energy (potential plus kinetic energy) present in the translation motion of the single electron or proton is constant in time. This is not so surprising, because without radiation there cannot be energy losses.

Sometimes however, at unpredictable times, the particle does shed some of its energy. It does not do so in a gradual way, but by giving off a sudden "burst" of energy. This is accompanied by a sudden break of the radio silence, and because of the suddenness of the event, the radiation is also in the form of a burst of electromagnetic waves. The frequencies we are talking about in the case of an electron are high, in the region of light frequencies. Such a burst of energy therefore takes the form of a flash of light. The flash of light emitted is so sudden and of such a short duration that onlookers experience it as a discrete "energy packet" shot at them by the moving electron. Accordingly, such an energy packet is known as a "photon", or "a particle of light".

One would expect that the frequency (in other words: the colour) of the emitted light would correspond to the frequency of the back and forth motion of the electron within the potential well. In the case of a macroscopic antenna that would be so. Not so in the quantum world: the colour of the emitted light is related in some way to the amount of energy that is being shed by the electron. Strangely, we thus find a definite frequency in the radiation, but there is nowhere any charge or part of a charge that is oscillating with this frequency. It is as if the radiation comes out of nothing, there is no obvious root cause for it. Apparently, in the realm of quantum mechanics almost nothing happens according to our expectations. And that same story goes on.

After having shed in this abrupt way some of its energy, the total energy in the translation of the electron of course has decreased. And here is again something strange: the new energy level is not arbitrary, but can have only certain fixed values. A marble in the *macroscopic* world can have *any* energy, which is evident from the fact that its excursions within the bowl very *gradually* become smaller when it loses energy by frictional losses. In contrast, the electron within its potential well cannot make excursions of any arbitrary magnitude. Only discrete values are permitted by nature. The magnitude of an electron's excursions is restricted to certain discrete values with *no gradual transitions in between allowed.* An electron's possible energy levels within a potential well are said to be "quantised", and the permitted levels are in fact very precisely prescribed by the laws of quantum physics, laws that are completely unknown in the classical physics of every day.

If a marble rolling inside a bowl would display this kind of behaviour we would be puzzled, to say the least (Fig. 2.4). First of all, the marble would for a long time not lose its energy, which might be seen by observing the amplitude of its motion, i.e. the height to which is climbs up the wall of the bowl during every cycle of its motion, or to use the earlier terminology: by observing the magnitude of its excursions. It would seem as if there were no losses due to dissipation: no air resistance, no rolling resistance, not even a sound would be heard because even the emission of sound would be an energy loss.

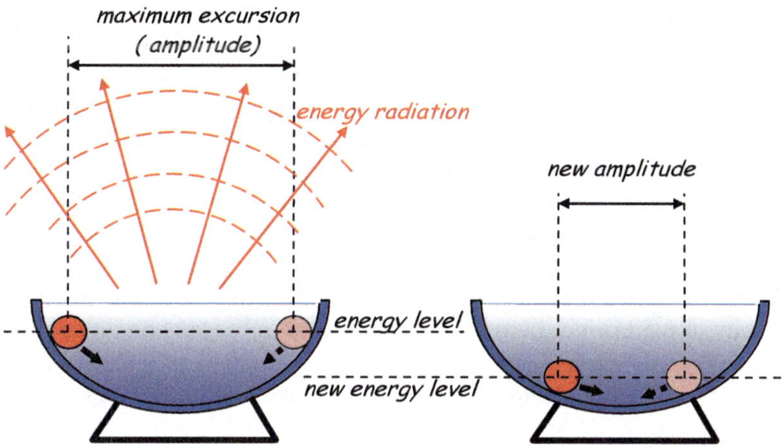

Fig. 2.4 Energy loss by emitting lumps of energy. Quantisation of the energy levels

And then suddenly, at an unpredictable moment, we would hear a thunderclap and at the same instant would observe that our marble no longer reaches the same height as before. And what is more: we could, using a felt tip, mark the new level reached after the thunderclap, and do it again after the next explosion of energy, etc. We would find that another marble in a bowl standing beside it would show *exactly the same energy* levels. The only difference with the first bowl would be that the thunderclaps coming from each bowl would come at different times. As precisely defined as the energy levels are, as unpredictable are the times: sometimes there would be a long time between the separate bursts of energy, so much so that we could easily put the kettle on and drink some tea. And at other times, the thunderclaps would follow each other rapidly, without any predictable intervals.

We would certainly call this "magical behaviour" in the case of a real marble. Yet, this is how marbles at atomic scale (or rather: electrons) behave.

The story of strange behaviour is not finished yet. One would expect that the process of shedding energy, although it is unconventional, would nevertheless finally result in the situation where the electron would come to rest in the centre of the potential well. Once again: not so! There is a minimum energy level, the so-called "zero-point energy" at which we will never see any more energy-bursts, no matter how long we wait.

To speak metaphorically in macroscopic terms: the marble never comes to rest completely in the bottom of the bowl. It forever keeps on "wriggling" a little bit in the bottom part of the bowl, but never gives up all of its remaining energy.

Let us now return to the situation where the electron (or using the metaphore again: our marble) still has plenty of energy, before it has emitted all these light flashes. In the macroscopic world we are used to see a smooth motion, the marble gradually accelerating and decelerating between the high points where it reverses its motion and the lowest point where it has maximum velocity.

What we see in the quantum world is again something completely different. It appears that there are certain, fixed points in the potential well where we hardly ever "see" the electron, whilst in between these points the electron is frequently spotted. It is as if the electron is "frightened" away from certain "forbidden" points.

When a macroscopic marble would show such behaviour we would interpret it as a very irregular velocity. We know that the marble is rolling from one side of the bowl to the other side, so that it has to pass all the points on the contour of the bowl, including these forbidden points. If we nevertheless see that the marble spends a much longer time in the regions between the forbidden points and hardly any time in the areas directly around these points, we will interpret it as a wild variation of the velocity of the marble. Before it reaches such a forbidden point it decelerates—it "hesitates" so to speak—and then takes a run to race past the "danger point" at very high speed. This is followed by a new hesitation before it takes very quickly the next hurdle (Fig. 2.5).

We humans do the same thing when for instance at night the lights on our bicycle have failed again and we see further on the road policemen on watch. We race past the "danger point" hoping that we will not be noticed. This is indeed a rational action because thanks to our high speed we are dwelling only a short time in sight of the patrolling police officers and can therefore hope that they are exactly at the most critical moment looking somewhere else. By racing past we have increased our chance not to be spotted.

The interpretation of "forbidden points" in terms of a wildly variable velocity must not be taken literally in the case of an atomic particle. Caution is needed if such a macroscopic picture is applied to the atomic world, and much more will be said about it in Chap. 3 and later again, in Chap. 10. For the time being, the picture of alternating "hesitations" and "sprints" is useful to remember the *phenomenology* of the atomic marble's behaviour.

We can make a different picture of this phenomenon. Let us take a camera and shoot at random times snapshots of the irregularly moving marble. Afterwards collecting all the photographs and printing them on the same frame, we will see that

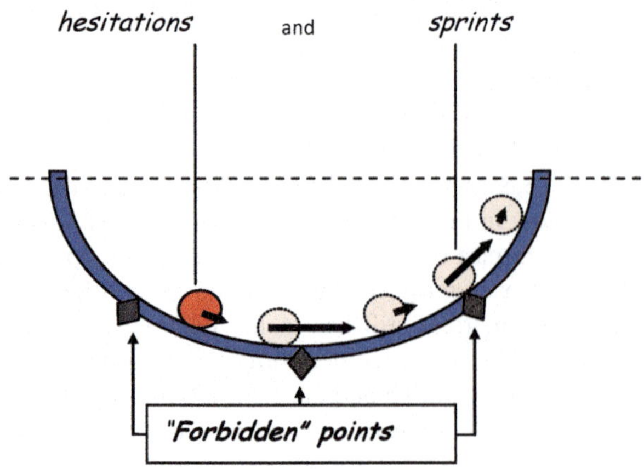

Fig. 2.5 The "forbidden points" in an atomic sized potential well, and the phenomenologic (macroscopic) interpretation in terms of jumps of the particle (not to be taken literally!)

the marbles on the photo's are not evenly distributed. Take for instance the vicinity around a "forbidden point". When the marble is in this vicinity it is racing to pass as quickly as possible the forbidden region. We get hardly the chance to catch it on randomly taken photographs, so short is the time it spends at these points. When making random snapshots of the marble that is racing past a danger point on the bowl surface, the chance to get a picture of it is much smaller than at the places where the marble is hesitating and has a small speed. Consequently, when overlaying all the pictures taken, we will find many pictures of the "hesitating" marble and just a few or none at all of the marble speeding past the forbidden points.

As will be seen later, in quantum mechanics it is customary to present results in the latter form, the picture one gets by taking snapshots at random. In fact, this is nothing else than determining the *chance* to find the marble at different points along its trajectory. Of course, in more dignified publications such information is usually presented in the form of graphs. They have the posh name "position probability density graphs" but really give the same type of information as contained in the picture of overlayed snapshots. In the next chapter we will have the opportunity to show several of such graphs, as calculated by using Schrödinger's equation (Fig. 2.6).

And now the most extreme behaviour of the marble at atomic scale. Alas the plot of this part of the story has already been given away in the introduction so that the surprise effect is lost a bit. Nevertheless it remains a remarkable thing: the marble sometimes jumps over the rim, out of the bowl! In the macroscopic world we would never expect this to happen, just because the marble has initially a given amount of energy which can only decrease, and which is insufficient ever to reach the rim of the bowl. After all, when we laid the marble in the bowl, it was somewhere *inside* the bowl, and we would not expect it to jump out of the bowl on its own accord, it simply does not have enough energy to ever come higher than where it started. At least, that is true in the macroscopic world (Fig. 2.7).

In the world of atoms an elementary particle which is trapped *inside* a potential well (so that it becomes an oscillator) is sometimes suddenly found *outside* the well! It would seem to have violated all the laws we know about conservation of energy.

And here we come back to our earlier introduced cyclist freewheeling down a hill and rolling up the next hill (still freewheeling). If he would not have sufficient energy to pass over the top of that next hill, and if we would nevertheless find him in a valley further on where he never could be whilst freewheeling, we ask

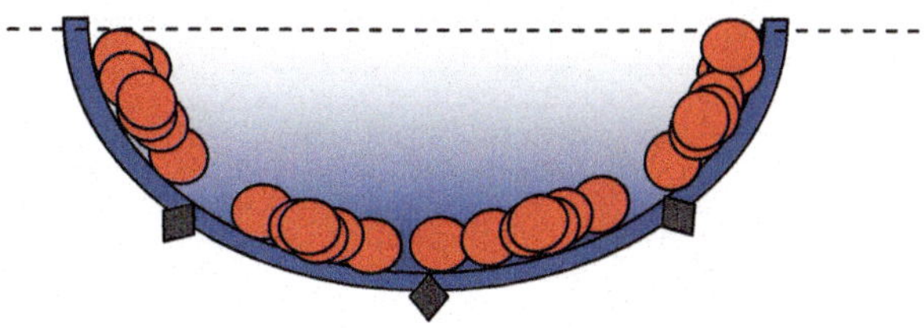

Fig. 2.6 Overlayed snapshots of the marble within a bowl with forbidden points

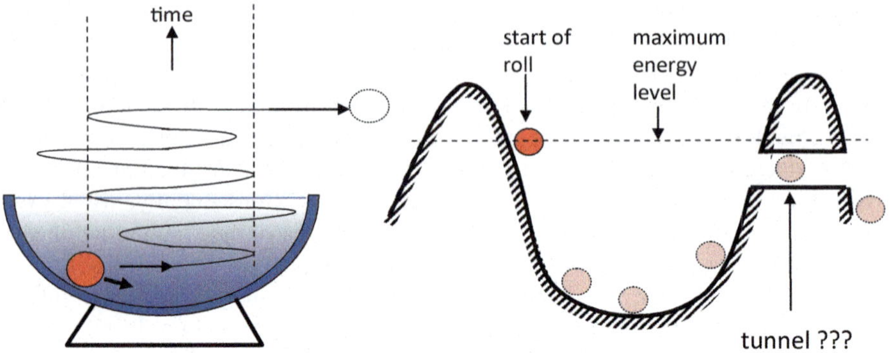

Fig. 2.7 The origin of the expression "tunnelling effect" for the energy-wise impossible escape of particles from their potential well

ourselves probably: "was he cheating and added some human energy by pedalling, or was there a tunnel through the hill, so that he has avoided a further—impossible—climb and sneaked *through* the hill instead of *over* it?

By analogy the strange, impossible behaviour of an electron which is sometimes found in places where it energy-wise cannot be, is called "the tunnelling effect". Although the phenomenon is impossible to understand (at first sight, but read on), it is very real. A special type of microscope can function only thanks to the tunnelling effect of electrons, and is therefore called "tunnelling microscope".

In conclusion: when consulting our two different books on mechanics (classical mechanics and quantum mechanics), the books tell us entirely different things about exactly the same situation, i.e. the same types of oscillator. And not only the stories are different, nature itself behaves entirely different according to whether we have in our oscillator just a few electrons or a great many of them. No wonder that we need completely different theories when we want to describe what we see happening, and want to make further predictions. The same model can only have different outcomes if different "laws of nature" are assumed to be applicable.

The aim of this book is however, to show that a *deformable* droplet of charge inside a potential well will show most of this weird behaviour, even if we analyse it using the "normal" laws of nature.

Chapter 3
The Transition Region as Predicted by Quantum Mechanics

3.1 Scaling All the Way Down Gives Quantum Results

In the chapter "introduction and outline" it was announced that in this book the strategy would be to start in the "normal" world, the macroscopic world at a human scale in which the classical laws of physics reign, and then try to find out whether hints of quantum-like behaviour would be found by making the scale smaller and smaller. In order to keep things as simple as possible, we would only look at one particular situation, viz. the one-dimensional oscillator such as a children's swing, a marble rolling in a bowl, or a charged object moving between stationary charges repelling it and thus forming a so-called potential well. The plot was given away already: it was disclosed that we indeed get quantum-like behaviour if the charged object in the oscillator is a *deformable* droplet of charge, instead of a rigid object. Surprisingly, an extrapolation of this kind of macroscopic oscillator fully down to atomic scales even yields the De Broglie formula, *including* the correct numerical value of Planck's constant.

All this is new stuff, not found in the present-day text books, and to remain on the safe side it must be considered as "tentative mental pictures about quantum phenomena". Before proceeding in more detail than was given in Chap. 1, it is wise to summarise the existing, *established* knowledge about the connections between quantum mechanics and classical physics. Such connections between the two worlds are found in the theory of quantum mechanics only, nothing will be found about this subject in books on classical mechanics. The following is thus restricted to a description what happens when we do an *up*scaling from quantum sizes to human scales, and what in this limit will be found according to quantum theory. It is the opposite of what will be considered in the later chapters.

T. van Holten, *The Atomic World Spooky? It Ain't Necessarily So!*
DOI 10.2991/978-94-6239-234-2_3

3.2 Upscaling: The Correspondence Principle of Quantum Mechanics

One of the dogma's of quantum mechanics, introduced by Bohr and often emphasised by him, is the so-called *correspondence principle*. The principle has to do with scaling *up* from quantum scales to macroscopic scales. It is thus the mirror of what we try to do in this book. The correspondence principle says that quantum behaviour will become "normal" behaviour if we assume the scale (or rather: the energy) to be sufficiently large. After having read Chap. 2 "Strange behaviour at quantum scale: the full list" you will agree with me that quantum behaviour indeed is *not* normal. It is therefore quite a harsh requirement imposed on the theory that it has to predict a behaviour which from "weird" will gradually become "normal" the larger the scale. Nevertheless, Bohr always insisted on checking every part of quantum theory, by investigating if it predicts this macroscopic limit correctly. In the present chapter we will give a few examples of such an extrapolation from quantum behaviour at atomic scales to "normal", often called "classical" behaviour.

What is more, a less well known theorem exists (at least: less well known in popular scientific books on quantum physics) which shows how Newton's dynamic law can be derived from the entirely different equation of Schrödinger. It is called "Ehrenfest's theorem". Why it is so unknown is really a mystery to me, in my opinion it must be considered one of the central theorems of quantum mechanics. Even the name of Ehrenfest is hardly or not at all known by the general public: one almost never sees streets named after him. Which is a shame, in my-humble-opinion.

3.3 The Parabolic Potential Well

The parabolic potential well will be featuring often in this chapter, as well as in the rest of this book. As a preliminary step it is therefore useful to introduce this particular situation more carefully.

What a potential well in general is, was explained earlier by showing the following, very sketchy figure (Fig. 3.1):

It showed a charged object which is mobile and can move in the space between two other, stationary charge distributions. The stationary charges exert a repelling

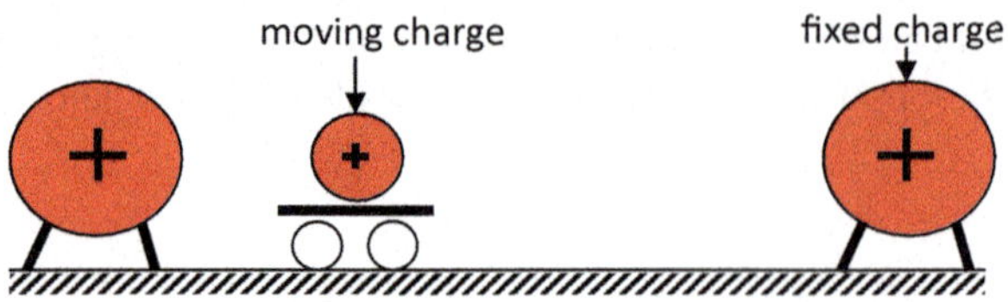

Fig. 3.1 Charge inside general potential well

force on the mobile charge, driving this charge back to the middle with a force that is larger the closer the moving charge approaches one of the stationary charge. The forces are of an electrical nature in this model. Such electrical forces are in the technical literature described by a so-called electric *potential field* in the space between the fixed charges. There is no need here to further explain how such an electric potential is exactly defined, or what it physically is. It is sufficient to know that the electric potential is an alternative, sometimes more convenient way to specify the magnitude of the electrical forces. The only reason to mention the "potential field" here is that it explains the name "potential well" for the arrangement of charges as sketched.

Of course, the specific configuration of the fixed charges will determine how fast or how gradual the force on the mobile charge will increase when it approaches the edges of the free space that is allotted to it. In order to visualise how the force on the mobile charge changes with its position within the well, the following types of picture are often used (Fig. 3.2).

This kind of picture strongly suggests the analogy with a bowl in which a marble is rolling back and forth. The shape of the bowl can be chosen such that any kind of force variation is simulated. In the first example the bowl is shaped like a parabola, and the marble then experiences the same force variation as a charge within a "parabolic potential well". The combination of a mobile charge and a parabolic potential well is also often called a *harmonic oscillator*. The parabolic well is taken in this book as "typical" for oscillators at an atomic scale, for instance if an electron is vibrating around a certain position, but cannot escape.

The rectangular shaped bowl is useful too as an analogy for certain kinds of potential well. A ball rolling inside the rectangular bowl will not experience any force, until it hits one of the walls. At that moment there is a sharp peaked, very

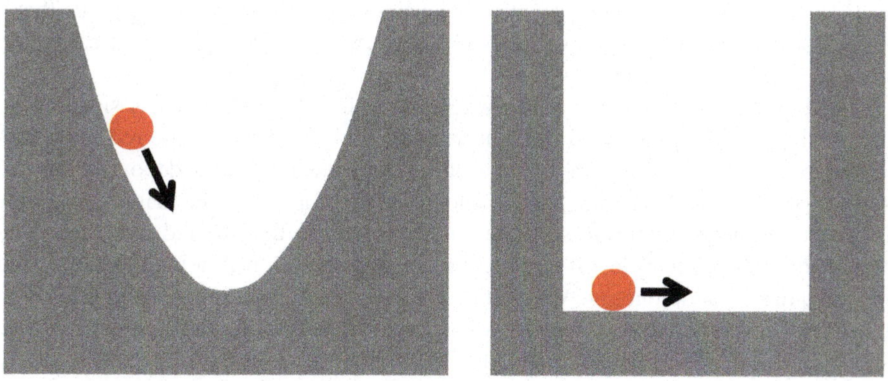

Fig. 3.2 Schematics of different types of potential well (*left* the parabolic well, *right* the box), based on the analogy with the marble inside a bowl

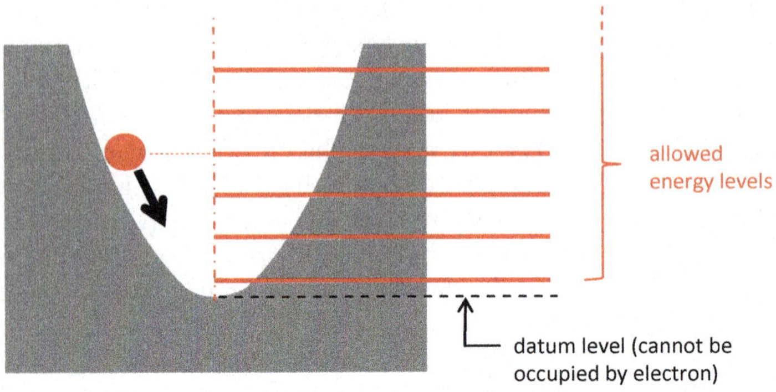

Fig. 3.3 Energy quantisation in a parabolic well

large force, kicking the ball back. In the electrical case we call this a potential *box* instead of a well. It represents the situation of an electron which is free to move, for instance inside a slab of conducting metal. But it is not allowed to escape from the metal, and experiences a very high "potential barrier" bouncing it back into the metal the very moment it wants to escape from the slab.

In the preceding chapter the phenomenon of quantisation of energy was mentioned. If an electron is confined to move within a small space from which it cannot escape (ignoring for a moment the possibility of "tunnelling" away), it cannot have any arbitrary value of energy. The electron is said to be "bound", and a bound electron can only have certain, discrete energy levels. We like to use—as in the above shown pictures—the visualisation which is based on the analogy with a marble and bowl. In this kind of visualisation the energy quantisation is shown in the way shown in Fig. 3.3.

The allowed energy levels are indicated by the red horizontal lines. Remember that a lower bound for the energy exists, the so-called "zero point energy", so that a particle at its lowest energy level does *not* have zero energy. This is also called the *ground state* of the particle.

The figure shown above is a nice way to visualise the energy levels since, if we deposit in the analogous bowl the marble at the height of one of the red lines, this would be the total energy of the marble. When the marble is deposited on the surface of the bowl it does not have a velocity at that moment. The height represents its potential energy, but that is all the energy it has, so that the red line shows the *total* energy of the marble. Of course, the marble starts to roll down the slope, and then acquires kinetic energy. When it has arrived at the bottom all its energy is in the form of kinetic energy, but the total energy during this whole motion is preserved and keeps the original value. The consequence is, that wherever it is at any instant of time, its total energy is still characterised by the height where it started its roll, i.e. by the red line corresponding to the point of release.

3.4 The Potential Box

Let us now have a look at the potential *box*, as shown below in Fig. 3.4. Again the same system of visualisation can be used, by drawing the analogous square bowl:

In this case the marble is all the time rolling on the floor of the bowl, so that it does not have potential energy. The only form of energy is here kinetic energy. The symbolism of the vertical stack of (red) lines to indicate the allowed energy levels is now less self evident. The level of the lines corresponds to the height obtained if all the kinetic energy would have been converted into potential energy. One can in this case imagine that the marble "thinks" it is a jet aircraft during a flying demonstration. The marble fancies that it like the aircraft can suddenly pull up from a horizontal high speed flight into a vertical, so-called "zoom" flight, thereby converting all its kinetic energy into height (potential energy). After such a "zoom" flight the marble would find itself at the height of one of the red lines. Because only the energy levels of the red lines are allowed, the initial kinetic energy before the hypothetical vertical pull up is quantised too.

Comparing the two Figs. 3.3 and 3.4, one symbolising the parabolic potential well and the other the potential box, we see that the red lines (the allowed energy levels) are at quite different places. Their distribution in the vertical direction is different. This is not due to carelessness by the draughtsman of the figures (me). The difference in the distribution of allowed energy levels is indeed significant, and shows that the energy quantisation is strongly influenced by the shape of the well.

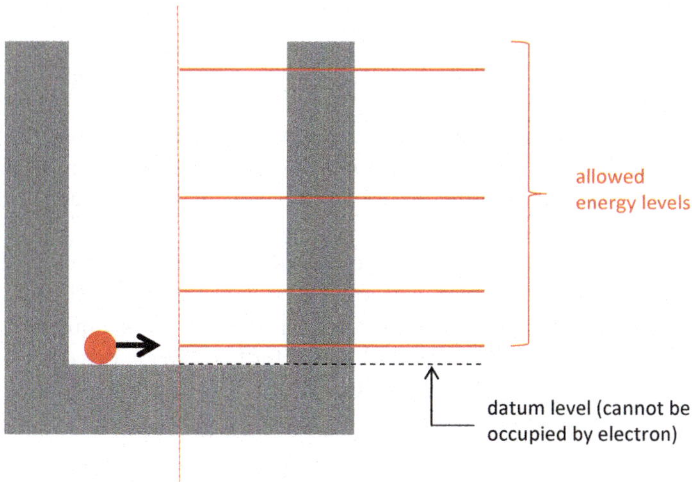

Fig. 3.4 Energy quantisation in a potential box

3.5 A Continuum of Energy Levels at Macro Scale

The correspondence principle now says that for very high energy levels one should recover the classical situation. What is meant by "very high energy level" is that we consider the hundredth or thousandth red line, or something like that. Obviously, the figures do not extend this far, they just show the situation at the lowest levels. The hundredth level would be far outside the edge of the paper. Now, looking at Fig. 3.4, one would at a first glance conclude that the gaps between the allowed energy levels become larger, the higher up we are in the diagram. Would this mean that in the macroscopic world we would deal with very noticeable gaps, and therefore with a very prominent quantisation? This is contrary to our experience, and certainly not in agreement with Bohr's correspondence principle! The point here is, that the *relative* magnitude of the gaps is the thing that counts.

The energy levels in a potential box can be calculated (see later chapters), and form a series 1, 4, 9, 16, 25, … etc., compared with the lowest level (the "zero-point energy"). The magnitude of the gaps increases too, but less fast: they form the series 3, 5, 7, 9, … etc. *Relative* to the amount of energy itself, the gaps diminish in importance the higher the level. It is seen that high up in the diagram of Fig. 3.4, at macroscopic scales, the gap between the successive red lines is almost negligible when compared with the absolute value of the energy. A macroscopic observer can hardly "see" on the scale that is relevant for him that in reality the energy levels are discrete: the energy levels form almost a continuum. Compare it with the screen of black and white television sets, where the image is in reality built up from separate horizontal lines. There are so many writing lines on the screen and they are so closely packed together, that for the eye of the viewer a continuous picture appears with smooth grades of brightness. Likewise, at high energy levels of a particle in a potential well it would seem as if any arbitrary energy is allowed. Just as we humans experience the macroscopic world.

Another, nice example of how the quantum properties seem to disappear with increasing scale is the following. If we compare a narrow potential well with a wide one, i.e. if we increase the dimensions of the well, we see what is sketched in Fig. 3.5.

In the upper part of the figure we see a narrow potential well with its allowed energy levels. In the lower part, we see that for a broader well the separation between the energy levels has become much smaller. It is not difficult to imagine what happens if we continue to stretch the well further: the energy levels will come closer and closer, until we finally are no longer able to discern between the different levels. Again the correspondence principle has worked correctly.

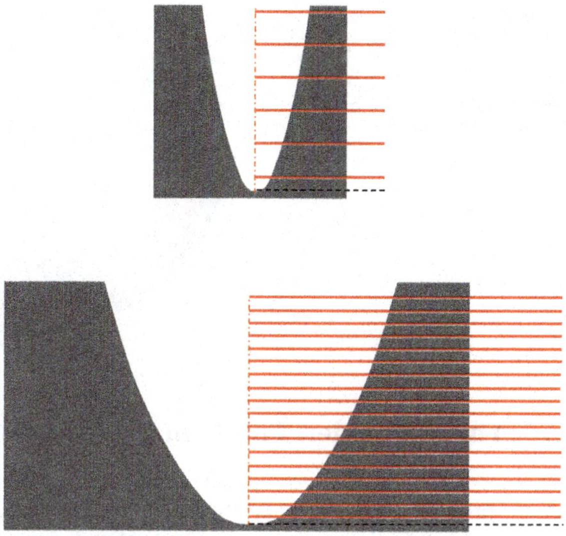

Fig. 3.5 energy levels of a narrow and a wide potential well

3.6 Where Came Figs. 3.3 Through 3.5 from?

So far, we have quoted the results of quantum theory, as found in the textbooks about the subject. Quantum theory is so complicated and so mysterious, that for the moment you just have to accept that these are the results. Now you may not believe this, but you will be able to derive these energy levels yourself, if you have the perseverance to read on at least to Chap. 9 of this book! The theory of the deforming droplet of charge leads to exactly the same results as sketched in the figures shown, but its basic ideas are definitely simpler than the conventional quantum theory. It may not be such light reading as a romantic story, but it will probably not demand from you more attention and alertness than a difficult detective plot, or a play by Shakespeare full of names and persons to be remembered. The basic reason why the ideas behind our droplet theory are so much more comprehensible is, that they come from the normal physical phenomena that you are used to in daily life. In Chap. 9 for instance, no more than a few lines of text and an explanatory diagram are needed to quantify the spacing of the energy levels in a potential box.

Therefore, you will not be left in the dark about the background of everything that has been told until now. But this background will be explained using the "deformable-droplet-theory", whilst the present chapter is about quantum theory itself. Right at this moment you will have to accept the correctness of everything on authority, the authority of the textbooks on quantum mechanics. The same applies to the remainder of this chapter. Sometimes, the author could not restrain himself from making his own comments, but these places will be clearly marked.

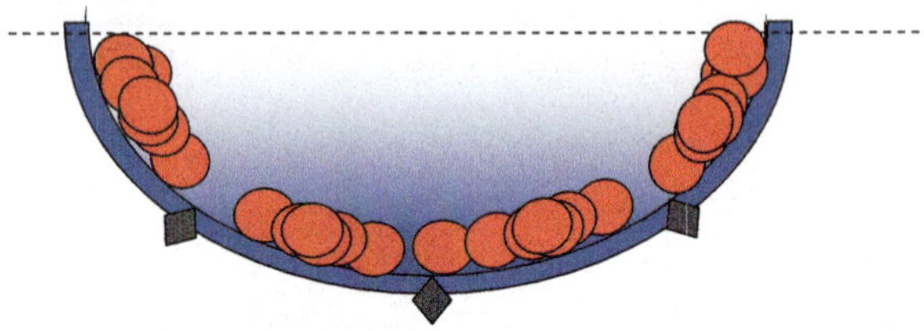

Fig. 3.6 An atomic marble in a bowl, "photographed" at random times, shows the whereabouts of the particle (position probability)

3.7 The Forbidden Points in an Atomic Bowl

Let us proceed with more examples of the correspondence principle, and next consider the "riddle of the forbidden points" on the path of a charge inside a potential well. To refresh everybody's memory (including mine), a sketch is repeated that was shown in Chap. 2.

We picked one of the allowed energy levels, and observed a particle having exactly this energy. Again the well is symbolised by its analogon, the bowl and marble. The broken line near the rim of the bowl indicates the energy level, since it is drawn through the point from which the marble was released. Actually, an insider will immediately see that the chosen energy level is the fourth one that is allowed, and if you read on you will be able to confirm this in a few moments. The atomic marble is rolling from side to side and back again, but it does so in a rather erratic way. We inspected its behaviour by taking snapshots at random times, and collecting these shots in the above shown Fig. 3.6. What appeared was, that we saw the atomic marble very often in certain parts of the bowl, and almost never at the points we baptised as "the forbidden points". Apparently, the chance to find the marble is not the same everywhere.

3.8 The Probability Density Curve

In textbooks on quantum mechanics this property of quantum oscillators is presented in a slightly different form, although the principle is not much different from the picture given above. In these books the probability to find a particle in a certain place is shown in the form of graphs. Here is one for example (Fig. 3.7).

This example concerns the second energy level in a potential *box*. Because it is such a simple graph (simpler than the graph which would represent the above

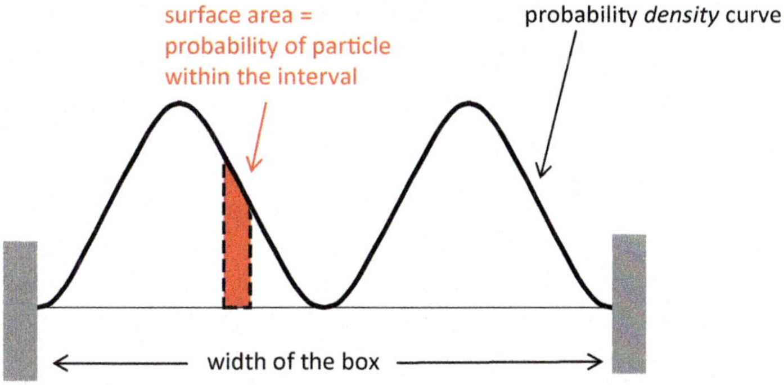

Fig. 3.7 Probability density to find a particle at certain positions within a potential box (second energy level)

sketched bowl with three forbidden points) we will use it to explain how this kind of graph must be read.

The horizontal line and the grey blocks are a representation of the potential box, the grey blocks symbolizing the potential barriers and the line between them is the straight path of a particle moving back and forth between them. The horizontal direction thus gives the *position* of the particle, no matter at which *time* the particle is in a certain position.

The black curve gives information about the *chance to find* the particle at a certain place. If in a certain region the curve rises high, the chance to find the particle there is also high. Comparing with the composite picture made out of a large number of randomly taken snapshots (Fig. 3.6) the highest parts of the curve thus correspond with the dense clusters of pictures of the marble. Where the curve is lower, we had less "hits" with the camera.

You would think that reading this kind of graph is thus easy: choose a point of the well, and read from the vertical axis the corresponding probability to find the particle at this point. Unfortunately, matters are a bit more complicated than this. The curve gives in actual fact the so-called *probability density*. Without going into all the mathematical details, this means that the graph should be read as follows. We take a small interval on the horizontal axis, and formulate the question about the chance to find the particle as: how large is the probability that the particle is *within this interval* at some point in time?

This probability is now given by the *surface area* below the curve, inside the chosen interval. The procedure is schematically shown in the above given example, where the red area gives the sought probability. In a few moments it will become clear why it is important to remember this property of the probability curve, or rather: the probability *density* curve.

3.9 The Meaning of the Forbidden Points in the Bowl (Or: The Zero-Points in the Probability Density Curve)

The curve shown as an example in Fig. 3.7 is typical for the probability densities determined by solving Schrödinger's equation. Again it shows that quantum mechanics has a lot of mysterious things in store for us. Take for instance the strange point in the middle of the potential box. Here the curve descends to zero. Sometimes this is interpreted as a point that cannot be passed by a particle moving inside the potential box. However, this conclusion ignores the explanation given above how to read the graph: one can only conclude that in an interval *around* such a zero-point the chance to see a particle is very *small*, the chance is not absolutely zero.

Nonetheless, it is a difficult issue indeed. The question that is immediately raised by the existence of zero-points is: if around such a point the probability to find the particle is very small, in other words: if there is hardly a chance to ever see it there, would that mean that the particle almost never passes this point? That would imply that the particle almost never steps over from the left side of the box to the right side, or the other way round. It would for a long time be dwelling in one of the two "compartments" of the box, and stay there most of the time. And what would then be the difference with a box of half the width? Very little. But an actual box of half the width would have a different quantisation of its energy levels! There is clearly a problem when we try to visualise the actual physics implied by the curve of Fig. 3.7.

We might have the hope to gain a better understanding, by invoking Bohr's correspondence principle. The principle says that the macroscopic situation should be gradually recovered by the process of increasing the energy of a system step by step. It means in this case that considering the higher energy levels in a potential well, we could perhaps get a clue how these strange zero-points are experienced in the macroscopic world. For the purpose of a comparison of the quantum results with the macroscopic world, the parabolic potential well is better suited than the box. The reason is, that the macroscopic equivalent then consists of an oscillator like the children's swing or other well known oscillating systems. The next section will therefore deal with the parabolic well.

3.10 The Probability Density of a Harmonic Oscillator

First of all it should be remarked that the number of zero-points in a potential well is not always one. If there are more of them, obviously they cannot all be in the middle of the well, it depends on the energy level. This is shown in the following figure (Fig. 3.8), copied from a physics textbook written by M. Alonso and E. J. Finn, titled "Fundamental University Physics". The figure comes from Volume III on "Particle and Statistical Physics". I will more often "steal" figures

The harmonic oscillator 73

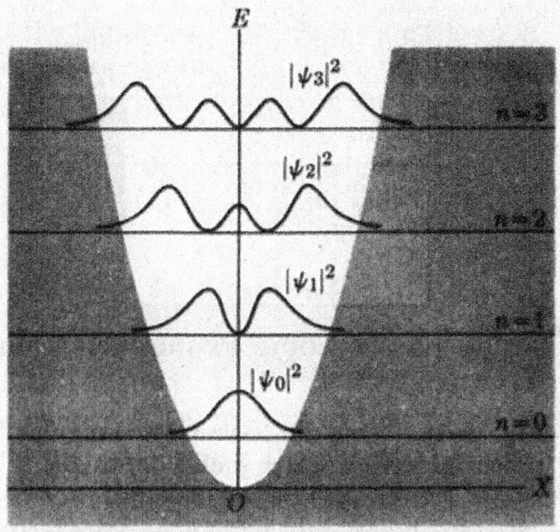

Probability densities corresponding to the first four energy levels of a harmonic oscillator.

Fig. 3.8 The harmonic oscillator, copied from Alonso and Finn: Fundamental University Physics

from this set of books, because of the clearness of Alonso and Finn's pictures. What they did here, was to determine solutions of Schrödinger's equation for a parabolically shaped potential well (or they in turn took these solutions from others), and sketched them in the form of one of their brilliant figures.

You will recognize the parabola giving the variation of the electrical potential, to be compared with the contour of a macroscopic bowl in which a marble is rolling. The horizontal line labeled "$n = 0$" indicates the lowest allowed energy level, or what is called the *ground state* of the electron which possesses the zero-point energy (note that the term "zero-point" here has a different meaning than in the previous section). Successive lines "$n = 1, 2, 3$" indicate the next higher levels. The curved lines drawn in the figure show the corresponding probability densities. Clearly, their shape depends very much on the particular energy level considered.

What can also be seen nicely is the tunnelling effect, mentioned already in previous chapters: we see that the chance to find the electron outside the contours of the well is not zero. The probability density curves extend a little bit outside these contours. The chance to find the electron in these regions is not large, however. No matter how interesting the tunnelling effect is, we will ignore it for the moment, and leave it to a later part of this book.

Inspecting Alonso and Finn's figure, we now understand how we were able, with a quick glance on an arbitrary density curve, to tell which energy level was associated with the particular curve. It is seen that the number of zero points is equal to the number, which gives the energy level. No mystery there anymore. Please do remember this property of the probability density curves, because it is an essential piece of information in the story of the following few alineas.

The questions we want to answer are: "how would the probability density curves look in the case of a macroscopic oscillator?" and "do they have a similar shape as the quantum curves?". And finally the crux of the matter: "is the required correspondence between the quantum results and the macroscopic result satisfied in the limit of high energy levels?".

3.11 The Macroscopic Probability Density Curve

To determine how a macroscopic probability density curve would look like, consider a macroscopic oscillator such as a children's swing (or the bowl and marble, or whatever). As a proud mother or father you are standing next to the swing on which one of your offspring is showing off. Fortunately you brought your camera, and you take lots of pictures. It is rather difficult to catch your child exactly at the most scary, highest points of the swings, so you take lots of snapshots at random times, hoping that there will be a few good photographs amongst them. At any rate, the blessing of the modern digital camera is that you can easily get rid of all the other shots. What you will find, coming home and studying all these snapshots, is that many more photo's show your child near the high positions than in the middle of a swing close to the ground. Your child will be delighted that his/her courage to swing so high is emphasized by your collection of randomly taken snapshots. The reason is of course, that in the high points the swing was slowing down, and even momentarily stopped before reversing its direction. Due to the small speeds, the swing thus dwells a relatively long time near the endpoints, and the chance to be caught by your camera is large. The reverse is true in the middle of a swing: due to the high speeds there, you will find relatively few shots of the swing when it is soaring past the middle.

The Fig. 3.9 shows the resulting chance to find the swing somewhere during its motion. The red curve goes to infinity at the endpoints, as the swing gradually comes to a stop.

Does that mean that the probability to find the swing in the endpoints is infinitely large? No, it does not. Remember that the curve is the probability *density*, the probability itself is given by the surface area beneath the curve, within a certain interval on the horizontal axis. It can be shown that this area does not go to infinity in the endpoints, as should be the case of course.

What we have drawn is only one curve, because macroscopically the probability distribution does not depend on the energy. If your child is swinging higher, and thus has more energy, this does not affect the *relative* variations of the velocity. The

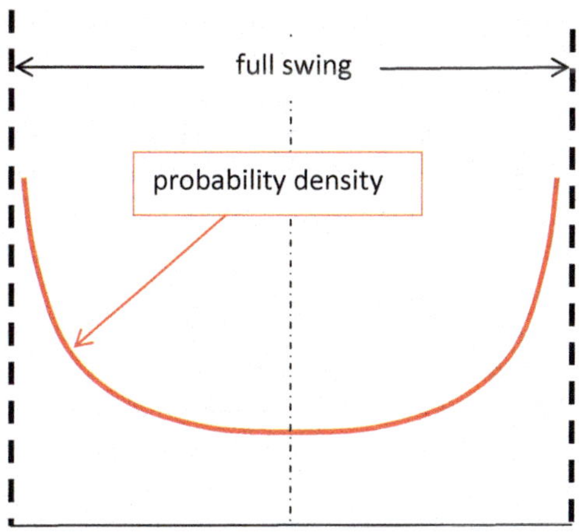

Fig. 3.9 Probability density for a children's swing

percentage of time that your child is in a given interval remains the same, if we take the interval also as a given percentage of the full swing.

Perfectionists will object and will say that this is not entirely correct. True, it applies only to low or moderately high swings. Think of the possibility that the swing is going so high that it goes "over the top". Rather do not think about it if it concerns your own child. But independent of whether it is your own child or—preferably—another's, in the case of going over the top the red curve would look differently from the one drawn above. We prefer to exclude extreme cases, and then have sufficient information from just the one curve shown above.

3.12 Is There Any Correspondence Between the Quantum Probability and the Macro Probability?

If we now compare with the probability density curves predicted for an "atomic swing", there is no similarity at all, see Figs. 3.8 and 3.9. Let us first of all look at the quantum curve for the ground state (the lowest energy). This is almost diametrically opposed to the curve of the macroscopic swing! The atomic swing has its *maximum* of the probability curve in the middle of the swing, the macroscopic swing has a *minimum* in the middle. If we would interpret the latter in terms of velocity, like we did for the macroscopic swing, it would mean that an electron oscillating back and forth would have its *lowest* velocity in the middle. Very strange indeed, and almost the mirror of what we macroscopically find.

One should be cautious, because the interpretation in terms of velocity is typically a macroscopic interpretation of the probability curve. It is kind of tricky to apply such a macroscopic interpretation to the quantum world, and most physicists would call it absolutely forbidden. In a later chapter, when we have seen how macroscopic phenomena may gradually "degenerate" into quantum behaviour during downscaling, it will be shown that this macroscopic interpretation of the probability curve is not entirely wrong, but it is incomplete, it is not the whole story. For now you may choose yourself what sort of pictures you like to have in the back of your mind when looking at these figures. We will come back to it.

Shifting our attention to the probability density curves at higher energy levels (as shown in Fig. 3.8), we observe that the character of the curves drastically changes compared with the ground state. But it is not possible yet to see any similarity with the red curve of Fig. 3.9 drawn for the "human harmonic oscillator".

But let us not be too hasty with our conclusions. In the book by L.I.Schiff: "Quantum Mechanics" the following Fig. (3.10) can be found. It concerns the 10th energy level of a quantum harmonic oscillator. What is shown here is the probability density for the quantum system (the fully drawn wildly oscillating line) in comparison with what is in the caption of the figure called the "classical oscillator", in our story the "human swing".

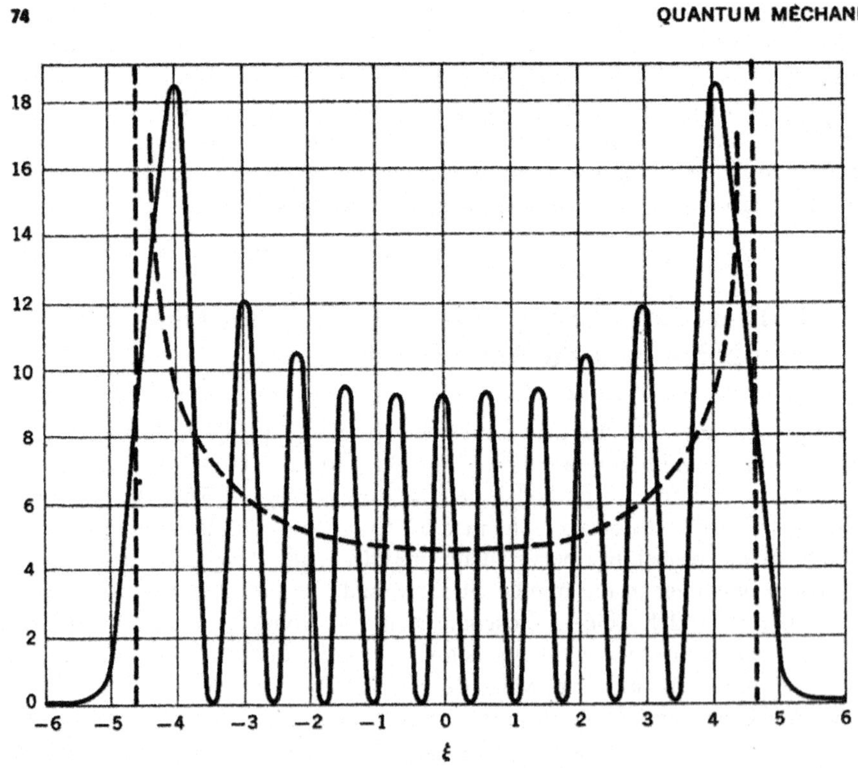

Fig. 3.10 The position probability density for the 10th energy level, compared with the macroscopic probability. Copied from: Schiff, Quantum Mechanics

It now appears that the classical curve coincides pretty well with the *average* of the wildly oscillating curve of quantum theory. The expectation is justified that if we go on and on, increasing the energy level for instance to the hundredth or higher, the classical curve is very accurately replicated by the average of the quantum curve. It would seem that the correspondence principle has worked again. Even so, some doubts remain, at least in the author's mind. Let me tell you about the worries I—secretly and in private—still have.

3.13 Once Again: These Nasty Zero-Points!

Is Fig. 3.10 really a confirmation of Bohr's correspondence principle? What now follows are a few private thoughts of the author, so there is no need to take them very seriously. The problem in my view is, that even if we continue to extrapolate to the human scale, the quantum curves still show these wild oscillations *including their zero points*. The extremely fast oscillations will average out because there are uncountably many of them, and they will not be seen on the human scale. But if a zero-point would be a barrier impeding or restricting the motion, increasing the energy would lead to uncountably many of such barriers. This not only applies to a single electron in the hundred-thousandth or so energy state. Quantum mechanics is just as well applicable to large bodies. I can pull my child, who is sitting on a swing, towards me to give him/her a start. But if all these quantum barriers are real they would inhibit the further motion, and the swing would be hampered or even stay in its place, my child hanging in an awkward position somewhere high above the ground and hardly moving.

There are books on quantum mechanics that offer another solution to the problem of the zero-points. I quote the—highly recommended—book by K.W. Ford, titled "101 quantum questions", in which it is summed up what sort of interpretations can be found: "Well, there are some things in quantum mechanics that we just have to accept whether we find them reasonable or not. In this case you can't really think of the particle making its way back and forth between the walls. Instead, at every instant, it occupies all the space in between. The particle, when in that particular state, *is* a wave. It is spread out."

Do we have to interpret the macroscopic limit as: our children are dissolving into a bunch of waves when they do a bit of swinging and want to pass all these barriers thrown up by quantum theory? Or alternatively and only a bit less bizarre: is my child "tunnelling" through all these difficult-to-pass points? Personally, I do not like either of these interpretations. In fact, I find them hardly acceptable. My personal feeling is that there is still something mysterious surrounding the interpretation of the zero-points of the probability curves of quantum mechanics.

It will be shown that the theory of the "deformable droplet" leads to quite another interpretation of the zero-points than in quantum mechanics. These zero-points do occur in the droplet theory, and they are found in the same positions as predicted by quantum theory, but their interpretation is a lot less problematic. The droplet theory can in a conceptually much easier way satisfy Bohr's requirement of "correspondence". But that will come later.

Chapter 4
From Rigid Marble to Vibrating Droplet of Charge: Model Assumptions

4.1 Warning: Chaps. 4, 5 and 6 Are Rather Technical!

The two preceding chapters (Chaps. 2 and 3) were about "normal" quantum mechanics. From this point on we change the subject, and turn to the mental pictures proposed in this book about quantum behaviour. It was announced that the weird behaviour of particles at an atomic scale can be replicated by a classical oscillator: the assumption that the oscillating object is deformable instead of rigid does the trick. This bold thesis is step by step corroborated in the next series of chapters, and it is tried to do so as systematically and completely as possible without going into the mathematical details.

However, the author would understand it if readers are a bit impatient and want to know whether it is worthwhile at all to read the next bunch of pages with all these systematic explanations. These readers might jump directly to Chap. 14, skipping for now the systematic story. Chapter 14 is limited to only the *results* of the analysis of a vibrating droplet of charge moving inside a potential well, without the arguments to justify these results. The chapter has the character of a list summarising the weird quantum behaviour as described in Chap. 2. Each of the magical phenomena described there will be accompanied by a mental picture based on notions of everyday physics. An example of this, already given in the overview of the introductory chapter, was the explanation of the tunnelling of atomic particles. In Chap. 14 *all* the magic phenomena of Chap. 2 will be gone through and will be ticked off. Later, the impatient reader could always choose to come back to the full explanations, hopefully with a whetted appetite. Remember, this is *not* a "who dunnit" or "who will get each other" story, there is nothing which forbids you to jump back and forth through the chapters of this book!

T. van Holten, *The Atomic World Spooky? It Ain't Necessarily So!*
DOI 10.2991/978-94-6239-234-2_4

4.2 The Modeling Philosophy

Having shed the impatient readers in this way, and thus having freed myself of the urge to hurry on, the explanations can now be given as systematically as needed in my opinion. As a reader, when you get bored with it, you can still at any moment choose to page quickly to Chap. 14.

The first subject then to be dealt with is how the droplet of charge was modeled in the mathematical analysis. The kind of deformations we can envisage in general are shown in Fig. 4.1. These photographs show a drop of water which has been deposited on an oil surface. The drop is kept together by its own surface tension, and is free to deform according to its own will, hardly hindered by the oil surface. Of course, gravity will squeeze the droplet a little bit onto the oil surface. It is good to realise that our own droplet of charges is not affected by such things as gravity, and is even more free to deform than the water droplet shown. Even so, the water drop "wriggles around" quite a lot, bulging successively in all directions, and if you would look more closely you would see tiny vibrations resembling shivers going through it. This is really the kind of freedom that we ideally would like to represent in the model of the droplet of charge.

Note in the above the word "ideally". It will be obvious to the reader that it would be nearly impossible to incorporate into the mathematical model *all* the "degrees of freedom" we see in the waterdrop. The dynamics displayed even by a common thing like a water droplet are practically outside the reach of mathematical analysis. Let alone the dynamics of a droplet of charge, where the electromagnetic field gives rise to "self forces".

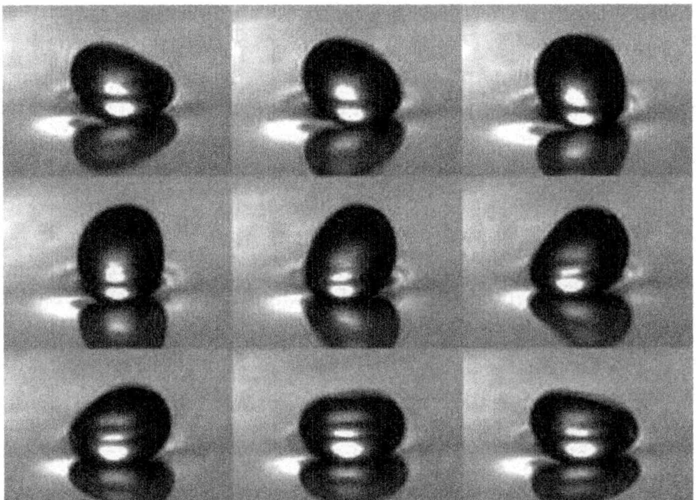

Fig. 4.1 A water drop lying on an oil surface (New Scientist)

In Chap. 1 it was already briefly described that one would have to start by taking into account the mutual repulsion forces between all the elements of the droplet (ultimately tamed by surface-tension-like apparent forces). But that would only be a start. One would also have to look at the electromagnetic radiation by each element, the influence this has on other elements and the recoil effects the radiation causes on the element itself. On top of all this, one has to account for the time delays of all these interactions due to the finite propagation velocity of electromagnetic waves.

No, we can forget about the *most general* dynamics of such a droplet of charge. For this reason the author has adopted a different modeling philosophy, and has asked himself: what is the *simplest possible* model of the droplet which nevertheless shows up the most essential differences with a completely rigid charged marble?

The priority is thus given to simplicity of the mathematical analysis. Of course, the model must allow deformations in shape and it must be consistent, i.e. it may not have features that contradict each other. These are the minimum requirements. Afterwards, we will try to judge from the results whether the simplifications were not too drastic. Such a—rather subjective—judgement will be based on whether the results seem to be physically plausible. It will also be based on an evaluation whether really the essential departures from the behaviour of a rigid marble have been caught. It is all rather subjective, and therefore open to criticism. The reader may rightly call the modeling more of an art than hard science. As a defense we may point out that this is true for most models used in physics, and always the "proof lies in the eating of the pudding": has the analysis yielded results in agreement with actual observations?

The author did his entire research whilst sitting behind a desk (or rather: the table in the sitting room which is the focus of the whole family life, from eating to colouring children's books and doing mathematical analyses), and never set a footstep in a laboratory. Actually, the author did spend a considerable part of his life in laboratories, but not for the present study. Nevertheless, we can compare the results of his paperwork with actual observations. The point is that uncountably many experiments have been performed in physics laboratories all over the world to check the validity of quantum mechanics. Quantum mechanics in its present form is the product of it, and can be viewed as the summary of all these measurements and observations. Therefore, if our theory of the deforming droplet of charge would indeed give the correct quantum results, then the conclusion would be that it correctly describes the real world, because the real world is described by quantum theory. No need for additional experiments, fortunately!

Of course, the present study covers only a small part of the whole of quantum theory, and we have to be cautious with our conclusions. More about that has been said in the introductory chapter, where it was stated that the ultimate test case is the double-slit experiment which at the present state is still outside the scope of the droplet theory. No more than nice mental pictures about quantum physical phenomena can be claimed to be the result from the studies presented in this book. What is important though, is to realise that these pictures at the very least are not contradictory to official quantum theory, and they therefore are not in conflict with experimental results.

4.3 One-Dimensionality

According to the above sketched philosophy we take as *prime* requirement that the model should allow a simple analysis. Only afterwards, after having done the analyses and having seen what are the conclusions, are we going to ask whether these results are plausible, or whether the model was too simple to really catch the essentials of the problem. This approach would seem to be a reversal of doing good physics, but it is pragmatic and it is successful as will be shown.

Taking into account this priority we can immediately say that the model should be kept one-dimensional. This does *not* mean that the shape of the droplet is one-dimensional, what it does mean is that we assume the *velocities* of its elements all to be in one direction. This leads to large savings in the effort to find the accompanying electromagnetic field, and therefore helps enormously to find the self forces on the droplet. The idea of one-dimensionality is shown in Fig. 4.2.

As explained in the beginning of Chap. 2, the assumption of one-dimensionality is not very exceptional. It is usual to start analyses of oscillators, irrespective whether they are classical or quantum oscillators, by taking the one-dimensional case. The object moving back and forth within the oscillator follows a path which can completely be described by one coordinate, e.g. the off-centre position of a mass acted on by spring-type forces. The velocity of the mass and its accelerations are in one direction, the same direction as the trajectory traced out by it.

If we now step over to a deforming droplet, we assume the same thing to apply to the individual *elements* of the droplet. One such element is drawn in Fig. 4.2, symbolised by the small square, whilst its velocity is symbolised by the red arrow. The position of the element, one second later, is thus as shown on the right hand side of the figure.

One special kind of element is the one in the centre of the droplet. More about what is exactly meant by "the centre" comes in the next section, for now it suffices to say that it is symbolised by the crossed circle. The motion of this centre will be called the *translational* motion of the droplet. For instance, the translational *velocity* of the droplet is in Fig. 4.2 symbolised by the red arrow between the two crossed circles.

Apart from the translation we have *pulsation*, symbolised by the stretching of the elliptically shaped contour. Needless to say, even the elliptical contour shown is meant just symbolically. In fact, in the theory no assumptions have been made

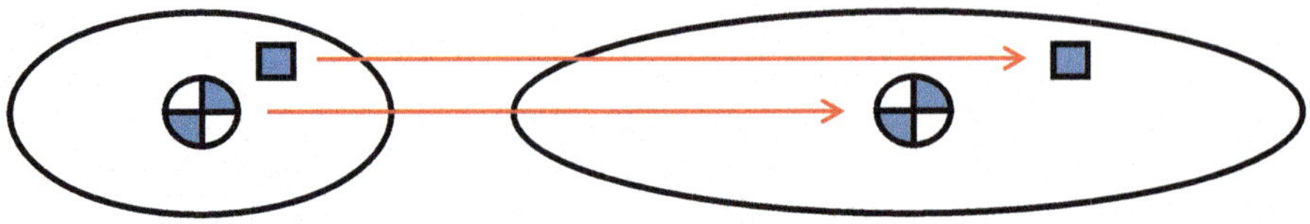

Fig. 4.2 The assumption of one-dimensionality: the velocities of all the elements are in one direction

about the precise shape of the droplet. It is only assumed that the elements which form the droplet, symbolised here by the small squares, can have a distance from the centre of gravity which varies in time. As a consequence, every element has a different velocity, differing from the speed of the centre. To be sure, these velocities are all in the same direction, despite their magnitudes being different.

4.4 Symmetry

As is suggested by the contours of the droplet sketched in Fig. 4.2, we as a next simplification assume symmetry, both in shape as well as in the distribution of the electrical charge. Actually, there are several symmetries. The first one is that the droplet has *rotational* symmetry around its length axis. Furthermore, there is left-right symmetry: the droplet is symmetrical with respect to an "equator plane", i.e. the left hand side in Fig. 4.2 is the mirror of the right hand side.

These symmetries cause that the geometrical centre point of the droplet (in Fig. 4.2 shown as a crossed circle) coincides with the centre of gravity. The term "centre of gravity" is used here, because it is probably well known to the general reader, and does not need a further explanation. Strictly speaking we should use the less well known term "centre of mass" but—although a subtle difference between these centres does exist—in the case at hand these differences do not play a role.

Similar to the centre of gravity, there is also a special point as far as the electrical charge is concerned, the so-called "centre of charge". One of the advantages of the assumed symmetries is, that all these different points coincide with the geometrical centre of the droplet. No further specifications are needed, no matter whether we are considering the electrical properties or the dynamical properties of the droplet. In most cases we will therefore use the term "mid-point" in the remainder of this book, unless it is really essential to make a difference.

4.5 Elongated Shape

The droplet is assumed to be elongated, with a shape varying between that of a rugby-ball and a long slender shape like a needle, with an average shape reminding one of an airship.

This kind of assumption, which excludes a spherically shaped droplet and certainly something like a thick disc, has an advantage when determining the electromagnetic field. The theory can then be considerably simplified compared with a "fat" configuration. The theory of long stretched out shapes is called "slender body theory" and facilitates simplifications to be made in the mathematical analysis. In the derivations of the appendix this advantage is fully used. But of course, the procedure involves some further approximations on top of the one-dimensional assumption. It is good to keep track of all the approximations introduced, so that we later do not draw unjustified conclusions, for instance if we by over-enthusiasm would tend to extrapolate results to "forbidden" shapes of the droplet such as spheres or discs. Which does not say that a droplet in the form of a sphere or a disc is to be ruled out as a physical possibility. It is just not included in the theory as it is at present.

4.6 Quasi-Static Deformations

This is not yet the end of the list of simplifications that have been introduced. The last one is, that only one dynamic "mode" of deformation is taken to be relevant. By "deformation mode" is meant the way in which the stretching and contraction is taking place. If there is an elongation of the droplet, we assume a proportional shift of the position of the separate elements. You can compare this with stretching a rubber band or a spring by pulling on the ends of it. If you do it calmly, without sudden shocks, all the elements of such a long spring move in proportion with the elongation between the ends. This is called the *quasi-static* mode of deformation, and this is the way we assume the droplet to stretch or contract.

In principle other modes of deformation are thinkable. For instance, if we do the stretching of a spring with a sudden pull, you will see dynamic effects. It is most clearly seen in the case of a long spring, as shown in the Fig. 4.3.

After a sudden pull on one end you can see a sort of longitudinal waves running through it. Do it symmetrically by pulling on both ends and then keeping your hands still, and the waves become standing waves. Such waves are the equivalent of the vibration modes of a violin string. The difference is that the elements of a violin string move in the lateral direction, i.e. they perform sideways motions. In the case of the spring the elements move lengthwise back and forth.

Another comparable type of waves is found in an organ pipe. Here too the particles of air oscillate in a lengthwise direction, and we find alternating regions of higher and smaller density of the air (see Fig. 4.4). In the figure it is sketched how the density variations look *in space*. From experience we know what the density variations do in *time*. They cause vibrations of the air in time, the kind of vibrations we describe as "sound" or as a "musical tone". The closer the bands of density

Fig. 4.3 Longitudinal waves in a long spiral spring

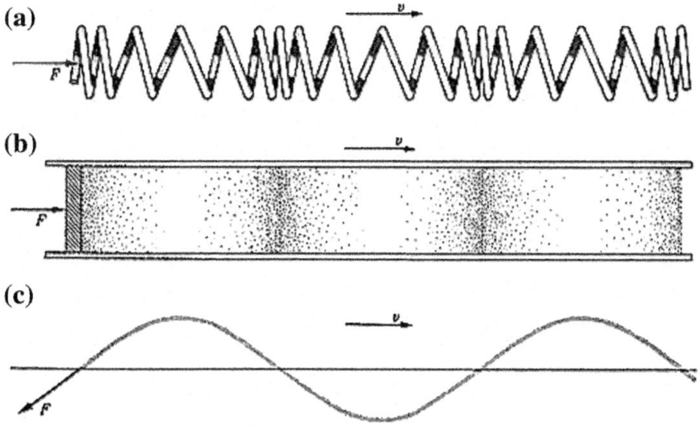

Fig. 4.4 Comparison of waves in a spring and in an organ pipe

variations are packed together in the pipe, the higher the associated tone. The different "modes" of vibration even can occur at the same time, so that the pattern of density variations becomes a superposition of several different patterns and represents different musical tones at the same time. It is called a ground tone with "overtones". The existence of these overtones gives a musical instrument its individual character, because the special mix of ground and overtones determines the "timbre" of the instrument, as experienced by the human hearing.

Likewise, in the droplet of charge a lot more deformation modes are in principle possible, apart from the "quasi-static" one that we take into account. The droplet's deformation could show successive bands of higher charge density alternating with less density of the charge. It would then look like the gas filled pipe in Fig. 4.4b. The final simplification of the model of the droplet is therefore, that we neglect all the possible "overtones" so that only the most basic deformation mode, i.e. the quasi-static deformation, is taken into account.

4.7 "The Proof of the Pudding Is in the Eating"

You may wonder if all these drastic simplifications are not going too far. Can the resulting model, stripped to the bone, still lead to anything useful? There is only one way to find out, and that is to explore the consequences. Consider the model for the time being as a "working hypothesis", just to see what comes from it. Later, after having seen the results, we can always decide to improve on it. The advantage of first awaiting those initial results is, that it is then hopefully more clear what are the most needed model improvements.

Chapter 5
Determining the Electromagnetic Self Forces

5.1 What Was Meant by "Self Force"?

What we now first of all must do is, to determine how the electromagnetic field around the droplet looks like. The procedure will be, to start with considering some arbitrary motion such as an assumed value of the speed of the droplet's mid-point, and assumed values of the acceleration and "jerk" (i.e. changes of its acceleration), without bothering how large they in reality will be. We just take some arbitrary values for them.

Next, we do the same for the pulsation motion. We take some arbitrary value for the elongation, and we assume how fast the elongation is changing. Let us call this the "pulsation velocity". The rate of change of this is also assumed, and is called the "pulsation acceleration", and similarly we define a "pulsation jerk".

Now the electromagnetic field is calculated. What this means is, that in every point around the droplet the strength and direction of the electrical attraction or repulsion forces are calculated, and how they change in time. Similarly, in every point the magnetic force and its variations in time are determined. The magnitude/direction of all these electrical and magnetic forces and how they vary in time will of course depend on the assumed values of velocity, acceleration, …., etc. up to the pulsation jerk. If you would have a look at the mathematical appendix (which I would dissuade you to do), you would see a set of formulae in which velocity, acceleration, etc. are the variables, so that by substituting their numerical values you would get numbers for the magnitude and direction of the electrical and magnetic forces. Of course, beforehand you should have picked a certain point in the so-called "field" around the droplet, where you want to know all these things. We say that this set of formulae is the specification of the "electromagnetic field".

Starting from the electromagnetic field we can do further calculations, e.g. how much energy is radiated by the moving droplet in the form of electromagnetic waves. In other words, we can find how large the effect is that the droplet has on remote objects. On the other hand the electromagnetic field can also be used to

T. van Holten, *The Atomic World Spooky? It Ain't Necessarily So!*
DOI 10.2991/978-94-6239-234-2_5

calculate how large the forces are on the elements of the droplet itself. These latter forces are what we call the "electromagnetic self forces".

There is a second task to be performed after all this. You will already have wondered how we can be sure that these self forces are in agreement with the motions we assumed beforehand. Well, we cannot. It is an entirely different, separate task to choose the values of the velocity, acceleration, etc. precisely such that the resulting self forces are in agreement with them. For instance, to take the simplest of these tasks: the calculation of the self forces leads to an expression that tells how large the force in the centre of the charge will be, amongst others how it depends on the assumed value of the acceleration. According to Newton's laws there is also a relation the other way round: the resultant force gives rise to a certain acceleration of the droplet. The problem is now, to find precisely that value of the acceleration for which both relations are in agreement with each other. How this task can be performed will be described in a later chapter, and will be ignored for the moment. The present chapter is merely about the electromagnetic field and the self forces, and how they depend on *arbitrarily* assumed motions of the droplet.

5.2 The Mathematical Approach and the Achieved Accuracy

If it would be attempted to write down, exactly and without any approximations, the electromagnetic field of the droplet with all its "modes of motion", this would lead to unmanageable mathematical expressions. This is so, despite the fact that in Chap. 4 already a lot of simplifications were introduced as far as the configuration of the droplet is concerned: the droplet was assumed to have certain symmetries and to have an elongated shape. Furthermore the kind of motion was restricted to one-dimensional motion of the elements, with the simplest possible form of pulsation only. Nevertheless there are still a lot of variables: the position in the potential well and the velocity, acceleration and jerk of the droplet's mid-point. Furthermore, we have to take into account the actual elongation as well as the pulsation velocity, pulsation acceleration, and pulsation jerk. It would be nice if we had some "ordering scheme" to discern between the effects that are really relevant and those that have a minor importance.

Besides that, it would be a waste of time *not* to make any approximations whilst determining the electromagnetic field. Take for instance points that are far removed from the droplet (the so-called "far field"). In these remote points we need to know the electromagnetic field for one reason only, viz. to calculate how much energy is radiated away to infinity by the droplet, and how much energy the droplet therefore loses every second. It is known that this "far field radiation" is mainly determined by the accelerations of the droplet's mid-point. All the other variables can only have a secondary influence, and these influences thus need not be determined with the highest accuracy.

Compare this with an analogous situation in astronomy. The gravity field of the earth is pretty complicated, it is not just the field of a uniform sphere of mass but it is noticeably influenced by the slight pear-shape of the earth and is even stronger above large land masses or mountain ranges. Artificial satellites flying relatively closely to the earth "feel" these variations of the gravity field, so that their path becomes somewhat twisted. On the other hand, the Moon is in the far gravity field. When calculating its path it does not make sense to take into account every minute perturbation of the gravity field like for instance that caused by the Himalayans.

In the case of the droplet of charge we want, on the other hand, to know the electromagnetic field close to the droplet (the so-called "near field") with a large amount of precision. High accuracy is here required, because this near field determines the self forces on the droplet. It would certainly not suffice to just take into account the acceleration of the mid-point and include only crude approximations for the other modes of motion. In the near field we cannot introduce the same approximations as were allowed in the far field. On the other hand, there is one aspect of the near field that is somewhat simpler than in the far field, viz. the retardation effects. You may remember what this was: it is the effect that every element of the droplet "sees" the other elements with some delay in time because of the finite velocity of light. It means that the influence on an element by other elements through their radiation is delayed. Any element is thus influenced by the position and the motion the other elements had some time ago. In Chap. 1 it was illustrated by relating the story of Lorentz's electron model that this so-called retardation effect may certainly not be neglected. In fact, it is the basic cause of self forces on the droplet. However, the time delays in the near field are much smaller than they are for instance in the far field, and this affords to make some approximations. The time delays in the near field are comparable with the time interval light needs to travel from one side of the droplet to the other, and are thus extremely small. How large an effect this delay can have also depends on how fast the variations are of the droplet's shape, compared with the time scale of the delays. A parameter characterising how fast these variations are is the ratio v/c, where v denotes a typical value of the velocity of the elements, and c is the speed of light. A possible approximation of the near field would be, to take all the terms proportional to v/c serious, but neglect smaller terms proportional to $(v/c)^2$.

A more accurate analysis then would be, that we do include terms of the order $(v/c)^2$, but neglect orders $(v/c)^3$, etc. In this way we have the freedom to choose ourselves how accurately the near field is calculated, whilst nevertheless simplifications can be made compared with an exact analysis.

The resulting situation is, that both in the far field and in the near field of the droplet approximations can be introduced without significantly detracting from the relevant phenomena. The problem is that the *type* of approximations is different, for instance the details of the droplet's shape and variations thereof are important for the near field, but less so for the far field. On the other hand, in the far field the full effect of retardation is of importance, whereas in the near field it may be approximated.

In the introductory Chap. 1 it was mentioned that a special kind of mathematical method would be used to determine the electromagnetic field and the self forces. Perhaps you remember the name "max-technique" which is short for a much longer, much uglier name that will not be repeated here. This technique is able to handle precisely the situation as indicated above.

The max-technique may be compared to using a set of geographical maps of a different scale. Our task of determining the electromagnetic field corresponds to not just reading the maps, we first have to draw these maps ourselves. The initial step is to draw a large scale map of the whole country, which does not show too much detail, but gives us the general overview. Most cities are shown on this map as just points, but their relative position with respect to other cities is clear. We see lines running towards these points. Imagining that these lines represent highways and railways along which cars and trains are moving, we might think that such a point on the map is a "singularity" where all these streams of vehicles converge, and disappear into some sort of "black hole", or are spewed out from seemingly nothing.

Now as a next step we direct our attention to one of these points of singularity. We want to see more detail. For this purpose we unfold a more detailed map covering just a part of the area of the large map. Or, more comparable to our task of map maker, we try to draw up a regional map. On a regional scale we see that cities are not singular points, but have some internal structure. We see how highways from outside the city are continued in the form of roads through the city, and how they connect somewhere in the centre with other highways in other directions. All this is achieved by choosing another scale of the map, or to say it differently: "by stretching the coordinates". Nevertheless, the information obtained by first having made the overall map is essential to understand the detailed map. For instance, a highway on the regional map stops at the edge of the map, and to see where it goes we need the larger overview. Usually, on a detailed map along the edges are the names of "destination cities" written, telling where the roads will bring you. But just the name of a city does not tell you very much, only the overview map can give more relevant information. In order to obtain correctly matching maps, where the smaller map is smoothly embedded in the larger scale map, some care is needed. In general it will require some back and forth working between the two maps to obtain a good and sensible connection.

If it is wished to see more details, the procedure of stretching the coordinates can be continued step by step. For instance, the next step could be to descend to the scale of a city map showing all the streets. We should be careful that such a street map blends in with the regional map, which in general will require once again a few iterations between these two maps. It might be that certain alterations we have to make to the regional maps even affect the first country map.

One could go on, depending on what the purpose of the maps is. Sometimes one would want to zoom in on the neighbourhood where you live. In principle it would be possible to go on and make matching maps of the interior plans of the separate houses, or even drawings of the pieces of furniture in your house and the pebbles in the garden.

The procedure to determine the electromagnetic field of a droplet of charge using the "max-technique" is in principle the same. The country map of the above given comparison would then correspond with the "far field" where the droplet itself is represented by a singularity. We get to see more and more details of the near field by stretching the coordinates in successive steps, all the time readjusting the successive approximations so that the stretched parts are properly embedded in the earlier determined field.

This readjusting is not a one way activity: making a further stretch on the detailed level may influence the whole field before everything is consistent. In the case of the map making this is in principle so, but that is theory only: a rearrangement of the pebbles in your garden will probably not affect the country map. Here there is a difference with the "drawing of the electromagnetic field". The scale of the most detailed maps (or: the amount of stretching of the coordinates near the droplet) is here characterised by the order (v/c), or the order $(v/c)^2$, order $(v/c)^3$, etc., as explained earlier. Every time we zoom in further, clicking the microscope to a larger magnification factor so to speak, one can be *sure* that there will be slight readjustments needed in the far field. This is not surprising: for instance stepping down from the level of detail O $(v/c)^2$ to O $(v/c)^3$ actually means that finer details of the motion of the droplet are taken into account, and these are bound to be "seen", even in the far field. It is one of the reasons why every step of further detailing requires increasingly more effort to work out mathematically.

There will thus be a strong urge to limit the amount of stretching to what is considered to be necessary from the point of view of accuracy. We stop the process of successive improvements as soon as we think that the accuracy achieved is sufficient. In fact, the analysis of the appendix includes effects of the order $(v/c)^3$, all the terms in the equations smaller than this were scratched out. Roughly speaking, terms as large as 1 promille compared with the leading terms are still included, smaller terms of the order of 0.1 promille are neglected. This is not too bad for a first order analysis, on the other hand it is not an accuracy to be particularly proud of. The choice of this kind of accuracy was based on two conflicting wishes. As pointed out, there is a strong urge to stop the successive refinements of the electromagnetic field as soon as practical, in view of the amount of time and effort it takes to work out the mathematics. On the other hand, we want to have a consistent theory. The chosen accuracy means that the radiation resistance on the droplet, i.e. the recoil force due to sending electromagnetic radiation, is correctly taken into account. Including the effects of order $(v/c)^3$ guarantees that the energy absorbed by the radiation resistance is in exact agreement with the energy which is found in the far field in the form of radiated energy.

5.3 What Kind of Self Forces Are Found?

The significant self forces thus found will be summed up now, including brief discussions from the physical point of view, leaving the mathematical aspects to the appendix.

1. **Electromagnetic mass**. This phenomenon was already mentioned in Chap. 1 as one of the conclusions Lorentz arrived at when he analysed his electron model (which was *rigid* in contrast to the present deformable model). He derived his theory more than a century ago, using a model of the electron that has been abandoned nowadays. The model consisted of a rigid sphere of some sort of isolator material, on the surface of which electrical charge was deposited (look at Fig. 1.23 for a sketch). The forces between the isolator core and the charges are sufficient to keep the charges fixed on the spherical surface. One might say that the charges are "glued" on the core, thus preventing an explosion of the assembly, and making the model consistent.

In some books his model is—curiously—called a deformable model of the electron, just the opposite of the above used description as a "rigid" model. This contradiction is confusing, but can nevertheless be explained easily. Lorentz, contrary to contemporary scientists working on electron models, took into account the relativistic effects as they were known at that time (before the unification, generalisation and cleaning up of existing ideas by Einstein). He assumed that an electron would become a bit shorter in the flight direction, due to what nowadays is called "Lorentz-contraction". A good description would therefore be, to say that Lorentz's model was "relativistically rigid", i.e. it could *not* deform under the influence of external or self forces, but it *was* subject to relativistic contraction.

If some amount of acceleration is assumed and if then the self forces are determined, Lorentz found from his calculations that his "electron" would resist the accelerating force. This resistance to acceleration appeared to be larger, the more charge he would assume. Since we call the unwillingness of a body to accelerate the "inertial mass of the body" the conclusion was that an electrical charge not only affects other charges, it also does something to itself: it represents an addition to the effective mass of the charged body. This addition to the mass was called the "electromagnetic mass". The idea was visualised in Fig. 1.24 and the accompanying text.

Let us denote electromagnetic mass by the symbol m_{em}. According to Lorentz's sums it is proportional to $(total\,charge)^2$, or in mathematical symbols: $m_{em} \div q^2$ where the symbol q stands for the total charge of the body, and the symbol $\div$ stands for "proportional to". We see that it does not matter whether the charge is positive or negative. The square of it is always positive, and the mass effect of the charge is always positive. Both positively charged particles such as protons as well as negatively charged electrons experience an increase of their mass due to their electrical charge. An electrically neutral particle such as a neutron would according to this theory not experience such an addition to its mass, unless it consists of an overall neutral collection of positive and negative charges. In fact, it is nowadays thought that neutrons indeed have some internal structure, so that one cannot be sure about their electromagnetic mass (if any).

Another thing found by Lorentz was, that the electromagnetic mass would increase if he assumed the size of his "electron" to be smaller. If we call the radius of his sphere a, then the proportionality $m_{em} \div 1/a$ was found. Since the radius a is in the denominator, a large charged sphere will have less electromagnetic mass than a small one. Of course, in both cases we assume the same amount of charge: the words "large" or "small" indicate just the physical dimension.

Both proportionalities can be combined to read:

$$m_{em} \div \frac{1}{c^2}\frac{q^2}{a} \text{ (Lorentz's electron model)}$$

In this formula one of the proportionality constants occurring in the complete formula has been retained, viz. $1/c^2$. Strictly speaking it is a bit illogical to do this, since we are used to disregard such constants in proportionality relations, and in fact in the complete expression for m_{em} there are other constants as well. The reason to make an exception for the constant $1/c^2$ is that it shows how small the electromagnetic mass actually is. The square of the speed of light is an extremely large number, and here we divide everything by it!

Now about the electromagnetic mass of our droplet of charge. The situation is a bit more complicated, because the droplet is a deforming object. If we say that mass is a measure for inertia, i.e. that it is a measure for the resistance to accelerations, what do we mean by that if all the elements of a body have different accelerations? It is a question that has been answered long time ago, during the infant years of the theory of mechanics, and the answer is really rather abstract though we have got so accustomed to it that it does not "feel" abstract. As a solution the theoretical artefact of "centre of gravity" was invented, which is a purely theoretical notion but something that even every layman has become accustomed to. The concept is in daily use even by people who loudly (and often proudly) proclaim to be absolutely ignorant of the natural sciences (funny, isn't it, that so many people seem to boast about their ignorance in this respect, whilst at the same time they happily make use of really very abstract concepts such as "centres of gravity"?).

To make a rather long story short: a procedure has been devised by physicists to sum all the different forces on the separate elements of a body (these forces can all have different magnitudes as well as different directions). It is a technique called *vector algebra*, specially invented to combine forces into one "resultant force" (a highly artificial process with a rather abstract result, really). If on top of that we now imagine that all the mass is concentrated in that particular point which mathematicians have defined to be the centre of gravity, this centre of gravity will move as if it were a rigid body on which this resultant force acts. A nice example which most people have heard of is the exploding cannon shell. The centre of gravity of these hundreds of small pieces will follow the same path as before the explosion, and will be acted upon by the earth's gravity just as if the shell was still intact.

In the same sense we can talk about the "electromagnetic mass" of a droplet of charge, even though the droplet is rapidly changing in shape. Recall that we made an idealised model of the droplet, the shape of which is changing between a sort of rugby ball, an airship and a needle. Let us denote the—variable—length of it by the symbol s. From the theory in the appendix we then find:

$$m_{em} \div \frac{1}{c^2}\frac{q^2}{s}. \quad \text{(deformable droplet of charge)}$$

It is seen that this expression is very similar to Lorentz's result for the rigid spherical shell. At first sight the similarity would seem encouraging: if the great Lorentz found the same result, then our sums cannot be very wrong, one is inclined to say. But that conclusion is a bit premature. At the time Lorentz found his result, Einstein had not yet developed his theory of relativity. Later, when Einstein had published the first version of his theory called the *special* relativity theory, it appeared that Lorentz's results were *not* in agreement with Einstein's. It will be explained more fully in the last paragraph of this chapter, but let me give a clue what the discussion was about. The theory of relativity predicts that the lengths of bodies, like our length s, will shrink with increasing velocity. This is the so-called "Lorentz contraction". Increasing velocity therefore implies a growing electromagnetic mass, exactly as predicted by the special theory of relativity for any kind of mass. The problem however is found in Einstein's relation $E = m \cdot c^2$ which is *not* satisfied by the electromagnetic mass. As already announced, this will be explained later. Briefly, according to this famous equation, mass and energy are equivalent. There is therefore some amount of intrinsic energy associated with the electromagnetic mass. Now, this intrinsic energy can also be calculated along another route, using the theory of electrostatic phenomena. Alas, these two ways to calculate the intrinsic energy do not lead to the same result!

Some consternation amongst physicists was caused by this observation, and other deep thinkers like Poincaré had to come to the rescue. Poincaré suggested that the noted discrepancy had something to do with the forces Lorentz had assumed to keep his electron from exploding (the "glueing" of the charges on the isolator core in his model). Feynman, in his famous trilogy "Lectures on Physics", relates the story in his usual flamboyant style.

All this might be thought to be of historical interest only, because this special kind of model for an electron studied by Lorentz has been abandoned anyway. But .... in our case we run up against the same problem again, and we cannot ignore the matter altogether. The person who has come to our recue is somebody with the name Arthur D. Yaghjian, although he certainly does not realise it himself and cannot be aware of the existence of a "theory of a deforming droplet of charge" in need of rescue. Clearly, we cannot evade the duty to tell the whole story later in this chapter (the last paragraph), since it is really essential to clear up the riddles. For now we move on to complete the list of self forces.

2. **Radiation resistance**. Again a subject in which some riddles are buried. In itself the existence of radiation resistance would seem to be a clear-cut matter. We are not going to repeat all that has been said about it in the introductory chapter, which was quite a lot. The essence is, that radiation resistance is a recoil force, associated with the emission of electromagnetic waves by a charged particle. As a reaction to such an emission a recoil force acts on the particle that is sending

out the electromagnetic waves. It appears to be proportional to the "jerk" of the particle, i.e. proportional to the change of acceleration:

$$(radiation\, resistance) \div (jerk)$$

Note carefully that we are not talking about *velocity*, but about *acceleration* that is varying. The situation is clarified in the schematic of Fig. 5.1.

The formula that was found by Lorentz, and that also comes out of our own droplet theory contains of course more factors than only the jerk, the most important being:

$$(radiation\, resistance) \div \frac{q^2}{c^3} \times (jerk)$$

To be precise: in the formula it is the jerk (defined as the change per second of the acceleration) of the *centre* of the droplet. Once again we have shown how small it is by leaving in the expression one of the constants, i.e. the term $1/c^3$. Note that in

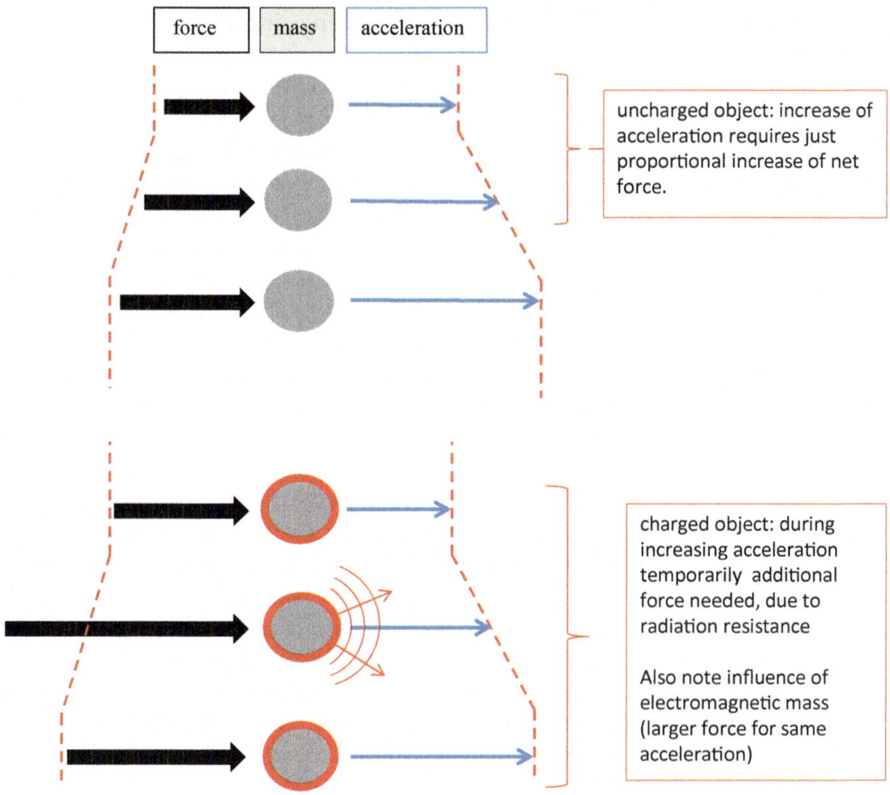

Fig. 5.1 Visualisation of radiation resistance

this case we are dealing with the *cube* of the speed of light. For the rest, the formula is what we would expect: the square of the charge sees to it that it does not matter whether the charge is positive or negative.

The above given formula came out of Lorentz's theory, and is also the result from our own theory of the deforming droplet of charge. These two theories are quite different, and follow different lines of reasoning, but the results seem to confirm each other nicely. The theory given in the appendix of this book has an additional feature, which is not mentioned in Lorentz's work (as far as I know). It concerns the possibility of a "non-radiating" state. In this special condition, imagine that the radiating charge is surrounded by a large sphere. It does not matter how large, it may be a sphere somewhere in outer space. The inside of the sphere reflects the radio waves back to the charge, and prevents that any radiation is coming out of the sphere. As a consequence the whole system of charge and the field inside the sphere together does not lose any energy by radiation, all the radiation energy is trapped inside the system. Probably a pattern of standing electromagnetic waves will be set up inside. The theory in the appendix about a droplet of charge predicts that the radiation resistance then vanishes, as soon as we assume the existence of such a "blockage" of the electromagnetic waves. This is in agreement with our physical feeling: no energy lost from the system, then neither a recoil force. This is confirmed by our droplet theory.

Later it will be discussed how this phenomenon has some bearing on one of the weird properties of quantum particles. Unlike macroscopic antennae, particles at quantum scale do not always radiate. During these times of radio silence they do not lose any energy, and are not subject to a force such as radiation resistance, despite the fact that there is plenty of jerk in their motion. Our theory at the very least does not exclude such a possibility of a vanishing radiation resistance, although the immediate question is of course: what then in the quantum situation gives the same effect as some sort of radiation shield around the particle? Obviously, this matter must be taken up in the later chapter about radiation.

The riddle that we spoke of earlier is a different one. It is a riddle which has nothing to do with quantum behaviour. There is something odd even if we consider purely the classical laws of radiation. Have a look again at the above sketched picture of what is meant by radiation resistance of charged objects, and let us now consider the energy balances. If the object is subjected to "jerk", the force causing it must be larger than we would expect without any radiation effects. It also means that this additional force pumps more energy into the charge than what is needed to just increase the kinetic energy of the particle. That is logical of course, because the force has to provide the power for the radio transmission as well, on top of the power necessary to increase the velocity. The extra force needed equals the radiation resistance, and we can easily tell how much extra energy it pumps into the charge to make the radiation possible. We can derive (do not worry if you do not find this immediately obvious, it is standard mechanical theory that you can safely accept to be true):

$$(energy\ (per\ second)\ supplied\ by\ a\ force) = (velocity\ of\ the\ object) \times (magnitude\ of\ the\ force)$$

We can make this general statement more specific, and conclude for the situation at hand:

$$(extra\ energy\ (per\ sec)\ absorbed\ by\ charge,\ subsequently\ radiated\ away) = (velocity\ of\ charge) \times (radiation\ resistance)$$

Finally, in this formula we can also make use of the fact that the radiation resistance is proportional to the jerk of the charge, resulting in:

$$(energy\ per\ second\ radiated\ by\ charge) \div (velocity) \times (jerk)$$

I hope you will find all these steps quite obvious. The problem crops up when we do a *separate* analysis of how much energy is found in the radiation far away from the particle. This is a piece of classical physics, which was first worked out by J.J. Larmor (1897). What he derived was an entirely different expression for the total radiation energy passing per second through a closed surface, far from the particle but completely surrounding it. He found that the energy which is present in the radiation is proportional to $(acceleration)^2$ of the charge:

$$(radiated\ energy\ per\ second) \div (acceleration\ of\ charge)^2 (\text{due to Larmor})$$

This is called Larmor's law of radiation, and it also comes out of our own theory of the moving droplet. This indicates the problem. At first sight, according to these expressions, the energy supplied by the extra force acting on the charge does *not* have the same magnitude as the radiated energy, because

$$(velocity) \times (jerk)\ \underline{\text{is not the same as}}\ (acceleration)^2!$$

Something is wrong apparently. Luckily, there is an escape from the problem. It was pointed out already by Larmor, that the kind of shaking of the charge he assumed in his theory was a special one, it is not entirely arbitrary. He assumed that the charge would move back and forth in a nice sinusoidal way. And this provides the solution to the dilemma, at least partially. In that special case of sinusoidal motion, we find that the *time averaged value* of $(velocity) \times (jerk)$ is exactly the same as the *time averaged value* of $(acceleration)^2$. Taken over one complete cycle of the smooth back and forth motion, the incoming energy which is absorbed by the charge from an external force is then equal to the energy that goes out in the form of radiation. On a *time averaged* basis the balance has been restored.

The fact that the energy balance comes out right on a *time averaged* basis does not say that *all* problems have been solved. Apparently, still there must be some kind of *short term* energy storage. Instantaneously, at each instant of time apart, the in- and outgoing energies are not balanced, so that a buffer must exist that during half of the cycle stores the energy surplus, and gives it up again during the second half of the cycle. This momentary surplus is called the "Schott-energy". What kind of buffer, handling the Schott-energy, might that be? It is usual to point at the electromagnetic field around the charge as the culprit. We say that a certain amount of energy is present "in the field". In fact, if we imagine that somewhere distant from the charge the energy in the radio waves is measured, this energy has been emitted some time earlier. There is thus some energy busy travelling in the space between the sending charge and our receiver. This amount of energy "in the field" can make up for the energy which seems to be lost or is superfluous at any moment.

Quantum physics offers a very vivid picture of the still somewhat mysterious energy residing in the space between the charge and our radio receiver. Mysterious, since there may be a complete vacuum there, whilst we always associate energy with some hardware: kinetic or potential energy *of a body*, acoustic energy *in a gas*, thermal energy in a solid or fluid medium, etc. But energy in nothing?

There is a branch of quantum physics which deals with the processes of sending or receiving electromagnetic energy by particles. More generally it deals with the interaction between matter and fields, whatever the nature of the field. This branch is called "*quantum field theory*", and several times in this book we will mention it because it can offer very down to earth views (or so it would seem at first sight) on certain mysterious looking phenomena. The picture of an electromagnetic field is in this theory based on the particle-wave duality of quantum mechanics. If a charge transmits electromagnetic waves, this is viewed as a stream of "light particles" or "photons" emitted by it. Each photon travels at the speed of light, and each photon represents a packet of energy (as well as momentum). From this point of view there is little that is mysterious about "energy residing in the field", it is just the energy of the photons that have left the radiation source but did not arrive yet at their final destination.

The reader will meet these "quantum field views" more often in this book, because of the very pictorial views quantum field theory offers on some phenomena. It is ironic that quantum theory contains so many mysterious, almost magical phenomena taxing the "physical feeling" (the reason why this book has been written), whilst on the other hand it sometimes can offer enlightening views on phenomena that are mysterious in classical physics.

3. **The "spring force" provided by the potential well**. Proceeding with the list of electromagnetic forces, the next one is straightforward, and for a change does not contain any riddles.

The droplet of charge finds itself within a potential well, which means that there are electrical forces which try to push the droplet back to some equilibrium position whenever it is displaced. This is a straightforward process. The electrical forces due to the potential well are equivalent to the mechanical springs in the so-called mass-spring analogue of oscillators in general. The situation was sketched earlier in a schematic which is repeated here (Fig. 5.2).

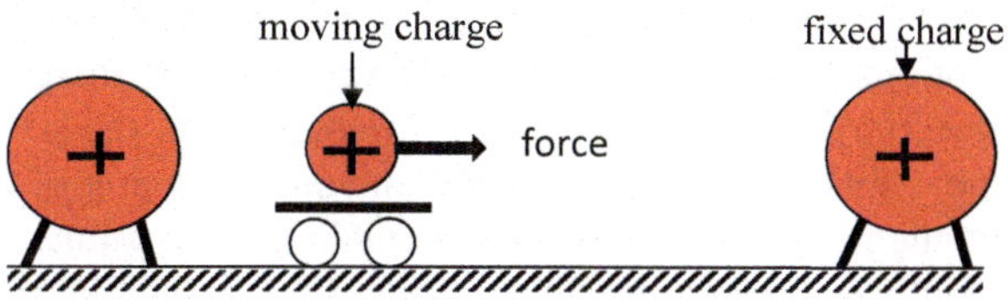

Fig. 5.2 The "spring" force provided by a potential well

In this schematic the charge moving within the potential well is drawn as some sort of marble, at least this is suggested by the sketch. In reality we are of course considering a deforming droplet of charge, and there is the problem that the elements of the droplet are situated at different places in the well, and therefore all experience a different force.

Any problems caused by the fact that the droplet has a finite extension and undergoes deformations can be solved as before, by the introduction of the concepts "resultant of the forces" and "centre of gravity". All the elements of the droplet may be at different positions within the well, but we can sum all these different forces into one "overall" force acting in the centre of the charge. As far as the motion is concerned, this totalised electrical force may be thought to work on a so-called "point mass", i.e. on an imaginary body placed in the centre of gravity, where all the mass of the droplet is concentrated. If it is phrased like this, it is again rubbed in that in reality all this is not so simple at all, although we have become accustomed to concepts like "centre of gravity" and therefore do not find the procedure to be very special. Anyway, the conceptual problems that lie behind this "summing of forces" and "summing of masses" and how to deal with these sums, are a part of very classical mechanics and are taken for granted here.

4. **Coriolis-like force**. At this stage we might think that we have obtained all the ingredients needed to calculate the behaviour of the droplet. We have a body with inertia (the "normal" mass of the droplet supplemented with electromagnetic mass), on which a "spring force" acts (the electrical force by the potential well) which tries to drive the body to the centre of the well but never is successful in doing so because, due to inertia, the equilibrium point is overshot every time that the body comes back to it. We also have a force (the radiation resistance) that damps out the motion, so that after a shorter or longer time the body will come to rest, having spent all its initial energy in radio transmissions.

This coincides exactly with the classical picture of the motion of a charge within a potential well. It is indeed correct in macroscopic situations, and the above listed forces are sufficient to describe what will happen in the world as we usually see it. We know from quantum mechanics however, that there must be more to it if we descend to the scale of atomic particles, since we then observe a quite different kind of motion. Quantum mechanics as such does not tell us anything about other, additional forces that could cause this difference. It just gives a prescription how to calculate this other kind of motion occurring at atomic scale, and it tells us to forget the classical "mass—spring—damper" picture.

On the other hand, the author of this book has argued that the weird kind of motion at atomic scale can be understood by imagining that a second "degree of freedom" plays a role next to the translation, the so-called pulsation. The fact that the—almost—periodic elongation and shortening of the droplet has some influence on the translation motion can explain a lot of things, as we promised to show. This of course implies that in our model there must be more forces affecting the translation of the droplet than just the above listed three. There must be forces that are carried over from this pulsation to the translation of the droplet. These additional forces apparently are practically negligible at macro scale, but have an increasing influence the smaller the droplet is. Therefore, we cannot finish the list at this point, and must continue with pointing out an additional force which affects the translation of the droplet but basically has its origin in the pulsation.

Such an additional force is indeed present, and it will be called a **Coriolis-like force**. Obviously, this is very much jargon-like language (the author is a bit ashamed to use it but does not know how to avoid it) and it needs further explication.

You may have heard about "Coriolis-forces" in connection with meteorological events. However, Coriolis forces are much more general and they occur in every mechanical system where objects are subjected to two different kinds of motion. Still, the author likes to use as an example the flow of air in the atmosphere, because of his background in aerodynamics.

In meteorology a question might be, what happens if a mass of atmospheric air flows from a high pressure area to a region with a lower pressure. Such pressure differences can be caused by differences in the warming up of the earth's surface by the sun radiation. What one would intuitively expect is, that the air would flow straight from the higher pressure to the area of low pressure. Actually, it does not, the rotation of the earth spoils the game. It is as if a cross-wind force is present, with the result that the air is going to circle around the low pressure region, instead of taking the shortest route to the middle of the depression. Everybody recognises these typical wind patterns when shown on the TV screen during the weather forecasts. They are the well known whirlpools seen by weather satellites in the form of spiralling cloud formations. An aerodynamicist would call them "vortices" in the atmosphere, a phenomenon which is closely akin to the well known vortex in an emptying bath tub.

The crosswind forces causing these phenomena are called *Coriolis forces*. They occur because the air in this example is subjected to two different kinds of motion: the motion towards the centre of low pressure, as well as the rotation of the earth (Fig. 5.3).

Coriolis forces not only occur in the case of the flows of air or water. A body subjected to two different kinds of motion (the insider will understand that we are talking about relative motions, so that I take the liberty to skip over the finer details) will always "feel" such additional forces, the origin of which is often not clearly visible at a superficial glance. The magnitude of these Coriolis forces depends on

Fig. 5.3 The devastating tropical storm "Katrina", illustrating the Coriolis forces leading to atmospheric vortices

the two velocities that are at stake: in an airflow with a velocity V, which flow is affected by the earth's rotational speed Ω, we find that this additional—or if you like: apparent—force is proportional to $V \cdot \Omega$. It is always the product of two velocities, in this particular case the wind speed and the so-called angular velocity of the earth. This should be considered to be the bare minimum of explanation of course: in the complete expression for the atmospheric flow other factors occur, such as the geographical position and orientation of the line between the low and high pressure areas.

Coming back to our droplet of charge: it too is subjected to two different kinds of velocity. The first one is of course the translation velocity V_{transl} associated with the motion back and forth between the sides of the potential well. The second kind of motion is the pulsation, for which we also can define a "speed of pulsation" V_{puls}. If the instantaneous length of the droplet is indicated by the symbol s, as we have done before, the associated speed V_{puls} is the *change of the droplet's length s per second*. Now working out the electromagnetic self forces as has been done in the appendix, it appears that we find an additional force in the direction of the translation (in this case it is *not* a force perpendicular to the direction of motion!). The additional force is proportional to the two velocities we have just defined: it is proportional to the product $V_{transl} \times V_{puls}$. Even though this force has an electromagnetic origin in contrast to the usual mechanical Coriolis forces, there are so many analogies (even deeper down in the mathematical derivation) that it is tempting to call it a *Coriolis-like force*. And that is what was done in the heading of this section.

The complete expression for this Coriolis-like force on the droplet of course contains other factors as well. An interesting finding is, that it is also proportional to an old acquaintance of us, viz. the electromagnetic mass m_{em} of the droplet. The final result can thus be summarised as:

$$(Coriolis\text{–like force in the translation direction}) \div m_{em} \cdot V_{transl} \cdot V_{puls}.$$

This additional force is the one that couples the translation and the pulsation of the droplet. It would be absent if the pulsation would not exist, and it would be very small if the pulsation would be weak. In a later section of this chapter it will be seen that this is precisely what happens if the droplet is assumed to be macroscopically large. The pulsation then becomes of such a minor importance, that the coupling practically disappears. Without such a coupling the droplet would perform a translating motion as if it were a rigid object, and what then is obtained is a normal oscillator with a "normal" behaviour, the sort of behaviour we are used to in the case of marbles within bowls and things like that. On the other hand, at atomic scales the coupling causes profound changes in the behaviour of the droplet, so drastic that it provides the justification to write this book.

This concludes the list of forces working in the direction of the translation of the droplet.

Some general remarks about the forces that affect the pulsation.

We now come to the second half of the list of self forces, viz. the forces that influence and drive the pulsation. Here you will immediately see a problem: the electromagnetic effects might have an effect to "squeeze" the droplet or let it expand, but can we really call these effects "forces"? The usual picture we have about forces is that they are actions that push or pull on something so that this "something" is going to accelerate or decelerate. In the case of squeezing or stretching something a combination of at least more than one force is at play. Think of squeezing a rubber ball between the thumb and the forefinger, both fingers are pressing in opposite directions. A squeezing effect is thus at least described by two opposing forces.

In the case of our droplet of charge the situation is even more complicated. It is not sufficient to have two opposing squeezing forces on the outside surface of the droplet. Recall that the droplet was compared with a shoal of fishes, where all the separate elements are free to move despite the apparent cohesion. *Each* element moves by influences felt by this element *itself* (Fig. 5.4).

Squeezing the *collection* of fishes requires that every fish *apart* is admonished to move a bit closer to the centre of the collection. In this respect the fishes in a shoal or the birds in a flock are more obedient to the rules of civilisation than passengers in an overcrowded train. Mostly the passengers on the balcony are packed together, whereas people in the aisles in the middle of the wagon hardly move to make more

Fig. 5.4 The analogy between a droplet of charge and a shoal of fishes. Dolphins attack and each element individually reacts, although some amount of coherence in the quickly deforming shoal is preserved

space. Apparently, the latter passengers do not feel the urge to squeeze. In contrast to the train, in the case of the droplet of charge we stipulated a model where stretching or squeezing would lead to a proportional movement of *all* the elements. Inherent in this model is therefore that *all* the elements, including the ones near the middle of the droplet, respond to a squeezing influence. We are thus talking of a very large number of forces, as many as there are charge elements, all of them of an electromagnetic nature, which cause the droplet to shorten as a whole. It is clearly difficult to speak about "the" force which squeezes or stretches the droplet.

Here we are rescued by an ingenious piece of physical theory thought up by Lagrange (1736–1813), a theory which was already developed very early in connection with similar problems in purely mechanical systems. Lagrange developed a theory for the motion of bodies where the variables causing the motion can be *anything*, no matter whether they represent straightforward forces, torques, couples of forces, or combinations of arbitrarily many forces. Even complexes of infinitely many forces are allowed in Lagrange's theory, and are actually often used in the practice of engineering calculations. When there are many forces acting on many different elements of a body, such a complex of forces can be handled in Lagrange's theory as if the whole complex were one single variable. Such a variable in Lagrange's theory, really representing an entire complex of forces, is called a *generalised force*. Similar to Newton's laws that give the relation between forces and the resulting motion, Lagrange derived formulae which give the relation between these generalised forces and the resulting motion of a body, even if that motion involves changes in the shape of the body.

Needless to say, these generalised dynamic equations are completely in agreement with Newton's fundamental laws. They can be viewed as a *generalised, alternative form* of Newton's laws. It really is one of the miracles of classical mechanical theory that such a thing has been found to be possible!

This concept of a generalised force will be used in our theory of the pulsating droplet. An electromagnetic effect that causes the droplet to squeeze or stretch is thus called a "generalised squeezing force". It is either positive so that it acts to give an elongation (in jargon: it is a generalised force *in the positive direction* of the elongation), or negative so that it gives a contraction.

Before proceeding with the list of generalised self forces found by the theory of the appendix, a few more definitions have to be agreed upon. Several of them were already mentioned, but let us give a systematic list for the sake of completeness.

- The variable length of the droplet is denoted by the symbol s and will be called the elongation.
- The change of s per second will be called the *pulsation velocity* V_{puls}.
- The change of V_{puls} per second will be called the *pulsation acceleration*, by analogy with normal velocities. In the following text no need crops up to use a mathematical shorthand notation for it, therefore no symbol will be attached to it.
- The change of pulsation acceleration could be called the *pulsation jerk*, but fortunately we will not need this at all, so forget it.

And now the list of generalised electromagnetic self forces which will affect the pulsation. The numbering system of the previous list will be continued so that in the later text it will be easier to refer to the particular force we there want to indicate.

5. **Generalised mass of pulsation,** which in the text will more often be referred to by the easier and more descriptive name **squeezing inertia**.

If for some reason there is a pulsation acceleration, we will find that the electromagnetic field acts in such a way that the squeezing will be opposed by the combination of forces on all the elements of droplet. It should be emphasised that this is a resistance against *accelerations* of the elongation, and it certainly should not be confused with anything like the force one feels when stretching a piece of elastic band. A better idea about the squeezing or stretching *inertia* can be obtained if we imagine a rubber band along which a large number of beads are fixed. If the beads are relatively heavy, for instance because they are large and made from gold, they influence the stretching of the band by their inertia, especially noticeable when we stretch the band by a sudden pull.

Not surprisingly, the squeezing inertia depends on variables such as the total amount of charge q, the length of the droplet s, the way the charge elements are distributed in the droplet, and also on physical constants like the speed of light c.

The expression for squeezing inertia will not be given here, because that would not really serve any useful purpose, but if you are curious how it roughly looks: it so happens that the way the squeezing inertia depends on all these variables strongly reminds one of the expression for the electromagnetic mass m_{em}. Not so strange after all, because the squeezing inertia can be viewed as a generalisation of the electromagnetic mass.

For instance, squeezing inertia is proportional to the square of the total charge, i.e. it is $\div q^2$, so that it is always positive no matter the sign of the charge itself. Another feature of the squeezing inertia is, that it becomes larger the more the droplet has contracted, i.e. it is $\div 1/s$ (compare the earlier given expression for electromagnetic mass m_{em}). And the reverse is of course also true: the squeezing inertia becomes small if the droplet is stretched out in a long Zeppelin-like shape.

The order of magnitude of the squeezing inertia is not enormously different from the value of the electromagnetic mass, it is again proportional to $1/c^2$. Nevertheless it is definitely somewhat smaller than m_{em}, typically by a factor of around 10. Of course, the precise difference depends on the details of how the charge is distributed within the droplet.

6. **Generalised radiation resistance** is the item expected next in this list. What sort of physical picture can we have about "generalised radiation resistance"? Recall the earlier treated radiation resistance that is associated with the back and forth motion of the droplet in the potential well. If the droplet due to its motion was emitting radio waves, it experienced as reaction a recoil force, acting like a sort of damping. Likewise, we would expect that the pulsation of the droplet will be accompanied by radio emissions, and that there must be a reaction force which will tend to damp the pulsation motion. This is what we would expect and what we would call "generalised radiation resistance"

But you can forget about it, in the present theory it does not occur. In the following a few words will be said why we expect such a force to be negligibly small. What is however much more interesting is, what the consequences are of the practical non-existence of any radiation resistance accompanying the pulsation. These consequences are profound! So, we have a lot of things to tell you about something that is not there. Still, I have the hope that the following discussion is more enlightening than most conversations about nothing.

First of all: why is it expected that the phenomenon of radiation resistance is not relevant in the case of the pulsation? The simple answer is, that energy radiation itself is unimportant in the case of pulsation. Any recoil effects associated with radiation are therefore unimportant too. In fact, we can easily explain why it is that the energy lost by radiation is negligibly small if it is coming from a droplet that only pulsates.

Such a droplet is shown in Fig. 5.5, its elongation becomes periodically larger and smaller than a certain equilibrium length. If we consider a small charge element of the droplet, e.g. somewhere near the bow of the "zeppelin" (you may choose yourself what you consider to be the bow and stern of the zeppelin, either on the right side or on the left) then such an element will move forward and backwards

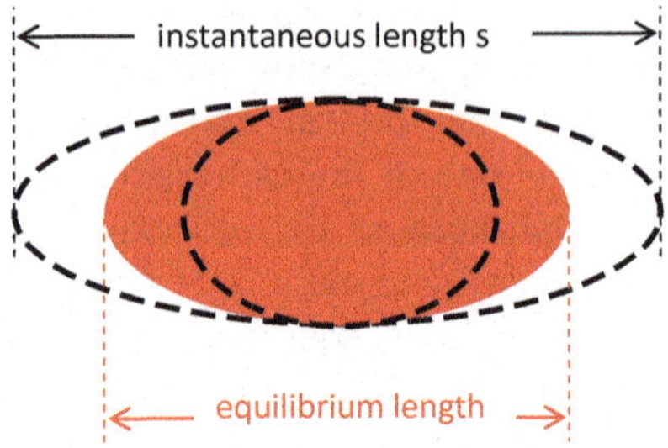

Fig. 5.5 Pulsation around an equilibrium length

during the pulsation. This in principle does not differ from the motion of an element when the droplet as a whole is translating within the well. The crucial difference with the translation motion is, that in the case of the pulsation there is always another element, symmetrically positioned near the stern of the zeppelin, which is moving in the opposite direction.

Seen from a distance, i.e. in the so-called "far radiation field" the radio transmissions from these two elements tend to cancel each other. Not quite completely, but the cancellation is sufficient to weaken significantly the radio signal that can be detected far away from the droplet. For this reason the pulsation contributes far less to the total energy loss by radiation than the translation. Quantification of the effect shows that the energy loss by pulsation is so small that it is negligible within the scope of a theory which is only accurate up to and including effects of the order $O\left(\frac{v}{c}\right)^3$, which is the accuracy pursued here. If we would improve the theory, so that it would include smaller effects by "clicking" the mathematical microscope to the next larger magnification level (recall the description of the "max-technique" in the beginning of this chapter) this would reveal effects of the order $O\left(\frac{v}{c}\right)^5$. In such a higher-order analysis the pulsation would be found to be a non-zero contributor to the total loss of energy by radiation. The "c" in these expressions is the speed of light, hence the factor $\left(\frac{v}{c}\right)^5$ is extremely small indeed! For the present, we may safely conclude that the shedding of energy by the droplet is practically limited to the effect of translation accelerations only.

What can be concluded from this? The pulsation is "non-dissipative", it does not lose energy in the form of radiation, at least not within the scope of the present theory. Must the conclusion be that the pulsation energy has a given value that will never change and forever will remain the same? No, there is another way by which the pulsation can lose its energy, that way is by transferring its energy first to the translation, so that it from there can be converted into radiation. How such an energy transfer can happen is the subject of a later item in the list of generalised forces, number 8.

This particular, indirect way of shedding the energy of the elongation/pulsation mode has rather profound consequences. To explain it, have a look at a children's swing, such as the bronze figure shown in Fig. 5.6. About the artistic value of this small statue one may have different opinions, but one aspect of it is very realistic.

Fig. 5.6 Bronze statue on grandma's mantelpiece, showing the most effective—though not the most clever—way to energise a system by the interaction with another system

The boy apparently knows that the most effective way to put energy into the swinging motion of his sister is, to give the push near the middle of the swing. The problem of the real world is of course, that doing so the "energy source" will be hit painfully during the next swing back. Most parents, if they had to do the job, would be wiser and would stand back a bit. But nevertheless, they too will be careful to give the push asymmetrically, i.e. the maximum push is given *after* the maximum excursion of the swing towards them. It means that they exert the force mostly *in the direction* of the motion, at least if they want to please their children by making the successive swings larger. In graphical form, the process is sketched in Fig. 5.7.

As you see, there must be a shift between the curve representing the swing position and the curve showing how large the exerted force is, if it is wished to pump energy into the motion. The maximum of the push must come after the maximum of the swing position. In jargon, one says that there must be a "phase-difference" between these two curves.

Now returning to the pulsation of a droplet of charge: if energy has to be transferred from the pulsation to the translation (or back), there *must* be a phase-difference between these two types of motion. If there is no phase difference, then the pulsation cannot get rid of its energy because another road for the energy does not exist if direct radiation is practically impossible.

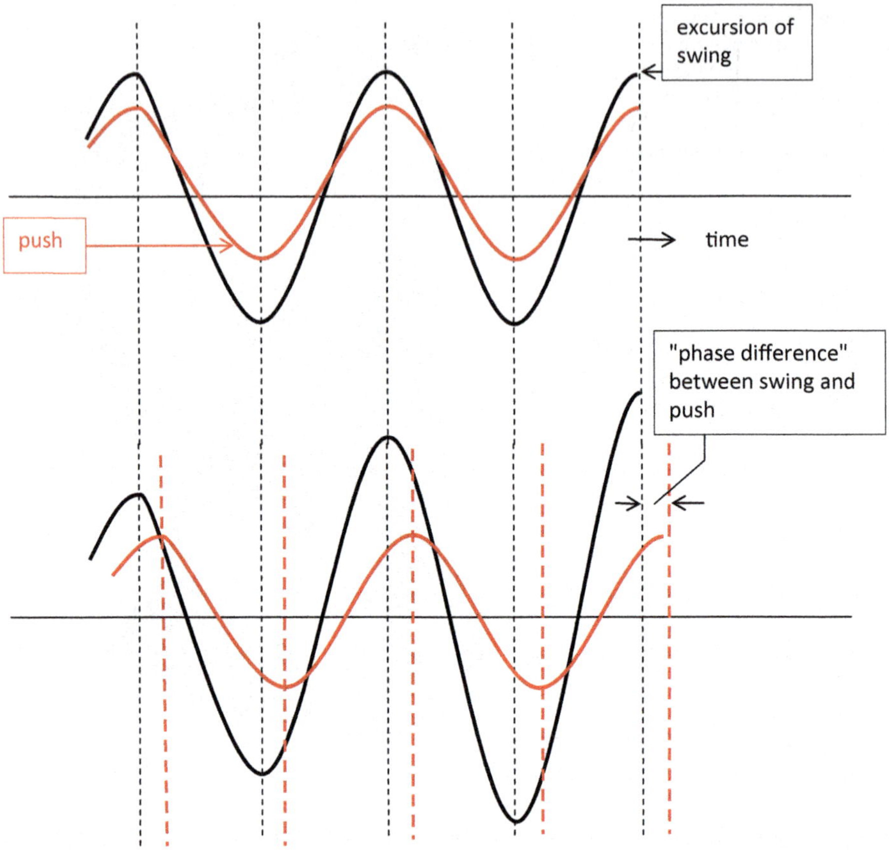

Fig. 5.7 A phase difference is necessary between the swings and the pushes in order to pump energy into the motion

This is the reason why it was said earlier that the absence of energy loss from the pulsation directly by radiation has profound consequences. From the above given description of the processes involved it will be clear that there are strict conditions to be fulfilled, or no shedding of energy can take place at all (that is: energy shedding by the pulsation mode). Later, it will be shown that the required conditions are certainly not always fulfilled. To anticipate later conclusions, here is the reason found why a droplet which flies freely through space, like a free electron following a straight unobstructed path, is not able to emit photons, even if it has sufficient energy to do so. Why a droplet within a potential well on the contrary does have the ability to emit photons—be it only when it is at certain places of the well—has the same root cause.

7. **Electrostatic repulsion** between all the charge elements in the droplet, leading to the tendency to explode. The generalised force associated with this will often be called the **explosion force** or **explosion tendency**.

The existence of such a force hardly needs any further explanation. All the charge elements within the droplet repel each other, and they eagerly want to move out of each others vicinity. Obviously, how large the tendency to elongate is, does depend on the actual, instantaneous shape of the droplet.

The generalised explosion force, i.e. the tendency to elongate the droplet by the internal repulsion forces, is found to be proportional to $1/s^2$ where s is the length of the droplet. This is illustrated in Fig. 5.8, where it is assumed that the droplet is pinched between the thumb and forefinger to keep it at a constant length.

The more the droplet is compressed in length, the harder the fingers have to press. The pressure needed to overcome the electrical repulsion (disregarding the cohesion forces) rises progressively the shorter we want to make the droplet, this is what is being expressed by the proportionality $(\textit{explosion tendency}) \div \frac{1}{s^2}$.

On the other hand, the needed force drops off fast when we allow the droplet to stretch, but the electrical repulsion forces never become completely zero except in the purely theoretical case that the elongation is infinite.

It must be emphasised that we are here talking purely about electrical effects. As has been explained in Chap. 1 the complete model of a droplet of charge comprises other forces as well: a sort of surface-tension-like force has been added to make the model consistent, which force (also a *generalised* force) acts in the same direction as the thumb and forefinger in Fig. 5.8. These additional model assumptions are treated in a separate chapter (Chap. 6), since the present chapter deals only with the self forces caused by electromagnetic effects. Let us therefore first conclude the list of generalised forces affecting the elongation of the droplet, before taking up the matter again in a separate chapter.

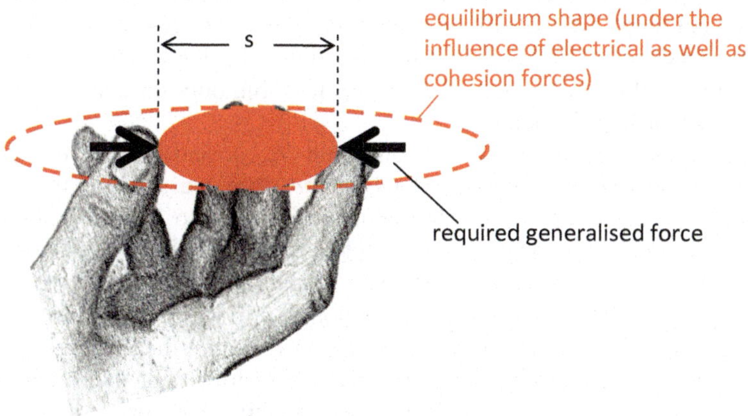

Fig. 5.8 The droplet must be pinched to maintain the length s

A last remark that later will appear to be of importance concerns the influence of the total amount of charge in the droplet. The reader can easily work out for himself what this should be: the explosion tendency cannot depend on the sign of the charges in the droplet. Positively charged droplets will have all positive elements, and negative droplets will contain all negative charge elements. The mutual repulsion between the elements is just as large, no matter the sign of the charges. Therefore, the formula for the explosion tendency of a droplet must depend on the total amount of charge in such a way that the sign of the charge does not matter. You will immediately suspect that there will be a proportionality of the form

$$(\textit{explosion tendency}) \div (\textit{total charge})^2$$

and that is confirmed by the mathematical analysis of the appendix. In conclusion, let us finally combine the proportionalities found into one expression:

$$(\textit{explosion tendency}) \div (\textit{total charge})^2 \times \frac{1}{s^2} \ (\textit{excluding cohesion effects})$$

8. **Coupling force from translation to pulsation**

In item 4 the "Coriolis-like force" was discussed, which drives the translation whereas the pulsation influenced how large it is: a coupling from the pulsation to the translation.

It is to be expected that also a "mirror force" will exist, which works the other way round: a force driving the pulsation whereas the translation has an influence on how large it is. Well, conscientiously working out the electromagnetic equations (as done in the appendix) reveals that there is such a force indeed, a *generalised* force in this case.

The influence of translation on pulsation does *not* take the form of a sort of "counter-Coriolis-like" term. It is quite different, and looks more like a relativistic effect (do not worry, all this will be explained). Under the previous item 7 about the explosion tendency of the droplet, it was stated that this tendency is proportional to $1/s^2$. Actually, this is only half of the story. It is true only for a stationary droplet that has no translation velocity.

If the droplet has a velocity of translation V_{transl}, and if that velocity is not negligible compared with the speed of light c, then we see that the explosion tendency is reduced somewhat. The reduction is never so large that there is any chance of completely getting rid of the explosion risk. But it may be sizeable if the ratio V_{transl}/c is not too small: the explosion tendency appears to be proportional to $\div(1 - V_{transl}^2/c^2)$. This represents a reduction factor which is almost equal to 1 when the translation velocity is small (and as a *reduction* factor it then can practically be ignored therefore), but it becomes more important the larger the ratio V_{transl}/c is. A droplet having a velocity approaching the speed of light would, because of this factor, lose almost entirely its tendency to explode.

Combining this with the earlier result about the effect of the elongation of the droplet we thus have:

$$(electrical\,explosion\,tendency) \div (total\,charge)^2 \times \frac{(1 - V_{transl}^2/c^2)}{s^2}$$

The formulae become bigger and bigger! I promise that this one will be the ultimate to which we go, and if you think it helps, you may think away the factor $(total\,charge)^2$, because for the present discussion it is unimportant.

From the above given expression it is clear that the explosion tendency will never disappear entirely, since no material body can go so fast as the speed of light, and neither can our droplet of charge. Anyway, there is an undeniable effect of the translation on the elongation of the droplet. The velocity effect therefore represents a coupling force from the translation to the pulsation.

How to understand this effect of the translation velocity? For an explanation we must make a diversion to the special theory of relativity. Do not fear to be bothered here with a complete discourse about this famous piece of thinking by Einstein. Just a few of his conclusions will be mentioned, without any derivation or justification. What he considered was the situation that we look at something which is moving past us at a high speed. One of his conclusions is, that we see it shrink in the direction of the motion. A body that is passing us at high speed is seen to contract, the more so the higher its speed is relative to us. The apparent contraction has nothing to do with any force that would squeeze the thing together, there is no such force. The contraction is just a consequence of the way we measure lengths. Einstein showed that the system we use to do measurements can never be set up such that these apparent contraction effects are absent, they are unavoidable. The contraction is usually called "Lorentz contraction", in recognition of Lorentz's work on electromagnetism, in which field the effect was already known before Einstein. Einstein generalised this, and showed that this consequence of relative velocity on length measurements is always present, not only in electromagnetic systems. Furthermore, a really revolutionary addition by Einstein was, that something similar happens to time measurements: they too are "polluted" by relative motions, the so-called "time-dilation". It should be realised that Lorentz contraction and time dilation are not one-sided. As an observer we see things flying past us contracted, but that also applies the other way around: an observer within the other system, when looking at us, will see us "squeezed" in the direction of the velocity we appear to have relative to him.

Let us go back to our droplet of charge. Previously we concluded that the droplet behaves like a kind of spring. If we want it to maintain a given constant length we have to apply an external force, like squeezing a spring between our thumb and forefinger.

As visualised by Fig. 5.9, if the droplet has non-zero velocity the Lorentz contraction already makes a contribution to shortening the spring, and it does so

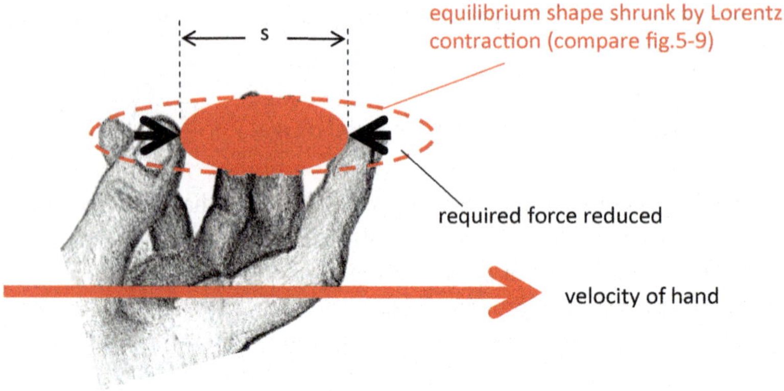

Fig. 5.9 Reduced explosion tendency due to velocity

without involving any force. The rest of the compression has to come from a real force, but the squeezing force "between the thumb and the forefinger" does not have to do the whole job. This is what is expressed by the above given formula which said that the droplet's urge to elongate diminishes a bit, if the droplet has some velocity. It is as if the Lorentz contraction acts as an additional squeezing force. However, this is a tricky way to interpret the formula, because we just had stated that Lorentz contraction is not due to any force. The final *effect* due to relativity is thus to be conceived as an "apparent" force, not a real force.

5.4 Conservation of Energy, Equipartition, and Scaling Effects

In this paragraph several of the above given results will be repeated, reconsidered, and especially *related* to each other, so that all the possible information is obtained that can be "wrung" from them without going so far to combine them into equations of motion and solving these equations. Setting up and solving complete equations of motion will be the task of later chapters.

We have seen that in the equation for the translation a coupling term occurs:

$$(Coriolis - like\,force\,working\,on\,the\,translation) \div m_{em} \cdot V_{transl} \cdot V_{puls}.$$

It is a force driving the translation, but of course it is not the only driving force, the Coriolis-like one works in cooperation with other forces. How large it is and in which direction it works is amongst others determined by the pulsation velocity V_{puls}. The pulsation is therefore one of the controls steering the Coriolis-like force.

Because the force also feeds or taps energy into/from the translation, the pulsation is one of the controls for the flow of energy into/from the translation. The pulsation is standing at the tap of the translation energy and is controlling the energy flow, so to speak.

The mirror is a generalised force occurring in the equation for the pulsation:

$$(explosion\,tendency) \div (total\,charge)^2 \cdot \frac{(1 - V_{transl}^2/c^2)}{s^2}$$

We may give a "mirror" description of it, just by reversing the words translation and pulsation in the above text. The—generalised—force of the last formula is driving the pulsation, but of course it does so in cooperation with other forces. What is special about it is, that it represents a coupling between translation and pulsation. The force occurs in the equation for the pulsation but how large this coupling term is, is determined by the translation velocity V_{transl}. The translation is therefore one of the controls steering the explosion tendency.

Because this force also feeds energy into the pulsation or taps energy from it, the translation is one of the controls for the flow of energy into/from the pulsation. The translation is at the tap of the pulsation energy.

The physical mechanisms represented by the Coriolis-like force and by the relativity effects (i.e. the effect due to the reduction factor $(1 - V_{transl}^2/c^2)$) seem to be completely dissimilar. After all, Coriolis effects were known already in good old classical mechanics. In contrast, the concepts and rules of the theory of relativity are much younger and seem to be completely separate notions. It may surprise you therefore to hear that if we work out how large the energy flows actually are, we find the following:

If an amount of energy is leaving the translation (through the action of Coriolis-like forces), we find that at the same time an exactly equal amount of energy is added to the pulsation (through the effects of relativity). And vice versa.

Really surprising! Although entirely different mechanisms are at work, they work in such a way that nowhere any energy is spilled. Apparently, the law of *conservation of energy* is obeyed, despite the fact that entirely different mechanisms are at work to pump energy into or out of the two "degrees of freedom" assigned to our droplet model. This is reassuring, it tells that the model we made was not *over*simplified, the proposed model at least is consistent with the law of conservation of energy.

We see that—under the right conditions—energy can be mutually exchanged between the two dynamic modes assumed in the model. In general we expect that the energy flow is back and forth between the two modes. Earlier it was pointed out how we by this "energy sloshing" between the modes may understand the tunnelling effect. Confining our attention to the translation, we see that the energy in it is fluctuating around an average value. Temporarily it will be larger than the average value, at other times it will be smaller. If the droplet is moving inside a potential well we are able to notice this energy fluctuation by observing the maximum excursions of the droplet. Sometimes these excursions are larger, sometimes smaller than average. It may thus happen that the droplet reverses its motion at a point which is further off-centre than where it started its motion, with the result that the droplet can even escape over the edge of the potential well: we would interpret this as "tunnelling", the droplet has arrived at a place where it could not be on the basis of its *average* translation energy.

The phenomenon of tunneling is associated with temporary energy fluctuations of the translation motion. We can also expect to see another effect caused by what was called above the "sloshing of energy", an effect that slowly develops in time. It is the tendency to arrive at so-called *equipartition* of energy in the long run. It is well known in classical physics that different systems, brought in close contact with each other, will tend to equally share the total energy available. The total energy shared by two modes of motion will be divided so that each gets half, although it may take some time before such a balance of energy has been achieved, it is a long term effect. An example is found in the molecular motion in a gas: the molecules fly around and are also spinning. It appears that on average the energy in their translation motion is equal to the energy in their rotation. Or to give another example: mix two gases, coming from two different reservoirs with a different temperature. After a while all the molecules in the mixture will on average have the same kinetic energy (i.e. they have the same temperature).

Such a long term tendency towards equipartition of energy may be expected in our case as well, in this case implying that the energy in the translation will become equal to the pulsation energy (on average of course, disregarding the shorter term energy sloshing).

The equipartition has interesting consequences. It can explain why a macroscopic collection of charges shows a different behaviour compared with a small "blob" of charge at an atomic scale. Here we come to the question why there should be a transition between "normal" physics and the realm of quantum phenomena. Finally! Perhaps you have been waiting for this explanation from the very time you started to read this book, but it is only now that an explanation can be given.

If the droplet, no matter whether it is of a macroscopic or atomic scale, is placed inside a potential well we will see a translation motion back and forth between the two points where it reverses its direction. The motion may not always be a smooth sinusoidal one, it can be chaotic, but we can always see how much energy there is in it by looking at these extreme points. In these "points of reversal" momentarily the velocity is zero (meaning that the kinetic energy is zero there), and the energy is completely in the form of potential energy. The further the reversal points are away from the centre of the potential well, the more potential energy (and thus total energy) the translation possesses.

The energy present in the translation is thus expressible in terms of the position of the extreme points of the translation. We will use the following relation derived by old fashioned school mechanics (note however that the expression below is strictly speaking valid only for a parabolic well):

$$(\textit{total energy in translation}) \div (\textit{charge of droplet}) \times (\textit{off} \\ -\, \textit{centre position of reversal points})^2$$

The force exerted on the droplet by the electrical field of the well is proportional to its charge, so that it not unexpected that in the expression above we also find a proportionality with the charge of the droplet. Why the energy would be dependent on the *square* of the excursions is a rather technical matter, explained in school books on mechanics. But unexpected it is not: the off-centre position of the droplet is on the right side of the well given by a positive number, and mathematically we count it as negative if the position of the droplet is on the left side. The occurrence of the square of the position thus anyway prevents that we would find different values of the energy in the two sides of the well.

The expression above remains valid in case radiation occurs, when the total energy is gradually lost. We see the reversal points of the motion then gradually shift closer to the centre.

For the energy present in the pulsation a similar expression can be written down. We can conclude how much energy is present in the pulsation by observing the maximum amount of lengthening the droplet assumes relative to its equilibrium length. There is an essential difference compared to the case of translation though, and this is how the pulsation energy grows with the total charge of the droplet. In the above we have already seen that the generalised force describing the explosion tendency depends on the square of the charge. It is to be expected (and confirmed in a later chapter) that the cohesion force which "tames" the explosion tendency scales in the same way with the amount of charge. Therefore, the formula in this case is:

$$(\textit{total energy in pulsation}) \div (\textit{charge of droplet})^2 \times (\textit{maximum lengthening})^2$$

On comparing these two expressions we see the important difference between them: they depend in a different way on the total charge of the droplet. The energy in the pulsation is much more sensitive to total charge, due to its dependence on the *square* of the charge.

Suppose that we have the state of equipartition of energy, i.e. that we have on average the same energy in the translation and in the pulsation. Combining the two expressions shown above then results in:

$$(\textit{charge of droplet}) \times (\textit{off} - \textit{centre position of reversal points})^2 \\ = (\textit{charge of droplet})^2 \times (\textit{maximum lengthening})^2$$

Now, comparing two droplets, one with a large total charge, the other with a small charge, we see that the pulsations of the large charge will be relatively smaller. The more charge a droplet contains, the smaller the pulsations (relative to the translation excursions). For very large, macroscopic amounts of charge, the pulsation will become completely unimportant.

There is thus a clear scale effect, which is the explanation why there is a transition scale where, coming from the large scales of the human world, we gradually begin to see the typical quantum behaviour where droplets of charge are influenced profoundly by their second degree of freedom. To be sure: both degrees of freedom exist at all scales, but their relative importance is very much scale dependent. At macro scales the influence by the pulsation may be ignored completely.

5.5 The Last Riddle to Be Solved: How Can Electromagnetic Mass Be Made to Comply with the Theory of Relativity?

It was mentioned under item 1 of Sect. 5.4 about electromagnetic mass that there is still something of a riddle in the expression for the electromagnetic mass. This is so in the old formula derived by Lorentz as well as in our own formula, derived in the appendix. Let me first explain what the problem is.

In both Lorentz's model of an electron as well as in the model we have made of a droplet of charge, the charge is assumed to be *divisible*. In both models the total charge is smeared out inside a certain volume, in Lorentz's model in the form of a layer covering a sphere of isolator material, and in our model in the form of a continuous sort of "fluid" filling an airship-like contour with variable length.

Inherent in this assumed divisibility is, that it costs energy to form such a droplet or shell of charge. First look at this matter the other way round. Imagine that the cohesion forces keeping the droplet or shell intact are suddenly removed. To speak in the colourful language of Feynman: what would happen if the "rubber bands" keeping the droplet together would suddenly break? Well, these "rubber bands" prevented an explosion, so that cutting them would mean that the impending explosion now takes place. All the separate charge elements of the droplet would fly outwards under the influence of their mutual electrostatic repulsion forces. They want to get away from all the other elements, as far as possible.

What this theoretical exercise shows is, that a lot of energy would be freed during such an explosion, energy that was present in the droplet when it still was a droplet.

In reverse, if we would imagine that the droplet has one time been formed by compressing a great many charge elements together into a small volume (again a purely theoretical exercise) a lot of energy must have been put into it. We needed forces on all the separate charge elements to bring them together, and these forces would have to do a lot of work. Therefore, the droplet represents stored energy.

Now, we made a model where cohesion forces would hold the thing together, and you may wonder if these cohesion forces would not represent just as much energy, but then of a negative sign. Not necessarily so! These additional forces could be considered as a sort of lock. We have pressed too many clothes into a trunkcase with a lot of effort, and click the lock shut to keep the lid closed. As soon as the lock is reopened the contents bulge out again. Apparently, there is energy in the closed trunk, waiting to come out again. This internal energy is *not* neutralised by any energy needed to shut the lock, because the lock itself does not really require much energy to close it. Or to use Feynman's metaphor, we have expended energy to form the small blob of charge, and the rubber bands are wound around at a small expense of energy, without detracting much from the energy waiting inside the blob to be released.

In conventional electrostatic theory a formula is found which tells how much energy is represented by a collection of charges. We can thus calculate relatively easily how large the energy is that is inherent in a "blob" of charges. We can perform this type of calculation for the inherent energy in Lorentz's electron model as well as in our own model of a droplet of charge. Let us call the outcome of such a calculation $E_{el.static}$, i.e. the potential energy stored when forming the configuration of charges against the electrostatic repulsion forces.

According to the special theory of relativity, energy and mass are equivalent. Quantitatively their relation is given by the famous formula $E = m \times c^2$. One can thus continue the above calculation, to find the mass that is equivalent with the internal potential energy, it is:

$$m_{el.static} = E_{el.static}/c^2$$

Along quite a different route we also did a calculation on the *inertial* mass due to electromagnetic effects, this was called the "electromagnetic mass" m_{em}. Now, one would expect that these two masses should have the same value. Thus, it is expected: $m_{em} = m_{el.static}$.

However, at the time Lorentz did his investigations on his electron model, he found different values for $m_{el.static}$ and m_{em}! This did not particularly worry him, because Einstein had not yet formulated the general law $E = m \times c^2$. Only a few years later however, after Einstein had published his work, physicists became alarmed. Everybody then expected that electrostatic mass (inherent mass due to formation energy) should be equal to electromagnetic mass (inertial mass), but already the results were there: these two masses were not equal! In fact, one can show that electromagnetic mass m_{em} will *always* be larger than electrostatic mass $m_{el.static}$. In the case of Lorentz's charge configuration they differed by a factor $4/3$. If we do the calculations for our droplet of charge, which has a different configuration, we find a factor 2. In other words, the electron model we are investigating in this book always has an electromagnetic mass *twice as large* as the mass corresponding with the formation energy.

How to explain this "riddle of the factor $4/3$" as it was called in Lorentz's time? Another famous mathematician and physicist, Poincaré, offered the solution of the problem: he proved that it has to do with the binding forces, or in our metaphore: the energy required to lock the lid of the trunkcase. He calculated that the riddle would be solved if one took into account the forces that bind the charge on the surface of the isolator (in Lorentz's model). Accordingly, the "forces due to the elastic bands" were later baptised "Poincaré forces", a posh name suggesting that we now know the nature of these forces. Actually, what binds charges on an isolator has become clear in the later developed atomic theory, so that we now know what sort of force the "Poincaré force" really is. At least, this is so in Lorentz's model which has a core of isolator material.

Feynman uses the indication "Poincaré force" more generally, even if it is not at all clear what kind of force the charge configuration binds together. Agreed, he used this name with a wink: he discusses these "rubber bands keeping the thing together", and suggests that by using the name "Poincaré force" physicists try to conceal their ignorance about what such forces could be, or the doubts whether such forces really exist at all. If the emoticon had already been invented, the books by Feynman would be full of symbols like;-).

In our case of the droplet of charge we have exactly such a situation: we are forced to assume that there is "something" binding the droplet together, otherwise the model cannot be made consistent and we would have to end this book here and now. In the first chapter of this book it was discussed that a cohesion effect in the form of some sort of surface-tension-like force would not be unreasonable, and the arguments for such a working hypothesis were there given extensively. We will come back to these matters in the next Chap. 6, where it is shown that the form of surface tension for the cohesion is almost unavoidable. Despite this definite choice for a surface-tension-like force, in the remainder of the present chapter we will preferably use the more neutral name "cohesion", because it covers both the Poincaré forces of Lorentz's model as well as the "rubber bands" assumed for the droplet of charge.

Can any kind of cohesion force, whatever its nature, now explain the fact that $m_{el.static}$ and m_{em} are unequal? Relatively recently (1992), a book has been written by the author Arthur D. Yaghjian, with the title "Relativistic Dynamics of a Charged Sphere". He proposes the following solution. He argues that charge does not exist on its own, it is always coupled to mechanical mass. There exists not only electromagnetic mass, but on top of that also "normal" mass, which Yaghjian denotes as "bare mass" m_{bare}.

The total mass of a blob of charge, felt as inertial mass which can be measured in the laboratory by accelerating this blob, is thus equal to

$$m_{measured} = m_{em} + m_{bare}$$

This is the only mass that can be measured, it is impossible to perform any experiment which discerns between the two types of mass in the right hand side of this formula. Only their sum can ever be measured. We should apply the theory of relativity to what can be measured, which means that the energy intrinsic in a blob of charge is

$$E = m \cdot c^2 = (m_{em} + m_{bare}) \cdot c^2$$

This is the result about the intrinsic energy that can be obtained from experiments in the laboratory. Now, if the intrinsic energy determined by electro*static* theory is found to be

$$E = m_{el.static} \cdot c^2$$

there need not be a contradiction, if we take

$$m_{el.static} = m_{em} + m_{bare}$$

The "bare" mass bridges the gap between $m_{el.static}$ and m_{em} and it can thus explain the discrepancy. This seems logical, and this relation between electrostatic mass and electromagnetic mass seems to be quite innocuous. It would appear that we have found an easy way to get everything in agreement with the theory of relativity. Except….: we know that electrostatic mass invariably comes out of the sums to be smaller than electromagnetic mass:

$$m_{el.static} < m_{em}$$

The line of reasoning above seemed so logical, but it implies that the mechanical mass m_{bare} is negative! Earlier we called the bare mass loosely the "normal" mass of an object, to which we are used, as distinct from electromagnetic mass. But how "normal" is *negative* mass?

Is it realistic at all? OK, the total mass that can be measured is still always positive:

$$m_{measured} = m_{em} + m_{bare} = m_{el.static} > 0$$

So we will not find in the real physics of the laboratory strange things like objects that accelerate in a direction opposite to the force we apply (this would be the behaviour of an object with negative mass). Still, the concept of a negative component of the total mass is an audacious one! After having proposed this hypothesis, Yaghjian stops any further philosophising about the matter. So let us ourselves ponder somewhat more about it.

The concept of negative mass becomes a little less weird, if we carefully keep in mind the last formula: what we are able to measure is still always a positive mass. The negative component in the positive sum could therefore be called a "mass deficit". And formulated this way the negative component is less strange. Mass deficits occur more often, they may be the result of chemical reactions as well as of nuclear reactions. A well known example is shown in Fig. 5.10.

The figure essentially gives information on how heavy the nucleus is of the atoms of different chemical elements. What is called "mass number" on the

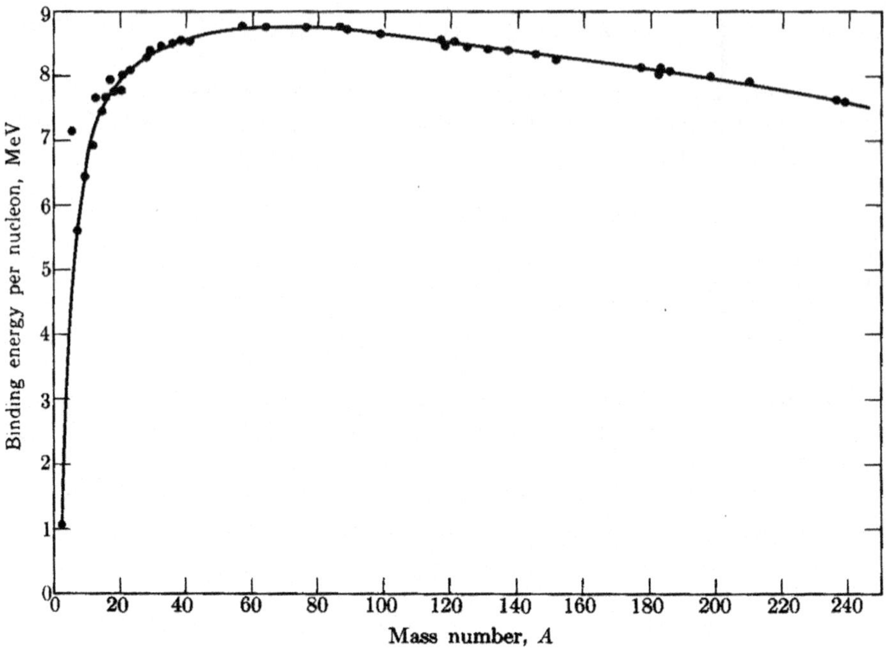

Fig. 5.10 Binding energy of atomic nuclei. The figure may be read to show the mass deficit of the united nucleus with respect to the summed masses of its components. Figure taken from Alonso and Finn: Fundamental University Physics

horizontal axis indicates how many particles (protons and neutrons) a nucleus contains. For instance, the element Hydrogen has a nucleus consisting of only one proton, it thus has a mass number of one. The next chemical element is Helium, which has a nucleus consisting of two protons and two neutrons, giving a mass number of four.

Now, the word "mass number" is a bit misleading. It suggests that this number also gives information about the mass of the nucleus, which it does not. Protons and neutrons have almost the same mass, so that one would expect that the nucleus of helium is practically four times as heavy as the nucleus of Hydrogen. In other words, one would expect the mass of Helium to be equal to its mass number, if the mass would be expressed as "so and so many times the mass of Hydrogen". Well, this is not true.

Imagine that we try to make a Helium nucleus out of separate nucleons. We have two baskets, one containing a lot of protons, the other one filled with neutrons, and pick with a set of tweezers two protons and two neutrons. First of all, when we try to assemble them it will cost some effort to bring the two protons together because of their mutual repulsive forces. Having done that, the next step is to "forge" them together, so that they will form a unit. In other words, after having pressed our clothes into the trunkcase, we now have to close the lid on it and lock it. At this moment a funny thing happens: by the very act of uniting the four nucleons we cause the assembly to lose some weight. We can notice it by weighing the Helium nucleus: it will be seen that the mass of the total is *less* than the sum of its components. Mass has been lost in some way, which can be formulated as we did before: there is a mass *deficit*. The mass is in fact not lost, it has been "transformed" into energy. The act of uniting the components leads to the release of energy, and certainly not a little bit because the exchange ratio between mass and energy is given by the formula $E = m \times c^2$. In the case considered here we should substitute for the "*m*" in the formula the observed mass deficit.

We can illustrate the process by an ugly example. In a hydrogen bomb an amount of hydrogen is heated to an extremely high temperature by a "normal" atom bomb, used as a sort of match to set off the fire. The high temperature causes that the separate protons, having a large kinetic energy, can collide with each other despite their electrostatic repulsion. At the moment they are thus brought into intimate contact with each other and with neutrons, they can fuse to form a helium nucleus. This fusion releases vast amounts of energy, much larger than the energy put into the reaction by the "match".

The energy released by fusing nucleons to form a new unit is called "binding energy". The binding energy is a way to measure how tightly the newly formed nucleus is bound together. If we would ever want to separate the thing into its individual components again, we would first have to supply this binding energy, before a separation is possible.

Now back to the figure: along the vertical axis the binding energy is shown (per nucleon). It is seen that elements with mass numbers around 60 are the most stable, because they require the largest amount of energy to separate them into single nucleons. In reverse: during their formation the largest amount of energy will be released. Or put in different words again: the mass *deficit* is largest around the mass number of 60 (somewhere in the vicinity of iron in the periodic table of elements). It is also the reason why one can make bombs out of heavy materials like uranium as well: splitting their nuclei leads to fragments, and the total mass of them shows a mass deficit compared with the original nucleus.

Let us return to the story of a droplet of charge. It was argued that the difference between electrostatic mass $m_{el.static}$ and the electromagnetic mass m_{em} indicated that there is a mass *deficit* found after having compressed the total charge into the small volume of a droplet. This mass deficit is a measure for the binding energy, it is thus directly associated with the cohesion forces. The conclusion is almost exactly the same as Poincaré's analysis of Lorentz's electron model: according to him the difference between $m_{el.static}$ en electromagnetic mass m_{em} was associated with the binding of the charge on the surface of the core. In our case, although the charge configuration of a droplet is different, and although there is a different binding mechanism, the consequence of assuming "something" which binds the droplet together implies that we must have a difference between $m_{el.static}$ and m_{em}. Because the values of both these masses can be calculated using the theory of the appendix, we can even quantify the effect of the cohesion forces. Funny, isn't it? Using electromagnetic theory we can tell something about cohesion forces that have been assumed (for now) to be of a *non*-electromagnetic nature. One suspects a deeper meaning of this fact.

Anyway, conceptually the "riddle" has been solved.

Chapter 6
A Brief Excursion to Sub-atomic Scales: The Electron, the Muon and the Tau Particle

6.1 Cohesion Forces: Real or "Apparent"?

The title of this chapter, mentioning muons and other strange things, might be confusing unless you already read about the subject in Chap. 1. What is the connection between these three elementary particles and our subject? For hasty readers who did skip parts of Chap. 1 (quite possible, I myself am one of those people who in the bookshop just scans a few randomly opened pages to get a taste of the book): the relevance of muons and tau-particles for our subject will become clear in a few moments, but do not worry, we are still talking about the same subject as in the last two chapters. In Chap. 4 a simple model of a droplet of charge was made, and in the previous Chap. 5 it was summed up what sort of electromagnetic self forces work on it. The conclusion was that the list of forces in Chap. 5 cannot possibly be exhaustive. Something is missing if we want to have a consistent model. The missing force is "something" (perhaps real elastic bands?) needed to prevent an explosion of the droplet of charge. We must avoid that the droplet in our one-dimensional theory will stretch out indefinitely, from a rugby ball, via a Zeppelin-shape, to a needle, etc.,.

In the model of Chap. 4 there are no electromagnetic forces that would be able to take on the role of an "anti-explosion device". That does not mean that no electromagnetic effect whatsoever would be able to restrain the droplet from stretching without limit. It only means that the kind of motion we allowed in the model does not provide such a restraining force. In fact, a friend of the author has the strong feeling that allowing certain rotations of our droplet could give the desired effect, purely by electromagnetic forces. This is not the road chosen in this book, because it would go against the "design philosophy" of striving after the utmost simplicity of our model. It was decided in the beginning of Chap. 4 that an attempt would be done to work out the consequences of the *simplest possible* model, before any further refinements would be investigated. Adding more degrees of freedom such as rotations would certainly not comply with such a strategy.

T. van Holten, *The Atomic World Spooky? It Ain't Necessarily So!*
DOI 10.2991/978-94-6239-234-2_6

In line with the chosen strategy we are thus compelled to assume (for now) that additional forces capable to stop an explosion must be sought in other than the electromagnetic effects allowed by the model of Chap. 4. The immediate and obvious question is: what could be the nature or origin of such forces? There are several roads that can be chosen at this point of the story.

The first choice could be: let us not bother at all about this question about the physical nature of the cohesion. We could just *state* one extra model assumption, viz. that there indeed exists some kind of cohesion, and that we are for now not concerned with its nature. It would be just a working hypothesis, and we are going to work out the consequences of this "Deus ex Machina". Only if the assumption would lead to meaningful—or at least interesting—results, then the time would come to delve more deeply into a possible physical explanation. This last subject, to ponder about the nature of the cohesion forces, could even be left to other people. In that case the subject would not crop up any further in the present book.

Although it would be quite a defensible and perfectly respectable approach, this is not the choice that has been made in this book. It would not fit in with the character of the author, who never can be stopped musing about things he does not understand. You as a reader are of course allowed to take the above indicated path of thinking, if you are not satisfied with the other choices that could be made. Nothing stops you from making your own choice.

However, there are other choices possible. One of such possibilities has been proposed in Chap. 1. To briefly summarise it: the concept of a "cloud of charges" was shown to be compatible with usual quantum mechanics ideas, if we consider it as a time-averaged view of a "zittering" electron. It was mentioned that one of the conclusions from the usual quantum theory is, that an electron is always jittering a bit. About the *cause* of this so-called "zitter" one may have different theories, but most physicists agree that the phenomenon is real. Within a very small volume an electron jumps around from place to place, in a discontinuous manner. The time-scale of this phenomenon is of the order of 10^{-21} s. We could restrict ourselves to an approximate theory where the *relevant* time-scales are more human-like. Let us say a theory where only phenomena are taken into account that are slower, e.g. on a scale of 10^{-15} s, one million times as slow. In that case the individual jumps are no longer visible, and the electron would be "seen" in such a theory as a small smeared-out cloud.

A well-known analogy is the theory of thermodynamics versus kinetic gas theory. In kinetic gas theory the fast motions of all the individual molecules in a gas are considered, to derive from them the properties of gases. In thermodynamics, on the contrary a more phenomenological approach is chosen: the individual molecules are ignored and a gas is treated as a continuum. The macroscopically measured properties of gases such as pressure, temperature, etc. are then taken as the starting point, and their mutual relations are studied. In a way, thermodynamics thus deals with averaged quantities. Thermodynamics and kinetic gas theory do not lead to different conclusions—they are perfectly compatible theories—but they differ with respect to the scale chosen to be relevant, depending on the purpose for which the theory is developed.

In thermodynamic theory certain quantities are defined that do not have a place in kinetic gas theory. Take for instance "pressure", which is typically a macroscopic property of gases. Comparing with kinetic gas theory pressure is the average effect of collisions between molecules mutually or between molecules and the walls of the vessel in which the gas is contained. In kinetic gas theory itself the quantity "pressure" is meaningless: to speak about the pressure of a single molecule is evident nonsense. Likewise, it makes no sense to speak about the "entropy" of a single molecule. However, on the averaged scale of thermodynamics one *must* add laws about the entropy, or else the macroscopic theory would not correctly describe the real behaviour of a large collection of molecules.

Something similar is the case with the zittering electron: it randomly jumps from place to place, and its average position may therefore slowly drift away, but there is no reason why this drunkard's walk would have a tendency to vehemently explode. On the other hand, when the drunkard's walk is time-averaged so that the view of a diffuse cloud of charge is adopted, then a 'blind" application of the laws of electrical repulsion would lead to an—unrealistic—explosion tendency. We *must* add a new quantity, the "cohesion" (like entropy had to be added in thermodynamics) or else the time-averaged theory would not correctly describe the real behaviour. Cohesion is therefore not a part of the theory at the smallest time-scales, but it is a natural and unavoidable concept when these smallest scales are—artificially, and only for the purpose of making a manageable model—averaged. In Chap. 1 cohesion forces were baptised "correction forces" or "apparent forces". The nice consequence of this approach is, that Maxwell's laws may be applied without restrictions (as was done in the previous Chap. 5), be it under the strict conditions that the time-averaged viewpoint is supplemented with these apparent forces called "cohesion forces".

Now, the problem is that it does not suffice to conclude that a time-averaged theory requires additions to make it consistent, like thermodynamic theory requires additional laws about entropy. One has to be more specific or else no quantification will be possible. What exactly is the form taken by the cohesion? How can it be described mathematically? In Chap. 1 it was just stated that in this book the cohesion will be treated as if it were a kind of surface tension effect. Some justification was given, but that had the character of a guess. In the present chapter the matter is taken up once again, and is considered more deeply. The conclusion will be that the form of a surface-tension-like force is almost inescapable. This conclusion will be based on more fundamental considerations than just "a hunch" like it was described in Chap. 1.

It should be emphasised that this certainly does not mean that the surface tension is a "real" force, no more than pressure, temperature, entropy, enthalpy, etc. are "real" from the point of view of a single molecule. Surface tension does not exist within the scope of un-averaged theory. Quantities like pressure, cohesion, etc. must be understood to be artefacts, the introduction of which is necessary to make the phenomenological, time-averaged theories consistent. But as such, *within the scope* of such a time-averaged theory, we can work with the surface-tension-like force *as if it were* physically real. This is the essence of all the "apparent" forces we use in physics, such as centrifugal forces and Coriolis-forces in mechanics or “apparent mass” in unsteady fluid dynamics. Once they have been introduced, we can henceforth work unrestricted with the usual laws. It is also the fundament on which Chap. 5 rests, where the electrodynamic self forces were determined by the full unrestricted application of Maxwell’s laws, as if nothing special was the matter.

A final remark must be made about the “zitter” of electrons. In theoretical physics there is more than one way to derive it. One way to give an explanation for the phenomenon was used in Chap. 1, where it was “explained” by the view of virtual particles popping up and disappearing in vacuum. There are people who abhor from this kind of visualisation. These people may be soothed when it is realised that the reason *why* there is zitter does not really matter for the present discussion. The only thing that matters is, that there *is* zitter, no matter the cause.

6.2 How to Model the Cohesion?

We want to calculate how cohesion of the droplet is going to influence the pulsating motion. How large is the cohesion force? It will undoubtedly depend on how far the droplet has been stretched. The cohesion will work out differently for a rugby ball shape than for a needle shape of the droplet. Obviously this question about the dependency on the shape of our droplet is difficult to answer, considering the fact that the cohesion is either just a working hypothesis or an “apparent” force. What can be done however, is to have a look how physicists tackled a similar problem, in connection with the cohesion forces working on the nuclei of atoms. This problem is not dissimilar to ours, because the nucleus of an atom is a collection of protons and neutrons. The protons with their positive charges mutually repel each other, and it is not immediately clear how the nucleus can form a stable entity without the tendency to explode. The cohesion force keeping the nucleons (the particles in the nucleus) together is called the “strong nuclear force“, and its origin is unknown, or rather: it is one of the fundamental forces of nature, unrelated to the other known forces. It just exists, and cannot be related to any type of force known in the macroscopic world.

What the well known physicist Von Weiszäcker did as a first attempt to quantify the characteristics of the strong nuclear force was, to reason along the following line of thought. He said (I quote somewhat freely from the book by Alonso and Finn): if there is some sort of cohesion, then it must be accompanied by energy. The cohesion keeps the nucleus together, which means that one would have to supply energy from the outside to split it up again into its constituent parts. This is so-called “binding energy”. The binding energy will be proportional to the number of particles in the nucleus, so that we expect it to be proportional to the volume of the nucleus, or: $(binding\, energy) \div (radius)^3$, where the constant of proportionality is unknown of course. However, the nucleons close to the nuclear surface are less tightly bound than those inside, because they have fewer neighbours with which to interact. Thus we have to correct our first estimate of the binding energy by adding a negative term. Since this is a surface effect, the reduction term required must be: $(reduction\ due\ to\ surface\ effect) \div (radius)^2$, again with a different, but also unknown proportionality constant.

In this way he went on to add more terms, for instance one taking into account the kinetic energy of the nucleons, and another for the energy due to the electrostatic forces. The proportionality constants of all these terms were still unknown, but what could be done next is to compare with actual binding energies, as measured for lots of different chemical elements. The nuclei of different chemical elements are all composed of different amounts of protons and neutrons, so that a lot of comparison material is available. In fact, Von Weiszäcker used the information summarised in Fig. 5.10. In this way he could fit the unknown constants in his tentative formula, to obtain an expression that replicates the series of measurements as closely as possible. Having done that, one can henceforth extrapolate towards other characteristics of nuclei than only the binding energy, and predict properties that have not been measured. The theory thus built up, although semi-empirical, was very successful.

Let us try to follow a similar procedure in the case of the cohesion forces binding a droplet of charge together. Do not think it is implied that our cohesion forces are the same as the strong nuclear force. Not at all! The only thing that is imitated here is the *procedure* by which Von Weiszäcker obtained his results. In fact, our task is simpler, because effects of velocity and of electromagnetic forces have already been taken into account. It was the subject of the previous chapter. The only thing that remains is to consider the binding energy due to the effects not covered yet by Chap. 5. We may, as Von Weiszacker did, take this binding energy to be proportional to the volume of the droplet, diminished with a correction proportional to the surface of the droplet.

But if no more terms are needed than just these two (because all the other relevant effects were already covered by Chap. 5), then a considerable simplification is possible: we are then talking about cohesion effects similar to those that keep a drop of water together. Once again: it is not meant to say that physically there is any similarity, what is meant is just that the mathematical description of the surface effect is similar. In the case of water drops the cohesion effect is well-known to almost every layman, it is known under the name of *surface tension*.

Analysing in more detail this phenomenon of surface tension, classical physics discovered that there are in fact two separate effects. One is that the droplet is kept under a constant pressure by the surface tension, this average pressure depending on how much "skin" there is compared with the volume. The second one, applicable to non-spherical shapes, is that the force exerted by the "skin" becomes larger, the more this "skin" is curved. These two effects combine to give the two characteristics of a droplet that have been sketched in the figure (Fig. 6.1). In case you wonder why the droplet looks different than in other sketches: the blue colour suggests water. In the upper part of this figure it is sketched that the surface tension maintains an internal pressure in the drop, so as to minimise the drop's overall size.

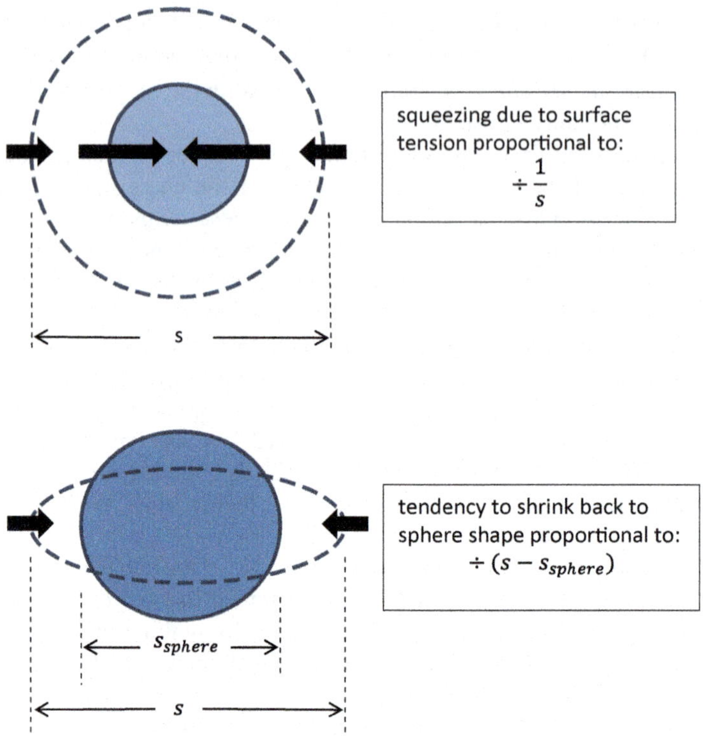

Fig. 6.1 The two effects of surface tension on a drop of water: internal pressure causing the droplet to minimise its overall size, and the tendency to minimise the surface area by striving towards a *spherical shape*

The surface tension becomes smaller, the larger the droplet is. This is easy to understand: a larger droplet has, relative to its volume, a smaller surface area. There is—relatively—less surface to give surface tension. This is the reason why very large water drops are not stable, the surface tension is unable to keep very fat drops together and they will fall apart into a number of smaller drops.

The lower sketch illustrates the well known property of a droplet that it always tries to take the shape of a sphere. At least, this is so in the absence of other forces. A falling drop of water experiences air forces due to its velocity, and will deform. The tendency to contract into a spherical shape is most clearly seen on movies of floating drops taken inside a space station. Under such circumstances there are no perturbations due to gravity and if the drop is floating, the influence of an airstream is absent too. The drop could be vibrating, but it does so around a pure spherical shape. The reason is that a sphere has the smallest surface area relative to its volume, compared to other shapes.

The striving after a sphere shape is clearly seen in soap bubbles too. In this case dynamic effects are also well known: during the blowing of the bubble we see large

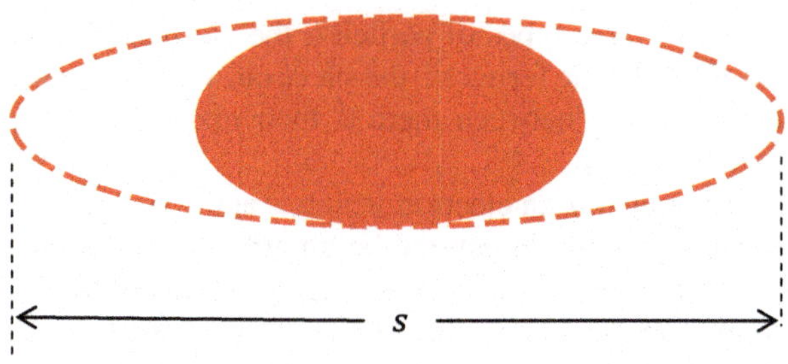

Fig. 6.2 Type of deformation assumed for droplet of charge is a combination of increasing volume as well as increasing slenderness → squeezing due to surface tension: $c_1 \cdot \left(s - s_{sphere}\right) + c_2 \cdot \frac{1}{s}$

departures from the spherical shape, but as soon as the bubble is coming loose from the blowing pipe, it contracts into a sphere. Often vibrations around this equilibrium shape are visible.

As indicated in Fig. 6.1, if we try to deform the droplet this tendency to go back to a sphere is—in the first instance—proportional to the amount of deformation.

Now, if it is remembered how we modelled our droplet of charge and how it was permitted to deform, it will be seen that this gives a combination of the above sketched effects. It is shown in Fig. 6.2: the deformation "mode" allowed in our model was just a lengthwise elongation in one direction. The "thickness" of the droplet was taken to be constant, only the length s is variable. This means that an elongation increases the volume of the droplet, and at the same time also gives a further departure from the spherical shape.

We had "squeezing tendency" expressed as a generalised force in the formal theory (as defined in Lagrange's mechanics), and thus find:

$$(\mathit{generalised\,force\,tending\,to\,reduce\,droplet's\,length\,due\,to\,surface\,tension})$$
$$= c_1 \cdot \left(s - s_{sphere}\right) + c_2 \cdot \frac{1}{s}$$

In this expression we have changed the proportionalities into equations, by giving the constants of proportionality the names c_1 and c_2. The result now looks quite impressive with these new symbols, but of course you may rightly ask "what is in a name?". The answer is: nothing new is represented by this "impressive" formula, it is still the sum of a few proportionality relations and no new information is obtained as long as we do not know anything about the actual value of these constants c_1 and c_2. By the way, you should realise that there are in fact *three* unknown quantities in the expression, since the value of s_{sphere} is not known either.

But recall the procedure Von Weiszäcker followed when he derived the properties of the strong nuclear force. He too was able only to determine a series of proportionalities. The key to making his expressions into quantitative relations was, to invoke the help of experimental data. He then could guess the value of his proportionality constants, so that his expression would replicate as closely as possible the experiments.

In our case we can try to do the same. What sort of experimental data could help us? Now finally we come to the explanation of the title of this chapter, which mentions three types of particles: the electron, the muon and the tau particle. They are all different forms of the electron, having identical characteristics except one: the difference between them is their mass. The "common" electron is the lightest particle.

The muon is an electron which is approximately 207 times as heavy, and is often found in cosmic radiation. An amusing story about it is that it provides one of the experimental proofs of the special theory of relativity. The muon particles are formed in the upper part of the earth's atmosphere due to collisions of extraterrestrial particles, coming for instance from the sun, with the atoms of the atmosphere. After having been formed the muon particles travel at a high speed, approaching the speed of light, towards the earth surface where they may be detected. The funny thing is, that muons are unstable and have such a short lifetime that one would expect them to vanish long before they can have reached the earth surface, i.e. long before they could be detected. The explanation is given by relativistic effects, in particular the so-called "time dilation" or "stretching of time at high speeds". Without going into the details of the theory of relativity, this means that from the standpoint of observers on the earth the time in the muon-system has slowed down, and it ages more slowly. From the point of view of the muon, the earth surface approaches so fast, that the thickness of the atmosphere appears very small due to Lorentz contraction.

The third type of electron, the tau particle, is even a lot heavier: its mass is something like 3477 times that of the lightest of the three, the common electron. It was discovered in artificial particle accelerators where nuclei are shot at and smashed into small pieces, releasing showers of debris to be studied by physicists.

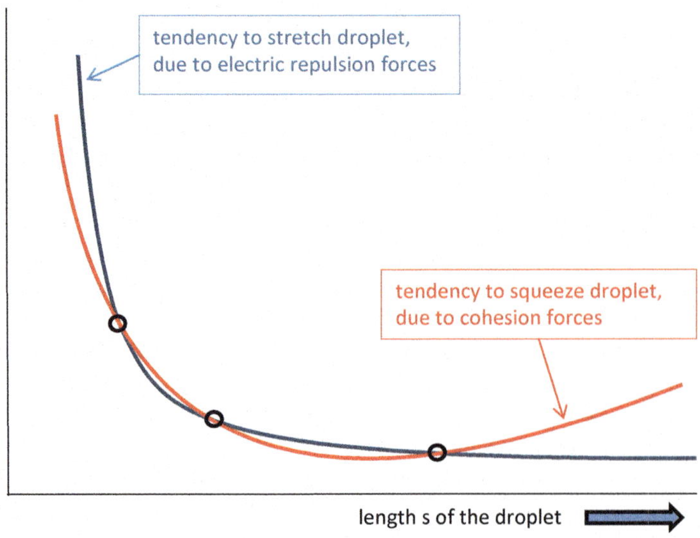

Fig. 6.3 How the tendencies to stretch and contract depend on the length of the droplet, sketched in one graph to show their relation

You probably wonder how this information is going to help us. Fortunately, it is not very difficult to explain. Have a look at the following figure (Fig. 6.3).

In this figure we have summarised everything we know so far about the tendencies of the droplet of charge to elongate or to shrink. In Chap. 5 we found electromagnetic self forces that always try to increase the length of the droplet. The associated generalised force was found to be proportional to $1/s^2$:

$$(\textit{tendency to \underline{expand} due to electrostatic effects}) \div \frac{1}{s^2}$$

And we have just determined the effect of surface tension, which squeezes the droplet and thus counteracts the electromagnetic effect:

$$(\textit{tendency to \underline{shrink} due to cohesion}) = c_1 \cdot \left(s - s_{sphere}\right) + c_2 \cdot \frac{1}{s}$$

Although we do not know yet how large the proportionality constants must be, one thing is clear: the shape of these curves will be quite different if they are drawn to show how they depend on the length s. In the above given figure the two curves are shown for some arbitrary values of the proportionality constants. The expansion tendency is shown by the blue line, and the tendency to shrink is shown by the red one. Clearly, their shape is indeed different, because the two counteracting effects depend in an entirely different way on the length of the droplet.

The consequence of the different shape of the blue and red lines is evident as well: the two curves will cross each other at several points. In fact, the rules of algebra tell us that one may expect *three* intersection points (sometimes imaginary intersections, but that is a technicality not relevant here). What does that mean physically? Easy: in an intersection point the tendencies to elongate and to shrink are equal, they balance each other. Such an intersection represents *equilibrium*, the length of the droplet could—in principle—remain constant once it has arrived in this equilibrium situation.

We can conclude that *three* equilibrium situations are possible, meaning that there are three different lengths the droplet could permanently maintain. If the droplet would have another length than one of the three equilibrium values, the expansion force would differ from the squeezing force. It would either be smaller or larger than the squeezing, which would result in a net—generalised—force which would cause a departure from the length it happened to possess. There are only three values of the length where this does not happen, and where the droplet can maintain a constant length. In principle at least: you will see in a minute that even under these equilibrium conditions the droplet will not stay quiet. But in principle the droplet could be content with its size if it has obtained the equilibrium length, and could keep it there.

The three equilibria where the droplet can maintain a constant length represent three different values of the *mass* of the droplet. To see this, let us recall the expression for the electromagnetic mass of a charged droplet:

$$m_{em} \div \frac{1}{c^2}\frac{q^2}{s}$$

The length is in the denominator of this expression. The equilibrium point that is the most to the left in the above figure thus represents a possible state where the droplet is heavy, heavier than in the other two guises it can assume. On the other hand, the rightmost equilibrium point is a possible state where the droplet has its lowest mass.

We now have made a bold assumption, and probably you already have an inkling what is coming. Perhaps you cannot really call it an *assumption*, it is rather the inescapable *consequence* of taking the droplet of charge to be the model of an electron. This model may be true or false, but if—for now—we have adopted it as a working hypothesis, then strict reasoning leads to the conclusion that there must then exist three types of electrons.

As we have seen, this is in agreement with actual observations in the real world. So, let us proceed on the path we have chosen to follow, even if it was chosen just for the purpose to satisfy our curiosity where it would lead. It leads us to the conclusion that a droplet of charge can take three guises, to be identified with the three guises of an electron in the real world. The previous Fig. 6.3 is now repeated in Fig. 6.4, with the new interpretation inscribed in it. Obviously, the figure does not have the correct scale. It should be seen as a rough indication of our ideas. For instance, if the equilibrium point in the middle is to represent the muon with its mass of 207 times that of the electron, the figure should in reality be such that the droplet's length should there be approximately 1/207-th of the length of the

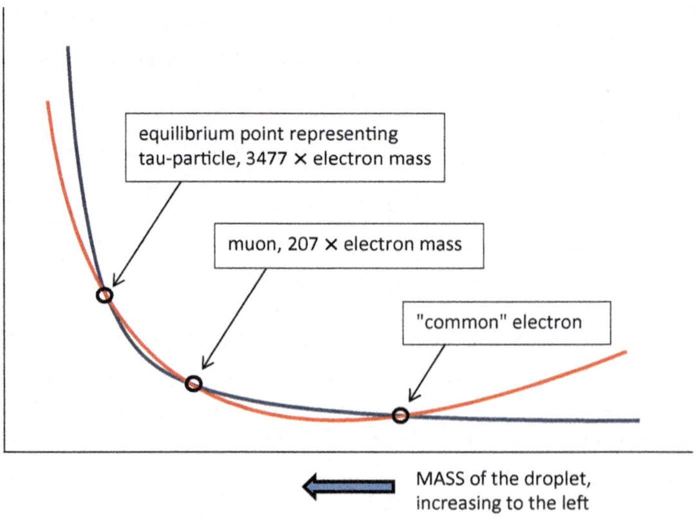

Fig. 6.4 Identification of the three equilibrium points of Fig. 6.3 with the three guises of the electron

"common" electron. We use the word "approximately", since the electromagnetic mass is not the only type of mass, there is also "mechanical mass". We have dealt with this at length in one of the last paragraphs of the previous chapter, discussing the matter of what was there called the "bare mass".

Anyway, the general procedure to proceed will now be clear to the reader: we will have to *deform* the red line, representing the cohesion force, in such a way that it crosses the blue line of the electromagnetic force at the correct points. "Correct" meaning, that the masses corresponding to these points coincide with the experimental masses of the real electron, the muon and the tau particle. Deforming the red line is done by appropriately choosing the values of the hitherto unknown proportionality constants in the expression for the cohesion force. The values of these constants will not be given here in the main text of the book, since they are not very meaningful. Hopefully the reader has understood the principle of the procedure, which has been worked out in detail in the appendix.

After having determined those values of the proportionality constants that are needed to make the droplet model a consistent model of electrons, we can then go on. Other properties of our model can now be calculated, no longer in general terms but quantitatively, and precise. It should never be forgotten of course that the model remains just a working hypothesis. Furthermore, the model may now be consistent but it is still approximate due to the many modelling simplifications and the approximate character of the mathematical analysis applied to it. Nevertheless, within the scope of the assumptions precise figures can be calculated. In the last paragraph of this chapter, you will read what sort of characteristics can then be quantified. The surprising conclusion is, that the theory of the deforming droplet of charge appears to be in *quantitatively correct* agreement with one of the foundational theorems of quantum mechanics (De Broglie's relation).

Before doing so, we will have to return to a discussion of what are other implications of "points of equilibrium".

6.3 Stable and Unstable Equilibria

In equilibrium the expansion forces and the squeezing forces balance each other. In principle this state can exist permanently, the droplet will not feel any urge to change its length. But that is only true "in principle". In actual fact one must always expect that there will be disturbances, no matter what causes them. A *stable* equilibrium means that any disturbance is counteracted by the forces in such a way that the droplet will automatically return to its equilibrium state. A stable equilibrium may be said to be robust, it can counter all sorts of attacks on it without disastrous consequences. Such a robustness is not always guaranteed, and *unstable* equilibrium states may also occur. How one can get a first impression about the stability of an equilibrium is sketched in the following figure (Fig. 6.5).

In this figure the rightmost equilibrium point of Fig. 6.4 (the point identified with the normal electron) is sketched in an enlarged form. It is assumed that

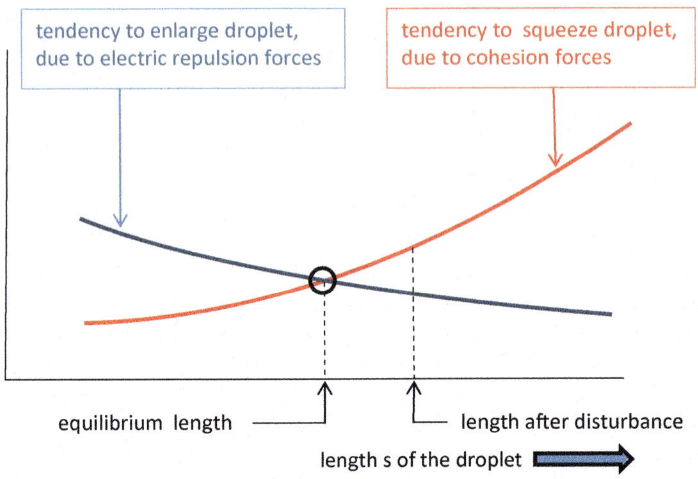

Fig. 6.5 Enlarged sketch of what happens after a perturbation of the length has occurred. Situation shown is that of the common electron

something has perturbed the length of the droplet from its equilibrium value. The resulting, new length is indicated by the broken vertical line to the right of the equilibrium circle. Let us now read off from the figure how the forces working on the droplet have changed by this perturbation. From the figure we can see that we are now on a higher part of the red curve, meaning that the contraction force due to cohesion has become larger. The reverse has happened to the explosion force due to the electromagnetic effects, it has decreased. In the disturbed situation we therefore see that the tendency to shrink has become larger than the tendency to expand, resulting in a net force trying to reduce the length. After a disturbance in length has occurred, automatically the droplet will be driven back, to reduce its length and return to its equilibrium.

A similar thing, but then in the opposite direction, will happen if the disturbance was such that it had reduced the length of the droplet. Once again we will find that the disturbance will be counteracted by the resulting changes of the forces. Any disturbance finally results in a return to the equilibrium: this is evidently a *stable* situation.

Actually, we should be cautious with this conclusion. The occurrence of a net, positive restoring force after a disturbance is *necessary* for stability, but it is not a *sufficient* condition. Even if positive restoring forces do exist, the equilibrium may still be unstable, as exemplified later when we consider the tau particle. Anyway, let us conclude for the moment that our model of the electron has a good chance to be stable, and that this is in agreement with the observed stability of real electrons.

In the case of the muon the roles of the red and blue lines are reversed, as seen in Fig. 6.4.

If we assume that a disturbance has happened, the net resulting force will be such that it tries to *increase* the disturbance. It will drive the droplet further away from its equilibrium point. One could say that there are negative restoring forces. General theory about the stability of dynamic systems now tells us that no rescue is possible whatsoever: this kind of equilibrium is *unstable*. Negative restoring forces always lead to instability, this cannot be repaired by any other dynamical effects. The conclusion is, that our model of the muon is always unstable. The real, physically occurring muon appears to be unstable as well. Remember the story about its lifetime which is so short that it is only thanks to relativity effects that we are able to discover the muons in cosmic radiation.

But wait! What is the matter with the tau particle? From the diagram (Fig. 6.4) of our tau-model it is clearly seen that after a disturbance we will have *positive* restoring forces, thus indicating the possibility of a stable equilibrium. In actual fact however, the physical tau particle is even more unstable than the muon. This would appear to be in contradiction with the stability considerations as explained. The point is, that we here encounter the provisos made earlier, where we said that positive restoring forces are necessary, but not sufficient to guarantee stability. What can make a system unstable, even if it is driven back to its equilibrium after a disturbance? Well, every motorist knows it.

What can befall a car driver is that he/she is momentarily distracted, looks up and finds him/herself half inside another lane. It is then, especially at high speed, a matter of not too briskly steering back, because a too brisk and too fierce correction could easily lead to *over*controlling, ending in an unintended swerve to the other side.

In aviation there is even a name for it, it is called PIO (pilot induced oscillations). If a sudden control correction is needed, the pilot must move the control stick in the correct direction but if he *over*controls, the aircraft could go into an ever increasing oscillation, it could in the worst case get completely out of control. Student pilots on—not artificially stabilised—helicopters often need many hours of training before they are able to keep their aircraft reasonably stable in hover, because helicopters are so sensitive they almost invite the inexperienced pilot to overcontrol them. Once in a PIO, the help of the instructor is needed or else the student pilot would go back and forth with increasing severity, he would see all the corners of the airfield, and would almost certainly crash. The PIO gives him the feeling that the more he tries —with clenched teeth—to fight the oscillation, the faster everything gets out of control. In fact "pilot induced oscillations" is a euphemism, one could better call it "pilot induced instability", the oscillations become larger and larger and get out of hand quickly. A saying by experienced helicopter pilots is: "do not move your stick, just *think* that you will move it, and keep your hands still".

In this kind of so-called "dynamic instability", the corrective action is in the right direction: after a disturbance a force is applied so as to go back to the desired equilibrium situation. But the correction is too large and almost always has a small delay, so that the system swings through the equilibrium and on the other side ends up further away from the equilibrium, after which the whole cycle starts all over again with an even more brisk correction.

To summarise: it is not enough to have corrective actions in the right direction, the corrections should be large enough to give any effect, but also should not be too large or else one gets "dynamic instability". This is a general rule applying to any sort of dynamical system, even to things like a country's economy. In the latter case, one often hears amateur economists say that the free market guarantees a balanced system, because it provides automatic counteractions for every disturbance. It may be true that any unbalance in the system automatically provokes corrective tendencies, this however certainly does not guarantee a stable system!

Now back to the tau-particle. Recall Fig. 6.4 from which it was concluded that the tau particle is represented by the leftmost intersection of the blue and red lines. Here we are in a region where both these lines go up fast. A warning was already given that the figure does not have a realistic scale. If we take into account the real ratios of the electron and tau masses (and therefore their length ratios) a simple sketch is almost impossible to make: the tau would be very high up in the figure, and would be situated practically on the vertical axis, with both the red and blue line almost vertical. The idea is sketched in the enlarged Fig. 6.6, but even this figure is certainly not drawn to the correct scale. Matters are much worse in reality than can ever be shown in a simple graph. What we see is, that even the minutests of disturbances would give rise to enormous restoring forces. It is typically the situation where the restoring forces can lead to *over* corrections. This particular model of the tau particle almost begs for "dynamic instability". Our model is therefore not

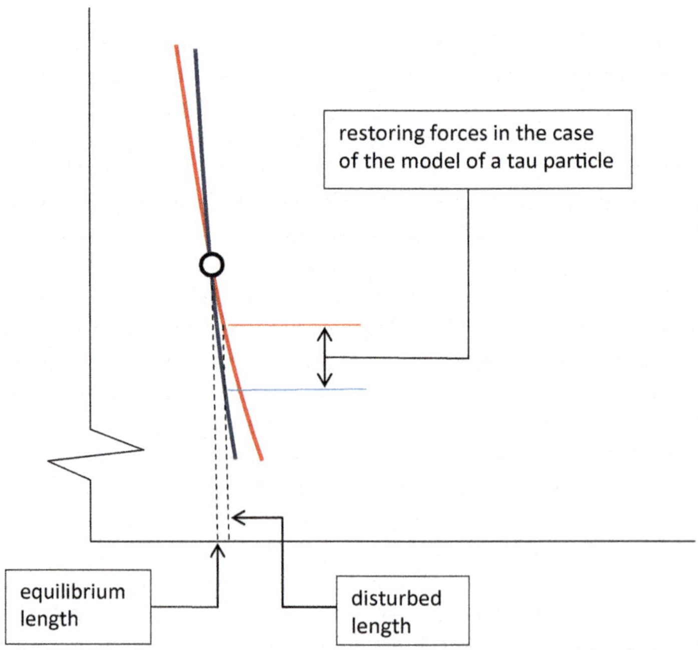

Fig. 6.6 The equilibrium situation of the droplet representing the tau-particle

in contradiction with the actual observations that real tau particles have an extremely short lifetime. Agreed, our model is a very speculative one. But it does not lead to conclusions in conflict with experimental observations.

6.4 Vibrations Around the Equilibrium

We now come back to some other properties of the droplet that can be quantified once we have fitted the unknown constants of proportionality to replicate experimental results. The conclusion at this point is that the model we have made of the electron can be stable. If there is for some reason a perturbation in length, restoring forces will counteract the perturbation and will drive the droplet of charge back to its equilibrium configuration.

Now remember another conclusion, mentioned in Chap. 5, that electromagnetic forces cause a droplet of charge to possess "squeezing inertia". This was a kind of generalised mass which comes into play if the length of the droplet is being changed. To be more precise: squeezing inertia is associated with what was defined as "pulsation acceleration", it is for length variations the same as the mass of an object is for translational accelerations.

Let us now look what the effect is, if after some perturbation of the length the restoring forces do their work. They drive the length of the droplet back to its equilibrium length, but at the very moment the droplet has acquired the "correct" length it has some "pulsation velocity" and by its inertia will overshoot the equilibrium. It is exactly the same as with all other kinds of oscillators: the result will be a pulsation vibration, i.e. a periodic lengthening and shortening of the droplet. It depends on the amount of damping, i.e. on how much energy is carried off per second, how long such a vibration will persist.

The pulsation of a droplet of charge is rather persistent, because loss of energy by radiation is negligible. This was one of the other conclusions of Chap. 5.

We can calculate the frequency of these pulsation vibrations. It would not be very enlightening to give you just a number of "so and so many cycles per second". It is a very large number you would see. A better impression about it can be gained by imagining that the droplet has a certain speed, and to have a look at the wavelength of the pulsations.

During the translation motion of the droplet one pulsation cycle starts at a certain point, and the next cycle at a place a bit further on. The distance between these two points will be called the "wavelength" of the pulsation. Now, let us take as an example a droplet that is moving with one tenth of the speed of light: $V_{transl}/c = 0.1$. The number 0.1 seems innocuous, it suggests that the speed is nothing special. And indeed, for an electron this is not a frightfully high velocity. But do realise that we are nevertheless talking about something like 30 000 km per *second*, which is an almost impossible speed in the human world.

Now, the frequency of the pulsation is so high that the wavelength of the pulsation amounts to only 0.00024 times the length of our electron model! As we have seen before (Chap. 1), the pulsations are accompanied by fluctuations in the speed of translation. We see the droplet irregularly moving forwards, hopping as it were, as if it was taking part in a sack race. The length of these separate hops is well within the contours of the droplet. In fact, the hops are very much smaller than the length of the droplet, by the same factor 0.00024. If our model in any way represents something which is comparable with real electrons, we are then talking about "hops" having the scale of quarks. It gives a feeling for the extremely high frequency of the pulsation, that even at these incredibly high speeds the distance between two consecutive cycles of the pulsation (or between the hops of the translation) is still incredibly small!

In fact, such a high frequency completely falls outside the technical possibility to measure it. A pity, because we can discard any hope to ever determine whether this piece of theory makes any sense. At least, a check on the theory of the deforming droplet is impossible by *directly* measuring the predicted pulsation frequency.

Fortunately, prospects to verify the theory of the pulsating droplet are not as bleak as one would fear from the above given numbers. This is thanks to the fact that the frequency is *velocity dependent*. Recall that in Chap. 5 we found that the electromagnetic "explosion force" depends on the velocity of the droplet. It was item number 8 in our list of forces and was described by an expression which has, for this book, a record length:

$$(explosion\,tendency) \div (total\,charge)^2 \cdot \frac{(1 - V_{transl}^2/c^2)}{s^2}$$

Here we saw that the explosion tendency becomes smaller, the higher the speed of the droplet is. Now, the balance between on the one hand this electromagnetic explosion force and on the other hand the cohesion force determines at which values of the length we find an equilibrium. The equilibrium length of the droplet, and therefore its mass, therefore both depend a bit on the velocity. We try to avoid too much detail of explanation, but I think you will not really be surprised to hear that the restoring forces after a perturbation of the length also depend on the velocity. It all works out such that the pulsation frequency varies with the speed of the droplet. A *shift* of frequency is noticeable if a particle at rest ($V_{transl} = 0$) is going to move with a speed V_{transl} which is non-zero. How large this shift is, is given by a proportionality relation:

$$(frequency\,shift) \div V_{transl}^2$$

Or: the frequency shift is proportional to the velocity squared. From the quantification in the mathematical appendix it appears that the constant of proportionality is very small. So small indeed that, even if it multiplies the large "velocity squared" in the last formula, the resulting frequency shift is still not very large (relatively speaking). In a few moments it will be shown what sort of magnitude it has. It will then be seen that the frequency *shift* lies well within the measuring possibilities. The frequency *shift* is something that belongs to the scale of the human world, whereas the pulsation frequency itself is something belonging to an entirely different sub-world.

Before giving some numbers about the frequency shift, we will have to answer a preliminary question: how can we measure a frequency *shift*? The problem is that nothing material is vibrating with a frequency that is equal to the *shift*. The droplet is vibrating with a certain frequency if it is stationary. We could, if ever we could measure it (which is impossible), write down this value on a slip of paper, and then compare it with the frequency we (cannot either) measure if the droplet has acquired a certain velocity. This is always the way to find out about a change of something: compare two different measurements. But, if we are observing just one moving droplet and have no comparison material with a stationary one, there is no measurable characteristic that gives any information about how the frequency was shifted when the droplet started its motion.

The problem would be solved if, apart from the moving droplet, there are other droplets in the vicinity that do not move. Remember that the pulsation causes radiation (although not the type of radiation which travels to infinity and would cause a loss of pulsation energy). The frequency of this radio transmission is the same as that of the pulsation. If we would put a radio receiver very close by, so that it can detect the field of both a stationary droplet and a moving droplet, we would hear a "howling" noise, much like the noises coming out of antique radios when two stations are not sharply separated. Mixing two frequencies close to each other gives interference between the two signals, and results in a tone with the *difference* frequency of the two signals. In Chap. 1 already a lot has been said about such interference phenomena: the interfering signals cause *beats* in the received signal.

This same phenomenon could serve as an observational tool to check by experiment the theory regarding the pulsation of a droplet of charge. The howling sound we would hear if listening in on the combined radiation field of a stationary and a moving droplet is nothing else than the sound of beats. The train of beats would have a frequency equal to the frequency *shift* due to velocity.

To give a feeling for how large this beat frequency is, the table below has been calculated using the theory of the appendix. In this table we have taken several values of the velocity of an electron, given as a ratio of the droplet's velocity compared with the speed of light. It is the range of velocities electrons may have inside an atom, although $V_{transl}/c = 0.5$ is rather extreme.

In the second column the wavelength of the pulsation λ_{puls} is given. We have defined earlier what is meant by this wavelength: it is the distance between successive pulsation cycles, when we see the electron passing us. To avoid the use of very small, rather meaningless numbers the wavelength is given as a fraction of the length of the droplet: $(wavelength\,of\,pulsation)/(length\,of\,droplet)$ or in short notation λ_{puls}/s. In fact, you will probably recognise the first number 0.00024, which was given before as an example to show how small this wavelength is.

v/c	λ_{puls}/s	λ_{beats}/s
0.1	0.00024	16 960
0.3	0.00071	5 653
0.5	0.00118	3 392

In this second column you see that the wavelength of the pulsations increases with the velocity of the droplet. A larger velocity will have two effects. Dominating is the fact that the droplet will travel a longer distance before the next cycle of the pulsation comes on. This first effect is slightly counteracted by the fact that the frequency of the pulsation becomes larger with velocity. This tends to make the wavelength slightly smaller, because the next pulsation now comes somewhat earlier, and the droplet has travelled a slightly smaller distance. But as is seen in the table this second effect is very small, due to the smallness of the frequency shift due to velocity.

The third column gives information about the frequency of the beats, if there is interference because the fields of a stationary charge and a moving one are mixed together. Again it is expressed in the form of a wavelength. This wavelength is the distance travelled by the moving particle between two successive beats. The wavelength is again brought in relation with the size of the droplet, the numbers in the third column give the calculated value of $(wavelength\ of\ beats)/(length\ of\ droplet)$ or symbolically: λ_{beats}/s.

There are two striking differences between the second and third column. First of all, the values of λ_{beats}/s are larger than those of λ_{puls}/s, not just a bit but by many orders of magnitude. This reflects the fact that the frequency of the beats is very much smaller than the pulsation frequency itself, their frequency equals the small frequency *shift* due to velocity. The corresponding wavelengths have now become of a "human scale". A typical value around 10 000 times the size of the droplet is comparable with the dimensions of an atom. So, it may still be a bit exaggerated to call it the human scale, but what is meant is that we here have arrived at a scale within reach of measurements. By considering the wavelength of the beats instead of those of the pulsations themselves, we have suddenly returned to a world where technical possibilities exist to perform measurements. We can have the hope that a verification of the theory of the pulsating droplet could become feasible!

The second striking feature of the numbers in the third column of the table is, that we see the wavelengths *decrease* with velocity. This is not really surprising, since the frequency of the beats is proportional to $\div V_{transl}^2$. The reduction of the wavelength due to an increasing frequency has now become dominant over the stretching of the wavelength directly due to velocity, which is only proportional to

$\div V_{transl}$. In fact, a little bit of puzzling shows that the wavelength of the beats must be inversely proportional to velocity:

$$(wavelength\, of\, beats) \div \frac{1}{(velocity\, of\, droplet)} \quad \text{or} : \lambda_{beats} \div \frac{1}{V_{transl}}$$

This expression reminds us of work by the physicist De Broglie. Remember that De Broglie was one of the first people to lift a tip of the veil of quantum phenomena. He proposed that every particle is accompanied by some sort of wave phenomenon, although the nature of this wave is unclear. He called them "matter waves". These matter waves accompanying a moving particle have a certain wavelength, just as we defined in our case a wavelength of the pulsation and of the beats. De Broglie, after deep thinking, proposed the relation

$$(wavelength\, of\, matter\, waves) \div \frac{1}{(velocity\, of\, particle)}$$

or:

$$\lambda_{DeBroglie} \div \frac{1}{V_{transl}}$$

Seeing this, one becomes anxious to know how large the two constants of proportionality are in the two formulae for λ_{beats} and $\lambda_{DeBroglie}$ respectively. Would both expressions perhaps try to tell us the same thing? Well, as remarked before, at this stage we are able to calculate precise figures (now that we have substituted in our theory the mass ratio's of the electron, the muon and the tau particle). And here follows the "pièce de resistance": if we take the droplet to have the charge of an electron, the constant of proportionality appears to be the same as in De Broglie's relation!

The theory in the appendix thus affords to calculate the actual wavelengths of the train of beats, and they come out practically the same as the wavelengths of De Broglie's matter waves! (apart from a small discrepancy of 1.5 % which is to be expected in an approximate theory). In other words:

$$\lambda_{beats} = \lambda_{DeBroglie}\,!!$$

We talked earlier about the possibility to experimentally check whether the theory of the droplet of charge has some validity. We would ideally like to measure if a real electron indeed pulsates, but had to conclude that such a direct verification is impossible. However, the thing we would be able to measure is the frequency of the *beats*, i.e. one of the derived consequences of the pulsation. A practical experiment would thus have to look for confirmation that a real electron indeed is accompanied by some sort of wave phenomenon having the same wavelength as the beats that are predicted by the droplet theory.

It now appears that such an experiment is unnecessary: it has already been done. Our prediction of the beat frequency and how it depends on the velocity of a particle is confirmed by quantum theory itself, where it was long known that any particle is accompanied by waves, waves which happen to have the correct wavelength.

It is quite another matter of course, whether the *interpretation* of matter waves as an interference effect due to pulsation does have any reality. This has not been proven at all by the above equality. But neither the opposite is proven: the two theories are not conflicting. Therefore, at the very least we may use all this as a "mental picture" of matter waves.

This remarkable result shows that there "must be something" in the theory of the droplet of charge. It is intriguing that such an unconventional model of an electron, analysed by the classical laws of nature, would give De Broglie's relation which is always considered to be reserved strictly to the domain of quantum mechanics. It gives us an excuse to not immediately throw away all our theories, and continue our investigations.

If several subjects in this last section were described too fast or were treated too superficially for the taste of the reader, do not despair: in the following chapters we will come back to several matters more thoroughly.

Chapter 7
Dynamics of the Droplet: A Pictorial Representation of the Equations of Motion

7.1 What Is an Equation of Motion?

To start with: what is an "equation of motion"? You can get two answers to that question, one short—and probably incomprehensible unless you already knew what is meant by an equation of motion—and a longer story containing a historical example. If you might be satisfied with the short answer, the advantage is that you can skip this paragraph and turn straight to the one explaining the phenomenon of physical analogues. A good chance that you want to skip that paragraph as well. But do not skip the one after that, because new stuff is presented starting with the paragraph titled "7.5. A mechanical analogue of the pulsating droplet inside a potential well".

Now the short answer: an "equation of motion" is a mathematical summary of the forces acting on a body, combined with the laws of dynamics. The summary is so complete that all the essentials of the problem are included and a mathematician will be able to calculate exactly how the body is going to move and what its position will be at any instant of time.

This short answer does not say anything about the form such equations take, and what to imagine about the procedures followed. The next, longer description is intended to give a better impression about the techniques physicists and engineers apply, but do not worry: the few "formulae" occurring in this description are simple and probably already well-known to many people.

One of the first, and perhaps the most famous instance of setting up equations of motion and solving them is Newton's exercise to calculate the paths of planets around the sun.

T. van Holten, *The Atomic World Spooky? It Ain't Necessarily So!*
DOI 10.2991/978-94-6239-234-2_7

7.2 A Historic Example

Newton started from the fact that a body in motion wants to maintain a constant velocity. This is not so self evident as we perhaps think. Nowadays we have grown up with this idea, it is so familiar to us that we think it is natural. Actually, another idea, on which Aristotle's thinking was based, would perhaps be more in accordance with intuition. This idea entails that a body will always lose its velocity unless we help it by applying some force. After all, that is what we in actual life observe: a ball rolling over the floor will always lose its velocity, and in the long run always comes to a stop. In contrast to this Newton's starting point was, that a body persists in its motion if *strictly* left to itself. If a body changes its velocity, i.e. if it accelerates or decelerates or if it deviates from its straight path, this must according to Newton be the result of some kind of force. Why we always observe a ball to lose its velocity is due to forces like air resistance, rolling resistance, etc. If in an imaginary world such forces would be absent, the ball would go on and on without ever stopping.

The reverse of an accelerating body is less counter intuitive: if we would observe that the speed of the ball increases, we would intuitively search for the force causing this acceleration. It might for instance be the gravity if the ball rolls on an inclined plane. Anyway, nowadays we know that Aristotle was wrong, and that both accelerating and decelerating a body requires a force.

Newton subsequently quantified this idea and put it in mathematical terms. One of his laws states that the acceleration of a body is proportional to the magnitude of the applied force. If we apply a force twice as large, the acceleration will also be twice as large. If the symbol a is being used to indicate the amount of acceleration —Newton defined very precisely what should be understood by "amount of acceleration"—and the symbol F for the force, we have obtained the relation $a \div F$ (to be read as: "acceleration a is proportional to the applied force F").

The proportionality can equally well be reversed: $F \div a$ ("F is proportional to a"), now meaning that if we want the body to accelerate twice as fast, then we need a force which is twice as large.

Finally, such a proportionality can always be written as an equality, using a proportionality constant: $F = m \cdot a$. The proportionality constant m has a clear physical meaning, because we know that a more massive body needs a larger push to obtain a certain acceleration than a less massive one. Consequently, m is called the "mass of the body". Mass is thus a measure for the inertia of a body, inertia being the tendency to resist changes of its velocity.

The law $F = m \cdot a$ is very general, and applies to all sorts of situations, ranging from the act of raising an arm to planetary motion. For a specific situation we need more: one has to tell what exactly is the nature of the force, how it acts on the body, how large it is and how it varies in time. Apart from that it is necessary to know how large the mass m is in the case considered or, if the system is a combination of more parts, how large the masses are of the separate components and how these parts are arranged and connected with each other.

For the case of planetary motion Newton had to add the law of gravity to his general dynamic law $F = m \cdot a$. This law of gravity, hypothesised by Newton, says that in planetary motion the force on a planet is always attracting it in the direction of the sun, and that the gravity pull decreases the further we are away from the sun. He proposed that the force of gravity decreases with the square of the distance: $F_{gravity} \div \frac{1}{r^2}$. In this expression the sign "÷" again means "is proportional to", and r denotes the distance of the planet from the sun. If a planet (or another heavenly body like a comet) would be moved further out to a distance twice as large from the sun, the attraction by the sun would decrease to only one quarter of the original value. Alternatively: if the planet approaches the sun to halve the original distance, the gravity forces would quadruple.

By combining the law of gravity $F_{gravity} \div \frac{1}{r^2}$ with the general dynamic equation $F = m \cdot a$, i.e. by *specifying* the force, the description of the dynamics of a planet is complete. What we in fact have achieved is that wherever the planet is, we know how large its acceleration is in the direction of the sun. This appears to be sufficient, the rest is a mathematical trick to completely determine the position of the planet as it moves in time.

The dynamic law in which a certain force-law has been specified is called an *equation of motion*. Newton's great achievement was not really to think up the $F = m \cdot a$ law. The ideas underlying this law were gradually developed by more than one scientist (amongst others Galilei and Huijgens), although the existing body of knowledge was systematised by Newton, cleaned up and freed from misunderstandings. Even the exact definition of what we mean by "acceleration" and how it is related to "velocity" (so-called differential calculus) was invented independently by two different people (Newton as well as Leibniz). Newton's great achievement was, that he proved how the gravity force depends on the relative position of two bodies so that he could set up the equation of motion for planetary motion, and that he subsequently was able to go through the mathematics that predicted in detail the paths of planets inside the solar system. The outcome of the exercise is of course well known and has become one of the "saga's of physics": the mathematically calculated trajectories of planets were in almost exact agreement with Kepler's empirical "laws", which state that planets move in elliptically shaped orbits around the sun, and acquire a larger velocity when approaching closer to the sun, or lose some velocity when they are further out. Kepler's laws were found by studying a multitude of very accurate astronomical observations collected by Tycho Brahe. The agreement between Newton's theoretically derived orbits and the actual observations proved that the law of inertia $F = m \cdot a$ was correct as well as the inverse square law $F_{gravity} \div \frac{1}{r^2}$.

7.3 In Our Case: *Two* Equations of Motion Needed

In summary, an equation of motion states in mathematical terms the relation between accelerations on the one hand and position on the other. It may be that the forces on a body also depend on its velocity (think of air resistance) or on its jerk (think of radiation resistance). In that case the equation of motion states how the acceleration is related to the body's position, speed, etc. Anyway, having obtained the equation of motion we can calculate exactly how a body is moving, when using the proper mathematical techniques.

In the particular case dealt with in this book, I have simplified the motion of a droplet of charge to translation along a straight path, and pulsation (contraction and elongation) in that same direction. We thus have a system with two different kinds of motion, officially called two *degrees of freedom.* As a consequence we need to set up *two* equations of motion, one for the translation and one for the pulsation.

What was done in previous chapters was in fact, to determine the forces. First of all there is the force the potential well exerts on the droplet, which drives the translation oscillation in a similar way as gravity drives the rolling of a marble inside a bowl. Then we also considered the electromagnetic "self forces", such as the electromagnetic inertia force (it was called the electromagnetic mass) and the radiation resistance (that was the recoil force as a reaction to the emission of radio waves), both—so to speak—disturbing the translation. The latter forces depended on the instantaneous acceleration and jerk (the change of acceleration) of the droplet.

As far as the pulsation is concerned, an electrical force was considered such as the "explosion tendency" due to internal repulsion, the magnitude of which depends amongst others on the amount of elongation of the droplet. On top of that we determined—or rather hypothesised—the "squeezing" of the droplet by cohesion forces, also depending on the amount of longitudinal stretching of the droplet.

More forces appeared relevant, coupling the two kinds of motion. For instance, the explosion tendency was found to be influenced by the translation velocity. More of such coupling effects have been pointed out and have been quantified in the mathematical appendix. The situation is complicated, but what it comes down to is that we have obtained in these previous chapters a complete specification of all the forces (at least I hope that it is complete).

What remains to be done now is to combine all these forces with the dynamical law $F = m \cdot a$ (or with convenient alternative forms of this dynamical law).

This procedure results in the two equations of motion sought, so that henceforth by applying suitable mathematical procedures we can predict the behaviour of the droplet.

To avoid bothering the reader with mathematical expressions, the resulting equations of motion will not be given here in the main text. For curious readers who think they will be able to stomach some mathematics, the equations of motion are written down in a separate text block (to be found at the end of this chapter) together with a few explanatory lines.

Fortunately, there is an alternative to mathematics. It appears that there is also a pictorial way to present the equations of motion, using the concept of *analogies*. To explain what this concept entails and how it works, let us go back to some earlier examples.

7.4 The Concept of Physical Analogues

In the chapter "Introduction and outline" the simple oscillator was introduced by giving the examples of a marble rolling inside a bowl, a clock pendulum, a cyclist freewheeling through a valley, etc. Intuitively it was rather clear that all these systems would show the same kind of behaviour, although their actual appearance is completely different. A cyclist does not really look like a marble, but his behaviour does (at least when he/she is freewheeling up and down the two facing slopes of a valley). No one will really be surprised that all these systems show a comparable type of motion. Mathematically their similarity is even more clear: it turns out that the equation of motion for all of them is exactly the same. Systems described by the same type of equations are said to be *analogues* of each other.

Often the reverse is also true. Once we have set up the equation of motion for a particular system, we can then—sometimes to our own surprise—conclude that another system must have the same behaviour. Even if the similarity was not intuitively obvious, we conclude it by comparing their mathematical description. Consider for example the next figure, on the left showing something like the balance of an old fashioned mechanical wrist watch.

In essence the balance consists of a body which can rotate around a vertical shaft, and is restrained by a torsional spring. The body has inertia in the rotational direction, as is symbolised in the figure by a pin radially sticking out, with a mass at its end. It turns out that this set-up is mathematically described by the same equation of motion as the clock pendulum sketched on the right. Seen in time, the balance's swaying to and fro is entirely comparable to the swinging of the weight of the clock. The rotation of the balance is measured by an *angle* and the swinging of the pendulum could be given by a *distance* from the midpoint, nevertheless they are analogues both performing what has earlier been called "sinusoidal" motions.

The forces driving the systems towards an equilibrium point (which point is "overshot" during each cycle so that the motion does not stop) are the torsional spring and the gravity force respectively. Therefore, the spring force and the gravity are also called to be "analogues" in this case. In the same way the relevant inertias of the fly and the pendulum are said to be analogues of each other.

Practical differences do exist of course. Both systems can be used to measure time, and actually both are being used for that purpose. However, the clock pendulum is inconvenient to use on board rolling and pitching ships, whereas the system using a torsional spring is much less sensitive to disturbances of this type, and can be made to be practically insensitive. In fact, in 1714 the English admiralty wrote out a contest to invent an alternative for the pendulum clock, because on board ships it was too inaccurate for navigational purposes. The winner of the contest was John Harrison who, after some struggles with the government's accountants, won the considerable prize of £20 000 by developing a clock based on a mechanical spring. This story contains a useful warning too: analogues are useful only to a certain level. Mathematical models are—intentionally—simplified idealisations of the real world, and neglect all the details deemed irrelevant for the

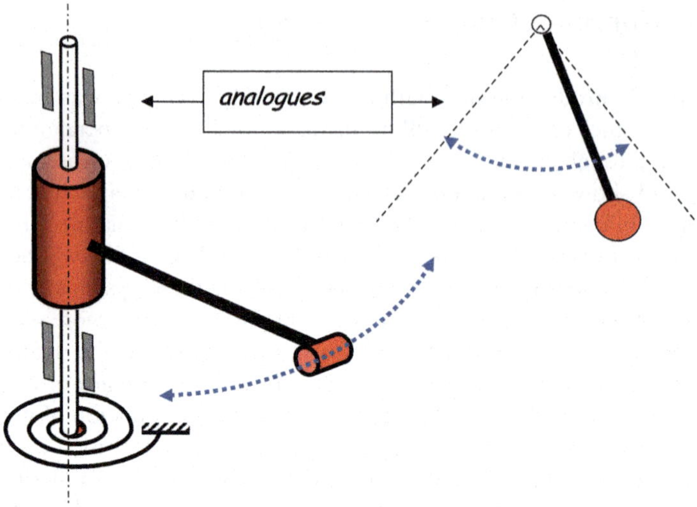

Fig. 7.1 The balance mechanism in a mechanical wrist watch and the clock pendulum are mathematically analoguous

special purpose for which they are set up. For some purposes theoretical models are brilliant tools, for other purposes they are useless: theoretical models should not be confused with the real world!

A system that at first sight is different again, is sketched in the next figure. It is a mass on a rod swinging up and down around a horizontal bearing, again including a mechanical spring which tries to drive the mass back to an equilibrium position.

Within the framework of a theoretical model it is no problem to assume that this contraption is placed on board a space ship so that gravity is zero. Under such circumstances the equation of motion describing the system appears to be identical to that describing the balance of a watch. It is just another simple oscillator with simple sinusoidal behaviour. Actually, this may be seen easily without setting up any mathematical equations, because we can obtain the system by just "switching off" gravity, tilting the oscillator of Fig. 7.2 over 90°, and finally noting that the schematic for the spring is different but essentially equivalent to the torsional spring in Fig. 7.1.

Fig. 7.2 Another analogue of the pendulum

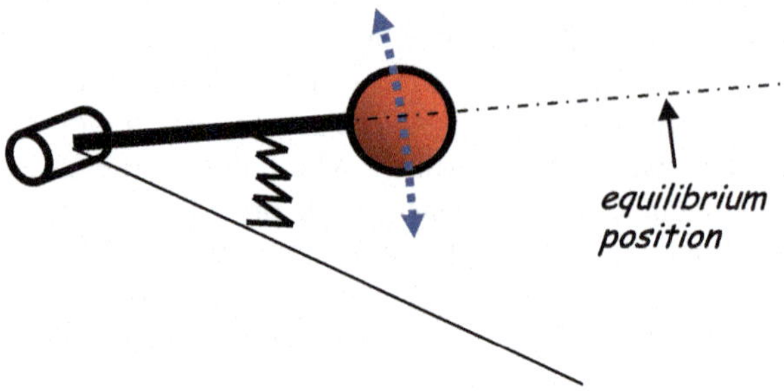

7.5 A Mechanical Analogue of the Pulsating Droplet Moving Inside a Potential Well

And let us now (at last!) come back to the motion of a droplet of charge, translating inside a potential well, and at the same time undergoing the deformations in length which were earlier baptised "pulsations". After having determined the equations of motion for these two degrees of freedom we can try to find analogues. This may help the understanding of the resulting motions of the droplet, at least if we are able to find an analogous situation for which we have an almost intuitive feel how it will behave. And fortunately, that appears possible. The next figure is the *mechanical analogue* of a pulsating droplet of charge moving inside a potential well (Fig. 7.3).

This mechanical analogue at first sight may appear pretty complicated, but on closer inspection it is nothing else than a combination of the previous two figures. The balance of the watch is modified, in such a way that it no longer has a fixed mass at the end of the radial pin, but instead has a hinge at this place so that the mass at the end of a second pin can move up and down. Compare it with the elbow joint in the human arm, with a mass kept in the hand. Because of the freedom to rotate around a vertical axis, the whole thing might be compared with a dancer standing on spitzen and performing a pirouette. On the other hand, one will not often see dancers with springs in the elbow joint and another—soft—spiral spring restricting their pirouette to just a few turns before it changes direction. No comparison is ever perfect, alas! The comparison with the dancer is useful, though, to explain some aspects of the dynamics of the analogon.

Note on terminology used in the remainder (Fig. 7.4).

The horizontal circular arc described by the "elbow joint" represents the excursions (the translation) of the droplet inside the potential well. For the corresponding motion of the mass, in a horizontal plane, the term *swaying* will be used. One single pass of the swaying motion from stop to stop will be called a *sweep*. If the torsional spring is present, a sweep in clockwise direction (viewed from above) is followed by a next sweep back in the anti-clockwise direction, etc. In principle such a sweep may cover several complete turns around the vertical shaft.

The "up and down motion" of the lower arm will be called the *nodding* of the motion of the mass. The mathematical analysis of the appendix shows that, in order to be a perfect analogue of our droplet of charge, the angle between the "lower and upper arm" must be chosen as 33.7° in the equilibrium position. Therefore, it is strictly speaking incorrect to use the term "up and down" motion of the mass, because the motion is skew: when going upwards the mass also moves closer to the vertical axis, and in reverse: going down implies a larger distance from the vertical axis.

The terminology used in the remainder of this book is clarified in the following Fig. 7.4.

End of note on terminology.

As already remarked, the analogon is just a combination of the two oscillators shown in the two previous Figs. 7.1 and 7.2. If the motion of the lower arm is

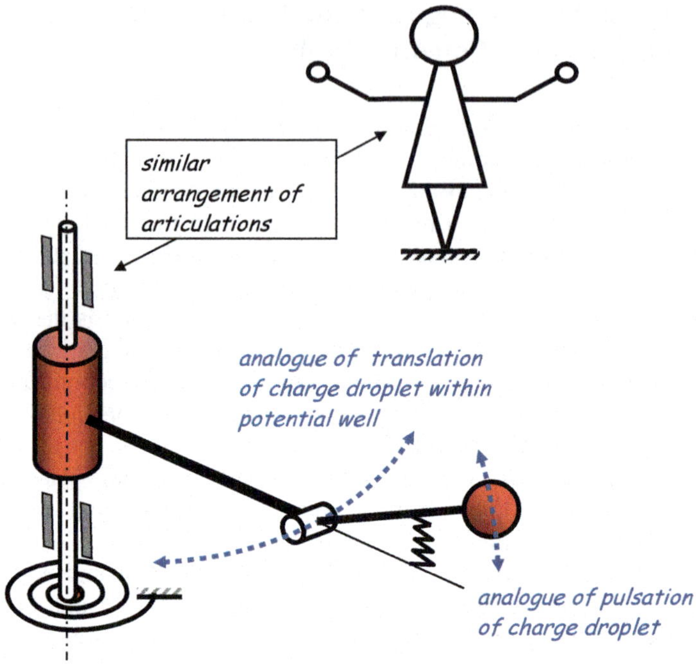

Fig. 7.3 The mechanical analogue of the pulsating droplet of charge moving inside a potential well

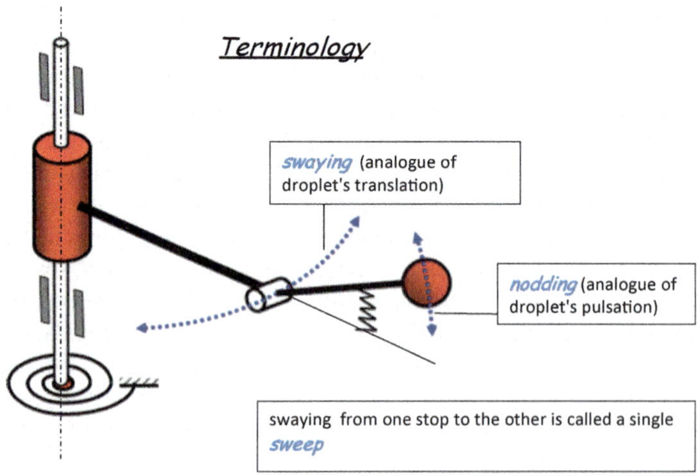

Fig. 7.4 The terms swaying, nodding, and sweep describing the motion

locked, what is essentially left is not different from the diagram of Fig. 7.1 for the watch balance. The resulting motion has already been mentioned, it is simply the motion of any simple oscillator: a smooth sinusoidal motion.

The same is true if we lock the whirling motion, i.e. if the dancer stops the pirouette and keeps the feet firmly on the ground. What we are then left with is a stationary support having a hinge which allows a mass to swing up and down under the influence of a spring (gravity effects are assumed to have been "switched off"). It is an arrangement essentially identical to the one sketched in Fig. 7.2. And again, the resulting motion is simple, it is a smooth sinusoidal motion.

Transferring these conclusions to the analogous situation of a droplet of charge: if there would—hypothetically—be *no* influence of the translation on the pulsation and if *neither* an influence the other way round would exist, then the translation as well as the pulsation would be rather uninteresting, just regular sinusoidal motions like any old oscillator.

However, it is important to emphasize that the frequency of the two oscillations is very different. In general the pulsation motion of a collection of charges is a high frequency phenomenon, in the macroscopic world known as so-called "plasma vibrations". On the other hand the translation oscillation inside a potential well has in comparison a very much lower frequency.

Therefore, in the mechanical analogon the spiral spring has to be chosen soft, so that an easy-going swaying to and fro results. The swaying motion may involve several complete turns, before the motion stops and reverses its direction. In contrast, the elbow spring will be very stiff, and the up and down motion of the hand (or better: the nodding) would be more comparable to the vibration of the leg of a tuning fork after having been struck. To put it differently: the swaying would be seen by our eyes whereas the up and down vibration would be heard by our ears and is too fast to be seen.

Building the mechanical analogon in the form of actual hardware and putting it in the laboratory for further study is subject to a few restrictions. As said before, one would be able to visually inspect the swaying motion. One would be able, using a stopwatch, to quantify the frequency of the swaying back and forth. Special kinds of apparatus, for instance a stroboscope or an oscilloscope, could be used for measuring the instantaneous velocity, even though the swaying speed is varying all the time. For people who are not sure what sort of instruments the stroboscope or oscilloscope are: do not bother, at this point it suffices to say that they are instruments to refine our visual observations, and easily quantify them (in both names the syllable "-scope" occurs, which is derived from the Greek word for looking; compare better known scientific tools like the microscope and telescope).

One would think that the same kinds of observation and quantifying would be available for the nodding motion. But in this case this is not so, the use of such equipment must here be *considered to be forbidden*. The reason is, that the nodding motion is the analogue of the pulsation of the droplet of charge. In the case of the actual droplet the frequency of the pulsation is extremely high, so high in fact that it lies outside the human possibilities to ever "see"or measure it. This has been pointed out earlier, in Sect. 6.4. To be a good analogon, we must then put a restriction on the kind of measurements than can be done on it in the laboratory: we must require that in the laboratory the frequency of the nodding cannot be measured directly. Neither can the nodding speed be detected by any means. Actually, in the laboratory the nodding would not be seen by the naked eye, but of course it would not be difficult to measure it, for instance again using a stroboscope or using acoustical equipment to measure the tone of the sound spread around the set-up. But, in order to respect the analogy, these kinds of measurements must be forbidden, the equipment needed for it must be banned from our laboratory!

In the next chapter an inspection of the behaviour of our laboratory contraption will begin. The investigation how to get from the analogon as much information as possible about the behaviour of a droplet of charge will then be continued throughout the Chaps. 8–10.

The present chapter ends with a textbox for the real enthousiasts, showing the form of the equations of motion in mathematical language. Once again: most readers will prefer to skip this. They will not miss anything essential for the understanding of the remainder of this book, the formulae are just written down for the benefit of the more curious, mathematically inclined readers.

Textbox: The Equations of Motion

The equations are given without derivation, for further details see the appendix.

Notations:

The electromagnetic mass m_{em} is given by $m_{em} = \frac{1}{c^2}\frac{q^2}{4\pi\varepsilon_0}\frac{\alpha}{s}$, where:

$c =$	velocity of light
$q =$	charge of the droplet
$s =$	elongation of the droplet
$\alpha =$	factor depending on nondimensional shape of the charge distribution
$\varepsilon_0 =$	vacuum permittivity
m_{em}	is variable through the elongation $s = s_e + \Delta s(t)$ where s_e is the equilibrium value

The equation of motion for translation is

$\dot{v} + \omega_z^2 \cdot z = 2 \cdot \frac{d}{dt}\left(v \cdot \frac{\Delta s}{s_e}\right)$ (excluding radiation resistance)

where $v =$ translation velocity in Z-direction, $\dot{v} =$ acceleration. The angular frequency ω_z is the natural frequency in the absence of coupling with the pulsation.

The inertia in s—direction (the generalized mass) is m^*_{em}, related to m_{em} by $m^*_{em} = \frac{m_{em}}{\sigma}$.

The pulsation equation is linearised in $\Delta s/s_e$ and reads:

$$\frac{\Delta \ddot{s}}{s_e} + \omega_s^2 \cdot \frac{\Delta s}{s_e} = \sigma \frac{v^2}{s_e^2} \cdot \left[2\frac{\Delta s}{s_e} - 1 \right]$$

where the angular velocity ω_s is the natural frequency of the pulsation in the absence of coupling effects. Its value is $\omega_s = \frac{c}{s_e}\sqrt{\sigma} \cdot \sqrt{\mu.\tau}$ where

$\mu =$ ratio of mass of muon w.r.t. mass of electron
$\tau =$ ratio of mass of tau particle w.r.t. mass of electron

Notes:

(1) In the form stated above a parabolic potential well is assumed.
(2) Why are there no terms with m_{bare}, the "mechanical mass"? The mechanical mass was related to the electromagnetic mass by "Yaghjian's hypothesis" (A.D.Yaghjian: Relativistic Dynamics of a Charged Sphere, Springer 2005). According to this hypothesis $m_{bare} \cdot c^2 = U_{es} - m_{em} \cdot c^2$ where U_{es} is the electrostatic energy, i.e. the energy needed to collect all the charge elements into the droplet. m_{bare} turns out to be negative, i.e. it is a mass deficit associated with binding energy by the cohesion forces.
(3) How do the ratio's μ and τ enter the equations? The pulsation equation is a linearised perturbation equation, and has to be accompanied by an equation determining the equilibrium around which the variations $\Delta s/s_e$ occur. This appears to be a third-order algebraic equation in s, which allows three different, real solutions, not necessarily all of them stable solutions. These solutions correspond with three different values of m_{em}, and lead to three values of the total mass. In the case of the electron it is assumed that the three possible droplet masses correspond to the electron, the muon and the tau particle.
(4) The particular form of the equations above is the simplest arrangement of terms, but not necessarily the most convenient. In the formula appendix several other forms are used.

Chapter 8
First Glimpses of Quantum Behaviour: Matter Waves

8.1 The Behaviour of the Analogon if There Is No Restraining Spiral Spring

Let us now study a few properties of the mechanical analogon, whilst carefully respecting the limitations just pointed out. The first step during a study of the contraption built up in our laboratory is, to leave out the spiral spring. Once we have given a whirling motion to the system, it will go on and on rotating in one direction without being stopped by a restraining spring. The system now is fully comparable with a dancer performing a pirouette. Remembering that it is an analogon of the droplet of charge, this means that we now, during this first step, are considering a droplet *outside* a potential well, i.e. a free-flying droplet.

The path of the rotating mass will be different, depending on how large the swaying speed is. This is schematically shown in Fig. 8.1.

If the swaying speed is very small, centrifugal forces are not strong enough to appreciably compress the stiff elbow spring. In the nodding direction, the mass will practically not become displaced from the equilibrium position and the path traced out by the mass whilst rotating is given by the red dashed line.

When we have given a somewhat larger swaying speed to the mass, the centrifugal force is larger, so that the mass is moved outwards and down, against the opposing action of the elbow spring. The path of the mass becomes as shown by the blue dashed line.

And finally, if the swaying speed would be very large, the action of the spring would be practically negligible compared with the strong centrifugal forces. The mass would then trace out a path close to the "equator plane", as symbolised by the drawn black line.

Recalling the analogy between the nodding and the elongation of the droplet of charge, the above is just the same as was discussed in Chap. 5. In Figs. 5.8 and 5.9 it was shown by the red dashed contours how the elongation of the droplet is

T. van Holten, *The Atomic World Spooky? It Ain't Necessarily So!*
DOI 10.2991/978-94-6239-234-2_8

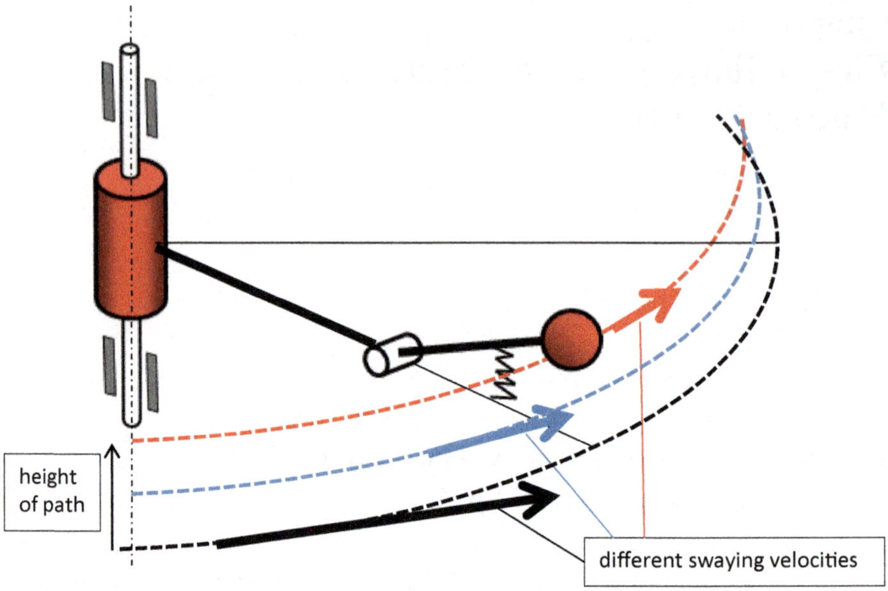

Fig. 8.1 No spiral spring. The height of the path traced out by the mass depends on the swaying speed, due to the centrifugal force on it

influenced by the speed of translation. The droplet shortens, the larger the velocity is (not considering any other squeezing effects).

As a next step, the mass is kicked and acquires a nodding oscillation (which is the analogue of the pulsation), on top of its rotation around the vertical shaft. We assume that the swaying speed is medium large, so that the resulting motion is given by the red curve in Fig. 8.2.

However, the oscillating red curve in Fig. 8.2 shows the motion just in principle, as if the nodding could be seen. But actually, it cannot be seen (as discussed earlier) because the nodding is too fast to be observed by the naked eye. And remember that special laboratory equipment to make it artificially visible was banned.

Figure 8.3 shows what will be observed when looking at the mass when it passes us: just a reddish blur, comparable with a fast turning bicycle wheel where the individual spokes cannot be discerned. The individual nods cannot be seen, and neither can their frequency be measured.

Earlier it was suggested that in the analogon the nodding motion perhaps can be heard, even if it is too fast to be seen. But this is probably too optimistic too (unless you have the hearing acuity of a dog): taking into account the relative frequencies such as occur in the case of the droplet of charge, sound emitted by the nodding would be ultra-sound, undetectable by the human ear.

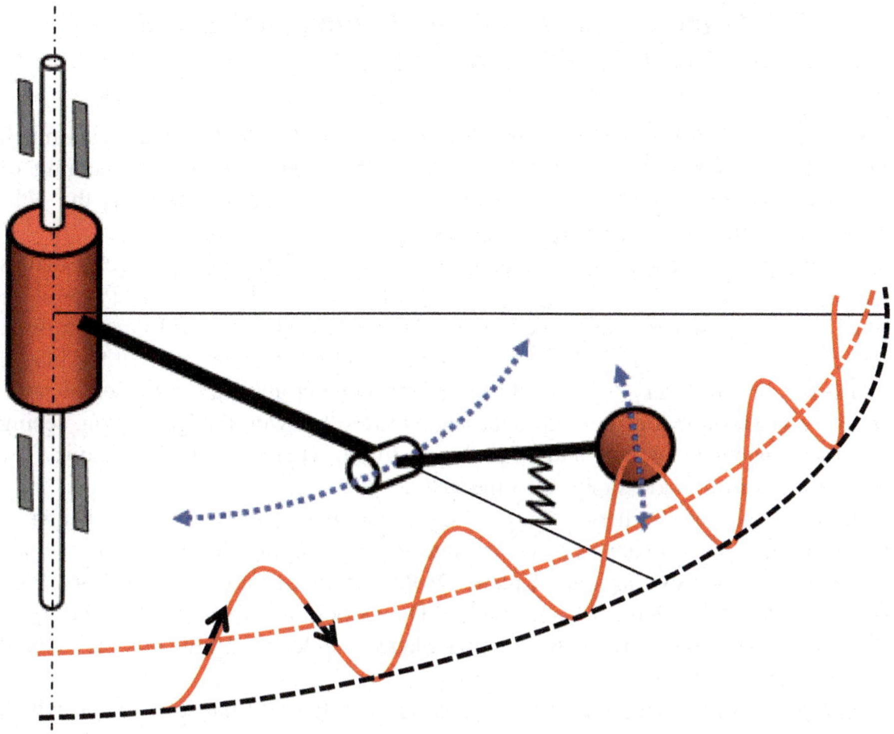

Fig. 8.2 Combined swaying and nodding (the analogues of translating and pulsating)

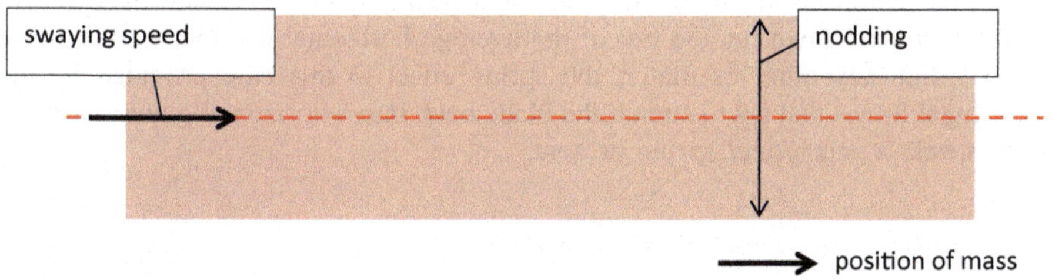

Fig. 8.3 View of the nodding mass passing an onlooker, taking into account the high frequency of the nodding

Fortunately there is a trick to get more information about the nodding than just the hazy picture of Fig. 8.3. In Chap. 6, Sect. 6.4, a similar trick was performed when dealing with the droplet of charge itself. It is a clever trick that does *not* violate the restrictions imposed by the banning of the normal laboratory toolbox. The procedure is simple, but requires a rather extensive explanation, to be given in the next two sections.

8.2 Why Is the Frequency of the Nodding Influenced by the Rotational (Swaying) Speed?

Proceeding with the explanation of the trick that can be used to extract as much information as possible from the laboratory experiment (despite the banning of proper instrumentation), we must make a large diversion. Surprisingly, the helicopter rotor may give a clue. It is sketched in Fig. 8.4.

In the sketch the rotorhead of a helicopter is shown with one of the rotor blades attached, of which schematically only a stub is shown. The real blade would be much longer, obviously. Such a helicopter blade is not rigidly fixed to the rotorhub. Quite the reverse, it is hinged in every direction so that, considering that the blades themselves are also very flexible, during flight they act almost like pieces of rope. Movies have been made showing helicopter blades that would frighten you, seeing how these blades bend, curl and twist during flight, whereas they at the same time are supposed to lift you safely into the air!

That they are able to lift the heavy body of the helicopter without the danger of collapsing is thanks to the large centrifugal forces, keeping the blades on average stretched in an almost horizontal plane. If the rotor does not rotate, do not even think of lifting the helicopter by its blades! They would just flap up vertically or—if it has stops to prevent this—the flexible blades would bend upwards and would surely break.

The centrifugal stretching when the rotor is at full speed has a by-effect, which shows up when we would give an upward kick to the rotating blade. I do not propose to literally kick a helicopter blade, but you can imagine that turbulence in the air, going through the rotor, will have that same effect by sudden changes of the aerodynamic forces. What we then see as a result is, that the blade is going to oscillate up and down, in and out of the average horizontal disc plane. The blade behaves like any other oscillator, the spring effect in this case provided by the centrifugal forces that try to stretch the blade back into a horizontal plane. There is, so to speak, a *centrifugal spring* present.

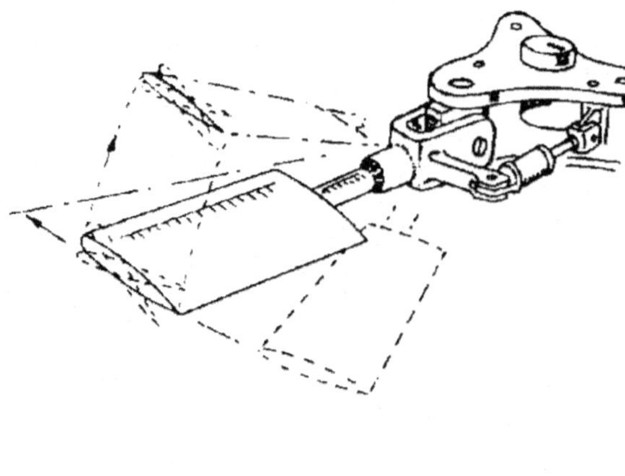

Fig. 8.4 Helicopter rotorhead. Hinges allow the blades to freely move up and down (so-called "flapping") and to sway back and forth (so-called "lead-lag" motion)

If the blade itself already had a mechanical spring in its hinges (there are many rotorheads in existence where the pure hinge is replaced by a soft bending element), the centrifugal stiffening adds a lot to the total spring stiffness, and often dominates it completely. One of the consequences of a stiffer spring in any oscillator is, that the oscillation becomes faster. In other words, a stiffer spring will cause a higher frequency of the oscillator. In a helicopter type where the total spring consists of two contributions, a mechanical part and a centrifugal part, we will see that a higher rotational speed (thus increasing the relative importance of the centrifugal stiffening) increases the frequency of the "flapping" oscillation. The influence rotational speed has on the oscillations of a helicopter blade is something that has to be calculated carefully by helicopter designers. It is vitally important to do this, to avoid that flapping oscillations give rise to resonances with other components of the aircraft, or that they even cause downright instabilities with complete destruction of the rotor as a consequence!

Returning to our mechanical analogon, in this arrangement an angle of some 33° between the upper and lower arm was "built in", around which an oscillation (the nodding) will take place. Due to this built-in angle it is not so easy to see that in this case the centrifugal stiffening effect is also working. But be assured that it is, the formulae in the mathematical appendix prove so. The spring in the elbow joint is assisted by the centrifugal stiffening effect, with the result that the nodding frequency increases if the rotational speed (the swaying speed) is higher.

The effect of the centrifugal stiffening can be put in the form of a simple formula. The expression will be shown here—despite the promise to use formulae as little as possible—because it may then be easier to see the connection with earlier chapters, in particular Chap. 6. Let us denote the frequency of the nodding by the symbol f, meaning that there are a number of f nods in every second. The nodding frequency increases when the swaying velocity (i.e. the rotational speed and centrifugal force on the mass) becomes larger. Let us now use the symbol f_0 for the nodding frequency corresponding to no rotation at all, i.e. when the swaying velocity $v_{swaying}$ is zero:

$$f_0 \leftrightarrow v_{swaying} = 0 \quad (\textit{definition of} \text{ ''zero - speed frequency''} f_0)$$

For simplicity this value of the nodding frequency will be shortly called the "zero-speed frequency". Now, if there is rotation and therefore a non-zero swaying velocity, the nodding frequency will become slightly larger (due to the centrifugal stiffening effect), or symbolically written:

$$f > f_0$$

To become more specific: the actual frequency at speed, compared with the "zero-speed frequency" is given by the relation

$$f = f_0 \left(1 + k \cdot {v_{swaying}}^2\right)$$

As you see, when the swaying velocity is zero, the formula tells that the nodding frequency reduces to the "zero-speed frequency". When there is a non-zero swaying velocity, the frequency becomes higher, and the amount by which it increases depends in a quadratic way on the swaying velocity. The fact that the *square* of the velocity enters into the formula should not be a surprise: you may remember from school that centrifugal forces (the forces that are here responsible for the change in frequency) are proportional to the square of the velocity. One can put all this in different wordings, by saying that the *shift* of frequency $(f - f_0)$ is proportional to the square of the swaying velocity:

$$(\textit{frequency shift due to swaying}) \div {v_{swaying}}^2 \quad (\textit{analogon})$$

And here you can once again conclude that our analogon faithfully reproduces the behaviour of the droplet of charge. In Chap. 6, Sect. 6.4. "Vibrations around the equilibrium" we were discussing the actual droplet of charge, with its pulsation and translation. It was stated that the pulsation frequency was influenced by the velocity of translation, and that the *shift* of the pulsation frequency depended as follows on the velocity:

$$(\textit{frequency shift due to speed}) \div {V_{transl}}^2 \quad (\textit{droplet of charge})$$

which is indeed the same phenomenon as we have now met in the analogon. Obviously, physically there is no similarity between a droplet of charge and a helicopter. This shows the remarkable power of mathematics, which is able to uncover such strange analogies.

8.3 The Analogue of "De Broglie's Matter Waves"

Let us continue the explanation of the trick how to wring more information from the hazy observation depicted in Fig. 8.3. For this purpose we ask the technicians of the laboratory to build a replica of the analogon shown in Fig. 7.3, and to install it near the analogon we already had. Now the following experiment can be done, schematically pictured in the next Fig. 8.5.

In the second analogon the swaying motion is blocked, so that here the mass is only performing a nodding motion. The frequency of the nodding is then equal to the above defined "zero-speed frequency". This is a slightly smaller frequency than the nodding frequency of the "machine" on the left, which performs both a nodding motion as well as a swaying motion.

Although it is outside the hearing range detectable by the human ear, there will be sound transmission from the analogon on the right to the one on the left (and vice versa). The combined sound field is a field caused by two different sound sources with slightly different frequencies, resulting in a so-called beat phenomenon. What "beats" are has been explained rather extensively in Chap. 1, where

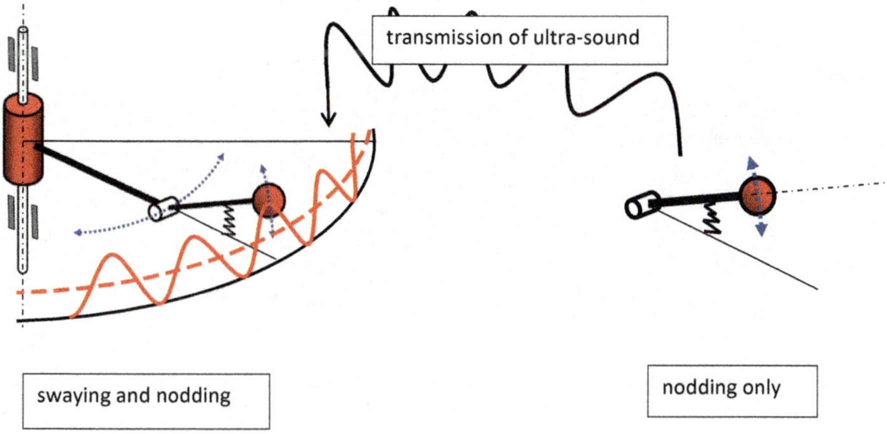

Fig. 8.5 Interference experiment. The ultra-sound of an analogon with blocked swaying is transmitted to the full analogon with non-zero swaying velocity

examples were given ranging from piano tuners to multi-engined aircraft. The reader is referred to this chapter, if he/she finds it useful to get some memory refreshing.

In the case at hand there is a new phenomenon as well. Sound waves are able to excite vibrations in other things, at some distance from the source where they come from. This is obvious, after all this is the essence of hearing: sound waves in the air let the eardrum vibrate, so that the stimuli on the nerve system can transmit the signal to the brain where it will be further processed and interpreted (unless something goes wrong in this chain of events, a failure which the author is often accused of by his children). Let us assume that in our laboratory experiment all the elements of the transmission chain are intact, so that the sound from the analogon on the right is correctly transmitted without meeting any obstructions. The result will be that the nodding and swaying mass in the left analogon receives an additional excitation, with a frequency which is slightly different from its own.

To make a long story short: it is not only the—inaudible—sound field that contains beats, the *nodding motion itself* will now display a beat phenomenon. The vague hazy picture of the nodding (Fig. 8.3), which was the only—and rather minimal—information that could be obtained from a single analogon-machine, will be reshaped in the presence of a second machine and in this way yields somewhat more—and significant—information. It is shown in Fig. 8.6. It may be well to emphasize that this new information is *not* obtained by directly measuring the nodding. The detective work that has produced the additional information did not violate any of the "laws" which had to be laid down in view of the fundamental "observability" limitations and the banning of certain toolboxes.

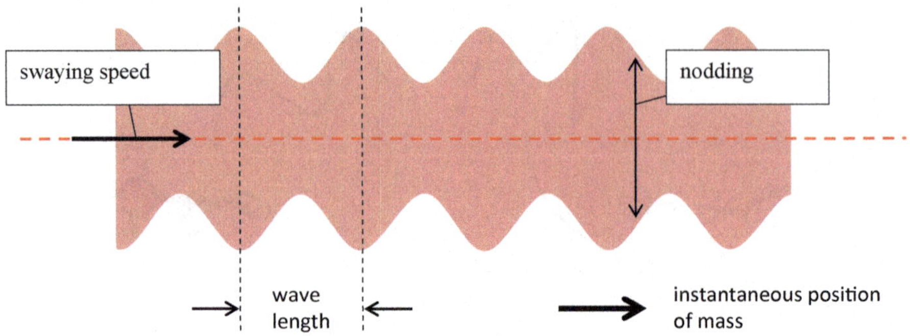

Fig. 8.6 Beats in the nodding, due to interference with "zero-speed frequency"

Considering Fig. 8.6, the individual nods are still invisible, of course. Still no nodding frequency can be measured, nor the instantaneous nodding velocity. However, what is visible and can be measured is the wavelength of the beats. And these wavelengths have a remarkable property: they shrink when the mass travels (is swaying) faster!

I am not going to repeat the long story in Chap. 1 about a truck with a hard-working man unloading pylons, it took too many pages already. But I hope you remember the first of the pictures, showing that the higher the speed of the truck, the longer the wavelength (Fig. 1.22, uppermost line). This is intuitively what we expect when a vibrating thing is moving past us: the distance travelled between any two of the vibration cycles will increase the faster this thing is moving (do not confuse this with the well-known Doppler-effect, which deals with the *frequency*, and certainly not with "wavelength" as it is defined here). If we further think about it, this first intuitive expectation may be wrong, as is shown in the other pictures of Fig. 1.22. The wavelength as defined in the figure also depends on how constant (or variable) the frequency is: if the frequency of putting the pylons down on the road changes *together* with the speed of the truck, then the distance between the pylons (i.e. the wavelength) may be anything. In the—very special—case that the frequency is proportional to the square of the truck's speed (Fig. 1.22 cont'd) then we find the result

$$(wavelength) \div \frac{1}{(speed)} \qquad (truck\ and\ operator)$$

i.e. the wavelength becomes *smaller*, the higher the speed of the truck! And the opposite is also true, of course. So that you do not have to page back, the relevant figure in Chap. 1 symbolising this result is here copied:

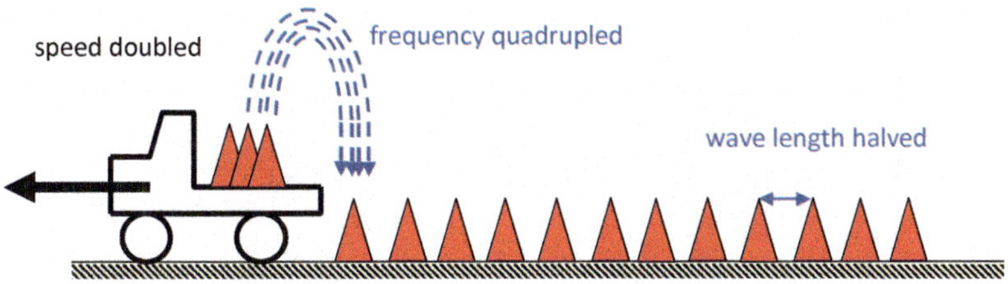

Fig. 1.22 (continued): the wavelength if the frequency depends quadratically on the speed

Now, this very special situation where the frequency is proportional to the *square* of the speed is exactly what we have in Fig. 8.6. Recall that the fluctuations of the width of the reddish area in this figure represent *beats*. Beats are caused by interference between two vibrations, and their frequency (i.e. the number of beats per second) is equal to the frequency *difference* of the two sources of vibration. The beat frequency in Fig. 8.6 is thus equal to the difference in the nodding frequencies of the two contraptions on the left and right side in the earlier figure Fig. 8.5. In shorter wordings:

$$(beat\,frequency) = (shift\ of\ nodding\,frequency\ due\ to\ speed)$$

This may be combined with the already mentioned proportionality:

$$(frequency\ shift\ due\ to\ speed) \div v_{swaying}{}^{2}$$

which gives:

$$(beat\,frequency) \div v_{swaying}{}^{2}$$

The latter relation tells us that we have in Figs. 8.5 and 8.6 exactly the same situation as in the just repeated Fig. 1.22 (cont'd). Consequently, again in short wordings:

$$(wavelength\ between\ consecutive\ beats) \div \frac{1}{(swaying\ speed)} \qquad (analogon)$$

The conclusion is of sufficient importance to be expressed in yet another figure, Fig. 8.7.

You may notice the remarkable similarity with De Broglie's formula concerning the matter waves accompanying elementary particles:

$$(wavelength\ of\ matter\ waves) \div \frac{1}{(speed\ of\ particle)} \qquad (De\ Broglie)$$

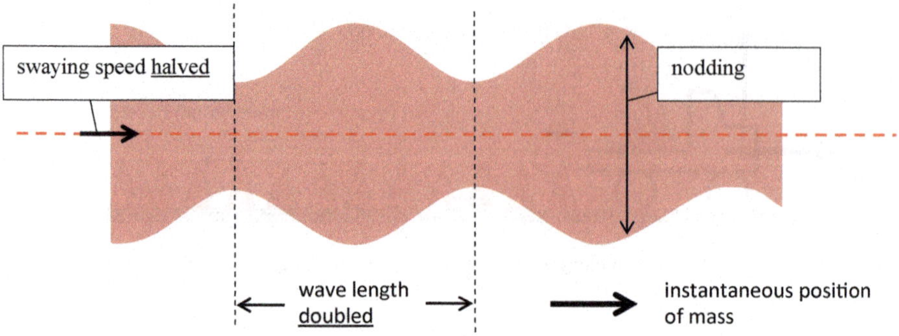

Fig. 8.7 Beats in the nodding, swaying speed halved compared with Fig. 8.6

It may be so that there is a clear similarity between the beat phenomenon in our laboratory set-up (Fig. 8.5) and the matter waves of quantum mechanics, this however does not mean that these two things are identical. In the above we just considered proportionalities, so that only more definite conclusions can be drawn if these expressions are first converted into real equality equations. This is done by writing the above formulae with appropriate proportionality constants. For an identification of the beats with matter waves one must require that in the following two expressions the proportionality constants K and h (i.e. Planck's constant) are the same:

$$(wavelength\ between\ beats) = \frac{K}{(mass) \times (swaying\ speed)} \quad (analogon)$$

$$(wavelength\ of\ matter\ waves) = \frac{h}{(mass) \times (speed\ of\ particle)} \quad (De\ Broglie)$$

In general, the constant K of the analogon is *not* the same as Planck's constant h. Naturally, because the mechanical analogon just represents the equations of motion of an arbitrarily large droplet of charge. The factors in the equations of motion have a different quantitative value for differently sized droplets. Now, the surprising result is that if the droplet is assumed to have the unit charge (i.e. the charge of one electron), then K becomes the same as Planck's constant h! Once again, it would be a waste of paper to repeat everything that has already been written in Chap. 1. Therefore, if you wish you can page back to this first chapter to find a separate text box where the complete formulae are shown. It is there found that K and h are—considering the approximations in the droplet theory—*practically* the same, with an "error" of 1.5 %.

The results thus found suggest the following conclusion:

The beats shown in Figs. 8.6 *and* 8.7 *are the analogues of matter waves.*

8.4 A Sack Race Caused by Yet Another Coupling Effect

In the story until now we met several coupling effects between the swaying motion and the nodding motion. Figure 8.1 showed how the swaying velocity influences the *average* nodding position of the mass in the analogon, through the action of centrifugal force. Furthermore, the comparison with a helicopter rotor (Fig. 8.4) showed that the *frequency* of the nodding is also influenced by centrifugal forces, so that there is a shift of frequency due to swaying speed.

Both these couplings work in the direction from the swaying to the nodding, i.e. the swaying influences the nodding. You may wonder if there is not any coupling the other way round, viz. from nodding to swaying. There is, and it is a very clear one. This new type of coupling was already tacitly suggested by the insert in Fig. 7.3 of a ballet dancer.

Evidently, the "general arrangement" of a dancer is very similar to the mechanical analogon of Fig. 7.3, except for the two springs, the spiral spring and the "elbow spring".

If the dancer performs a pirouette, i.e. when she is whirling at large speed around her top-axis ("she" and "her" being used because of the evident female schematic in the figure, not because the mechanics involved are gender-specific!) she can control the whirling speed by stretching her arms out or by retracting them. Stretching her arms in a radial direction will slow down the whirling. On the other hand, the whirling will speed up by flexing her arms, bringing the hands closer to the body and thus reducing her inertia around the vertical axis. These effects are even more pronounced if she carries weights in her hands, for instance lead balls.

This ballet trick, perhaps somewhat superfluously sketched in Fig. 8.8, is well-known of course. Something not often or never done on stage is, to rapidly spread and retract the arms in turn (sometimes it is done with the feet). This would result in an irregular whirling motion. What would be seen are brief moments of a high speed rapidly alternating with a lower whirling speed, and this repeated as long as the dancer can perform the rather fatiguing exercise with her arms (remember the presence of the lead balls, and do realize that the centrifugal forces can be quite large!).

A similar effect will happen in our mechanical analogon. Remember that in this chapter the spiral spring was removed, so that a full round and round whirling motion is allowed, like the dancer is performing during a pirouette. The elbow spring, however, has been kept in the analogon. If now the "hand-held mass" receives a push so that it is going to rapidly move up and down under the influence

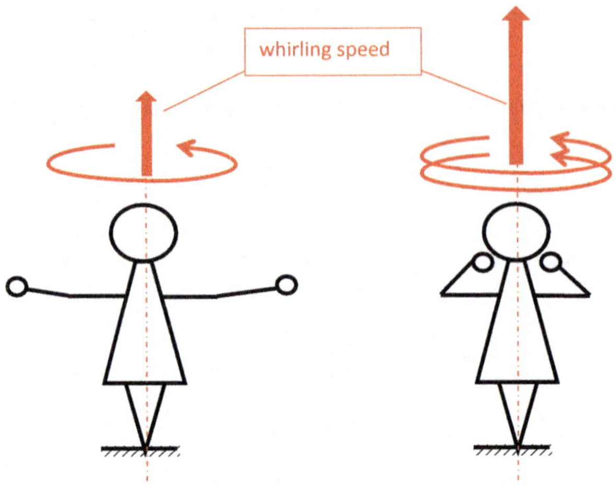

Fig. 8.8 Reducing the inertia around the top axis increases the whirling speed

of the spring in the elbow hinge, the rotation around the top axis becomes irregular, rapidly alternating between a higher and lower whirling speed. Remember that what is sloppily called an "up and down motion" is in reality skewed (around the equilibrium position of $\sim 33°$), so that the hand is at the same time moving closer to and further away from the body, thus changing the inertia around the top axis. This is exactly what also happens in the case of flexing the arms by the dancer, although in the analogon the nodding motion is less pronounced than sketched in Fig. 8.8. Even so, the nodding oscillation will cause "ripples" in the swaying speed.

One would like to picture this effect in a diagram like Fig. 8.2, but this is difficult. In this case it is more enlightening to go back to the droplet itself. The swaying is the analogue of the translation velocity of the droplet, and the nodding is the representation of the pulsation of the droplet. When there is a nodding oscillation which gives "ripples" in the swaying velocity, this process is the analogue of what is shown in Fig. 8.9.

The smooth translation is perturbed, and is "contaminated" by the much faster pulsation: the droplet passes through a sequence of small, rapid jumps whilst moving from left to right. In Chap. 1 the motion was compared with a sack race.

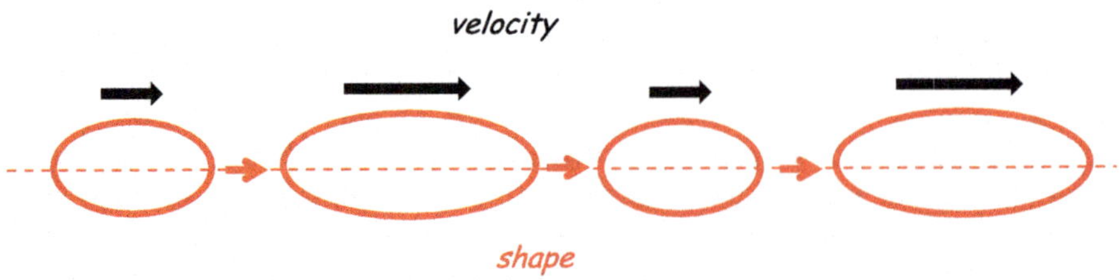

Fig. 8.9 Pulsation and velocity fluctuations of the droplet

Another possible comparison might be to think of the scrolling wheel of certain types of computer mouses. Large scrolling sweeps can be made by turning it smoothly, but the actual advancing of the small wheel is not really smooth, it is the result of a series of small "hops", even heard as separate "ticks".

In earlier chapters seemingly different explanations were given why such a sack race occurs, but at this point one can see that the different ways to describe the phenomenon all come down to the same basic cause. In Chap. 1 it was explained that pulsation is accompanied by variations of the mass of the droplet. It is the consequence of electro-magnetic mass, which is smaller or larger depending on the elongation of the droplet. The text continued rather superficially: "When the elongation of the droplet is large, its mass will be small, and vice versa. This is bound to have an effect on the velocity of translation, of course". True enough, but not very enlightening. Now that we have set up an analogon, the matter can be clarified somewhat further: the mass of the droplet corresponds with the inertia of the analogon around the top-axis. Variable mass of the droplet corresponds with the variable inertia due to the "spreading and retracting of the dancer's arms" (i.e. the nodding motion).

The phenomenon of the variable whirling speed of a dancer due to arm (as well as leg) movements was in the text above taken for granted, because almost everybody has seen it with his/her own eyes, either on actual stage or on the TV-screen. This kind of observation of course does not give an explanation *why* it occurs, from the point of view of the theory of mechanics. A deeper level of explanation was attempted in Chap. 5, dealing with the electromagnetic self forces of a droplet of charge.

In Chap. 5 something was said about so-called Coriolis forces, causing for instance the vortices in the atmosphere seen on meteo maps as large "whirl pools". The Coriolis forces were held responsible for this phenomenon, entailing that the masses of air and clouds which close in on the centre of a depression attain a gradually increasing whirling velocity. In other words, closer to the centre of a depression the wind speeds are higher (except in the centre itself, the "eye" of the storm, where the wind may be calm for another reason that will not be considered here). As you see, there is a striking similarity with the dancer: moving the body parts closer to the turning centre gives the body a higher whirling speed. This similarity is no coincidence, for both the dancer and the air masses in the atmosphere are subject to the same forces: Coriolis forces. Perhaps the reader would like to descend to the next deeper level of explanation, and would want to know where these Coriolis effects come from. Sorry, these curious readers must be referred to traditional school books on mechanics, this would go too far for the present discussion!

For now, the following can be ascertained. Electromagnetic self forces act on a droplet of charge which may be considered as "Coriolis-like forces". They couple the translation of a droplet to its pulsation. The mechanical analogon replicates this coupling faithfully, coupling the swaying motion to the nodding by Coriolis forces. The analogon thus gives a rather vivid visualisation why the droplet of charge is going to move in a "sack-race like" manner.

8.5 In Case of Wild Motion: Devastating Effect of the Couplings

Everything that was said up to this point about the behaviour of the analogon is only valid if the motions are not too wild. It has tacitly been assumed that kicking the analogon to start its "wobbling around" was done with some care, and did not involve really brute force. What would happen if we are less gentle?

Still all the effects mentioned above are present, in principle working out in the same way as described above. The difference is, that coupling effects are going to "sing round". To give an example: in the mechanical analogon the swaying motion was contaminated by a high frequency due to the fast nodding, the smooth easy-going pace of the swaying thereby being transformed into a "sack race". Conversely, all these "jumps" in the swaying cause a fast variation of centrifugal forces if the jumps are large enough. There is thus a feed-back to the nodding by the same kind of process as shown in Fig. 8.1, but in this case with a fast variation of the swaying speed.

It has something of a snake eating its own tail. The original "pure tone" of the nodding vibrations is now contaminated by the *ripples* in the swaying motion which were started by the nodding motion in the first place. And all this is carried over again and again, back to the swaying motion, back to the nodding, etc., etc.

If you have lost track by now, do not worry: that was the intention. The result is indeed an undescribable mess! Of the nice, smooth and regular sinusoidal motions where we started from, nothing nice and smooth is left after the coupling effects and feed-backs have done their devastating work. In the case of more gentle motions, this " singing round" of the disturbances is not significant, the feed-backs are too weak to play havoc. For gentle motions we keep the smooth sinusoidal variations of the nodding and swaying from which we start. In contrast, large motions run out of hand.

An acoustical analogue of the behaviour of our mechanical analogon is often experienced in lecture rooms. If the hall is reverberating and the quality of the electronic installation is not very good, the amplified sound of the loudspeakers feeds back to the microphone. A shrieking, ear piercing and deafening noise is the result, well-known to most people. Here literally the noise is "singing round".

8.6 Chaos!

In plain language the "messy" result can be called chaotic behaviour of our analogon.

It so happens that the term "chaos" is a mathematically defined idea as well. Often such strictly defined terminology in physics and mathematics has—irritatingly and confusingly—a meaning different from what a layman would understand by the same word. In this case, however, the meaning is essentially the same. The mathematically defined term “chaos” means in normal language “chaos”, lack of order.

But of course mathematicians always have the pretention that they will be able to say more about the concepts they have carefully defined. One of the conclusions drawn by mathematicians is that chaotic motion often arises when we are dealing with so-called “non-linear” equations of motion. This is not the place, nor is it useful to explain what non-linearity entails. But the more mathematically inclined reader may refer to the separate text box in Chap. 7 and will then after just a brief glance confirm that our equations of motion are indeed highly non-linear. From that point of view one cannot say it is really a surprise to find chaotic motion as a possible solution of the equations of motion for the droplet of charge.

To be slightly more specific than just a vague inkling what mathematical chaos means: it means that the droplet of charge will never in future come back to the same place with the same velocity, the same elongation in shape, the same pulsation velocity, and the same …. etc. (“etc.” meaning: whatever is an important factor in the equations of motion). Because, if it would return to an earlier position with all these factors the same, then it would proceed by repeating its past: it would at a later time come back to this same condition, and come back again, etc. In that case there would be periodicity in its motion, and “unpredictable motion” would be a misnomer, there would not be real chaos.

In books about the theory of chaos often the weather is given as an example of the most typical and best-known chaotic system. There is never an exact return to an earlier condition, all the flows in the atmosphere, the distributions of humidity, temperature, etc. measured at a certain instant of time are completely unique: these conditions have never existed before in history, and will never occur again. Weather is a nice example, because at the same time it shows that despite the unpredictability of all the details certain statistical phenomena are predictable. Think of the periodicity of the seasons, and on a smaller scale the phenomena observed during for instance a front passage. Such a passage shows a sort of standard pattern, though every front passage is different in detail. Apart from that, we also know that meteorological phenomena, though seemingly completely random, will nevertheless satisfy the ordinary laws of nature: the total energy in a certain part of the atmosphere must be in agreement with the energy radiated into it by the sun, energy losses due cooling by ocean water, losses by friction of the wind against the earth surface, etc., etc. Evidently, all these weather processes cannot be completely random, they must obey certain rules. This is certain despite the otherwise whimsical variations of the weather conditions. The conclusion is: “chaos” is not the same as “complete randomness”.

Likewise, it may be true that the motion of our droplet of charge is chaotic, erratic and unpredictable, that does not mean that no regularities at all can be discovered in it. First of all, the law of energy conservation is observed by the droplet. At a given instant of time the energy in the translation and the energy in the elongation of the droplet will have certain values. The next instant these energies will have changed, for instance the energy in the translation has decreased. We can be sure now that the energy in the elongation then has *in*creased by the same amount. Were it not so, energy would have been lost somewhere, which is impossible (we are for the moment considering the situation without energy *radiation,* i.e. the situation to which the mechanical analogon refers. If we would include radiation as well, then still the sum of translation energy, elongation energy and radiated energy would be constant). Apparently, certain correlations between all these chaotically varying quantities do exist. Once again: chaotic behaviour is not completely random.

Let us return to the chaotic behaviour of our mechanical analogon after a rough start-up. Figure 8.10 (upper part) shows a typical chaotic signal, representative of how the nodding motion would look like under these conditions. Due to the coupling between nodding and swaying (the pirouette effect of a dancer irregularly waving the arms up/down and in/out) the "looks" of the swaying are similar.

For the actual droplet of charge it means: pulsation and translation show the same behaviour as sketched in Fig. 8.10, both whimsically varying around an average value. The translation motion now reminds one of a "*drunkard's* sack race".

From the sketch of Fig. 8.10 one other thing is also confirmed: chaotic motion is not the same as completely random motion. A definite rhythm is seen in the signal, which in the sketch is made better visible by the addition of the grid of dashed vertical lines. The lower part of the figure shows the corresponding nodding motion if the start-up would have done more gently. The amplitude of the motion is then smaller, and the shape of the signal is much closer to a smooth, nice sinusoidal curve. The "rhythm" that is roughly discovered in the chaotic signal now contracts into a steady, constant repetition of the motion. One could say that the chaotic motion for smaller amplitudes "degenerates" into a sinusoidal motion. The other

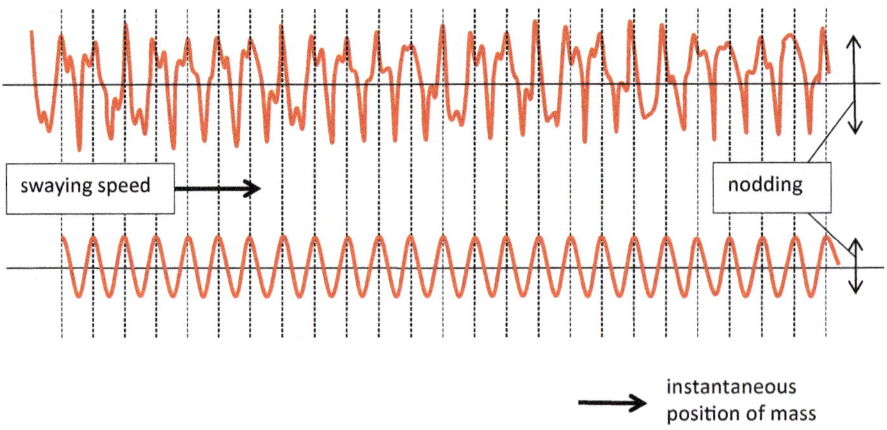

Fig. 8.10 Chaotic pulsation of droplet, represented by the nodding motion of the analogon. *Above* chaos fully developed. *Below* for small amplitudes "degeneration" into sinusoidal motion

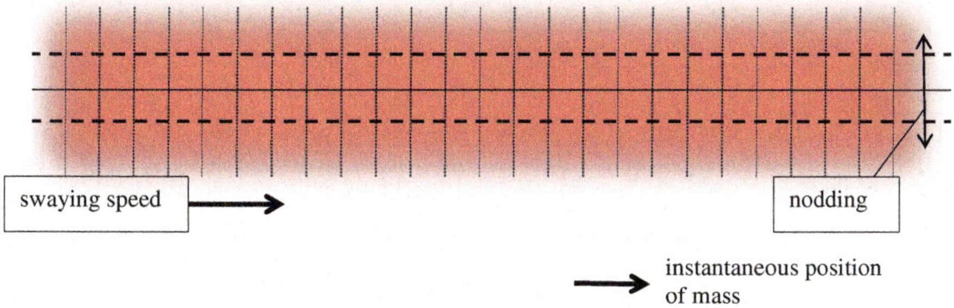

Fig. 8.11 Chaotic pulsation of droplet (represented by the nodding in the analogon), taking into account the high frequency of the pulsation. For small amplitudes the image shrinks into a narrower, sharp-edged band bounded by the *black dashed lines*

way round: one could imagine that the small amplitude, regular motion "grows grass" when its amplitude becomes larger.

Naturally, what is being shown in Fig. 8.10 is schematic. If one tries to imagine how the nodding motion in reality looks in the laboratory, one of the first things that has to be taken into account is the very high frequency of the nodding. As argued earlier, the individual nods would not be visible. A more realistic picture of what one would see is sketched in Fig. 8.11. The reddish coloured band with vague contours is what one would see in the case of the chaotic nodding. When at small amplitudes the chaotic nodding "degenerates" into sinusoidal motion, what would be seen is a narrower band with sharply defined contours. The edges then would be sharp, and located somewhere at the place of the black dashed lines.

8.7 Interference of Chaotic Motion with the "Zero-Speed Frequency"

Let us return to the experiment of Fig. 8.5, where the swaying and nodding of the analogon was mixed with the "zero-speed nodding frequency". Let us furthermore assume for now, just for the benefit of an explanation, that the nodding could be seen. If the motion is gently started the nodding and swaying would be seen as a near-sinusoidal motion with amplitude modulation. The amplitude modulation would resemble an AM (Amplitude Modulated) radio signal, where the high frequency oscillations represent the "carrier" wave, and the low frequency variations of the amplitude contain the significant audio information in the radio signal. The amplitude modulation in our case corresponds with the "beats" which are caused by interference between the frequency of the actual nodding and the "zero-speed nodding". It is shown in Fig. 8.12.

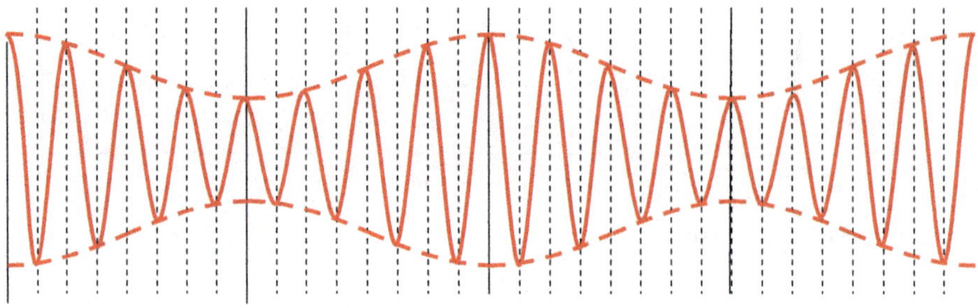

Fig. 8.12 Nodding motion with beats, in case of a gentle kick-off

Next the experiment is repeated, but now giving a really fierce kick-off to the nodding. The amplitude then becomes a lot larger, and due to the feed-backs (and the non-linearities) in the system a chaotic signal results. It is shown in Fig. 8.13. One could say that the neat signal of Fig. 8.12 "has grown grass". However, the chaos in not random: it still shows—roughly—the same rhythm as the fast near-sinusoidal oscillation of Fig. 8.12. When the chaotic signal is mixed with the "zero-speed frequency", we therefore still find the same beat pattern buried in the resulting interference pattern. It is schematically drawn in Fig. 8.13.

Figure 8.14 shows the image of the chaotic nodding with interference, when it is taken into account that the individual nods cannot be observed. The black wavy lines indicate the contours inside which the blurred pattern would shrink if there would be a degeneration of the chaos into sinusoidal nodding motion.

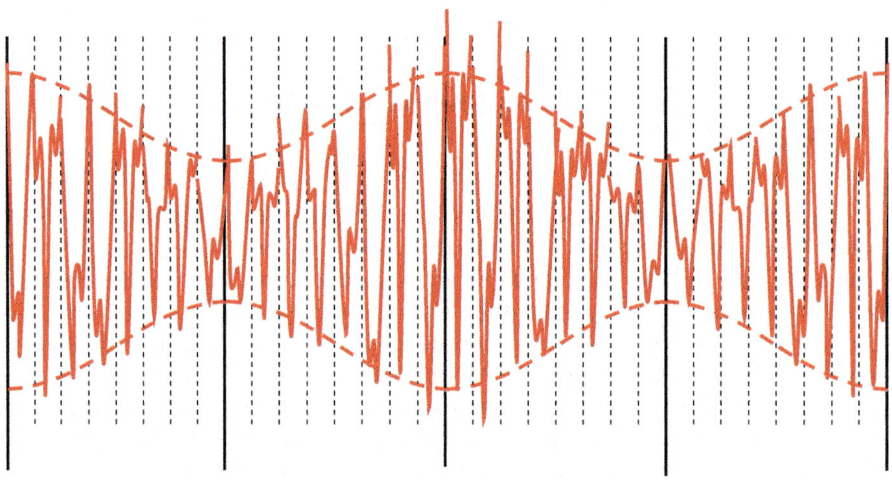

Fig. 8.13 Chaotic nodding motion showing a similar beat pattern as Fig. 8.12

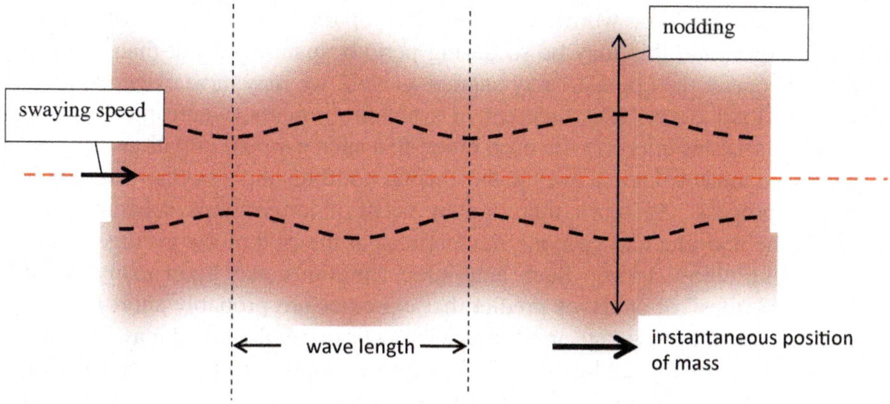

Fig. 8.14 Beats in the chaotic nodding, taking into account the unobservable individual nodding motions

8.8 Random Variations of the Chaotic Nodding Motion

In the last chapter it was made plausible that, even if there is chaos in the nodding motion, still a phenomenon like beats will occur, just like beats occur in a neat, sinusoidal signal. The beats are the result of interference, and imply a well-defined, periodic variation of the nodding amplitude.

In the present section it will be explained that there is also another type of amplitude variations, random rather than periodic and not caused by interference effects. We return therefore to Fig. 8.2, showing just the original analogon on its own, without the presence of a second one. We assume that the wobbling is so fierce that a chaotic motion occurs. What would then be seen was already pictured in Fig. 8.11. However, this picture is a bit too idealised, it shows what is called a *stationary* chaotic process, which not often will be realised. Mostly the signal will be somewhat more disturbed.

To explain this, think of weather variations, a typical example of a chaotic process. The lower curve in Fig. 8.10 could represent an idealised day-night cycle of the temperature of the atmosphere. It is the cycle such as corresponds with the heat influx caused by the sun during the day, and the loss of heat by radiation during the night. In practice, the temperature variations will be more like the upper curve in Fig. 8.10. Clouds come over and they influence the heat balance irregularly, whereas wind variations too make the temperature curve more chaotic. Nevertheless, in general a day-night pattern will still be clear, as is the case in the upper, chaotic curve.

Cloud cover, however, can also have longer term effects. Sometimes there are periods with an almost clear sky, for days on end, or in extreme cases during a week or even more. The heat of the sun during the day is then more intense, and the loss of heat during the night likewise. We will see that the temperature differences become larger. In other words, the amplitude of the chaotic signal of Fig. 8.10 becomes larger, and this may be so during periods spanning more than a few day cycles. Of course, the opposite may happen as well: we may have extended periods of a thick cloud cover. It is inherent in a chaotic process, where mostly a large number of variables interact with each other, that such *random* variations can occur.

Similar to meteorological events, the chaotic nodding and swaying need not be a stationary process. At times there might occur a longer-term accumulation of energy in the nodding and swaying fluctuations, at the cost of the average potential energy in the elbow spring. Such amplitude variations may span many nodding cycles. The "rules" of statistics predict that extremes are probably short-lived, and mild variations of the average amplitude will occur more often, and may last longer.

In an exaggerated form the not-quite-stationary chaotic nodding motion is shown in Fig. 8.15. In the caption of Fig. 8.15 these fluctuations spanning a large number of nodding cycles are called "random fluctuations", to differentiate them from the regular amplitude variations enforced by interference effects that were shown in Fig. 8.13.

Figure 8.15 on the right side shows still another incident that may occur. In exceptional cases the accumulation or depletion of nodding energy may be severe, the nodding showing a really extreme amplitude (large or small). What on the

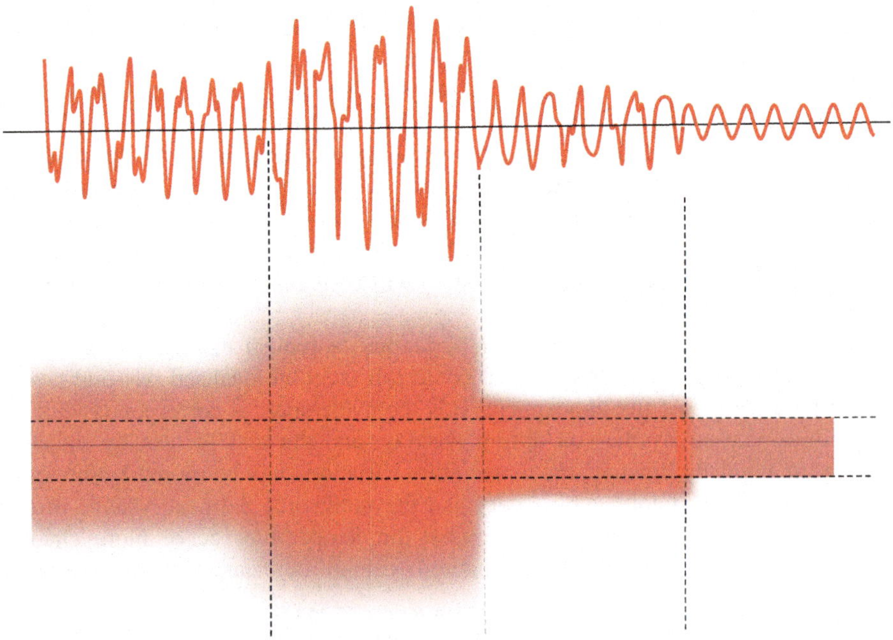

Fig. 8.15 Picture of the random long-term amplitude changes of the pulsation and translation, due to changes of the energy partition. *Lower part* shows same picture taking into account the high frequency of the pulsation. *On the right* degeneration into harmonic pulsation

right-hand side of Fig. 8.15 is shown is a special event: an extreme depletion of nodding energy may lead to such small amplitudes that we arrive at an almost sinusoidal motion. The chaotic character is then lost, it is in the caption of Fig. 8.15 called a "degeneration" of the chaotic motion.

Needless to say, we may translate all this to the droplet of charge, having a translation motion with irregular "hops", and a chaotic pulsation. In this case too we must expect that the amplitude of the "hopping" and pulsating can show random variations. These variations come on top of the regular variations (the beats) caused by interference effects. Sometimes, although seldomly, the random variations are so large that the chaotic character of the motion disappears. The motion then "degenerates" into sinusoidal motion, in other words the "drunkard's sack race" degenerates into "a regular sack race". The word "degeneration" is here perhaps not used in the most appropriate way, but it is hoped that the reader understands what is meant, and that he/she will recognise the phenomenon when it comes up again, later in this book.

Chapter 9
Energy Quantisation in Potential Wells

9.1 The Analogue of a Potential Well

In the previous chapter the motion of the "analogon-machine" was studied in case the torsional spring shown in Fig. 7.3 was removed. Our laboratory set-up thereby became the analogue of a free-flying droplet of charge.

What will be done in the present chapter is, to reinstall the torsional spring. The swaying motion then can continue during a few revolutions, but will be finally stopped because the spring winds up. The swaying motion will for a short moment come to a stand still, after which the direction of the motion is reversed. The swaying velocity, now in the opposite direction, gradually increases whilst the torsional spring unwinds itself. The velocity will reach a maximum value after which the rotation is slowed down again until it stops, and everything repeats itself. It is exactly the motion of the balance in a mechanical watch. This is the analogue of the back and forth translation motion of a droplet which is placed inside a potential well. In Fig. 7.4 concerning the terminology used, a new term was introduced: the word "sweep" indicates a single pass from stop to stop, no matter the direction of the swaying.

The magnitude of the swaying velocity is variable in time, having a maximum halfway during a sweep whilst momentarily reducing to zero exactly in the points of reversal. When we now recall the story about the helicopter blades in Sect. 8.2. where it was explained how the frequency of the nodding is affected by the swaying speed, you will understand that the nodding motion also feels the presence of the torsional spring. The nodding frequency will have a maximum value in the midst of a sweep, coinciding with the maximum velocity. The frequency of the nodding is lower near the beginning and end of a sweep. Exactly during the stops when the motion reverses its direction, the frequency of the nodding becomes the "zero-speed frequency". The very fast nodding would cause a field of ultra-sound in our laboratory, not heard by the human ear but perhaps noticeable by dogs. Dogs would

T. van Holten, *The Atomic World Spooky? It Ain't Necessarily So!*
DOI 10.2991/978-94-6239-234-2_9

thus hear a variable tone, it is as if the analogue is a musical instrument playing a sort of “vibrato”.

Because the sound field in turn stimulates other things to vibrate, the variable tone of the sound field may act back at our analogon-machine. In Fig. 9.1 it is shown schematically how this may happen. It is assumed that the analogon is placed in a normal laboratory which is not prepared acoustically. When sound waves are coming from it, caused by the nodding motion, these waves will be reflected by the walls of the room, and return echo’s back to our “machine”. Even if such echo’s do not directly return, they will bounce repeatedly against other walls or the floor and ceiling, and will fill the room with a so-called reverberating sound field. Of course, this is a well-known characteristic of concert halls too, or—easier noticeable—of large cathedrals: every enclosed space has its own characteristic reverberation. In particular large cathedrals are often not very suitable as music halls, because of the large reverberation effects.

What is visualised in Fig. 9.1 is, that the reverberation in our laboratory acts back on the analogon, and stimulates a secondary vibration of the nodding mass. This would not be very disturbing if the nodding takes place in one frequency. But, as we have seen, the nodding mass in this chapter emits a sound field with vibrato, a variable tone. There is a time delay between the primary nodding motion and the secondary motion induced by the returning sound waves. The returning sound waves therefore have a slightly different frequency than the frequency of the primary nodding motion: the returning sound belongs to the nodding as it was at a slightly earlier instant. In the situation of the analogon machine placed in a reverberating environment one must expect that the nodding motion is “contaminated”, the finally resulting nodding will show a mix of different frequencies. And this will lead to interference effects.

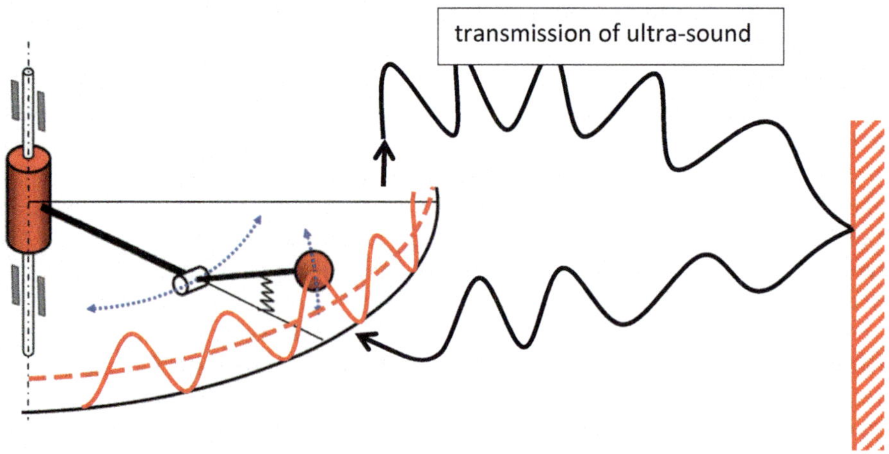

Fig. 9.1 Reverberation of ultra-sound, causing return of emitted sound with some delay, and feed-back of nodding history

The question that now immediately springs to mind is: is the reverberation effect a correct analogue of the electromagnetic interactions of the real droplet of charge? The answer, explained in the next section is: up to a point there is a certain analogy. In the real electromagnetic environment there is indeed a sort of reverberation effect, although the analogy with acoustics is not perfect. Therefore, the picture sketched above of the feed-backs due to the sound field is useful as a mental picture of what goes on but it would not be suitable for *quantitative* analyses. In fact, the sums in the appendix about the resulting interference effects were done in a different way. In Sect. 9.8. we will come back to it.

9.2 Does the Electromagnetic Field of the Real Droplet of Charge Show Reverberation?

The answer was already given, and comes down to: there is indeed a kind of reverberation (well: sort of). To explain it, let us go back to Chap. 5 where so-called "Schott-energy" was mentioned. This came up in relation with the translation motion, but equally well describes what happens with the radiation coming from a pulsating droplet.

The story went as follows, and an attempt to sketch what happens is given by Fig. 9.2.

If a charge is subjected to "jerk" (a change of acceleration), the immediate response of the charge is to pour radiation energy in the near field. Remember the metaphore of the "cries of complaint" by maltreated bus passengers? The complication is, that not all of this radiation reaches infinity. What is found at large distances is only a part of the originally emitted radiation. The part that is carrying off the energy to infinity, and that thereby causes the loss of energy from the total system (which includes the charge as well as the field), is proportional to

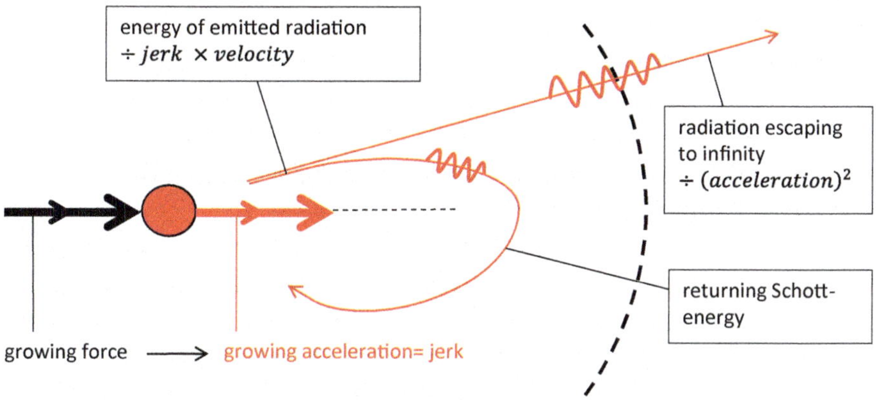

Fig. 9.2 Symbolic sketch of Schott-energy

÷(*acceleration*)^2, which is different from the initial radiation with an energy proportional to ÷*jerk* × *velocity*. Apparently, not all of the radiation in the near field propagates to the far field on a single trip to infinity. In fact, the field acts as a buffer to bridge the gap between the energy poured into the near field and the energy escaping in the far field to infinity. During periodic motions of the charge, the buffer absorbs energy during part of the cycle, and returns energy during another part.

The difference between what is originally radiated and what is found at large distances could be described as radiation that after a while returns to the charge. It is, as it were, a "home-sick" part of the radiation, that temporarily resides in the field and then decides that it does not like to travel to infinity (never to come back), but instead at second thoughts prefers to go back home. The energy carried by the home-sick part of the radiation was called the "Schott-energy", and is not definitely lost (from the point of view of the charge).

The fact that some of the original radiation leaving the droplet will later return to it, is somewhat comparable with the reverberation of acoustic radiation. Now, if the radiation is caused by the pulsation of the droplet, and if the pulsation has a variable frequency, then one may expect further complications. The part of the radiation that returns has the frequency the droplet had a few instants earlier. There is a difference between the instantaneous frequency of the pulsation and the frequency of the returning radiation. The motion of the droplet is disturbed by the returning radiation, and as a result the total pulsation will consist of a mixture of frequencies. Result: interference!

We thus see that, although the electromagnetic field is not the same as the acoustic field of the analogon, similarly complex interference effects may be expected. Our analogon-machine is not a mathematically correct analogue with respect to reverberation, but it may display similar effects as the droplet of charge. For a mental picture this suffices. Once again a warning should be given: it is no more than a mental picture, suitable to imagine how the droplet's past may influence its present motion. The strict mathematical treatment, however, does *not* follow this line of physical considerations, it shows that these "memory effects" must always exist if a droplet is trapped inside a potential well. They are inescapable to get a consistent theory, whatever the physical cause of such memory effects. Further remarks about this matter will be found in Sect. 9.8.

9.3 To Simplify Matters: The Potential *Box* Instead of a Potential *Well*

You may have obtained the impression that the nodding, including the feed-backs caused by the reverberation effects, has become extremely complex: the reverberation contains a lot of different frequencies, corresponding to the entire history of the nodding mass in our analogon. This impression is correct. Fortunately, to trace what is going to happen as a result, we can consider a simpler situation as a first step. This is the case of the potential *box* instead of a more general potential well. The box has already been introduced in Chap. 3, from which the following Fig. 9.3 is copied.

The situation in a box is that of an electron moving inside a piece of metal. The electron can move freely through the metal but cannot escape from it. Therefore, if it reaches the surface of the metal it is suddenly stopped and kicked back by a high potential barrier. In books about quantum mechanics this special shape of potential well is then usually called a potential *box*.

The mechanical analogon of it is shown in Fig. 9.4. The spiral spring has now been replaced by hard stops restricting the swaying motion. The swaying around the top axis would then take the form of a sweep with constant speed, followed by a hit against one of the stops, a rebound, and a constant-speed sweep back until the other stop is hit, etc. This kind of rough ride comes in the place of the much smoother sinusoidal swaying to and fro if a spiral spring would be present. As will be seen, nevertheless it is somewhat easier to get a feeling how the nodding motion in this case would look.

The reason why the box is simpler than a more general type of potential well is not difficult to see. The nodding frequency during a sweep from one stop to the other is given by the expression mentioned in Chap. 8:

$$f = f_0\left(1 + k \cdot v_{average^2}\right)$$

where the symbol f stood for the nodding frequency (i.e. the number of nods per second). This frequency has been raised, compared with the "zero-speed frequency" f_0, by the influence of the swaying velocity $v_{average}$. What the subscript "average" tries to convey is, that it is not needed to take into account the "ripples" in the swaying velocity caused by the nodding: substituting into the expression just the average value gives sufficient accuracy. In the case of the analogon representing a potential box the average swaying velocity is constant during a single sweep, it does

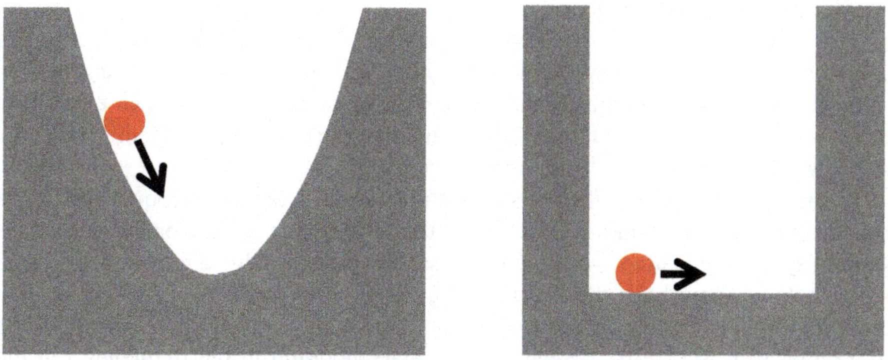

Fig. 9.3 (copy of Fig. 3.2). Schematics of different types of potential well (*left* the parabolic well, *right* the box), based on the analogy with the marble inside a bowl

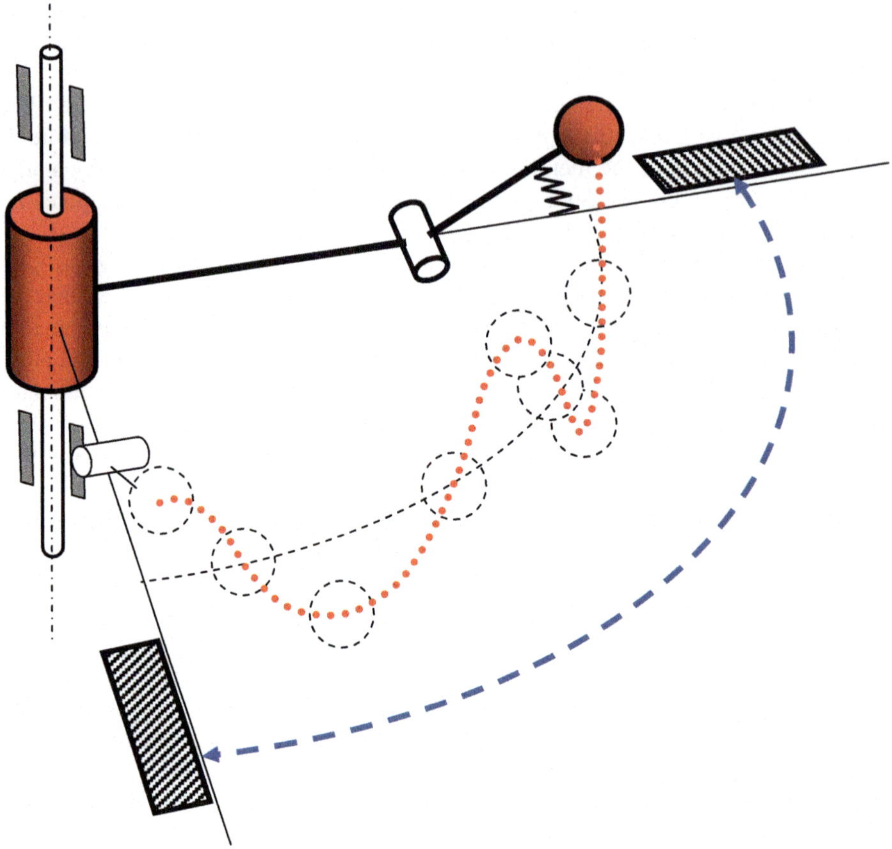

Fig. 9.4 The mechanical analogon representing a potential box

not vary in between the stops. The corresponding nodding vibration will thus contain just one single frequency.

This is true for the motion in between the stops. What exactly happens during the collisions and rebounds against the stops is more complicated, and is analysed in somewhat more detail in the formula appendix. Let it suffice to say that effectively there is a brief moment of stand still, and therefore a brief moment where the nodding takes place in the "zero-speed frequency".

In the case of the box there are therefore just *two* different frequencies to be considered, instead of the multitude of frequencies that would occur in a more general type of potential well. Here we see the reason why the box is so much simpler. If we now have the reverberation effect in our laboratory room, so that these *two* frequencies are mixed together, the resulting nodding motion is modified in a rather neat way. First we have the primary nodding in the frequency f at the speed $v_{average}$. The reverberation in the room will also contain sound emitted during the stops, which has the "zero-speed frequency" f_0. The reverberation sound will feed back on the nodding, so that the nodding also becomes "infected" with f_0, even if it is in the middle of a sweep. Finally, the nodding will show a mix of the two frequencies f and f_0.

You can easily guess what will be the result: interference leading to beats in the amplitude of the nodding. The nodding motion has been modified to look as sketched in Figs. 8.6 and 8.7. These latter figures were derived for the analogon *without* a spiral spring or stops, and the beat phenomenon was obtained because of the presence of a separate source sending the zero-speed frequency (see Fig. 8.5). We now see that the back and forth swaying representing a potential box does not need such an artifice: the source of the zero-speed frequency is provided by the analogon-machine itself, without the help of a second, auxiliary device.

The conclusion can directly be transferred to the droplet of charge. The pulsation of the droplet, when it moves back and forth inside a potential box must show beats in its amplitude.

The presence of the stops does more. It restricts the beat pattern to certain configurations, i.e. the position of the beats within the box is not quite arbitrary. To see this, it will be necessary to give somewhat more attention to the details of the bounces against the stops.

9.4 The Action of the Coriolis Forces During the Collisions with the Box-Walls

In Fig. 9.5 the hypothetical case is considered as if the nodding could be seen (in fact, any laboratory equipment that would enable this has been banned, as you will remember). The path of the mass approaching the stop is sketched as a drawn line. It bounces back, and the return path is shown by the broken line.

The collision is viewed as a small deflection of the wall, which yields slightly to the right and then springs back again. The flexibility of the wall is symbolised by the two vertical broken lines between which the front of the wall will be displaced during the collision. The force the wall exerts on the mass is symbolised by a spring force. Apart from the wall the rod which carries the mass may also have some flexibility. In the schematic of Fig. 9.5 the related spring force is also thought to be represented by the spring drawn behind the wall. The spring thus symbolises both types of flexibility. In the analogous situation of a droplet of charge bouncing against the wall of a potential box, one may take the wall to be infinitely stiff, whilst the spring force which reverses the motion is entirely provided by the deformation of the droplet. This does not change the schematic of Fig. 9.5, the principles remain the same.

Returning to the "analogon machine": the nodding motion during the collision with the stop is assumed to be upwards. Recall that the motion of the mass is not purely vertically up and down: in Fig. 8.2 is was attempted to sketch that a vertical motion is also accompanied by a "radial" motion: going *up* means that the mass at

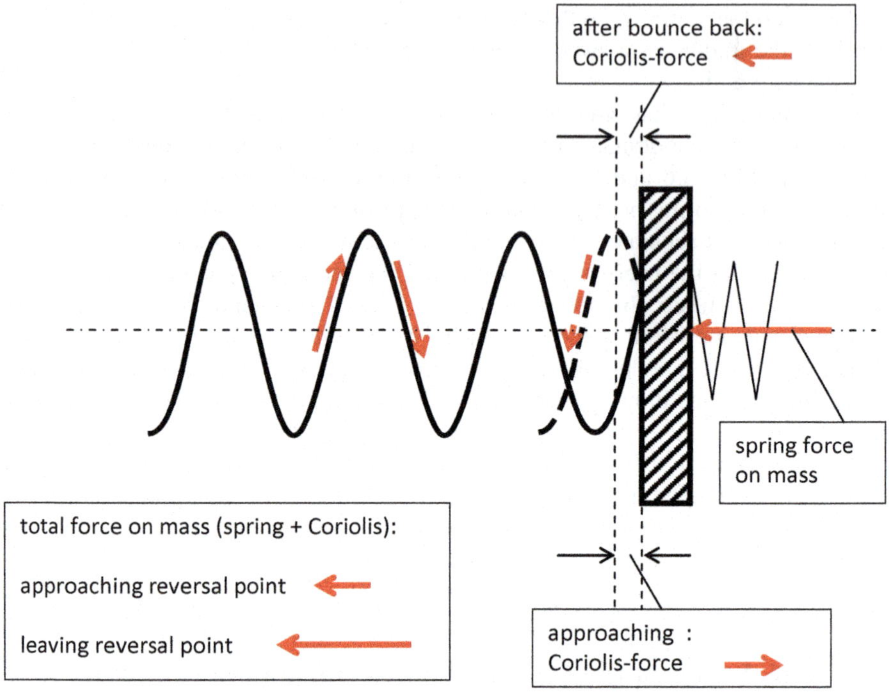

Fig. 9.5 Collision of mass with wall: force on mass is asymmetrical before and after the reversal point, due to Coriolis-forces

the same time moves *towards* the rotation centre, and this is the cause of Coriolis-forces. When the mass is close to the reversal point the Coriolis-force tries to increase the velocity, and is thus directed to the wall. After the reversal point the Coriolis-force acts in the opposite direction, it acts away from the wall.

The text in the lower lefthand corner of Fig. 9.5 summarises the total force on the mass during the collision. The forces due to Coriolis effects and the wall spring are here combined, to show that the force on the mass is not the same before and after the reversal point. We must expect that the swaying velocity is not the same either. In the assumed situation the mass will after the collision have attained a larger velocity (to the left) than it had before the collision (to the right). The forces may be compared to what happens to a tennis ball when it is driven by a racket (Fig. 9.6).

If we had considered another situation, where the nodding motion happened to be downwards instead of the sketched upward motion, then we would have found the opposite conclusion, where the mass would leave the wall with a smaller swaying velocity than it had before.

Fig. 9.6 The forces in Fig. 9.5 may be compared with the driving of a tennis ball by a racket. From http://web.mit.edu "tennis-ball-rebound"

In general, therefore, the swaying velocity may be expected to be changed by the collisions against the stops. In the case of the analogous droplet of charge remember that Coriolis-*like* forces exist, which means that the translation velocity within the potential well may slightly differ from one sweep to the next one. There is an exception though, and that happens when exactly at the instant of the collision the Coriolis-forces are zero. The nodding must then just have reached its highest (or lowest) point when the collision takes place. This is the situation as sketched in Fig. 9.7.

In Fig. 9.7 we have perfect before/after symmetry w.r.t. the hit against the stop. Therefore, the path of the mass after the rebound cannot be drawn separately: it exactly coincides with the approaching path, although the direction of motion is reversed. This situation will be called "path retracing". Before and after the bounce the swaying velocity (with the small ripples averaged out) is the same, apart from its direction.

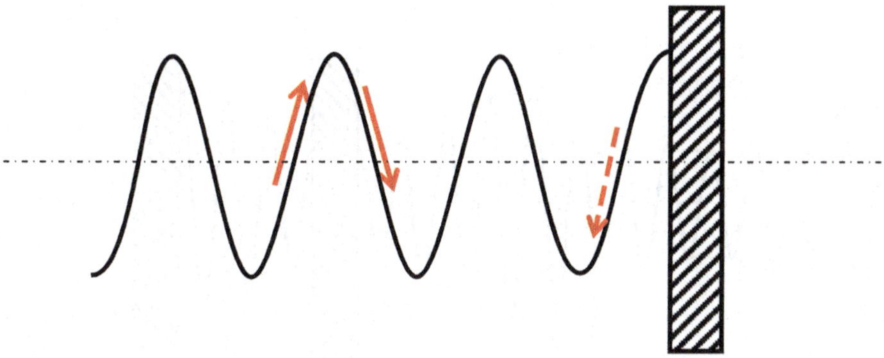

Fig. 9.7 The exceptional situation where the swaying velocity will not be changed by the collision against the stop

9.5 Quantisation of Velocity and Energy

As a next step, consider the situation where the nodding amplitude is not constant, but instead has a variable amplitude due to the beats which were discussed in Sect. 9.3. Again, there will be *only one* type of motion where the swaying velocity remains the same before and after the collision against the stop: this is the motion which has *before/after symmetry* with respect to the bounces. The requirement of symmetry now no longer concerns just the individual nods, it extends to the symmetry of the beat pattern w.r.t. the stops. This is clarified in Fig. 9.8. What is shown in Fig. 9.8 can again be called a motion with "path retracing". This path retracing appears to be the condition to ensure that the average swaying velocity—ignoring the short ripples—remains the same before and after the bounce against the stop.

If we finally consider not just one of the stops separately, but instead the entire box with stops on both sides and continuous sweeps from left to right and vice versa, it is clear that an average swaying velocity that is unaltered (in absolute magnitude) from sweep to sweep can occur for a strictly limited number of configurations. "Configuration" meaning here: the beat pattern and its length relative to the width of the box. The beat pattern must "fit" in the width of the box, otherwise the average velocity of the different sweeps will be varying.

If all the conditions have been fulfilled to make the sweeps identical (apart from the direction of motion, of course), this situation is called "stationary". In a stationary state everything periodically repeats itself exactly.

Figure 9.9 is a schematic of the configurations that allow a stationary state. It shows that the first possibility is, that half a wavelength of the beat fits between the two walls on either side. The next possibility is a full wavelength, etc. One obtains a series of $\frac{1}{2}, 1, \frac{3}{2}, 2, \frac{5}{2}, \ldots\ldots$ wavelengths in the box, where a stationary state occurs.

In words: the box must contain an integer number of half-wavelengths, in order to allow a stationary state.

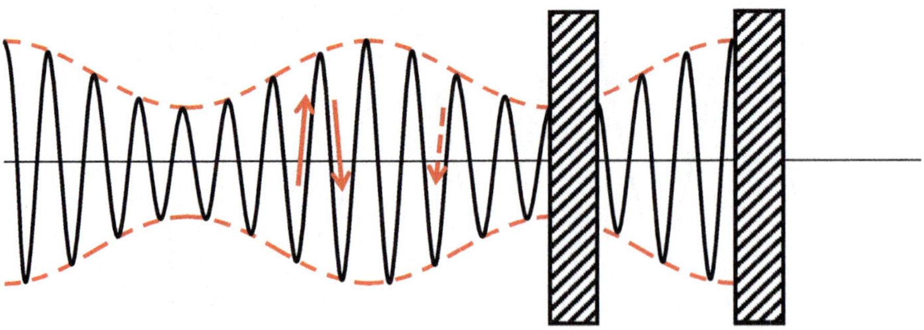

Fig. 9.8 Possible positions of the walls to obtain "path retracing", and thereby to ensure that swaying velocity before and after impact with the stops remains unchanged

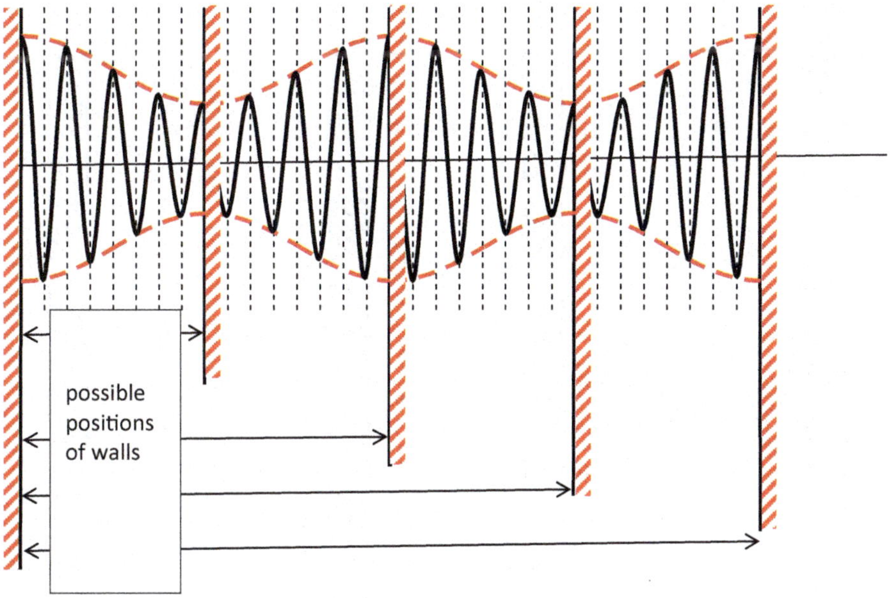

Fig. 9.9 The possible positions of the potential walls w.r.t. the pulsation (or: nodding), in a stationary state

The actual situation will be, that we have a given potential box, with a given, fixed width. Let us give the symbol a to this width. Figure 9.10 then shows the configurations that are able to sustain a stationary state.

In Chap. 8 the following relation was derived between the wavelength of the beats and the—average—swaying velocity:

$$(wavelength\, of\, beats) \div \frac{1}{(swaying\, speed)}$$

i.e. the wavelength becomes smaller, the higher the swaying speed is. We see that in Fig. 9.10 the pictograms on the second row thus represent a speed that is double the speed on the first, uppermost row. Likewise, the pictograms on the third row represent a speed three times larger than on the first row. This goes on, and what we find is:

$$(speeds\, allowing\, stationary\, state) = (1, or\, 2, or\, 3, \ldots) \times (smallest\, speed)$$

The kinetic energy corresponding to the average swaying speed is proportional to the speed squared:

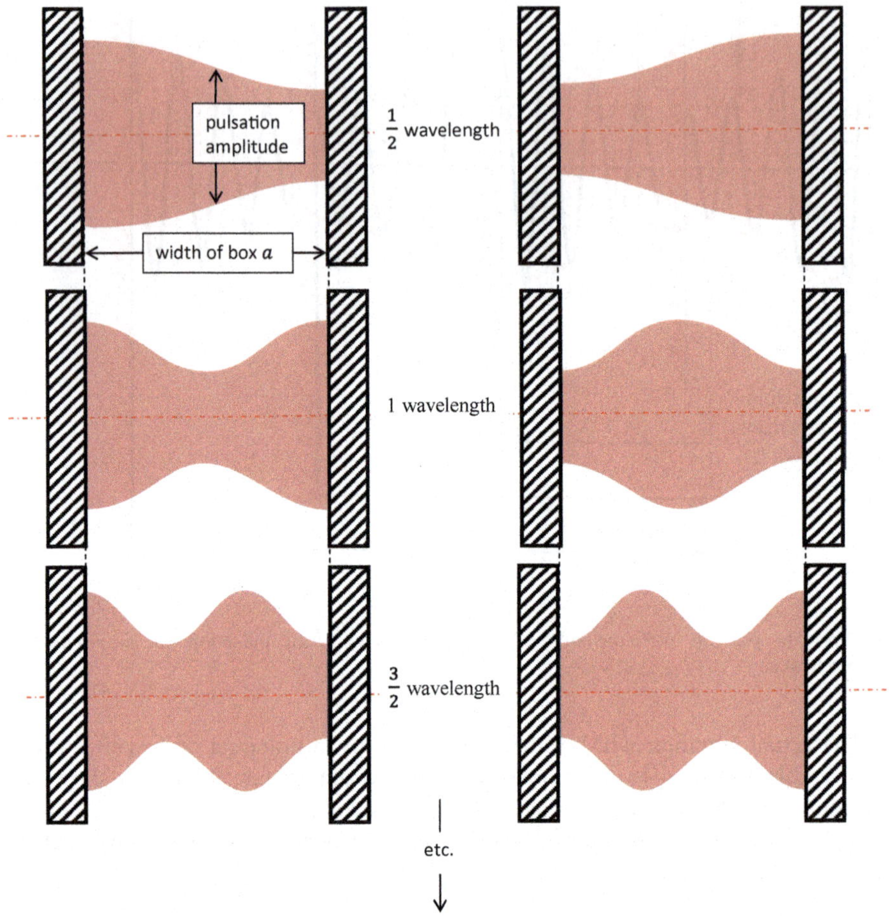

Fig. 9.10 The different possibilities to fit the wavelength of the pulsation beats in the box, in such a way that stationary states occur

$$(kinetic\ energy) \div (speed)^2$$

so that, combining the above two relations:

$$(kinetic\ energy\ corresponding\ to\ stationary\ states)$$
$$= (1, or\ 4, or\ 9, \ldots) \times (lowest\ energy\ level)$$

In words: the kinetic energy of the swaying cannot have any arbitrary value if we want to have a stationary state. Stationary states can only be realised at discrete values of the kinetic energy. There is a "ground level" of the energy, corresponding to the uppermost configuration in Fig. 9.10, which is the lowest possible energy

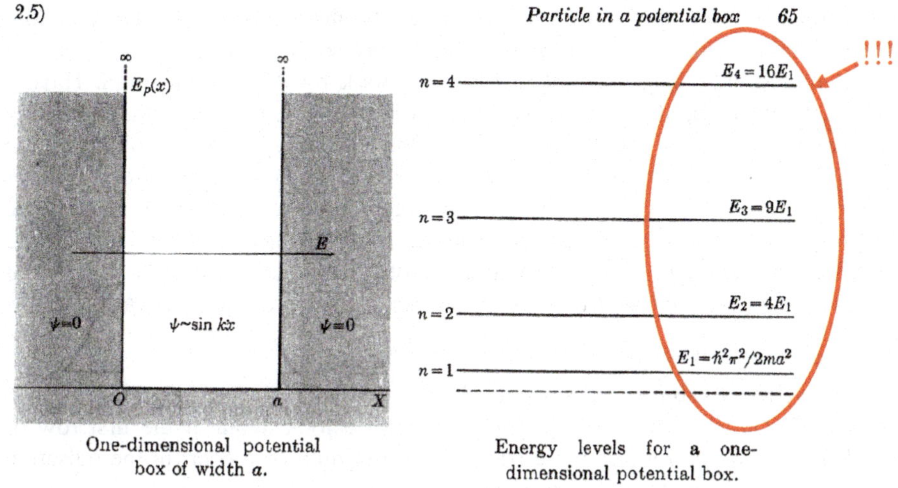

Fig. 9.11 The quantised energy levels in a potential box calculated in usual quantum theory by solving Schrödinger's equation. Figure copied from Alonso and Finn: Fundamental University Physics

which allows a stationary state. Higher energy levels are also possible, but only if they have an energy of 4 × (*ground level*),or 9 × (*ground level*), etc.

There is thus *quantisation* of the energy levels at least if we would insist on having a stationary state. More about this last proviso will follow later.

The conclusion about the quantisation is exactly in agreement with the prediction of usual quantum theory. This is illustrated in Fig. 9.11. The figure is reproduced from the physics text book by Alonso and Finn: Fundamental University Physics, volume III "Quantum Mechanics and Statistical Physics", page 65. The part of the figure to which I would like to draw your attention has been circled in red.

It is an amazing agreement, is n't it? Especially so, because Alonso and Finn here show the results from *quantum theory*, obtained by solving Schrödinger's equation for the case of the potential box. In contrast, our identical results had nothing to do whatsoever with quantum mechanics! We just took a droplet of charge and analysed its behaviour by the classical theories of Newton and Maxwell.

9.6 The Ground Level of the Energy

Cynics would call the excitement perspiring in the preceding section somewhat premature. They would point out that the agreement between the two theories only concerns the *ratio's* of the energy levels. What about the absolute levels, in terms of the energy quantified as "so and so many Joules"? Is there in this respect a similar agreement between both the theories, the droplet theory and the quantum mechanics? The answer is: yes! And it can be proven very easily.

Let us return to Fig. 9.11 taken from the book by Alonso and Finn. Have a look at the lower right corner of the figure, where you see the formula expressing how much energy is present in the first, lowest energy level. Looking at the formula more carefully (you as a reader are excused if you refrain from a closer look because of the vague reproduction of A&F's figure) and keeping in mind De Broglie's relation (see Chap. 8), one can see that it expresses the following: the lowest energy level of a particle within a potential box corresponds with *fitting half a wavelength of the De Broglie matterwave in the width of the box*. Or: $\frac{1}{2}\lambda_{DeBroglie} = a$ (lowest energy level).

This is a remarkable result, since it is identical to what we had found for our own droplet of charge. Turn a few pages back, have a look again at the first row of Fig. 9.10, and what you see there is: half a wavelength of a beat (in the pulsation signal) equals the width of the box. Or: $\frac{1}{2}\lambda_{beat} = a$.

To finalise the argument, it was established in Chap. 8 that De Broglie's matter waves have a definite relation with our beats. Wavelengths of matter waves and our beats both depend in the same way on the speed of a particle (or droplet). It means that droplet theory gives the same value for the lowest possible energy as quantum mechanics does. Not only that, all the energy levels are the same in both theories!

9.7 Why Would the Droplet Prefer to Be in a Stationary State?

The question is: what is so special about these stationary states? Intuitively, one suspects that the droplet would tend to move preferably in a stationary state, but why would it? Such an intuitive thought is triggered by what we know about vibrating strings of musical instruments. No matter how the string is brought in motion, after a short time (a so-called "transient") it will vibrate in the form of the ground tone and overtones that are allowed by the geometrical constraints of the instrument. Purely mathematically it can be worked out how this comes about. However, from the physical point of view, it is much harder to describe the process of these transients. One must then describe how waves of different wave lengths run along the string away from the place where the string has been plucked, how these waves are reflected at the end points, and how they interfere with each other after their return.

In the case of the droplet of charge (or for that matter: the case of the swaying and nodding mass) one of the mechanisms that would cause the droplet to be driven towards a stationary state is clear. It was pointed out in the text accompanying the Figs. 9.5 and 9.6 that in *non*-stationary situations (i.e. if there is no path retracing) the collisions with the walls cause changes of the—average—velocity. Now imagine that the droplet is shooting from wall to wall with a velocity that is slightly

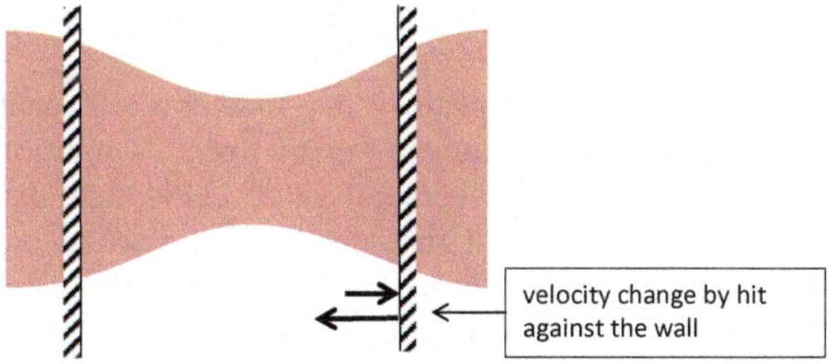

Fig. 9.12 The beat pattern does not fit correctly between the walls, due to a velocity that is too small (corresponding to a large wavelength of the beats)

smaller than one of the "allowed" values (i.e. "allowed" for a stationary state). It is sketched in the following Fig. 9.12.

It is one of the characteristics of the beats that they become longer when the speed of the droplet is reduced, and v.v. If it is assumed that the speed is too small, this implies that the beat is too long for the box, it does no longer fit correctly. This is what is shown in Fig. 9.12.

Clearly, in this situation there is no longer path retracing. The consequence is, that hits against the walls cause a velocity change (the "tennis racket effect" pictured in Fig. 9.6), in the case shown leading to a larger velocity. This in turn is accompanied by a shrinking of the beat pattern.

One can imagine that this effect tends to restore the configuration where the beat pattern exactly fits. One can therefore also imagine that this mechanism leads to *stability* of the stationary states: any perturbation away from the stationary state is counteracted.

Admittedly, the above given explanation gives no more than an inkling why stationary states will tend to be stable. For more details, it would be necessary to delve more deeply into the dynamics of the droplet during the bounces. This would go too far at this place, and the mathematical basis of the formula appendix should be worked out further than its present state. For now, as well as in the remainder of this text, the stability of the stationary state will be supposed to be assured.

9.8 A Further Remark on the "Reverberating Environment"

In Sects. 9.1 and 9.2 it was made plausible why the droplet is going to pulsate with a mix of frequencies, even if its *natural* vibration contains only one single frequency. For the mechanical analogon an acoustic phenomenon, the "reverberating environment" could be pointed out to cause a sort of memory effect. The "memory" in the experiment explains why different frequencies, coming from different events in the droplet's history, will be found in the final nodding motion.

In the case of the real droplet, similar "memory" effects in the electromagnetic field were identified. The so-called "Schott-energy" was pointed out, which is responsible for the memory effect, or to put it differently: for the electromagnetic reverberation.

In the formula appendix, the mathematical analysis of the droplet inside a potential well goes along different lines. A solution of the equations of motion *without* any external reverberation effects is there determined. The derivation does not take the mixing of frequencies as a starting point, therefore. The surprising conclusion is nevertheless, that the same mixing of frequencies "echoing from the past" is found, even though no *external* reverberation is assumed. It appears that the mathematical manipulations needed to solve the equations of motion themselves give rise to memory effects. One could, alternatively, say that the mathematics determine what the final outcome should be if the laws of physics are respected, irrespective of the physical details.

The mathematics therefore lead to the important conclusion that the interference effects leading to beat patterns will always occur, no matter the circumstances. This is an important supplemental conclusion to the present chapter. Think for instance of "Bohr's radio silence" mentioned in Chap. 1. Such a state of radio silence will not affect the interference phenomena and the beat patterns responsible for the energy quantisation.

Chapter 10
Still More (Non)Glimpses: The Droplet Is Sometimes Invisible

10.1 Magical Disappearances in Usual Quantum Mechanics

We now come back to Chap. 3, where something more was told about the usual quantum mechanics than was possible in the introductory Chap. 1. One of the magical things narrated there was the occasional disappearance of particles. Or was it just disappearance from sight? In the present chapter the behaviour of a droplet in this respect will be investigated, as revealed by the droplet theory. First a summary will be given of the issue at stake.

Figure 3.8 was taken from the textbook "Fundamental University Physics" by Alonso and Finn. It showed the so-called "Position Probability Density" curves as determined in usual quantum mechanics by solving Schrödinger's equation in a parabolic potential well. The same type of curves are drawn in Fig. 10.1, now for the potential box. They are the curves in the right-hand column of the figure, the lowest energy level on top, and the next energy levels underneath it. As you recall, these curves give information about the chance to find the particle at different places inside the box. The largest chance is found near the tops of the curves. And what was so mysterious: the curves at some places descend to zero, meaning a zero chance to ever observe the particle in these places! These mysterious points were in Chap. 3 called "those nasty zero-points".

A number of questions arose concerning these zero-points. For instance: if a particle will never be observed in such a point, does it mean that it never comes there? But, would that mean that the particle does not move from wall to wall inside the box, because it will find hurdles on its path that cannot be passed? In particular for the energy levels higher than the lowest: would a particle thus move back and forth along a small stretch of the box only, and is it destined to stay forever in a fraction of the entire width of the box? Most books shy away even from mentioning these conceptual problems. Other books state: "this is the mathematical solution of the Schrödinger equation. Do not try to form a physical picture. There are many

T. van Holten, *The Atomic World Spooky? It Ain't Necessarily So!*
DOI 10.2991/978-94-6239-234-2_10

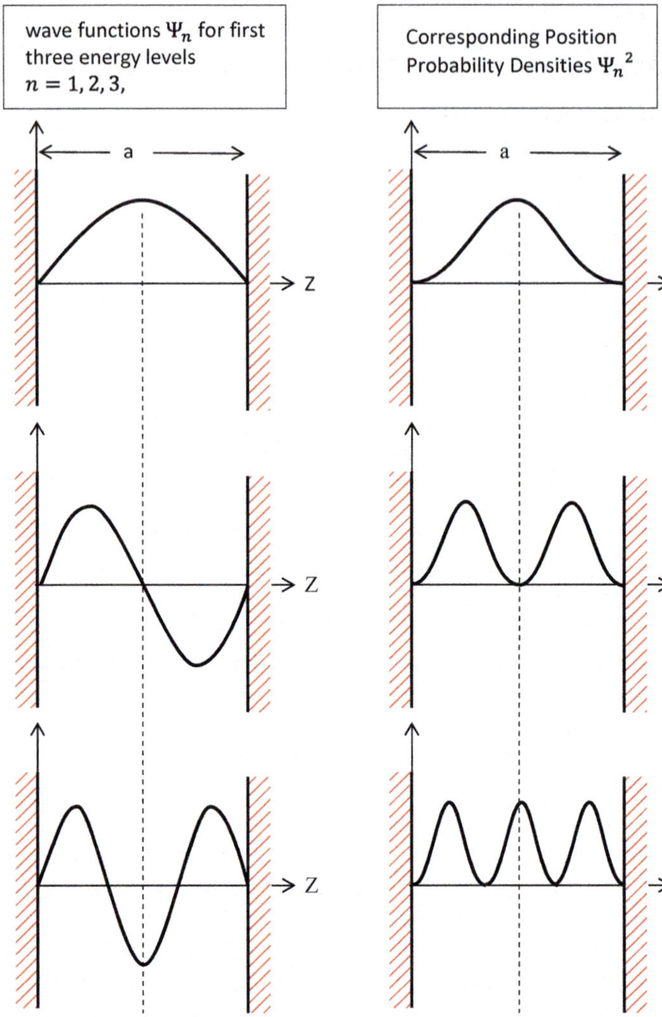

Fig. 10.1 Wave functions and Position Probability Densities in a potential box, first three energy levels

phenomena in the quantum world that cannot be understood by our human intuition, and this is just one of them. Therefore, accept the fact that the results are in agreement with experiments, however incomprehensible they may be". In still other books it is tried to give more explanation, and one finds: "In this situation the particle is no longer a thing as we are used to in classical physics, it really is a wave, and as such at any instant it is everywhere inside the well". The latter, admirably brave attempt to give an "explanation" makes matters worse in my opinion, rather than that it "explains" anything. The picture it sketches is close to the "world of Harry Potter".

Before we are going to propose an explanation based on the droplet theory, let us first complete the description of Fig. 10.1, where until now only the column of graphs on the right has been discussed. In the left-hand column the shape of the corresponding matter waves, so-called "wave functions", are shown as determined by solving Schrödinger's equation, and it is shown how they are fitted in the box in the case of stationary states. Do not try to form a picture what their meaning is, matter waves by themselves do not have a clear physical meaning. They are mathematical artefacts, semi-manufactured articles, and serve as an intermediate step to obtain the more meaningful graphs in the right-hand column. The curves in the right-hand column are determined by squaring the curves in the left-hand column.

Nevertheless, these wave functions are reproduced here because it is interesting to compare them with the patterns of beats, as determined by the droplet theory. Compare Fig. 10.1 with Fig. 9.10, where the beat patterns were shown for stationary states in a potential box. What then appears is, that there is a definite relation between these wave functions and the beat patterns. In places of the box where the amplitude of the pulsation does not change very much, the wave functions have a small value, around zero. At places where the rate of change of the beats is large, we find maxima of the wave functions. This feature is illustrated in the following Fig. 10.2, where the example is given of the second energy level.

Figure 10.2 is rather convincing to prove that our own beat patterns and the wave functions do have a relation of some sort. Mathematicians would say that the wave functions are the negative derivatives of the beat patterns. However, this is only correct for the potential box, the relation is more complicated in other types of potential wells.

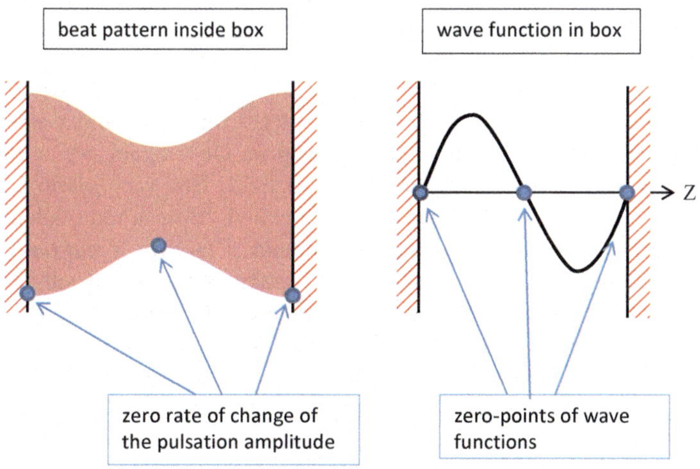

Fig. 10.2 Relation between the beat patterns of droplet theory and the wave functions of quantum mechanics

Now, do realise that the zero-points of the wave functions coincide with the more meaningful (but certainly not less mysterious) zero-points of the "position probability density curves" in the right-hand column of Fig. 10.1. They are the points where, according to quantum mechanics, the chance to observe the particle is zero.

It immediately raises the question: could it be that the beat patterns also tell something about the chance to find the droplet in a certain place? And if so, why would that be? The answer is: yes, they do. But why that is so requires a lot more explaining, which will take the rest of the chapter. From you as a reader some effort is therefore requested. But the reward will be, that a tip of the veil can be lifted from "the riddle of the zero-points".

If you are not up to the rather complicated full explanation, there is an alternative to be found in the next two sections of this chapter. This is a more heuristic reasoning, i.e. it is an explanation with gaps in the logic, it is more intuitive but nevertheless makes it plausible why there must be "zero-points" and it gives a feeling for what is happening when the droplet passes these special points.

10.2 What Determines the "Visibility" of a Droplet of Charge?

Whether a droplet of charge can be observed depends on its size. A large, macroscopic droplet containing much charge is easily "seen". Just observe the electrostatic field if it is at rest, or the electromagnetic field if it is moving. Simple laboratory equipment will do to tell where the droplet is. Of course, if the droplet is moving fast we run against the problems pointed out earlier, e.g. in Chap. 3. As in the case of the children's swing (see Fig. 3.11), we need fast registration equipment to "catch" it when its run from place to place is swift. But this is not an unsurmountable obstacle to pinpoint the instantaneous position of macroscopic blobs of charge.

At the other end of the scale, when the droplet is tiny or even just possesses the unit charge, it is comparable with a single electron. Once again, we can try to find its position by observing the electromagnetic field. But very often, there is no electromagnetic field at all! In Chap. 1 it was called "Bohr's radio silence": if the droplet is moving back and forth inside a potential well in a *stationary state*, it keeps the same energy during a long time. Constant energy implies that no energy is lost (or added) by radiation, and consequently we have no means to tell where the droplet is. The droplet is completely "invisible" in a stationary state.

It is only at the moments when the droplet loses or gains energy that communication with the outside world is restored through the emission or absorption of radiation. We can "see" the droplet when it transitions from one stationary state to another, for the rest it is invisible.

To answer questions about the visibility of the droplet, we thus have to investigate the conditions which make a transition between stationary states possible. Could it be that the zero-points in a well where the particle is never observed, indicate that in these points the conditions are never fulfilled to initialise an energy transition? Let us have a better look at the conditions required for leaving a stationary state.

10.3 A First Clue About the Conditions to Be Satisfied for Radiation

A first clue about what makes spontaneous radiation possible can be obtained from the inverse question: under what circumstances do we know for certain that spontaneous radiation never occurs? Easy: consider a free-flying electron when there is no interaction with the outside world. No collisions with other particles, and neither "collisions" with electromagnetic radiation coming from somewhere else. Such an electron will not spontaneously radiate. In the case of a free-flying charge there are no forces acting on it, by definition. The electron will fly at a constant velocity, without accelerations or jerks, and it therefore does not radiate. In a later chapter, Sect. 12.10, it will be discussed that another impediment exists too, if we require that both the energy balance as well as the momentum balance should be satisfied at the same time. No, an *undeformable* charge in free flight definitely does not radiate, and this is in agreement with experimental evidence.

At this point we become extremely concerned whether the model of a droplet of charge can really be a valid model of an electron. A droplet has been shown to perform a "sack-race", even if it is in free flight without external forces acting on it. It is thus subjected to continuous accelerations and jerks. We therefore expect, in accordance with classical electromagnetic theory, that it will give off radiation, even in free flight. This essential difference with the "normal" electron would thus seem to invalidate the droplet model of an electron.

Now, having read Chap. 1, the reader would perhaps say that this is not true. Was is not argued in Chap. 1 that the sack-race of the droplet would normally be chaotic, a "drunkard's sack race", which prevents any radiation? This is true, but recall Fig. 8.15 near the end of Chap. 8. In this connection it was argued that under exceptional circumstances a so-called "degeneration" of the chaotic motion could happen, resulting in an ordered, sinuoidal motion. It was a statistical matter, such a degeneration would be a rare event occurring as the result of an extraordinary coincidence of circumstances, something like a heat-wave in mid-winter. Nevertheless, statistics do not absolutely rule out such rare events, they may occur. A degeneration from a chaotic to an ordered state would make radiation possible. We must conclude that a droplet even if it is in a free-flying condition could radiate (be it under exceptional circumstances), *unless there are other causes to block* the start of radiation.

There is indeed another lock on radiation, apart from the chaotic motion, and its nature will be explained in the later Sects. 10.4 through 10.6. This second lock will be found to be effective in free-flight conditions. The consequence is, that the droplet theory for now has been saved: under the special conditions of free-flight the droplet theory is still in agreement with the experimental evidence!

Now another observation may be made: if the droplet moves inside a potential well, then it *can* jump from one stationary state to another, shedding energy by radiation. So, what is it that does make such an essential difference between free-flight and bound motion?

Comparing the free-flying droplet with the droplet inside a well, the obvious difference is the occurrence of beats. Under the conditions of truly free flight, without any interactions with the outside world, the interference experiment of Fig. 8.5 using a second source of ultra-sound does not apply. In true free-flight the amplitude of the pulsation is constant after a degeneration of the chaotic state has occurred. In contrast, inside a potential well there will always be interference effects leading to beats, i.e. to variations of the amplitude. This was the result from the mathematical analysis: beats always occur inside potential wells, no matter whether there is radiation or not. Figure 10.3 summarises the difference between free-flight

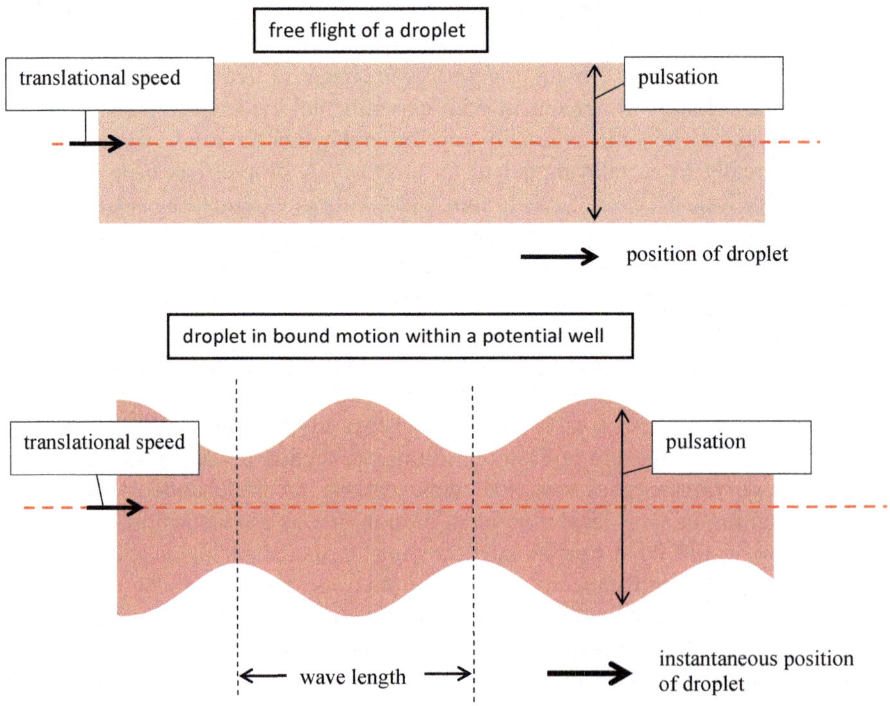

Fig. 10.3 The only difference that can be found between free-flight of a droplet and bound motion within a potential well is the occurrence of beats

and bound motion. In the motion of the droplet no differences can be found other than those shown in Fig. 10.3.

Pondering about Fig. 10.3, another feature strikes the eye. In the vicinity of the maxima and minima of the beat pattern, *locally* the amplitude of the pulsation hardly changes. When we restrict our field of view to the immediate surroundings of the maxima and minima, no difference at all can be discerned with the free-flight situation. On the basis of this locally existing similarity, one would expect that starting up radiation in or near these extremes of the beats is impossible, just as initialising radiation is impossible in the free-flight situation. And such a conclusion is exactly in agreement with Fig. 10.2. The blue points in Fig. 10.2 indeed coincide with the places where according to usual quantum mechanics the electron is invisible!

There may be readers who are already satisfied with this "explanation" of the occurrence of these mysterious zero-points, and they are excused if they skip the remainder of this chapter. The interpretation according to the droplet theory of the "Position Probability Density" curves of quantum mechanics is symbolised by Fig. 10.4.

The droplet is moving back and forth in the potential box, travelling the whole stretch between the walls. The motion of the droplet is not hampered by the zero-points, and a droplet may happily continue its journey right through these "forbidden points", which are not forbidden at all. The only thing is, that in and near these zero-points radiation cannot be started up, like radiation is inhibited in free-flight. Therefore, the droplet will never be seen there, it may pass the "forbidden" points but is temporarily invisible.

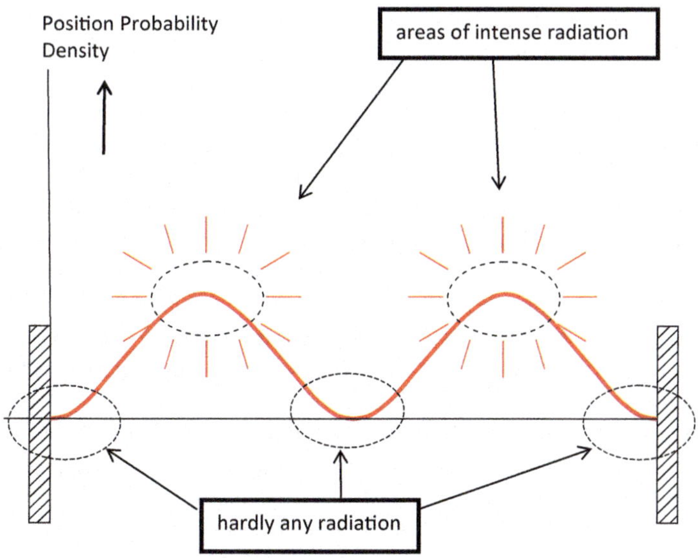

Fig. 10.4 The "droplet view" of the "Position Probability Density" curves of quantum

Will the droplet always radiate when it is far removed from these points of invisibility? No, remember what was earlier said about the usual chaotic state of its motion. The chaotic state *always* blocks radiation. It is only during the rare occurrences that there has been a degeneration into a more ordered motion that we have a *potentially* radiating state. Even so, the *actual* start-up of radiation will then have to await favourable conditions. And such conditions do not exist everywhere within the potential box. The start-up of radiation (and the possibility to observe the droplet) is restricted to places well away from the zero-points.

What is still needed, however, is an explanation why the starting up of radiation is blocked in a free-flying condition, and why in situations where the pulsation amplitude is changing fast, the conditions for radiation are more favourable. This has to do with the energy balance within the droplet, and it is quite complicated. Once again, if readers want to unhook at this point, they are forgiven! However, what follows is rather crucial in the complete story, because ultimately it determines whether we may consider the droplet of charge to be a good model of an electron (at least: in this respect).

10.4 Balance of Energy Within the Droplet

The energy which is present in the droplet comes in several "flavours". We have energy in the mean elongation, in the pulsation around this average, and in the translation. The purpose of what now immediately follows is, to become more specific and to get rid of any vaguenesses in this broad statement. After that, we will come back to the main question about when the conditions are favourable for radiating and when they are not.

Let us build up step by step the picture how the energy in the droplet is partitioned, by looking first at Fig. 10.5, which is partially a copy of Fig. 8.1. The figure shows the effect of three different values of the swaying speed. The centrifugal force on the mass causes it to move outwards and down, against the force by the spring in the elbow hinge. The compressed spring thus contains potential (stored) energy whenever the mass has a swaying velocity. By itself the swaying motion represents kinetic energy, so that in this situation the total energy has two components: kinetic as well potential energy. The analogous droplet of charge would similarly contain energy in two forms: the kinetic energy in the translation is accompanied by potential energy due to a change of its elongation compared with the "zero-speed" equilibrium length. The partition of energy is thus very clear, as long as there would be no nodding (or pulsation) oscillations.

The next step is shown in Fig. 10.6, where it has been assumed that the mass has additionally an oscillation in the nodding direction. The average position of the spring is again given by the blue, dashed circle in Fig. 10.6. The energy in the nodding has now become larger compared with Fig. 10.5. This is clear when we consider the mass when it just passes the blue broken circle on its way up or down. At such a moment the potential energy in the spring is just the same as in Fig. 10.5,

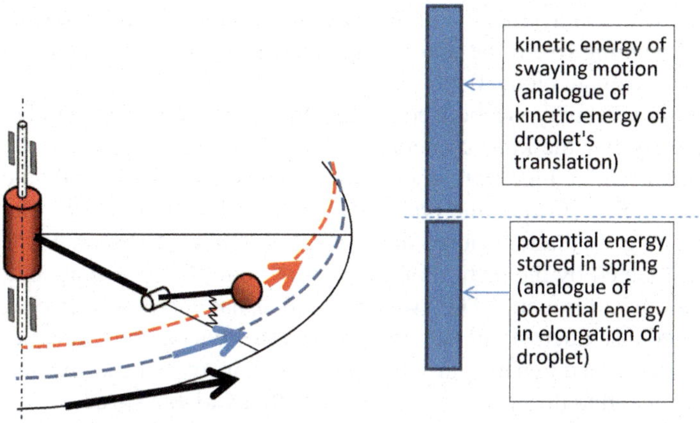

Fig. 10.5 The two kinds of energy: kinetic energy of the swaying motion and potential energy stored in the spring (analogues of kinetic energy in the translation of the droplet and potential energy in the elongation)

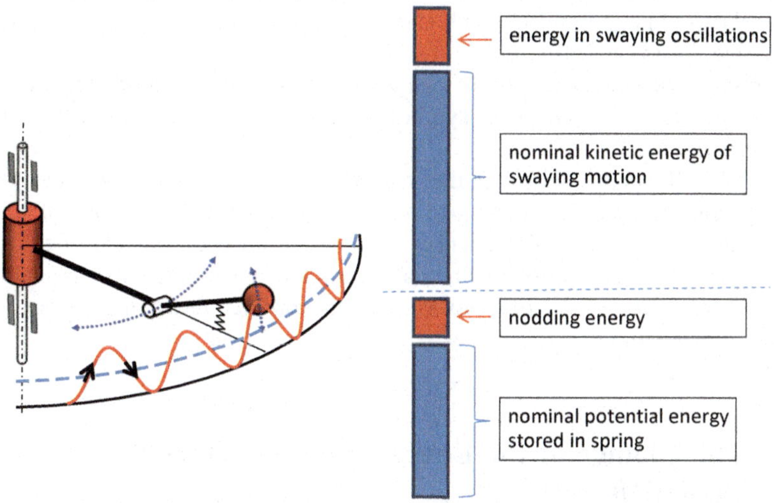

Fig. 10.6 Time-averaged break-down of energy if mass is oscillating in nodding direction. *Red blocks* indicate additional energy compared with Fig. 10.5

but there is additional kinetic energy associated with the up- or down velocity. In Fig. 10.6 this is symbolised by the small red block in the lower half of the energy partition diagram.

Definition.

In what follows the—rather artificial, and perhaps somewhat clumsy—nomenclature will be used to call that part of the potential energy in the spring which corresponds with the average nodding position of the mass: the *nominal potential energy* in the spring (or, if we are talking about the droplet: the nominal potential energy in the average elongation).

The additional energy symbolised by the red block in Fig. 10.6 will be called the *nodding energy* (or: the kinetic energy in the pulsation).

As symbolised in Fig. 10.6, the kinetic energy associated with the swaying velocity also increases when the nodding oscillation has been added. The reason is that the nodding oscillation causes the translation motion to become a "sack race" as earlier explained by the picture of a dancer periodically spreading and retracting the arms during a pirouette. The swaying motion has the character of an average velocity with fast fluctuations superimposed. The fluctuations have as a consequence that the kinetic energy in the swaying has increased too. An example can explain this: imagine that you drive your car with a variable speed, which is half of the time 80 km/h and the other half 120 km/h. The kinetic energy associated with the average velocity of 100 km/h is some factor times 100^2. The actual kinetic energy, averaged over the considered time period, is larger. It is the same factor times $(\frac{1}{2} \times 80^2 + \frac{1}{2} \times 120^2)$ which is larger than 100^2. Due to the quadratic dependence of kinetic energy on speed, the excursions of the speed to the high side have a relatively large impact, which is not compensated by the slower rides to keep the average speed on 100 km/h.

Therefore, the "sack-race like" swaying motion contains more kinetic energy than a steady motion, and this is symbolised by an extra red block in Fig. 10.6.

Definition.

In what follows we will call the kinetic energy associated with the average swaying velocity the *nominal energy of the swaying* (or in the case of the analoguous droplet: the nominal energy of the translation). The additional kinetic energy when there is a velocity fluctuation around the average will be called the *energy of the swaying oscillations* (or for the droplet itself: the energy of the translation oscillations).

10.5 The Change of the Energy Partition During Radiation

In the mathematical appendix it has been analysed what happens if the droplet of charge sheds some energy by radiating. The result is at first sight quite remarkable. It appears that radiating away energy goes at the cost of the two red blocks in Fig. 10.6 only. If the radiation would continue for a very long time, the pulsation around an average value of the elongation would asymptotically go to zero, i.e. the pulsation would diminish until after an infinitely long time only a constant elongation of the droplet would remain. The corresponding "sack-race" would asymptotically become a steady motion without any noticeable jumps in the translation.

Now recall item 6 in Chap. 5 concerning the "generalised radiation resistance" (which did not exist). The gist of it was, that the pulsation itself cannot lead to a type of radiation that carries off energy to infinity so that energy is lost forever. Only the fluctuations of the translation motion are capable of emitting radiation escaping to infinity, with a consequent definite loss of energy by radiation. It means that pulsation energy *can* diminish, but in a complicated two-stage process. The energy of the pulsation first has to be transferred to the translation oscillation, and from there it can be converted into radiation energy (at least into the type of radiation which is able to carry off the energy to infinity). The transfer of energy between these two dynamic modes, i.e. from the pulsation to the translation oscillation, takes place through the process symbolised by the dancer with flapping arms (Fig. 8.8).

Finally, you will perhaps remember the rather ugly bronze statue shown in the picture of Fig. 5.6. On purpose the author had sought out a picture that by many people would be esteemed rather low in its artistic value (this is a subjective opinion of course: granny liked to have the statue on her mantel piece), the express purpose was that it would be impressed on the reader's memory. The statue was used as an illustration that energy transfer from one oscillating system to another is possible only if there is a so-called "phase difference" between the oscillations. The oscillations of the two systems must be slightly out-of-step with each other, or else no energy transfer is possible. In the case considered here, during a radiating condition the pulsation must be slightly out of step with the translation fluctuations, or else the pulsation energy cannot be transferred to the translation, in order to be converted into radiation energy.

The mathematical analysis confirms this. A condition for shedding the droplet's energy in the form of radiation is, that during these circumstances the pulsation and the translation fluctuations are *not* completely in step with each other. This is a *conditio sine qua non.*

And here we have found why a free-flying droplet cannot radiate. It may happen to leave its chaotic state, so that the pulsations and the velocity fluctuations become less erratic and indeed perfectly sinusoidal. But in free-flight these two types of oscillation are *always* in step with each other. Therefore,we have still *not* fulfilled the conditions that are necessary to start up radiation. The free-flying droplet is simply incapable of ever going over to a radiating state. In this respect the droplet of charge thus behaves the same as a free-flying electron.

10.6 Radiation *Can* Be Started up at Certain Places Inside a Potential Well

Figure 10.3 (lower half) concerns the beats occurring if the droplet is bound inside a potential well. "Beats" is just another word for the amplitude variations of the nodding (or: pulsation). It should be realised that, through the process depicted in

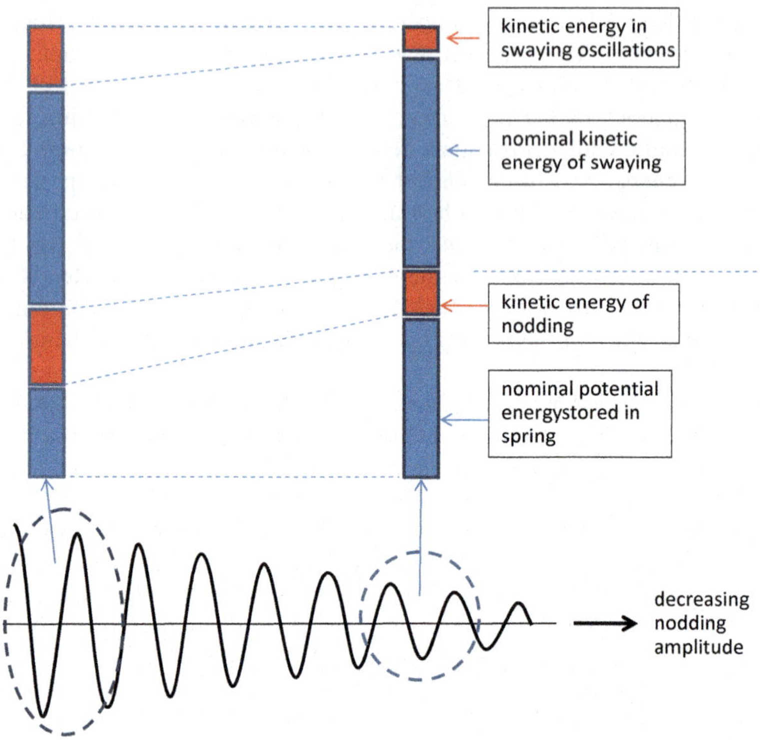

Fig. 10.7 Decreasing nodding amplitude in beat: energy partition changes. Constant total energy, and constant average swaying velocity assumed

Fig. 8.8 (the action of Coriolis-forces) there are similar beats in the swaying fluctuations (or: the fluctuations of the translation velocity). Energy-wise, the beats thus imply a periodic variation of the height of the red blocks in Fig. 10.6.

When the droplet is somewhere in between the "walls" of the potential box no external forces act on it. This implies that the *average* velocity does not change as long as it does not come in contact with the walls. Neither can the *total* energy change. When the red blocks become smaller and larger due to the beat phenomenon, this must go at the expense of the nominal potential energy of the spring (or: the nominal potential energy in the elongation of the droplet). The situation is clarified by the diagram of Fig. 10.7.

The periodic rising and falling of the energy in the velocity fluctuations due to beats means that now energy is exchanged between the upper red block in Fig. 10.6 and the lower red block. Such an energy exchange requires that a "phase difference" exists between the pulsations and the velocity fluctuations: they cannot be exactly in step with each other. Here we have a phase difference that is *enforced* by the occurrence of beats. In other words, the conditions of a droplet within a potential well open up the possibility that also radiation can be initiated. The possibility to start up radiation is not present everywhere in the well, the conditions conducive for radiation are only found at places where the phase difference happens to be in the right direction. The favourable conditions are certainly *not* present at places where the beats have their maxima or minima.

In summary, the above given arguments explain why a droplet of charge can enter a radiating state when it is moving inside a potential well, in contrast to the free-flying droplet. It also explains why at certain positions in the well radiation will never be started up. It has been shown that these positions, where the droplet cannot radiate and is thus invisible for the outside world, exactly coincide with the positions of the zero-points of the "position probability density" curves of usual quantum mechanics.

Chapter 11
Schrödinger's Equation

11.1 The Crucial Question: Is All This True for a General Potential Well?

If you page back and browse through the Chaps. 9 and 10, you will see that invariably the potential *box* was considered. It was chosen because of its relative simplicity. The question is of course: "might it be that the droplet's behaviour, which seems to conform so well with the results of quantum mechanics, has a *purely coincidental similarity* with real quantum behaviour, occurring just in the case of the box?".

It could all be a coincidence, happening solely in this simplified case. Who can tell how the droplet will behave in more complex situations, for instance in a potential well of a more general shape? Would then still be found that its behaviour conforms to the results of quantum mechanics?

This will be the subject of the present chapter.

11.2 The Parabolic Potential Well as Another Example: Fortunately OK

I can imagine that you feel some anxiety now, fearing that you have read so many pages for nothing, only to come at this point to the conclusion that the potential box is just a lucky exception, and that more general cases cannot be "explained" by the droplet theory. Fortunately, at least one other example can be given of a type of potential well where the droplet theory can be worked out and still leads to the same results as usual quantum mechanics. This is the parabolic well (in textbooks often called the "harmonic oscillator"), and its characteristics will be considered in the present section. The sections after that will then make it plausible that agreement

T. van Holten, *The Atomic World Spooky? It Ain't Necessarily So!*
DOI 10.2991/978-94-6239-234-2_11

between quantum mechanics and droplet theory may be expected *in general*, although it is obviously impossible to treat every type of well individually.

In the present context an important feature of the parabolic well is that it allows a mathematical solution in closed form, both in the droplet theory as well as in usual quantum mechanics. This is rather exceptional, most problems in quantum mechanics require more than "just a piece of paper and a table of mathematical functions" to yield solutions. The solution offered by droplet theory when it is applied to the parabolic well, is shown in the mathematical appendix. And it agrees nicely with the solution mentioned in quantum mechanics text books. Although I am sure most of the readers are not going to leaf through the appendix, this knowledge may take away most of their above sketched concerns. Another good point about having obtained a solution in closed form is, that the mathematical solution greatly helps to describe in non-mathematical terms what is the fate of a droplet of charge when it is moving inside such a parabolic potential well.

Let us start with a description of what is meant by a "parabolic well". The description will be helped by Fig. 11.1, which is a copy of Fig. 3.3 in an earlier chapter.

In a parabolic well the electrical force on a droplet of charge is non-zero (in contrast to the box), and the force on it depends on the position of the droplet. The variation of the electrical force is similar to the variation of the gravity force on a marble rolling inside a parabolically shaped bowl. This leads to the visualisation used in Fig. 11.1.

The force is zero in the middle of the well. If the droplet is at another place, it feels a force driving it towards the middle of the well, and this force is larger the further the droplet is away from the middle of the well.

The average velocity of the droplet ("average" meaning that the fast velocity fluctuations induced by the pulsation are ignored) varies in the same way as the velocity of a marble rolling back and forth in a bowl. The velocity is large in the middle, whereas the droplet slows down the further it moves away from the centre,

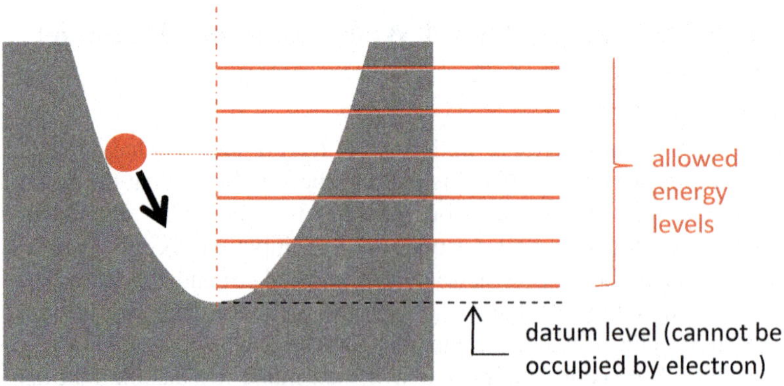

Fig. 11.1 The droplet of charge in a parabolic potential well, apart from its pulsation comparable with a marble in a parabolically shaped bowl

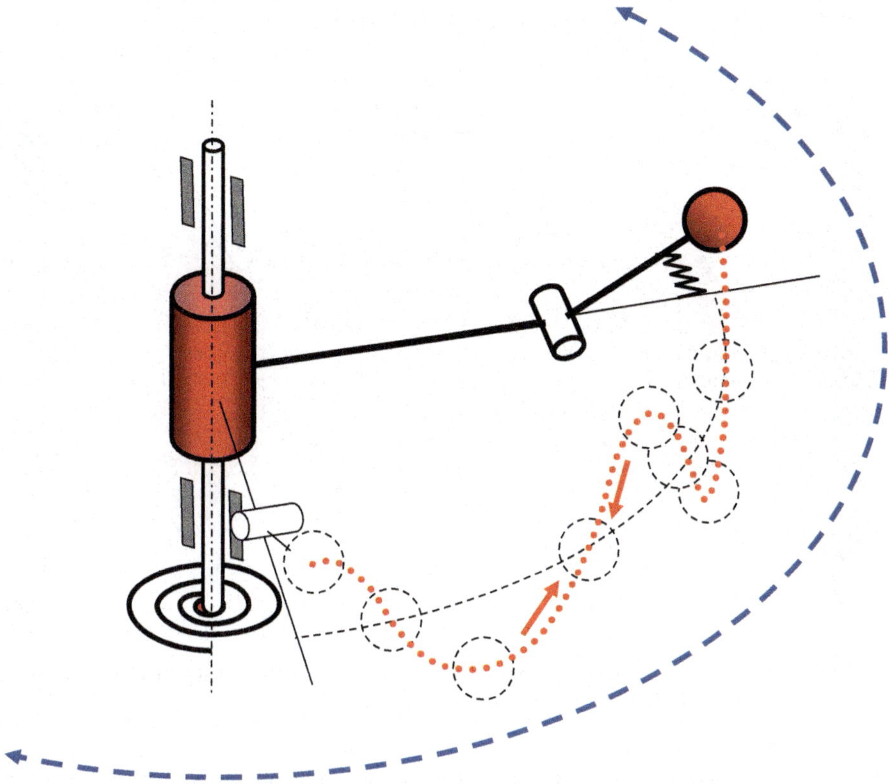

Fig. 11.2 Visalisation of the translation and pulsation of a droplet of charge by the swaying resp. the nodding of a mass. The spiral spring gives the same force variation as the electrical force inside a parabolic potential well

until it comes to a stop at a certain point after which it moves back to start a new journey towards the middle of the well (only to overshoot it at high speed).

The equivalent of the parabolic potential well is shown in Fig. 11.2 in the form of a swaying and nodding mass, where the swaying motion is restricted by a spiral spring. Compare this figure with the earlier Fig. 9.4, which showed the analogon of a potential *box*. The difference between these figures is that the hard stops in Fig. 9.4 are now replaced by the spiral spring, making the swaying motion more smooth.

Almost superfluous to say: the trajectory of the mass (the red dotted curve) is just a rough symbolisation of the nodding motion. The nodding is actually a very fast up and down motion, comprising uncountably many cycles in a single sweep between the "walls" of the potential well.

It should be stressed that what is symbolised in Fig. 11.2 is a *stationary* situation with so-called "path retracing". The mass retraces its steps, no matter whether it is swaying from left to right or in the opposite direction. This is symbolised by the red arrows in the figure.

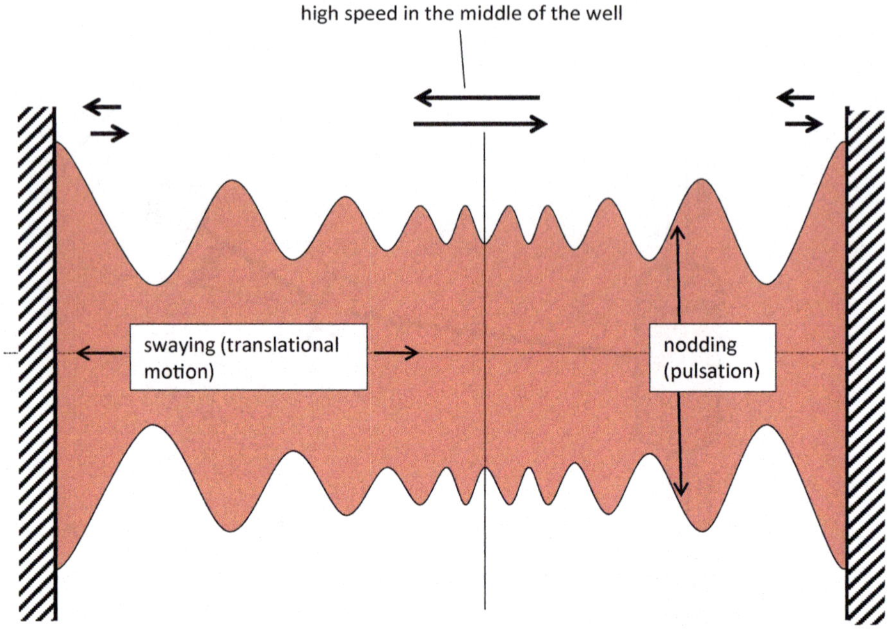

Fig. 11.3 Schematic of pulsation pattern with beats inside a parabolic potential well (intermediate high energy level), visualised by the analogon of Fig. 11.2

Yet another picture of what goes on in a parabolic potential well is given by Fig. 11.3. From the mathematical analysis it follows that again beats occur in the pulsating of the droplet (or what is the same: slow amplitude variations in the nodding motion of the analogon). In Fig. 11.3 the individual nodding cycles are not shown, they are symbolised by the hazy, reddish area to suggest that they cannot be seen on this scale due to their high frequency. However, the beats occurring in the nodding motion have a quite different, much larger scale so that they are visible without difficulty. For the sake of clarity an intermediate energy level has been chosen for the sketch, so that the number of beats is not too small, which helps the further discussion.

In this latter figure it is again indicated by the arrows that we are here considering a stationary state. The beats are exactly repeated every time the droplet travels back and forth between the points where it reverses its direction of motion.

What is immediately obvious from the sketch is, that the waves of the beats are shorter near the middle of the well than they are near the "walls". This is in agreement with the variation of the swaying velocity of the mass (or equivalently: the droplet's translational speed). The swaying velocity is high in the middle. Next, recall that the droplet's behaviour conforms with De Broglie's relation in quantum mechanics: the beats are shorter the higher the droplet's speed and vice versa. This was expressed by the proportionality

$$\lambda_{beats} \div \frac{1}{v_{average}}$$

In the middle of the well the droplet has the highest speed, just as the marble in the parabolic bowl, and the beats become shorter. On the other hand, near the sides of the well the droplet is decelerating to a stop. Exactly in the point of reversal of its motion the speed is momentarily zero, and according to the above given formula, the beat pattern there would show a large wavelength, even approaching an infinitely large wavelength in the end points.

Although this description of the droplet in a parabolic well is intuitively quite understandable, it does raise the question: what do we mean by "the wavelength of a beat *in a particular point of the droplet's trajectory*"? It sounds like a contradictio in terminis: wave*length* suggests something that has an extension over a certain distance, whereas a single point is the opposite of a distance. This dilemma will be taken up in the next section of the present chapter. Having answered the question, it will appear that it is not difficult to proceed and "derive" Schrödinger's equation in the section following the next one. We will find that the behaviour of the droplet is governed by this famous equation, just like the particles in quantum theory.

For now, let us return to Fig. 11.3. Obviously, the beat pattern is more complicated in a parabolic well than it is in a potential box, due to the variable wavelengths. Still, the same requirement is applicable as in a box to have a stationary state. The pattern of beats should exactly be repeated during each pass of the droplet between the walls, no matter in which direction the droplet is moving. In the analogon this behaviour would be called "path retracing", i.e. the swaying and nodding mass must tread the same path during each sweep. This is by definition what is meant by the word "stationary": although the word suggests differently, it does *not* mean "motionless", a stationary state indicates that the motion *pattern* remains the same, and does not depend on time.

In the case of the parabolic well the rather complicated beat pattern shown in Fig. 11.3 must thus be fitted in the width of the well. And this is only possible for certain discrete energy levels, hence the quantisation of energy. That is, there is quantisation if we insist on having a stationary situation, or if the droplet has a "natural desire" to do so; see Chap. 9, Sect. 9.7. The quantisation of the energy levels found in this way by the mathematical analysis of the appendix is shown by the red horizontal lines in Fig. 11.1. Quantitatively these lines are in complete agreement with the quantisation levels found in quantum mechanics for the parabolic well! On the one hand this may be reassuring, but on the other hand it may nevertheless surprise you: if you page forward to Fig. 11.12 you will there see that the so-called "wave functions" of quantum theory have an entirely different shape than the beat patterns in Fig. 11.3. This difference shows up especially clearly near the "walls" of the well. If these quite differently shaped wave-functions of quantum mechanics are fitted in the width of the parabolic well, again this is only possible for certain energy levels. And now it is found that these energy levels are exactly the same levels as we have found in the case of the droplet theory, despite the different shapes of the curves! Once again: in a way it may be considered to be reassuring, but it is surprising too or even somewhat disturbing. At the very least an explanation is required. In the later sections of the present chapter it will be made plausible why it is not so surprising, really. Anyway, the result is as promised: the quantisation according to droplet theory is the same as in quantum mechanics.

11.3 What Is "Local Wavelength"?

Let us now first continue with the more technical question: what should be understood by the "wave-length" of the beat pattern, when the wave length is variable and is different in each point of the wavy boundary of the reddish area in Fig. 11.3? Phrased differently: what is a *local* wavelength?

What we intuitively do when speaking about a "local wave length at a certain place" is, that we imagine the curve to be pieced together from small segments, each segment cut from an appropriate long sinusoidal wave. This idea is schematically shown in Fig. 11.4. For each of the small pieces which together make up the whole curve, we can define that its "wave length" is the wave length of the sine curve from which it has been cut. It is a rather intuitive idea, but in this way we at least have a (sort of) definition of what can be meant by a "local" wavelength.

Now the inverse problem: if someone would cut a piece from the curve with the continuously varying wavelength, gave it to us, and told us that it originally had come from a pure sine wave, would we be able to reconstruct from which sine wave

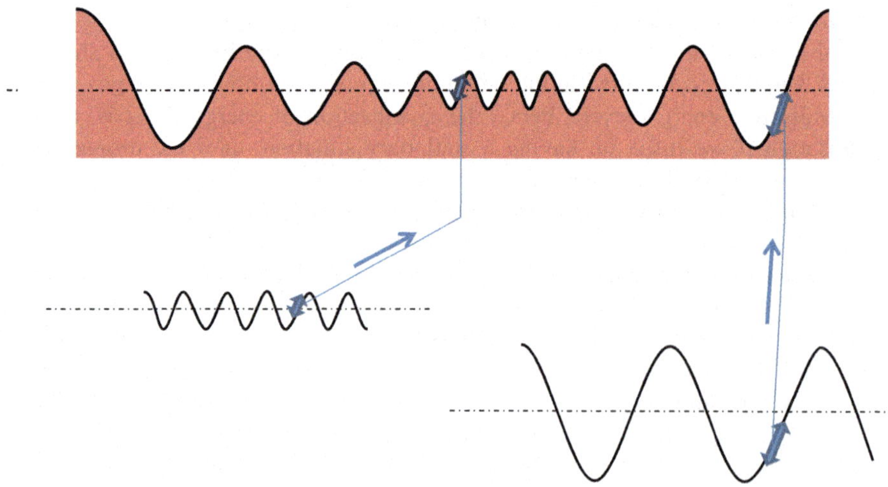

Fig. 11.4 The curve with variable "local wavelength" is imagined to be pieced together from segments, each segment cut from an appropriate sine wave

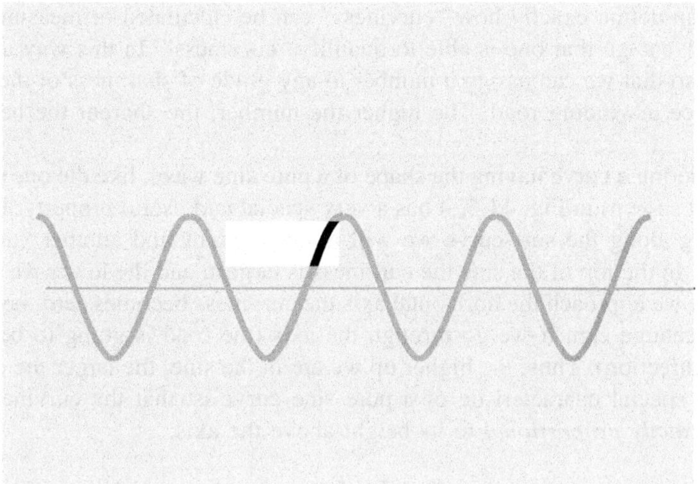

Fig. 11.5 Only a small piece of the entire sine wave is visible. Can we reconstruct the whole wave from just this small part?

it came and what the wavelength of that sine wave was? Perhaps at first sight the problem would seem to be of academic interest only. It is typically the type of problem only mathematicians like to spend their time on, but in a few moments you will see that the question is relevant for the subject of this chapter.

The problem can be formulated in slightly different words: if we would mask a sine curve, so that only a small piece of the entire wave would be seen and the remainder is blanked off, could we still determine how large the wavelength is of the wave from which the piece has been cut? The situation is shown in the picture, Fig. 11.5. We assume that only the black piece of the curve can be seen, the rest is invisible. Can we from the shape of that small piece reconstruct how the complete curve looks like?

It is like a paleonthologist who tries to reconstruct a complete prehistoric animal from just one bone he has found during an excavation.

In our case mathematicians have found a reliable reconstruction method. They start to have a look at the "curviness" of the piece of wave they have been given. The word "curviness" is not an existing word in English nor in mathematics, but it is found in some books about quantum mechanics (see e.g. the book by K.W. Ford: 101 quantum questions). The word was invented to avoid another word, viz. "curvature" which is already defined by mathematicians, but means a different thing. "Curviness" is a metric for how sharp the bends in a curve are. The larger the curviness, the sharper the bend. In normal language we would call it the curvature, but unfortunately this expression has already been claimed by mathematicians for another thing. For the mathematicians amongst the readers: the word "curviness" is here used for the second derivative of a function which oscillates around a zero mean value. For readers who do not "speak mathematics" it is sufficient to know that we can define exactly how "curviness" can be calculated or measured, which has the advantage that one is able to quantify "curviness". In this way a metric is available so that we can assign a number to any grade of sharpness of the bends in for instance a winding road. The higher the number, the sharper the bend in the road.

Considering a curve having the shape of a pure sine wave, like the one shown (or rather: not shown) in Fig. 11.5, it has a very special and useful property. Evidently, proceeding along the sine-curve we will at every point find another value of the curviness. In the top of the sine the curviness is largest, and the lower we come, the smaller. If we approach the horizontal axis the curviness becomes zero, because it is going to change sign if we go through the axis (the road is going to bend in the opposite direction). Thus, the higher up we are in the sine, the larger the curviness. Now, the special characteristic of a pure sine curve is, that the curviness of any piece is *exactly proportional* to its height above the axis:

$$(curviness\ in\ a\ point) \div (height\ of\ the\ point\ above\ axis)\ (for\ a\ sine)$$

If we divide the number indicating how sharp the bend is (i.e. the curviness) by its distance from the axis, we always find the same value, it does not matter which part of the sine wave is taken. Let us call the ratio $(curviness)/(height)$ the *relative curviness*.

For a sine-wave the relative curviness is constant, it is the same in each point of the wave. The whole of the sine is thus characterised by one number, i.e. the relative curviness in all its points. Its actual value will of course depend on the particular shape of the sine-curve. This is pictured in Fig. 11.6.

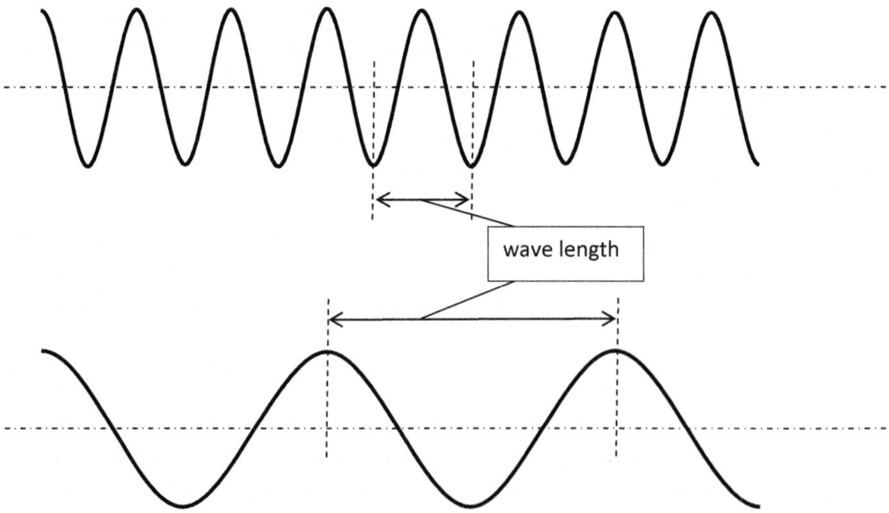

Fig. 11.6 The relative curviness, i.e. (local curviness)/(local height), is the same in each point of a given sine-curve. Obviously, its value is smaller if the wave-length is larger

If the sine is long stretched out, i.e. if it has a long wave-length, then the relative curviness is small, and vice versa. How this characteristic number, the relative curviness, depends on the wave length is a simple relation:

$$\textit{relative curviness} = (\textit{curviness})/(\textit{height}) \div \frac{1}{\lambda^2} \quad \text{(pure sine wave)}$$

where the symbol "÷" stands for "proportional to", and the symbol λ indicates the wavelength. The formula says once again the same thing: a long stretched out sine wave, with a large wavelength, is characterised by a small relative curviness. And vice versa of course.

Now back to the mathematician to whom the picture of Fig. 11.5 was shown, and who had the task to reconstruct the rest of the—invisible—curve. There are three things he/she would need to know about the piece of curve that is visible: (1) the height of the piece above the horizontal axis, (2) the slope in the middle of the piece, and (3) the curviness of the piece. By combining (1) and (3) the wave-length of the invisible sine curve is immediately known, because together they give the relative curviness. The extra information (2) about the slope is needed—and sufficient—to tell how large the amplitude is of the "mother curve", and from which place of the mother curve (the so-called "phase angle") the piece has been amputated.

Having learned how to play this game in the case of a sinusoidally shaped mother curve, we can play the same game for any other curve, *if* we assume that this curve—of relatively arbitrary shape—has been pieced together from segments cut from sine curves. Schematically this is shown in the next Fig. 11.7, resembling Fig. 11.4 but with the blue arrows reversed.

For each piece of the variable wave-length curve the relative curviness can be determined, so that it is known from which type of sine-curve it was cut. The "local wave-length", defined as the wave-length of the corresponding sine-wave, is now known from the formula given above.

The nice thing is, that mathematicians know how to determine the slope and the relative curviness from just a small piece of the whole wave. They are even more clever than that: they are able to calculate or measure the relative curviness when the piece they have been given is *infinitesimally* small. From just a piece which is near to nothing they can reconstruct the whole curve! Like you would offer a paleonthologist not just a bone, but even less: a small splinter of it, and would ask him/her to draw a picture of the whole beast.

In the next part of the story it will appear how useful it is that mathematicians are able to play games like this.

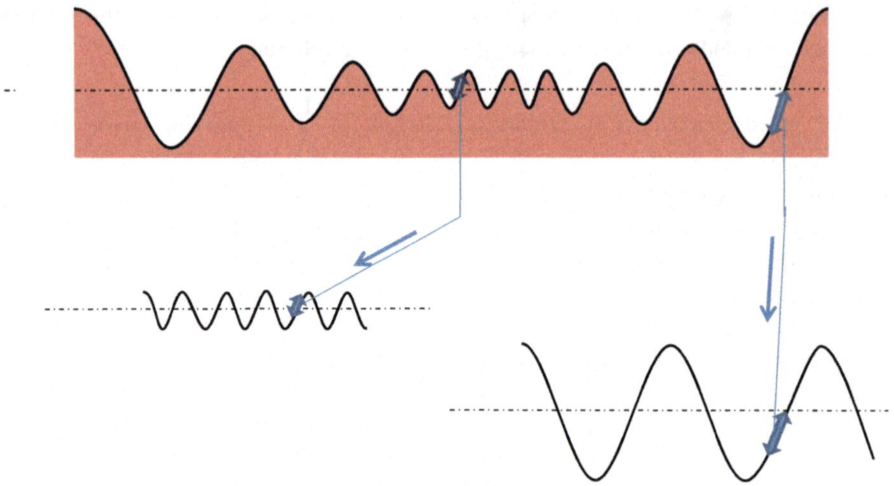

Fig. 11.7 A curve with variable wave-length: for each piece the relative curviness can be calculated or measured, so that it is known from which type of sine-curve it was cut. The "local wave-length" is defined as the wave-length of the corresponding sine-wave, and is now known

11.4 Schrödinger's Equation!

Everything that has been said about how we can assign a variable wave-length to a curve (and calculate the local wave-length in each point) was rather general. It applies to any type of curve, as long as the curve reminds one vaguely of a real sine-curve: it has to wave up and down around some average value.

But remember where the question "what as a matter of fact is a local wave-length" came from. The question came up when considering Fig. 11.3, where the beat pattern was shown in a parabolic potential well, i.e. the variations of the pulsation amplitude when the droplet is moving back and forth in such a well. It was rather intuitively surmised that the wave-length of the beats would be small in the middle of the well and large near the edges, based on the fact that in the middle the speed of the droplet is high, whereas the beats comply with De Broglie's formula:

$$(local\, wavelength\, of\, beats) \div \frac{1}{(average\, speed\, of\, droplet\, at\, this\, point)}$$

Now, in the preceding section it was explained what must be understood by "local wavelength". It was the wavelength of the sine that in the point considered could intimately nestle itself against the curve of beats. "Intimately nestling" was made possible by choosing a sine that has the same relative curviness as the curve of beats in the considered point. And finally, a quantified relation was given in

Sect. 11.3, connecting the relative curviness in a point of the beat-curve with the so-defined local wavelength:

$$(relative\, curviness\, in\, a\, point) \div \frac{1}{(local\, wavelength)^2}$$

Now combining these last two formulae:
Schrödinger's equation!

$$(relative\, curviness\, of\, the\, beat - curve) \div (local\, speed\, of\, droplet)^2$$

Perhaps you do not believe your eyes reading the exclamation above the grey box. In fact, there is a good reason to print this inside a box: it really is nothing else than the famous equation by Schrödinger, the equation that describes all of quantum mechanics! It is as simple as that! I should add a nuance to the excitement: it is Schrödinger's equation by which *stationary* states can be calculated in quantum mechanics, in one dimension. There is another form of the equation too, which describes what happens when there is a transition from one stationary state to another. That, however, does not concern us here, it is the subject of a later chapter. For now, we conclude that this famous equation crops up in our droplet theory, and describes the beat pattern found in the pulsations of a droplet of charge. Needless to say: in usual quantum mechanics there does not exist something like "pulsation of particles", and Schrödinger's equation there describes the form of the wave functions, i.e. the shape of the matter waves accompanying particles.

Of course, the form in which the equation is stated here is our own format, you will see it in the textbooks always written out in a different form. This may be the case, but be assured that Schrödinger's equation, no matter the format used, is nothing else than the expression in the grey box above. In a separate textbox at the end of this chapter it is shown how the above given formula can be written in terms of the usual mathematical symbols, so that we obtain the formula that you will find in all the physics books. The textbox is only meant for readers who think they can digest the more formal mathematical expressions. An advise to the rest of the readers: do not read the textbox! It does not really add any new thought or idea. What is more: we can easily in words explain the steps that Schrödinger took to arrive at the usual form of what is nowadays called "Schrödinger's equation".

There are three steps necessary to come to the usual form of the equation, starting from "the grey box".

1. First of all, of course the descriptor "local relative curviness" was formulated using mathematical symbolism. There is nothing special about it, it is a part of the

toolbox of mathematicians since the 17th century, when the branch of mathematics called "differential calculus" was invented and developed by Newton, Leibniz and others coming after them. It is purely a technical issue, and we will leave it to the textbox.

2. Second, and also needing hardly any further justification: the expression in the grey box, which is in the form of a proportionality (written with " $\div$"), is written in the form of an equation (written with "="). This is done by plugging in the constants of proportionality.

3. Schrödinger saved the users of his formula time, in other words: his formula was streamlined and was made as efficient as possible for the user. The point is that in our expression above the velocity squared $(v_{average})^2$ of the droplet was used ("average" here means: ignoring the fluctuations induced by the pulsating). The kinetic energy of the droplet E_{kin} is proportional to its velocity squared, so that the expression in the grey box above may be formulated alternatively:

$$(relative\ curviness\ of\ the\ beat - curve) \div (instantaneous\ kinetic\ energy)$$

In a potential well of a given, known shape such as the example parabolic well, the value of the kinetic energy velocity is known immediately as soon as we have specified in which point of the well our droplet is momentarily situated. Take the parabolic well as an example, and have a look at Fig. 11.8 where this kind of well is again represented by a parabolic bowl.

The total energy of the marble E_{tot} is given by the level from which it started its roll, because here the velocity was still zero (no kinetic energy) and the total energy was therefore equal to the potential energy E_{pot}. After having released the marble it rolls down the slope and acquires velocity. In other words: it exchanges its potential energy for kinetic energy E_{kin}. During the roll the total energy is conserved, so that we have

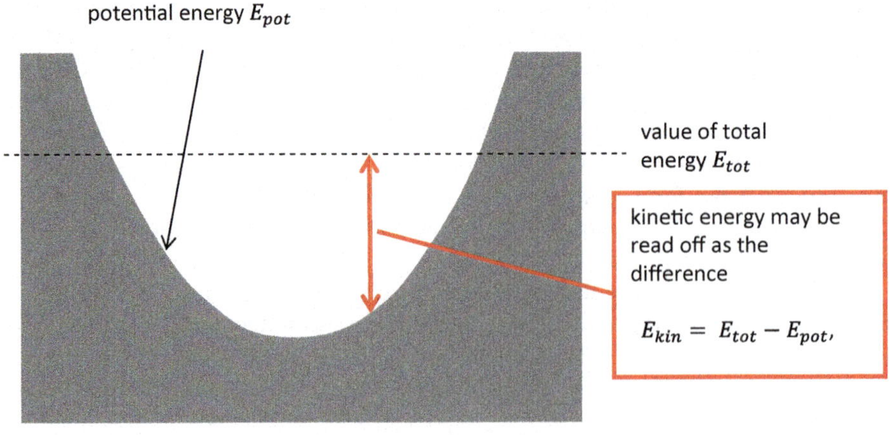

Fig. 11.8 Wherever the droplet is in the well, its kinetic energy is known at each point as the difference between total- and potential-energy: $E_{kin} = E_{tot} - E_{pot}$

$$E_{tot} = E_{pot} + E_{kin}$$

At each point where the marble is instantaneously found the potential energy E_{pot} is known, because this is given by the height of that point. We can in each point of its roll thus calculate how much kinetic energy the marble has, by using the above equation in an inverted form:

$$E_{kin} = E_{tot} - E_{pot}$$

Schrödinger avoided that the user of his formula had to take all these steps, by directly substituting all this in his equation. The expression in the grey box above can thus be written in the form:

$$(\textit{relative curviness of the beat} - \textit{curve}) \\ \div (\textit{total energy minus local potential energy})$$

And that is all. Quantum theory is simpler than you thought, isn't it? Unfortunately this derivation of Schrödinger's equation—"heuristically" as it is—is not given in any of the textbooks the author has knowledge of. But that can be his own fault of course, not knowing all the textbooks in existence.

11.5 How Is Schrödinger's Equation Used?

Let us stay with the parabolic potential well, to give an example. We assume to know the energy level, i.e. the total energy of the droplet. We come back later to this, because you may—rightly—wonder how we would know this beforehand. In fact, we do *not* know how large E_{tot} is, at this stage we can only make a wild guess. But let us assume that someone has stuck his head out, and provided us with a first guess. The way to proceed is then as follows. It is sketched in Fig. 11.9. The shape of the potential well tells us what the potential energy is at any position of the droplet within the well. As shown in the figure, the difference between E_{tot} and E_{pot} may be read off from the figure, at any place in the potential well where we wish to know it. According to Schrödinger's equation, we thus know at any place the value of the relative curviness of the beat curve we want to reconstruct. Let us say we want to reconstruct for instance a beat curve like the one shown in Fig. 11.4.

Now recall: relative curviness was defined as

$$\textit{relative curviness} = (\textit{curviness})/(\textit{height})$$

In other words, Schrödinger's equation tells us *how sharp the bend of the curve is, in each point* of the curve. A mathematician then has sufficient information to reconstruct the whole curve of the pulsation beats, spanning the entire width of the well. The work of the mathematician is pretty well comparable to that of a gas fitter

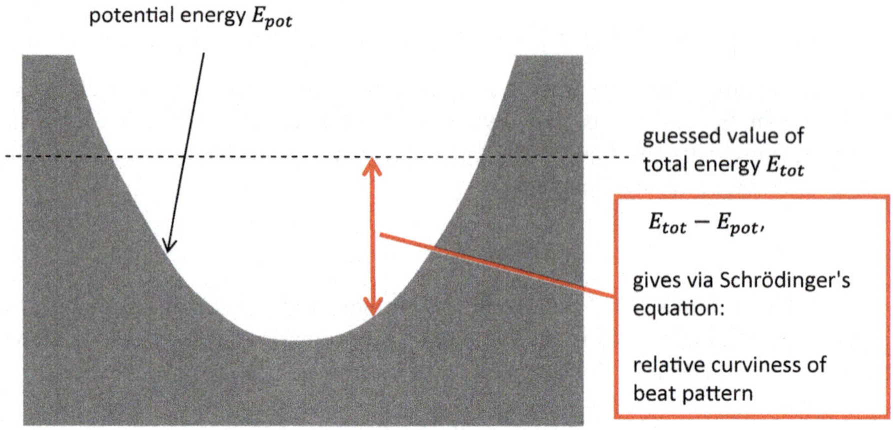

Fig. 11.9 First step to calculate the beat pattern inside a parabolic potential well: guess a value of the total energy

who has the task of laying a gas pipe from the cellar where your gasmeter is placed, to the new cooker you had just installed somewhere else in the house. He would start to install a T-shaped connection in the main pipe near the meter, and then he would proceed by preparing a lot of pieces of pipe, bent in the correct shape. Reading from his installation drawings he would cut small pieces, bend them individually, and afterwards would fit them piece by piece together to obtain the whole gas connection.

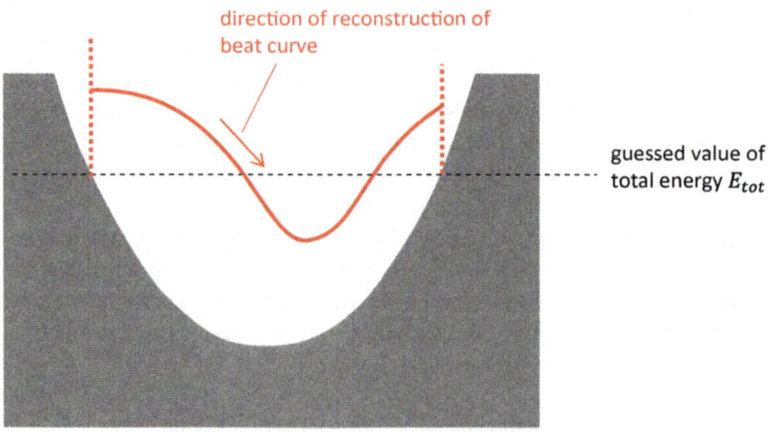

Fig. 11.10 Using the available information about the local curviness, the beat pattern is reconstructed, and it is checked whether the assumption about the total energy was correct

Likewise a mathematician would start for instance on the left side of the well. He/she starts with the first piece of the beat curve, which must have zero slope (because of our requirement of path retracing). Then, using Schrödinger's equation as a prescription that tells him/her how sharp the bend must be in the next piece, this next piece can be joined, and so on. A schematic picture of the procedure is given by the next Fig. 11.10. This time we take as an example a low energy case, with a small number of zero-crossings of the curve, again for the sake of clarity.

11.6 Quantisation of Energy

And then it appears that something has gone wrong! Reaching the right end of the potential well, we see that here the reconstructed curve does not have a zero slope, as we would want it to have. Schrödinger's equation cannot be blamed for this mishap. It could not know that zero slope is what we wanted, it has just faithfully prescribed the bends of the curve that are in agreement with the energy difference $E_{tot} - E_{pot}$. It is our own fault, because we apparently made the wrong guess about the total energy E_{tot}.

Now begins a process of trial and error. If we would assume a higher value of the total energy, in each point the difference $E_{tot} - E_{pot}$ would be larger, and the bends in the curve would become sharper everywhere. A higher total energy would thus shorten the reconstructed curve, which works in the right direction. What also helps is, that a higher total energy gives somewhat more space between the ends of the

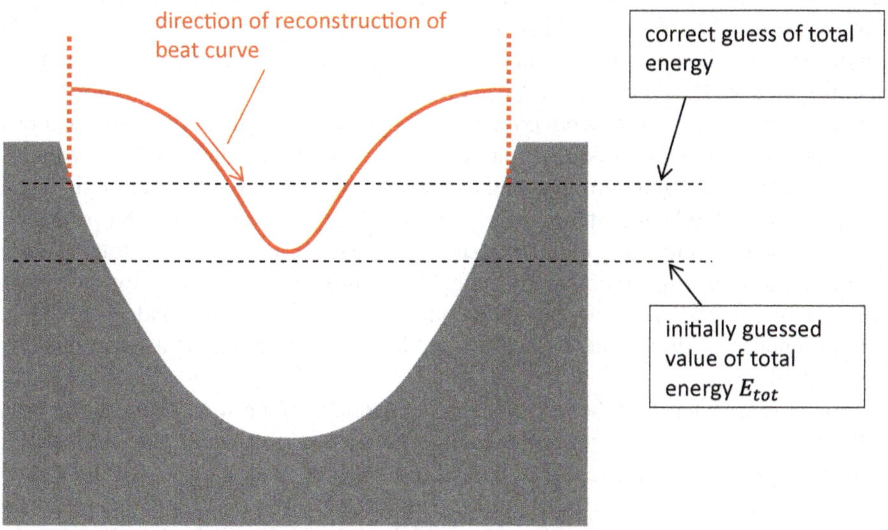

Fig. 11.11 The trial and error process to find the total energy which gives the correct boundary condition

potential well. After a few trials we would then be able to achieve the desired result, as sketched in the following Fig. 11.11.

It is a process of trial and error, like "shooting in on a target" by a piece of artillery to reduce in steps the "miss distance". What also becomes clear is, that the energy of the droplet in a potential well of a general shape is quantised: the energy must have a very special value in order to achieve the right conditions for path retracing. Of course, there may be more of such special energy values, but then one has to assume that the targeted curve has more zero-crossings. More zero-crossings automatically imply that the curve has sharper bends everywhere, and in order to get agreement with Schrödinger's equation we will have to assume a higher level of the total energy. But again, this higher level is also quantised: if we have for example a curve with ten zero-crossings, there is only one particular value of the total energy that can give the desired end conditions.

11.7 Do Less Cumbersome Methods for Reconstructing the Beat Pattern Exist, Other Than Trial and Error?

In a few exceptional cases this whole process of reconstructing curves and "shooting in" can be done by a streamlined analytical approach which does not involve any trial and error but arrives straight at the desired result.

You will understand that in the ordinary quantum theory the problems are very similar to those described above. The reconstruction of a so-called "wave function" in a potential well (see the examples on the left side of Fig. 10.1 or 11.12) must also be based on the use of Schrödinger's equation. The difference is, that the conditions that are required in the end points of the "wave function" are not the same as the conditions we require for the curves of the beat pattern. Because it is not the intention to give here a treatment of quantum mechanics theory, we will not go into the question what sort of end conditions apply to the wave functions inside a potential well. The end conditions are definitely different, which explains why the wave functions of quantum mechanics will unavoidably have a different shape compared with the beat patterns in our droplet theory. Nevertheless, the procedure to reconstruct the curves is the same: start by guessing a value of the total energy, do the pipe fitter's job with Schrödinger's equation as the recipe how the pieces should be bent, and check whether the end of the pipe connection indeed gives the correct junction with your new stove. If not, do it all over, based on a better guess of the total energy.

You may wonder if all these clever mathematicians in the world have never been able to invent a different, more efficient way to do the job. A real pipe fitter would abhor from the idea to make a pipe connection, only to find out if it was the correct one after having completed everything. Fortunately, in a few simple cases it is possible to find analytical approaches which go straight to the correct answer, without a tedious process of trial and error. In a very few exceptions such an

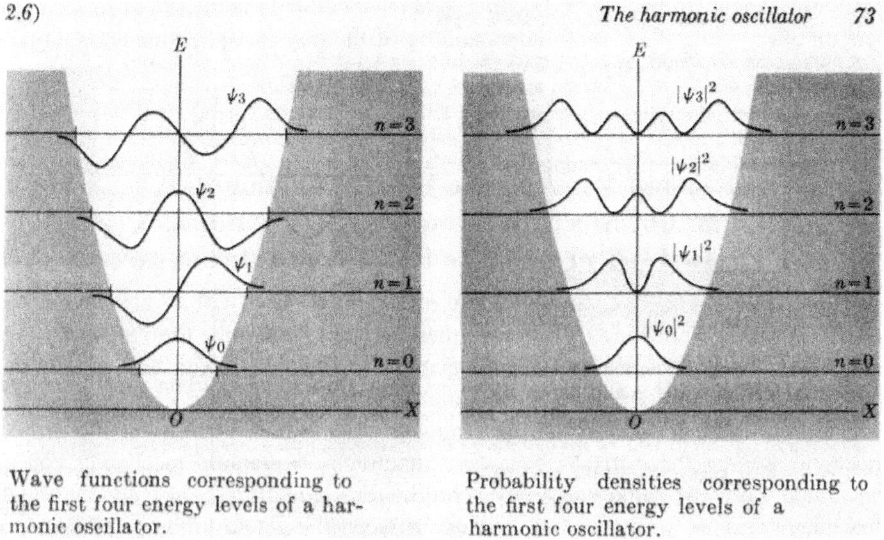

Fig. 11.12 Quantum mechanical solutions of Schrödinger's equation in a parabolic potential well. *Left* the wave functions. *Right* the square of them, representing the Position Probability Density. Copied from: Alonso and Finn, Fundamental University Physics

analytical solution can sometimes be found just using pen and paper, only assisted by a mathematical "bible" which gives information on all sorts of mathematical functions. If that would be possible for all the problems in physics, it would be great. But alas!

An example of such an easy case is the potential box. To find its solution completely analytically, and in closed form, is a piece of cake even for mathematicians who only did the beginner's courses. Incidentally, the derivation of the results for the potential box in Chap. 9 was done in a different way, and did not involve an attempt to solve Schrödinger's equation (after all, in Chap. 9 we were not yet familiar with that equation). However, we could have done it easily by solving Schrödinger's equation, and of course would have found the same answers.

A next level in difficulty is the parabolic well. A parabolic potential well requires a mathematician who at least has followed a few more advanced courses.

A still higher level of craftmanship and experience is needed if it is wished to unravel the architecture of the simplest atom, the hydrogen atom. It can be done analytically, in three dimensions, finding all the possible energy levels and the corresponding shapes of the clouds of electrons around the nucleus, but it is not "an easy piece".

If an analytical solution in closed form is not possible at all—and that is in the majority of the cases—then physicists may revert to approximations, or to number crunching on a large computer. Or they give up and content themselves with just qualitative conclusions. And why not? Whoever compels them to put exact numbers on everything, if the real understanding of the physical phenomena is already there (obtained from easier cases)?

11.8 The Kinship Between the Wave Functions of Quantum Mechanics and the Beat Patterns in Droplet Theory

Let us now finally return to the original question: why would the droplet theory give the same quantisation results as usual quantum mechanics? And why would this *generally* be true, for any type of potential well?

The answer is not difficult to give now. Both the beat patterns of the droplet theory as well as these mysterious wave functions of quantum mechanics can be calculated from the same equation (Schrödinger's equation). This already shows that there must be a rather close kinship between these two kinds of curves. For both types of curves the relative curviness is the same in each point of the potential well.

What has been seen above is, that the shape of such a curve is crucially dependent on the conditions in the end points of the well. An example has been given of the calculation of a beat pattern in a parabolic well, which showed that requiring a zero slope of the curve in *both* end points (i.e. satisfying the condition of path retracing in a stationary state) inescapably leads to quantisation (Figs. 11.10 and 11.11): this condition cannot be fulfilled for any arbitrary value of the total energy.

Now, the end conditions required in the case of wave functions are quite different from this. It may be seen in Fig. 11.12, taken from the textbook by Alonso and Finn.

On the left wave functions are shown for a parabolic well, the lowest one representing the ground state (i.e. the lowest possible energy). In total four energy levels for stationary states are depicted, with their corresponding wave functions.

By the way, what is shown on the right are curves obtained by squaring the wave functions. These curves have been shown earlier in this book (see Fig. 3.8), since they correspond to the so-called "Position Probability Densities", from which can be read off the chance to observe the particle at different places within the well.

Evidently, we can conclude that the beat patterns and the wave functions do have some kinship (because both comply with Schrödinger's equation), but they are still different members of a same family. And it is therefore still not certain that they would lead to the same quantisation of energy.

But there is another requirement, that *is* common for the beat patterns and the wave functions. To see this, have a look back at Fig. 11.10. Here a candidate beat pattern was shown which does comply with Schrödinger's equation, but is not able to satisfy the required conditions in the endpoints. During its "reconstruction" the conditions in the endpoint on the left (zero slope) were automatically correct, because this is where we started so that there was the freedom to do it correctly, right from the start. The job became difficult, because we also want to have the correct conditions (i.e. zero slope again) in the end point on the righthand side. That this requirement is satisfied is certainly not guaranteed automatically, and it compelled us to search for that specific value of the total energy that could give the desired result. In conclusion: it is the fact that we have to fulfill this *second* condition that causes the quantisation.

Now, the second requirement that was used until now might be replaced by another one, which is less specific but just as compelling. It is the requirement that in a potential well which is symmetric—like the parabolic one—one must have that the reconstructed curves (either the beat pattern or the wave function) are symmetric too, or alternatively: exactly anti-symmetric. For the "pipe fitter" this makes life easier, because he already will discover whether the connection is going to be correct when only half of the pipe connection has been completed. Nevertheless, the requirement of (anti)symmetry is just as good an indication whether the guessed value of the total energy was the correct one. The second condition, which was the cause of the quantisation of energy, has now been formulated in a more convenient way.

And what is more: this second condition is *the same* for the beat patterns of the droplet theory and for the wave functions of quantum mechanics. We now have arrived at a point where it is clear that these two (the beats and the matter waves) are really very close members of one family (though they are not identical twins). They are not only derived from the same equation (Schrödinger's one), they have to satisfy the same type of critical end condition determining the energy.

Agreed, this is still not a mathematically rigorous *proof* that for both the same quantisation of energy will occur. The less ambitious purpose of the present section was, however, to make it *plausible* that such a close kinship is to be expected. I hope I have succeeded in taking away your surprise that this close kinship exists. Rather the contrary, you would probably be surprised if it did *not* exist, and if the quantisation found in the droplet theory would be different from the one found in quantum mechanics. The fact that a full mathematical treatment of the parabolic well, such as given in the appendix, yields an identical quantisation in both theories, is hopefully a less mysterious result now. I hope you share this feeling with me, despite the lack of mathematical rigour of the argument (which you would not want to be bothered with, anyway).

Finally, the present chapter ends with the promised textbox, showing the heuristic derivation of Schrödinger's equation in mathematical notation. Ignore it if you wish, it does not tell anything different from what was already written down in plain language.

Textbox: Heuristic Derivation of Schrödinger's Equation (Time-Independent, One-Dimensional)

Let a sinusoidal curve be given by

$$f = a \cdot sin\left(2\pi\frac{x}{\lambda} + \in\right)$$

In the text the terminology "relative curviness" is used for

$$\frac{d^2f}{dx^2}/f = -\left(\frac{2\pi}{\lambda}\right)^2$$

and this relation is used to define what is meant by "local wavelength" for curves $g(x)$ other than sines. When $g(x)$ denotes a wave function, and if we assume De Broglie's relation to be valid locally, then:

$$\frac{d^2g}{dx^2}/g = -\left(\frac{p}{\hbar}\right)^2$$

where $\hbar$ is defined as $\hbar = h/2\pi$. We can substitute

$$p^2 = 2m\left(\frac{1}{2}mv^2\right) = 2m\left[E_{tot} - E_{pot}(x)\right]$$

After some rearrangement:

$$-\frac{\hbar^2}{2m} \cdot \frac{dg^2}{dx^2} + E_{pot} \cdot g = E_{tot}g$$

Chapter 12
On Radiation and Radio Silence; Interaction Between Charges and Radiation

12.1 Radiation Patterns Far Away and Close to the Radiating Charge

In Chap. 1 it was told that Maxwell, studying the then available experiments on electrical and magnetic phenomena, was the first to describe all of these phenomena in a comprehensive way in mathematical terms. He found, after an addition by himself to the empirical "laws", that four compact equations were sufficient to describe and summarise everything that had been discovered in this field of physics. These "Maxwell equations" were published by him in a famous book "A Treatise on Electricity and Magnetism" in 1873. As a good mathematician, he went on to derive from these equations other properties that were new, and that were confirmed by experiments only later. One of the properties of electric and magnetic fields he predicted was, that oscillating charges would be surrounded by a field of waves, so-called "electromagnetic waves".

The predicted waves would—very roughly—be comparable with the waves on a water surface, for instance caused by periodically dimpling a body into the water and retracting it again. Figure 12.1 shows the familiar wave pattern, in this case caused by falling droplets. The waves have the form of rapidly spreading circles.

Electromagnetic waves are more complex, and in most textbooks a picture will be found such as shown in Fig. 12.2. The waves consist of a mix of electrical and magnetic effects. If one would place a small test charge in a point where the wave passes by, the electrical charge would start to oscillate in the direction of the red arrows. If we would place a small magnet in the same point it also would oscillate, but in a direction perpendicular to the electrical charge (the blue arrows).

In a fixed point, we see both these waves running along us at the speed of light. This was one of the great discoveries by Maxwell: using his equations he calculated the velocity of the electromagnetic waves, and found practically the same speed as had come out of measurements of the speed of light. His conclusion was: light is one of the manifestations of the wave phenomenon predicted by his theory. It was

T. van Holten, *The Atomic World Spooky? It Ain't Necessarily So!*
DOI 10.2991/978-94-6239-234-2_12

Fig. 12.1 Waves on a water surface

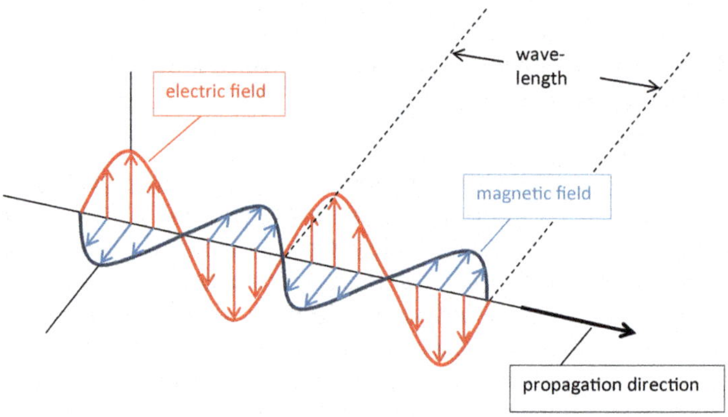

Fig. 12.2 Configuration of electromagnetic waves in the "far field"

later confirmed (for instance by Hertz's experiments) that there are other manifestations as well, such as the "radio waves" we nowadays are all used to.

The charge that is causing the waves needs an oscillating external force on it to be kept in a continued motion. In other words, the charge needs a constant feed of energy to continue its motion. The logical conclusion is, that the electromagnetic waves carry off energy, and that we see in a point of the field a "flow of energy" passing us. One can calculate that the energy that is passing every second through a unit area perpendicular to the propagation direction of the waves (the so-called flux of energy), is related in a simple way to the electrical and magnetic forces in the field:

$$(energy\,flux) \div (electrical\,force) \times (magnetic\,force)\ (Poynting)$$

This relation, derived by the physicist Poynting, may explain something needed later. It concerns the colour of light as perceived by our eyes. "Colour" is a measure for the frequency at which the electrical and magnetic forces are varying. Consider once again a *fixed* point in space, such as for instance the origin of the coordinates drawn in Fig. 12.2. When the waves are passing this point, what will be seen is that the strength of the electrical force is varying in time, between a positive maximum value, through zero, to a negative minimum, and so on. The magnetic force does the same. The frequency of light is defined as the number of such cycles passing us per second. Now, our eyes are a sort of frequency counters, they can see the difference between high frequency waves and waves with a lower frequency. However, instead of passing this information to our brains in the form of "so and so many cycles per second", the information is first translated by using a colour-scale. High frequencies of light are perceived as "bluish", and lower frequencies as "reddish". Physicists are able to tell which frequency exactly corresponds with every colour in the rainbow spectrum perceived by our eyes.

Now, the energy flux according to the above given formula is the product of the electrical and magnetic force. It therefore varies in time too, between a maximum value and zero (it does not become negative). It is simple mathematics—although the reader will not be bothered by the technical details—to calculate the frequency of the energy fluctuations, and it turns out to be *twice* the frequency of the light. Why it is relevant to keep this in mind will become clear later in this text (Chap. 13).

Although drawings like Fig. 12.2 are always shown in the textbooks to introduce the phenomenon of electromagnetic waves, they may give the wrong impression if this picture is the only thing which is remembered. The books are always careful to state that this is the configuration of the waves *far away* from the source, in the so-called *far field*.

The region close to the source, the so-called *near field*, in general is far more complicated. In fact, it is so much more disturbed that it hardly can be drawn in a simple sketch. Compare it with a pool which is being filled by a waterfall. Far away from the waterfall we may see the orderly pattern of waves in the form of concentrically widening circles. In contrast, right under the waterfall we see wild motions of the water. It is a turmoil of whirls, splashes and water shooting in all directions. The water velocity is different everywhere, and can certainly not be described as just waves emanating from one point. Instead you may see very complicated flow patterns, and there are people (like me) who cannot stop gazing at these intriguing "fields of flow".

This is what you should imagine too in the case of the near field of electromagnetic radiation. It is highly perturbed, certainly so in the case of a droplet of charge which is performing all these types of motion that we listed, ranging from just a steady velocity to an object speeding around under the influence of sharp "jerks", and at the same time contracting in spasms like a vomiting stomach. The electromagnetic field will be just as disturbed as the whirlpools under a waterfall. Something to be remembered later is, that it is not fruitful at all to have a mental model of the near field in terms of a "bunch of waves running criss-cross through each other, and interfering with each other". It may be that, purely mathematically, this might be a possible view of the near field. The actual physics, however, have no resemblance to such a picture of running waves. The reality lies closer to a picture where (again oversimplified) the near field more looks like a balloon inside which mainly standing wave patterns occur. Mainly, but not quite: the balloon surface is flexible, so that it transmits an energy flow to the world outside, the mean flux of energy originating somewhere inside, amidst the semi-standing waves. Forgetting entirely the picture of running waves may help later (in Chap. 13) to appreciate some properties of the near field and to find them less anti-intuitive.

12.2 Schott-Energy

The large difference in character between the highly disturbed near field and the smooth far field can explain one of the theoretical difficulties pointed out earlier. According to Maxwell's theory, the energy flux that is in the far field seen to flow away to infinity does not seem to be in agreement with the energy that is being fed into the oscillating charge. Far away, the energy flux is found to conform to the proportionality

$$(\mathit{energy\,flux\,to\,infinity}) \div (\mathit{accelerations\ of\ the\ charge})^2\ (\mathit{Larmor})$$

This is called Larmor's law, named after the scientist who first derived it from Maxwell's laws. In the mathematical appendix of this book the same result is derived, using a somewhat different mathematical approach. The approach of the present book was earlier explained (see Chap. 5), and was baptised "max-technique", to avoid the full, much uglier official name. There is a subtle difference, however, with Larmor's results. Larmor arrived at his expression by considering a source of the electromagnetic radiation (i.e. the oscillating charge in the centre) that is sinusoidally moving up and down. Often you will see in the textbooks the restriction mentioned that the above stated result is therefore strictly valid only for such sinusoidal motions. The derivation of the same result using the "max-technique" shows that we may forget this restriction: the result is always valid, no matter the history and future of the accelerations. The instantaneous flux at a certain instant corresponds with the instantaneous acceleration of the charge (of course, duly taking into account the retardation time, so that we mean the instantaneous acceleration at an earlier time).

Now, we can also calculate how large the energy input is by the external force that keeps the oscillating charge in motion. Already Lorentz had calculated how large such an external force must be so that it can counteract the self forces of the oscillating charge. First there is electromagnetic mass, which is in effect the inertial self force that resists accelerations. Part of the energy input by the external force is thus spent to increase the kinetic energy of the charge, exactly like a force overcoming the inertia of a normal mechanical body.

But Lorentz had also found a second self force, called the "radiation resistance". A lot already has been said on this subject, which will not be repeated. It is the recoil force on the charge by emitting electromagnetic waves.

Lorentz had calculated that a charge only provides such a recoil force, when it is subjected to "jerk", i.e. if there is an *increasing* acceleration or deceleration. The recoil force is not there if the charge experiences a constant acceleration. And most disturbing: we would be tempted to conclude that the radiation is not there if there is a constant acceleration. This is in direct conflict with Larmor's law! Most disturbing indeed.

These two forces, the electromagnetic inertia and the radiation resistance, both have to be counteracted by the external force, needed to sustain the motion. Summarising, and using a result from classical mechanics about the work done by forces:

$$
\begin{aligned}
&(\textit{energy made available for radiation by the force on the charge}) = \\
&\quad = (\textit{total energy absorbed from the external force}) - (\textit{increase of kinetic energy of the charge}) \\
&\quad \div (\textit{jerk}) \times (\textit{velocity of charge})\ (\textit{Lorentz})
\end{aligned}
$$

And this again appears to be wholly contradictory to Larmor's result. It was already mentioned that the problem is not unsurmountable in the particular case when the charge is indeed moving sinusoidally. In that case, it turns out that the two expressions given by Larmor and Lorentz are the same if we average over a complete cycle of the charge's motion. For a sinusoidal motion of the charge we thus find:

$$
\begin{aligned}
&\textit{average of}\,(\textit{energy flux to infinity}) \\
&\quad = \textit{average of}\,(\textit{energy available for radiation}) \qquad (\textit{sinusoidal motion})
\end{aligned}
$$

But the problem is more serious when we now recall the "max"-result that these two expressions are more generally valid, and are not restricted to sinusoidal motion of the radiating charge. Imagine for instance a charge that is in a constant accelerating motion, the speed is increasing all the time but it increases uniformly, which is the same as saying that it has a constant acceleration (see Fig. 12.3). According to Lorentz's result no radiation resistance would exist, and we only have to supply so much energy as corresponds with the increasing kinetic energy of the charge. The consequence would be that there is no energy surplus to "power" any radiation escaping to infinity. On the other hand, Larmor's result states that there would be a constant energy flux towards infinity! How to reconcile these conflicting results from both—perfectly sound—theories?

The answer is that the "near field" cushions the conflict between the energy input by an external force and the energy that is finally coming out of the system. The near field can act as a long-term energy reservoir, even with a large capacity to store and release energy. To return to the case of the charge with constant acceleration: there must always be a beginning and an end to this kind of motion. It is sketched in Fig. 12.3, from which it is apparent that very large jerks exist when the acceleration starts and stops. Mathematically working it out, it appears that the energy balance is restored when taking this into account. The total amount of energy fed into the system by the external force is then found to be exactly the same as the total energy

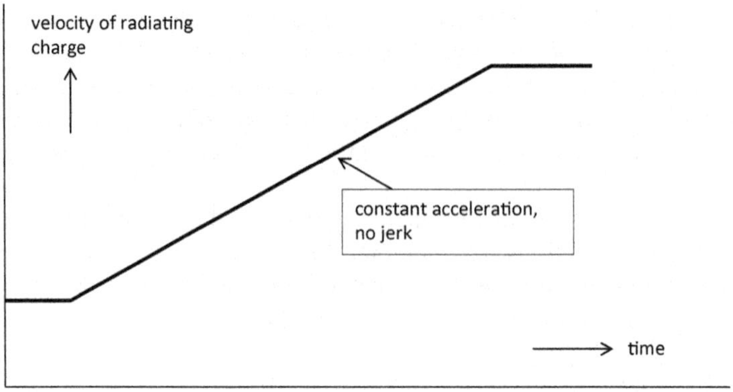

Fig. 12.3 Radiating charge with constant acceleration: a contradiction between Larmor's and Lorentz's theory?

radiated away to infinity. But the power input and the energy losses occur at quite different moments! The conclusion is that the field around the charge, and in particular the near field, acts as a large energy reservoir. The instantaneous unbalance between input and output is bridged by this energy reservoir. The near field apparently has a very large capacity to store energy, for a very large time. Recalling the above sketched mental picture of the near field containing standing waves, it is easily imagined how energy can be stored in the near field: it is in the form of the energy present in the standing waves.

At any instant of time apart, i.e. considering the situation instantaneously, we must conclude that a large amount of energy can circulate in the near field. In classical physics this is called the "Schott energy". Earlier, in Chap. 8 this effect was compared with the reverberation of sound in a music hall. The circulating electromagnetic energy in the field is not constant in time: in order to restore the *instantaneous* energy balances, the field sometimes has to supply energy to the flux of radiation in the far field, whilst at other times it has to transport energy back to the oscillating charge. At the instants when energy is coming back, the field helps to maintain the motion of the oscillating charge, since the action of the external force is then momentarily insufficient to do the entire job. Of course, at other instants this is compensated by absorbing energy from the charge, when the action of the external force would be more than needed to maintain the correct motion. A similar interaction with the far field radiation exists.

The near field thus bridges and cushions the instantaneous gaps between Lorentz's and Larmor's expressions. And something remarkable must be noticed: this bridging action is *instantaneous*, there are no delays between the supply of energy where it is needed and the depletion of energy at other places. It is as if the energy fluxes within the near field do not care about Einstein's speeding limit declaring the speed of light to be the highest achievable interaction speed between different events. We will come back to this aspect in a later chapter (Chap. 15), where it will be analysed more deeply.

12.3 The Observability of the Radiation Field

Let us turn our attention to the far field of a radiating droplet. Its structure is much simpler than the complex near field, have a look at Fig. 12.2 again. Unfortunately, there are complications here too. They are associated with the relevant time scales, determining what can be observed by a macroscopic onlooker and what cannot. The matter initially came up in Sect. 6.4. Some numbers were given about the actual frequencies of the pulsation, and it was concluded that "such a high frequency completely falls outside the technical measuring possibilities". This limitation was—in a way—circumvented by the later conclusion that the *shift* of this frequency due to the velocity of the droplet *could* be observed: mixing the pulsation frequency of a droplet at speed with that of a stationary droplet would give rise to beats, and these beats would have a so much lower frequency that they *can* be observed.

Later again, in Sect. 7.5 on the mechanical analogon, these considerations led to the banning of any equipment that could give direct information about the frequency of the nodding motion. And still later, in Figs. 8.6 and 8.7 it was visualised what would be seen when the nodding mass of the analogon is subject to a beat phenomenon. Figure 8.6 has here been repeated as Fig. 12.4. The individual nods would be "invisible", and the nodding motion is visualised by a vague reddish area. The beats are on the contrary clearly observable and measurable, they show up as the wavy outlines of the reddish area.

Now, the argument why some things will be observable and other things are not, rested in the above given description on the "technical impossibility" to measure these very high frequencies. This is not really a convincing argument. Who knows how measurement techniques will be refined in future? What is really meant is, that there is a *fundamental* limit to what can be considered to be observable. Perhaps the reader will remember from Chap. 1 what was the basis of the droplet theory. The droplet of charge, as an—unorthodox—model of an electron came from downscaling a macroscopic cloud of charges, so that the total charge would become the unit charge (i.e. the amount of charge a single electron possesses). A correction force (the so-called Poincaré force) had to be added to make it a consistent model

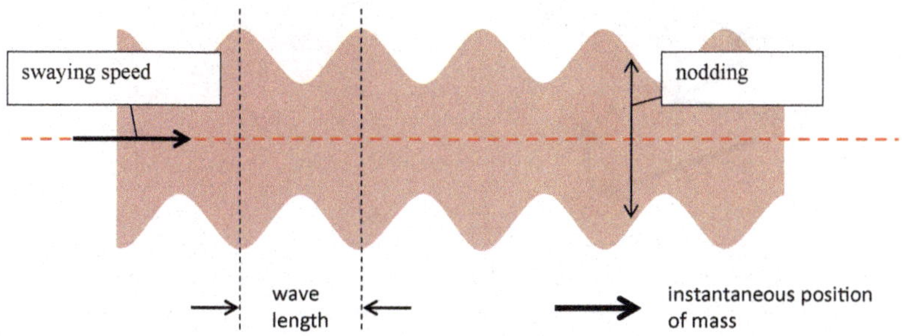

Fig. 12.4 (Copy of 8.6): Beats in the nodding, due to interference with "zero-speed frequency"

without the tendency to explode. The resulting, rather unusual model of an electron (i.e. a deformable droplet of charge) was proposed as a working hypothesis, just to study what would be the consequences of it.

However, another parallel line of reasoning was also followed, to justify the model assumptions, and make them appear to be less weird. It was shown that this unorthodox model is not necessarily incompatable with existing ideas about electrons. The reasoning was that we might arrive at a similar model by taking the droplet of charge to be a "time-averaged view" of a "zittering" singularity of the electromagnetic field. The idea was illustrated by Fig. 1.15, reproduced here as Fig. 12.5. Note that in the caption of Fig. 12.5 we do not talk anymore about a "zittering electron" but prefer instead to indicate it as the "zittering of the basic singularity of the electromagnetic field" (in other books this is sometimes called the "naked" electron). The reason is, that the time-averaged view may, depending on the equilibrium size of the "zitter volume" represent the physical electron, but also a muon or tau particle. This further precision could not yet be indicated in the original Fig. 1.15, since the three guises of the physical electron were introduced only later.

A time-averaged view implies that in the droplet theory certain—short—time-scales are taken to be irrelevant *within the scope of the particular theory developed*. At another place in this book the comparison has been made with thermodynamic theory as opposed to kinetic gas theory. Thermodynamics is a description of gases based on averaged properties of the gas molecules, in contrast to kinetic gas theory in which the motions of the individual molecules are studied in detail.

By adopting the model of a deformable droplet of charge we therefore have implicitly stated what time-scales are considered relevant, and which phenomena are so fast that they are not consistent with the droplet theory. In other words,

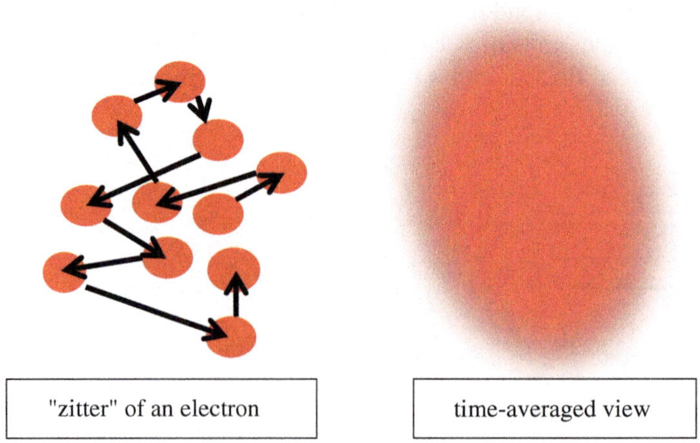

Fig. 12.5 (Copy of 1.15): Zitter of a singularity in the electromagnetic field, and the time-averaged view of it

implicit in the droplet theory there is a certain "cut-off" of the relevant time-scales. Certain events will be so rapid that they must be ignored, otherwise they would be incompatible and out of balance with the basic assumptions of the time-averaged theory. This is not to say that such rapid events do not exist. They just do not have a place *within the scope* of the adopted model assumptions. The existence of very rapid events may then be inferred through other kinds of observation or by deduction, but they cannot *directly* be ascertained. For example, the pulsation of the droplets is too fast to be "seen" directly, but its existence can be deduced from the beat phenomena associated with the pulsation.

So far about the observability of the droplet's behaviour itself. In the present chapter we must consider what are the consequences for the radiation field. At least, *if* there is any radiation at all. In Chap. 10 it was argued that the possibility to radiate is limited, only under certain exceptional circumstances the conditions will be right. First, the amplitude of the pulsations must happen to be small enough. Most of the time the amplitude is so large that it leads to *chaotic* pulsation, and this will block radiation. However, there are random variations of the amplitude, and sometimes—highly exceptionally—the amplitude may fall below a threshold where the chaotic motion "degenerates" into sinusoidal motion. This is a "*potentially* radiating" condition, but it requires more than that to actually enter the radiating state. The conditions for radiating can only be met fully if the droplet is moving inside a potential well so that beats occur in the amplitude of the pulsation. Only then at certain positions in the well all of the required conditions for radiating can be satisfied completely. This was the subject of Chap. 10.

The conditions to be fulfilled for radiating are summarised in Fig. 12.6. In the next Fig. 12.7 the electric and magnetic field strength around the radiating droplet are sketched.

The electric field strength has the same frequency as the droplet's pulsation. It thus fluctuates very fast around an average value zero. If we would try to observe it, we obtain a "null-result": the fluctuations are too fast to be observed, and only a time-average would be observed. But this time-average is zero! The electric field strength has become invisible due to its high frequency.

The same is the matter with the other component of the radiation, the magnetic field strength. When a small magnet is kept in the radiation field, it will not show any movement. The fluctuations of the field are too fast to be picked up by the device used to measure the field, and we will observe the time-averaged value of zero. Again, the magnetic field is invisible, just as the electric field. It would seem as if the pulsating droplet, although it is in a radiating state, does not have a radiation field!

Fortunately, there is one aspect of the field that is measurable and thus observable, and that is the energy flux. Its fluctuations are shown in the lower part of the Fig. 12.7. The energy flux is proportional to the product of the electric and magnetic field strengths, as was told earlier in this section. The frequency of the flux variations is higher still than that of the individual field components, it was mentioned before that the flux in fact always has double the frequency of the pulsations of the droplet. An attempt to observe the energy flux will always give a

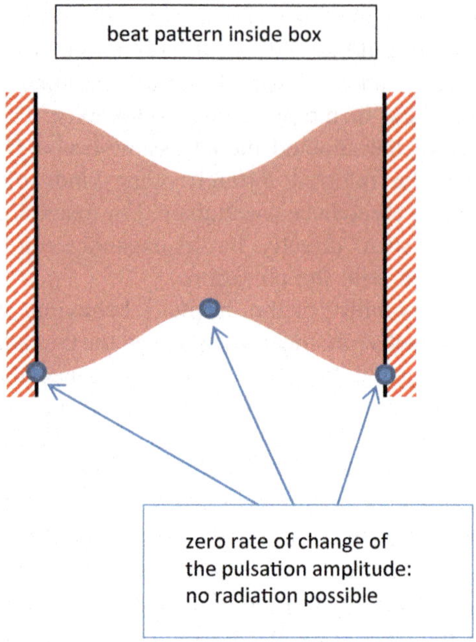

Fig. 12.6 Schematic showing the conditions enabling radiation: small pulsation amplitude (no fuzzy beat contours due to chaos), and droplet well away from the *blue* positions

time-averaged value, where the separate fluctuations of the flux are invisible. But averaging will not give a null-result in this case. The time-averaged value of the flux is shown in Fig. 12.7 by the red dash-dot line. The variations of this average have a much lower frequency (equal to the beat frequency of the pulsations) and are within the window of measurement possibilities.

If the droplet is in the radiating state, what can be observed is energy flux in the field, varying in magnitude with the beat frequency of the droplet. It is thus a strange form of radiation: just energy flux with a low frequency, and no supporting electric or magnetic fields! As if there are energy waves coming from the droplet, travelling away in isolation on their own, with the speed of light. It already "smells" a bit like the phenomenon in quantum mechanics called "the photon". In the next chapter all this will be investigated more deeply, to find out whether the droplet theory indeed can provide a consistent picture of photons.

What is shown in Fig. 12.7 bears some resemblance with AM-radio transmissions (Amplitude Modulated). There is a high frequency carrier wave, with a slow variation of its amplitude. If we would tune in on the correct frequency and directly feed the signal to an ear phone, nothing would be heard. The membrane of the ear phone cannot follow the high frequency, and responds only to the average value of the signal, which is zero. The audio signal can be recovered by first rectifying the antenna signal. This corresponds with the lower part of Fig. 12.7. Usually,

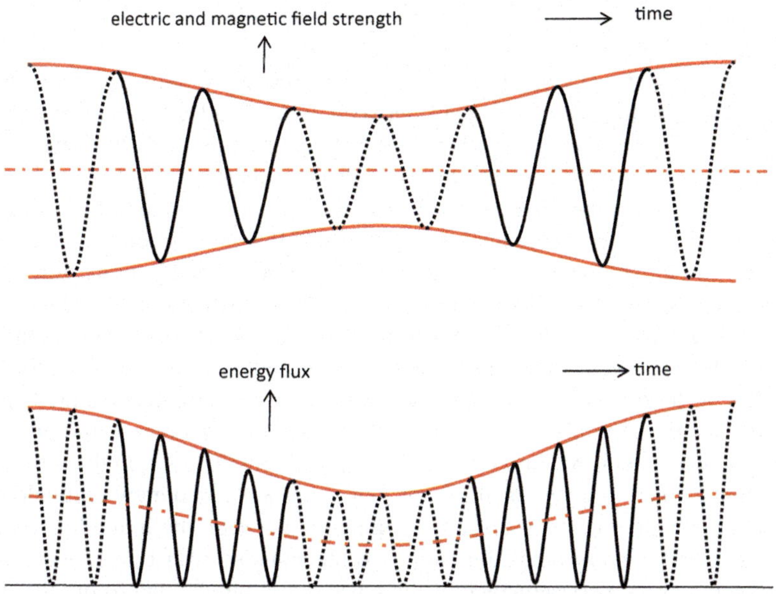

Fig. 12.7 Electric and magnetic field strength in the far field, both fluctuating around a zero average value with the frequency of the pulsation of the droplet. The energy flux in the far field is proportional to the product of the electric and magnetic field, and shows a variation around a non-zero average

rectifying in radio sets is done by a diode, so that the resulting wave shape is slightly different from the one sketched in Fig. 12.7. Nevertheless, squaring the signal, as is shown in the figure, is just as good a method of rectifying. If this would be done in the actual receiver, again the time-averaging act of the ear phone would make the buried audio signal observable.

12.4 Another Consequence of the Limited Observability: Red Is Red and Blue Is Blue

Let us return again to Fig. 12.7. It was argued that, due to the limitations of observability *within the scope of the droplet theory*, only the low frequency beat phenomenon in the radiation field can be "seen". We have a "visible" energy flow, without any "visible" supporting electric of magnetic fields.

If we had been dealing with a "normal" field, where everything would be "visible", we would call the frequency of the electric field the colour, as perceived by our eyes. The associated energy flux has double this frequency, but if we had done a measurement on the energy flux only, we would still call *half* of the measured frequency in the flux the colour of the radiation. In our case, where a measurement of the flux is the *only* way for an observation, we still have to interpret half of the flux frequency as the perceived colour.

Now, it should be realised that the observed frequency (and therefore: the perceived colour) is the result of beats between two ultra-high frequencies, which separately lie outside the observation window. The frequency of the beats is equal to the difference of these two interfering high frequencies.

The *same* beats could have been caused by *another* set of ultra-high frequency components, if these components have the same difference of frequency. The same colour would be perceived, even if the radiation field is actually quite different and contains quite different oscillations of the electric and magnetic field. As these basic oscillations are invisible within the scope of the droplet theory, we conclude that there may exist different radiation fields that nevertheless cannot be discerned from each other, as long as they give rise to the same colours. We perceive the "blue" of one field as the same "blue" of quite another field. For this reason the heading of this chapter was "red is red and blue is blue", expressing that within the scope of the droplet theory there are sets of different fields that are completely equivalent, and we do not have any means to discern a difference between the individual members of the set. Such *different* fields, belonging to the same set, must thus be considered as *identical*. This is a consequence of the basic assumptions underlying the droplet theory, and it is a property of the electromagnetic fields associated with this theory.

12.5 How Can Radio Silence Come About?

In the introductory sections of Chap. 1 is was narrated how Bohr had to assume that most of the time the electrons orbiting inside an atom do not radiate. This assumption was rather revolutionary, because it is in contradiction with Maxwell's laws. Or so it seems. Electrons orbiting the nucleus are in a state of constant acceleration (centripetal acceleration), and according to Maxwell's laws the acceleration would always be accompanied by radiation (see Larmor's formula above). But Bohr rejected this idea, and he was compelled to do so in order to explain why atoms are stable. It was one of the postulates by Bohr, i.e. the new set of rules and laws drawn up by him, that there should be "radio silence" most of the time.

In Chap. 1 the whole of Sect. 1.27 was devoted to an explanation why radio silence is not necessarily in contradiction with the classical laws of physics. This long story will not be repeated here, just a brief summary will suffice. It came down to pointing to the chaotic "drunkard's sack race" an electron is performing when it flies along its trajectory. The culprit is therefore once again the pulsation of our droplets of charge, which is the cause of the sack-race. The pulsation is a new feature found by the droplet theory, and absent in the conventional view on electrons. It explains why in droplet theory Bohr's "radio silence" can be derived from the classical laws, whereas Bohr had to introduce it in the form of a separate postulate.

The "drunkard's sack race" was connected with another phenomenon, extensively discussed in Sect. 1.27, which was the "starting-up delay" of radiation. The author of a relatively recent book, Yaghjian, had found that one is compelled to assume that such a delay exists, otherwise some of Lorentz's theoretical results would be completely anti-causal (as an example the phenomenon of "pre-acceleration" of electrons was given, the result that an electron is going to move *in anticipation* of a force applied later). By assuming a start-up delay of radiation, the anti-causality in Lorentz's theory can be removed.

The starting-up process of waves was considered in more detail in Chap. 1, and it was found there that starting-up of waves involves for a brief moment a combination of an in- and outgoing wave, until the wave has become fully detached from the radiating object. The sum of an in- and outgoing wave (of equal strength) means that momentarily there is no energy transferred to the field, and this explains the delay in energy radiation.

Finally, it was argued that an electron performing a "drunkard's sack race" (chaotic motion) can be considered as a charge that is permanently in a starting-up process. The associated delays overlap each other and prevent the normal radiation one would expect. This was the picture presented—somewhat tentatively—as a possible explanation of Bohr's radio silence.

12.6 Does Anti-causality not Creep in Again Through the Back Door?

The above given description is mainly a physical way to make plausible that a non-radiative state is possible that will be associated with chaotic motion of the droplet.

The point deserving special consideration is the use of incoming waves as an ingredient to construct the complete electromagnetic field. What must be reconsidered more thoroughly in the present chapter is this "sum of in- and outgoing radiation". It sounds rather logical that such opposite waves cancel the energy transfer to infinity, but there may be a flaw in the argument: incoming waves could themselves be the cause of anti-causality. This has to be checked carefully.

Let us systematically go through it again, starting with the left hand side of the sketch in Fig. 12.8. For the purpose of an easy explanation the radiating source has here been sketched as a pulsating sphere, an idealisation often found in acoustic theories. One could imagine the sphere as a small balloon with a variable internal pressure that is caused by a pump powered by an external energy source. One can easily imagine that the pulsating sphere periodically displaces the surrounding medium in- and outwards (in acoustics the air around the sphere). The displacements are carried outwards, first from the surface of the sphere to a shell of air which is in direct contact, and then every shell of air pushing against the next shell, and thus transferring energy from the sphere eventually to infinity. This is exactly

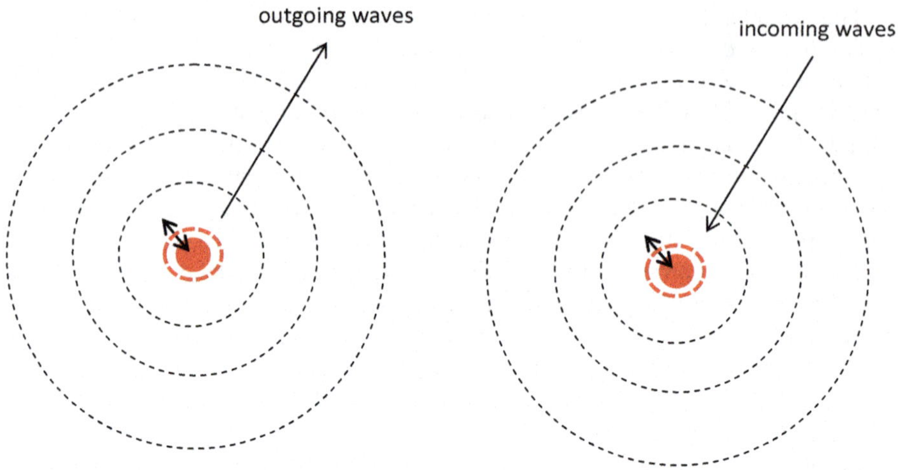

Fig. 12.8 On the *left* the usual picture of a radiating source. On the *right* another correct solution of the wave equation, usually rejected on physical grounds

the process as it is described by the so-called *wave equation* which mathematically predicts what is going to happen.

Now, this same wave equation has another solution, purely mathematically valid, but physically rather unlikely. This solution is sketched on the right hand side of Fig. 12.8. Here we have incoming waves: some energy source at infinity causes waves that converge toward the sphere as their meeting place. They have been "programmed" by the mysterious source at infinity in such a way that they exactly coincide with the pulsations of the sphere's surface. In itself this is already an improbable situation, but worse still: it can lead to anti-causal effects. If the frequency of the pulsation would gradually change in time, the source at infinity would have to anticipate this. It would have to know beforehand how the sphere will change its pulsation in future, because the incoming waves need some time (the "retardation time" as it was called earlier in this book) to arrive at the surface of the sphere. All this has led physicists to reject this second solution of the wave equation: purely mathematically it is possible but it cannot physically be realistic. Only the solution of outgoing waves is accepted, and this restriction is known under the name "Sommerfeld's radiating condition".

How then must the "summing of in- and outgoing waves" be interpreted that was mentioned in Chap. 1 in connection with the starting-up process of waves? One could say: "let us in this particular case forget about Sommerfeld's condition". If we by accident have stumbled on a physical situation where incoming waves do not lead to unrealistic conclusions, then we may use them. After all, mathematically there is no objection to this. And we do not worry any further, if it does fit the physics.

How would the picture then become? We accept for now that there is indeed a superposition of incoming and outgoing waves. The outgoing waves are powered by the pulsations of the sphere, but the incoming waves supply just as much energy back to the sphere. There is then no variation of internal pressure needed to maintain the pulsation of the sphere. The sphere can perform its pulsations without any pumping action, and without energy input to power the pump. In other words: the radiation resistance is zero, and there is nothing needed to counteract the (non-existing) radiation resistance. It explains why, during a start-up process of radiation which is being described by a brief moment of equally intense in- and outgoing waves, there is momentarily no radiation resistance. In other words there is a delay of the radiation resistance at the very beginning of the starting-up process.

So far, so good. What we are left with though, is an uneasy feeling about having to use incoming waves driven by some mysterious source at infinity. In order to resolve any remaining doubts, in the mathematical appendix the complete solution is determined of a droplet in the non-radiative condition, with the intention to find out whether the ideas developed above are tenable.

12.7 The Structure of the Non-radiative Field, as Found by a Mathematical Analysis

What is in the mathematical appendix considered is shown in Fig. 12.9. A free-flying droplet is assumed, pulsating and consequently performing its translation motion in the form of a "sack-race". The analysis that was done on this situation in earlier chapters carefully respected the conventional "Sommerfeld radiation condition", although it did not say so explicitly. Tacitly the existence of incoming waves was ruled out. In contrast, it is now tried to find a valid mathematical solution for the electromagnetic field where the restriction by "Sommerfeld's condition" has been ignored. Both types of allowed solutions of the wave equation are now being used (outgoing as well as incoming waves). The primary question to be answered is: is the mathematically allowed solution realistic in the physical sense? More specifically, are anti-causal effects absent, despite the use of incoming wave as an ingredient of the whole field?

The result of the—as yet—purely mathematical exercise has been sketched in Fig. 12.9, very schematically. What is found is, that a field of incoming and outgoing waves of equal intensity forms a field of so-called *standing* waves. Before a more detailed discussion of Fig. 12.9 is taken up, it may be wise to explain the phenomenon of standing waves somewhat more generally.

The phenomenon of standing waves is very clearly seen in the strings of musical instruments. Figure 12.10 is a picture of a vibrating string, and shows how the string can vibrate in several modes: ground tone and overtones. Mathematically, there are two different ways to analyse the vibrating string. The most straightforward approach is, to make use of our physical knowledge, obtained from

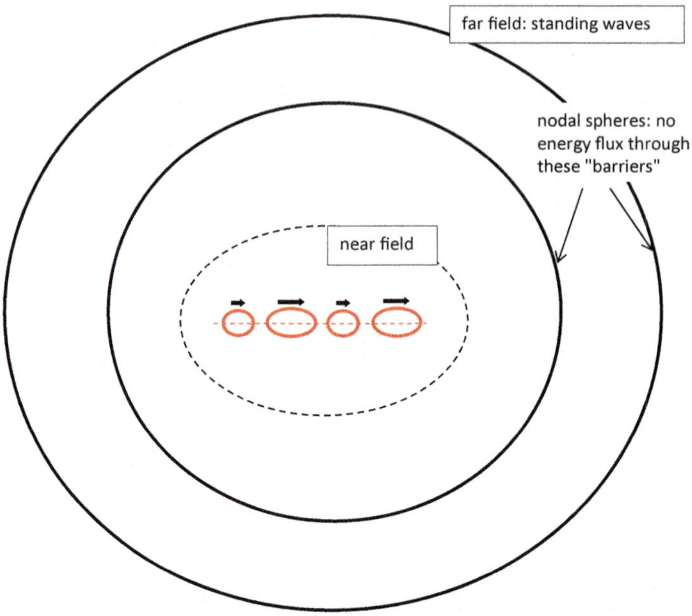

Fig. 12.9 The non-radiative field of a pulsating droplet performing a "sack-race" without radiation resistance. One finds a sequence of spherical surfaces through which the energy flux is zero

experiments like the one of Fig. 12.10. We see that a particular piece of the string is moving up and down, with a periodical (and we suspect: a neat sinusoidal) motion in time. Comparing two adjacent elements of the string, one notices that the motion in time is similar, but the maximum deflection is different (or: the amplitude of the motion depends on the position of the particular element). What is meant by "standing wave" is, that this spatial pattern does not depend on time: the places of maximum deflection do not move along the string but are stationary. There are even points where there is no up- and down movement at all, and these so-called "nodes" also keep a fixed position.

The known physics can to advantage be used as a priori knowledge in a mathematical analysis, to go straight to an "explanation" of the phenomenon and to make predictions about the effect of the length of the string, its mass, etc.

There is an alternative route, however. Although it is clumsy and hardly practical, this alternative is conceptually enlightening. The mathematical equation describing the motion of the string elements is the wave equation, the same equation that was mentioned earlier in this chapter. It predicts that waves will be running along the length of the string, waves in appearance very much like the water waves of Fig. 12.1! This is very strange indeed, and is entirely in contradiction with what our eyes see in Fig. 12.10. Of course, Fig. 12.1 depicted waves in a pond so that, to be better comparable, we should think of water waves in a long,

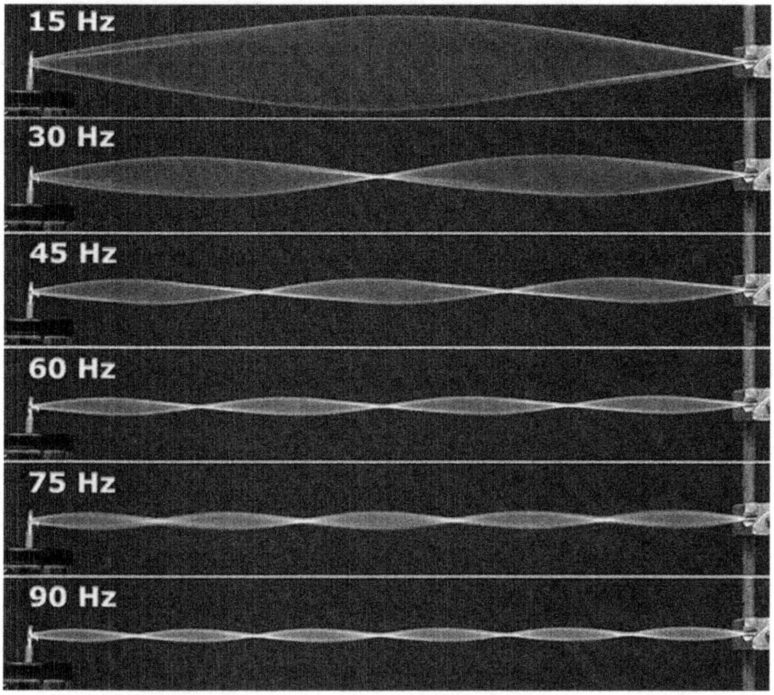

Fig. 12.10 Standing waves in the string of a musical instrument. From: https://fys.kuleuven.be

narrow channel. The problem is, that there are two different mathematical predictions, one telling that we should see waves running along the string, and the other that we should *not* see them running. Both predictions are correct, and the explanation is that the waves running along the string will meet the endpoints, and will be reflected to come running back. We get a superposition of several waves running in different directions, and the mix of them then leads to the standing waves as observed. This is actually what we wanted to show: the superposition of incoming and outgoing waves of equal intensity forms a field of so-called *standing* waves.

An example of standing waves in more dimensions is shown in Fig. 12.11. In this figure one of the possible vibration modes is shown of a thin plate, with its circular edge clamped. One may look upon it as the superposition of waves running away from the centre, and reflected at the edge. The reflection causes waves running inwards towards the centre so that the result is an interference pattern between the incoming and outward moving waves, again resulting in standing waves.

Similarly, the electromagnetic field formed in three dimensions by in- and outgoing waves of equal intensity must be expected to be a field of standing waves, although it is more difficult to draw in a simple picture than the examples given in one and two dimensions. Returning to Fig. 12.9, this is indeed what is predicted by the mathematical analysis. Incidentally, Fig. 12.9 is the picture predicted when a regular “sack-race” of the droplet is being assumed. In the case of chaotic

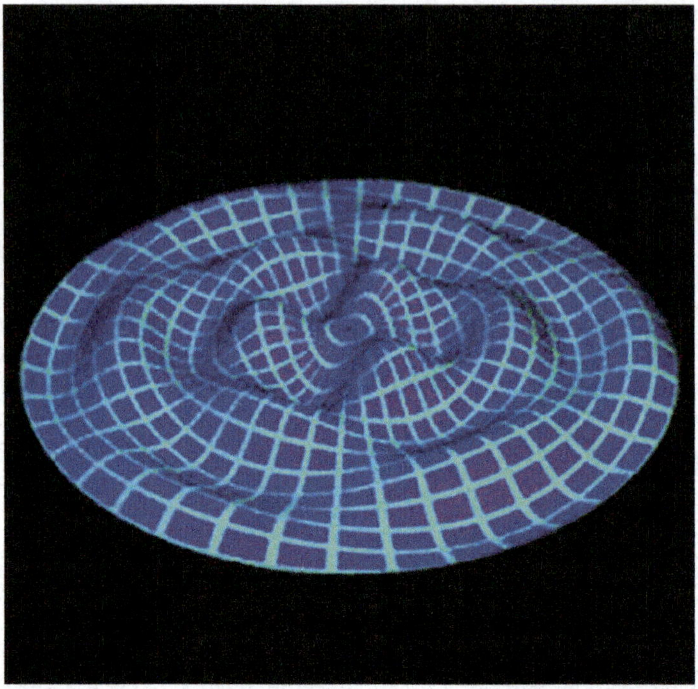

Fig. 12.11 One of the possible modes of standing waves in a circular plate clamped at the edges. From: www.mobiusinstitute.com

pulsations, we find the so-called "drunkard's sack-race", a less regular kind of jumpy procession of the droplet along its trajectory, however with a basic rhythm buried in it which is the same as in the case of sinusoidal pulsations. The sharply defined "nodal spheres" in the far field of Fig. 12.9 where no energy flux occurs from the inside to the outside, in the case of chaotic pulsations turn into the more "fuzzy" energy barriers shown in Fig. 12.12.

The conclusions we arrive at when scrutinising the mathematical results are threefold.

1. The patterns of standing waves shown in either Fig. 12.9 or 12.12 are mathematically valid solutions of Maxwell's equations for the electromagnetic field. At least mathematicians will not frown upon them. What about physicists?
2. Physically, there is no objection to accept the solution of standing waves. The unusual feature about them is, that fields of standing waves are normally associated with "hard" boundaries such as the clamping edge of a vibrating plate, or the hard lay-up points of a string. In the case at hand there is another kind of physical condition that forces the standing field upon us, which is the condition that the droplet does not shed energy. Remember that this was taken as the *starting point* of the mathematical analysis. The inability to lose energy is

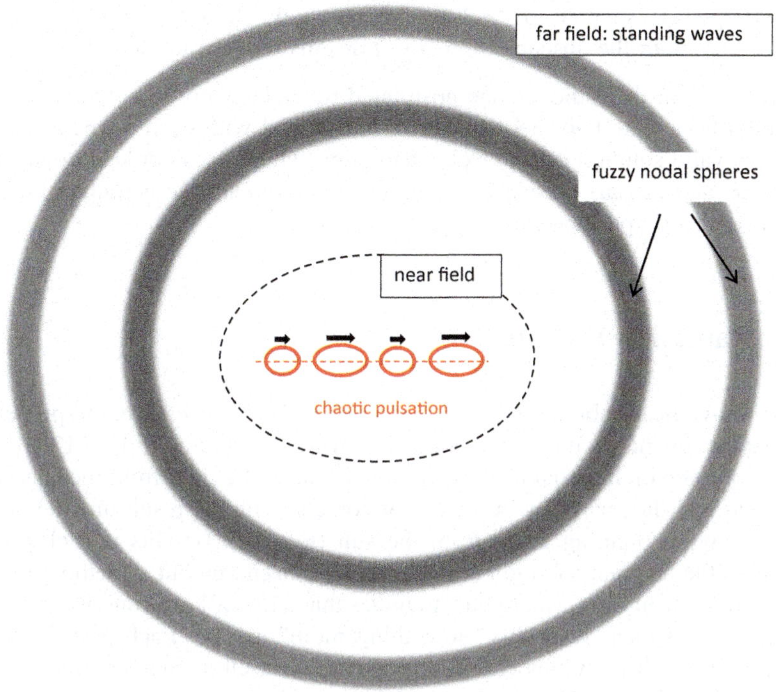

Fig. 12.12 The non-radiative field of a pulsating droplet performing a "drunkard's sack-race" without radiation resistance

a *physical* requirement, imposed on the ambiguous options offered by the mathematics. What we wanted to find out was, whether this state of "non-radiation" would or would not contain aspects of anti-causality. One might initially have had the fear that the admission of "incoming waves" (as one of the components of the mathematical solution) by itself could be the cause of another type of anti-causality. This fear does not appear to be justified, since the total solution (i.e. the entirety of the mathematical solution components taken in unison) is certainly not anti-causal: there is nothing unrealistic about fields of standing waves.

3. In the far field, it is evident that the condition of no energy-shedding is satisfied by the solution. There are barriers in the field that prevent energy escaping through them. No energy coming from the droplet would thus be able to reach infinity. At the starting end of the sequence of events, i.e. in the near field, there is another mechanism that prevents such an energy flux. The near field is the "cause" of self forces on the droplet. What is found in the mathematical appendix is, that *all* the usual self forces (see the list in Chap. 5) are there, all having their usual magnitude, except the radiation resistance. Radiation resistance is completely absent in the solution of the electromagnetic field, even if

the droplet is subject to "jerk". The near-field solution is thus in perfect agreement with the absence of energy transmission towards infinity.

In conclusion: the radio silence postulated by Bohr is not in contradiction with Maxwell's laws. The two theories can be reconciled with each other, by a mathematically valid solution of Maxwell's equations which does not lead to physically absurd (i.e. anti-causal) effects. The origin of the radio silence is found in the high frequency, chaotic pulsation of a droplet of charge.

12.8 Radiation Pressure

You will have heard about the phenomenon of *radiation pressure*, the property of light that causes the peculiar behaviour of comet tails pictured in Fig. 12.13. Light can exert a force on material objects. In this instance, the light from the sun pushes against gas and dust particles in the tail of comets, with the result that the tail of a comet is always pointing away from the sun (Fig. 12.13). This is a clear manifestation of the fact that radiation not only represents energy but something more. In the particle-view of light, where one imagines that a ray of light consists of a stream of photons, it is clear what this "something more" can be. Each particle of light behaves as a small body possessing both energy as well as momentum.

We met "momentum" before, it was defined as the product of mass and velocity of an object: $(momentum) = (mass) \times (velocity)$ or in short notation $p = m \cdot v$. In the case of a tangible object we can imagine that momentum is a metric for the impact a body may have when it collides with other bodies. A similar picture may

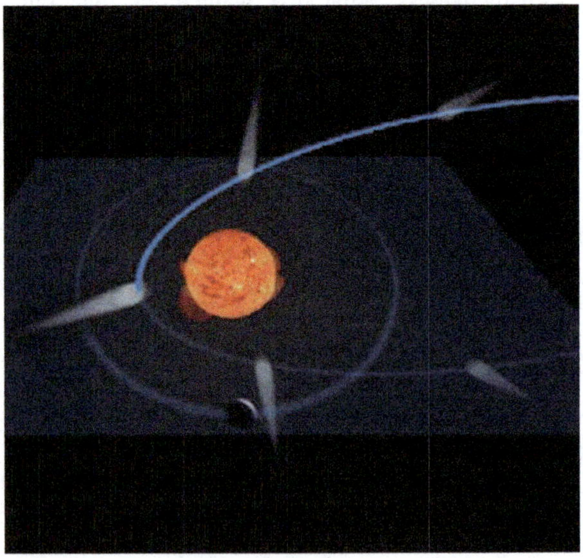

Fig. 12.13 Due to the pressure of light a comet's tail points away from the sun. From: www.telescope.org

be built up when dealing with photons: they collide with an object, and transfer part or all of their momentum. Such an event will be experienced as a pressure exerted by the radiation, just like the pressure on the walls of a vessel where molecules of a gas hit the inside of the container.

The wave-view of light, i.e. light radiation as a field of electromagnetic waves, offers perhaps more problems to imagine the pressure of light. The matter is complex, and no attempt will be done at this place to build up an intuitive picture. Let it be sufficient to say that radiation pressure is indeed covered by Maxwell's theory. This classical theory of electromagnetic radiation gives results in complete agreement with the view of radiation as a stream of photons. Some results from Maxwell's theory, needed below, are the following.

The first result is, that the momentum carried by the field (which can be transferred to objects placed in the field, and is then "experienced" as a force exerted on such an object) always has the same direction as the flow of energy in the field. Take for instance the simple electromagnetic waves depicted in Fig. 12.2. The propagation direction of the wave is indicated in the figure, and is here assumed to be to the right. It means that also the energy in the wave flows to the right. If the wave impinges on an object, the force experienced has the same direction, to the right. Even if such a "collision" has not taken place yet, we imagine that the radiation field carries momentum to the right, which *potentially* can be transferred to an object. According to Maxwell's theory the momentum in the field can quantitatively be related to the flow of energy. This relation can be captured in a simple formula, which will be reproduced here because it will be used further on in the text. Of course the next alinea with the formula, up to Sect. 12.9 may be neglected by readers who are satisfied to see just the general line of the argument.

Remember that we had a metric available to quantitatively express the intensity of the flow of energy. This metric was the energy *flux*, defined as the amount of energy which per second flows through a plane perpendicular to the flow, having a unit surface area.

Furthermore, the momentum in the field can be quantified in the form of "momentum density", defined as the total amount of momentum in a unit volume of the field. The quantitative relation between flux and momentum density is:

$$(\textit{momentum density in a point}) = (\textit{energy flux in that point})/c^2$$

which relation is accompanied by what already was mentioned: the direction of the momentum and that of the flux are the same.

In words: the momentum density is proportional to the flux. The proportionality constant is small, viz. $1/c^2$, where as usual the symbol c stands for the speed of light. It is seen from the formula that if we shine light on an object, the radiation pressure experienced by the object is really very small. Even so, radiation pressure is real and its presence cannot be ignored, however small it is. The reason is, that we now have to be careful that *two* conservation laws must both be satisfied: in any process there must be conservation of energy and at the same time the conservation of momentum. This is a much more severe restriction than if we had to deal only with the balance of energy. Therefore, the radiation pressure may be very small, it has a rather fundamental significance, as will be seen.

Before proceeding, the reader must be warned. The momentum which is present in radiation is so small, that it strictly speaking is not covered by the theory of the appendix. This theory includes effects of the order $(v/c)^3$ and chops off smaller terms, beginning with effects of the order $(v/c)^4$. This is not to say that terms of such small orders do not crop up at all in the theory. For instance, the momentum in an electromagnetic field comes correctly out of the theory. The problem is, that one cannot be sure that *all* the relevant effects of this small order systematically have been found by the present theory. For this reason, what follows must be considered as a purely *qualitative* discussion which is somewhat speculative.

12.9 Asymmetries in the Electromagnetic Field

In the following Fig. 12.14 it is assumed that a charge decelerates and thus radiates electromagnetic energy. For a start, we assume that the charge is moving inside an antenna rod. As will be seen later, this is a situation which is essentially different from a charge inside a potential well. Nonetheless the antenna is a good starting point because it indicates the problems arising if we require that both the conservation of energy and the conservation of momentum are satisfied.

In Fig. 12.14 the decelerating charge is shown in the centre, and the radiation field drawn around it is the *far* field, at large distances. The deceleration is caused by a force acting on the charge, which may be the sum of potential gradients in the antenna rod, radiation resistance (i.e. the recoil due to the radiation), and other

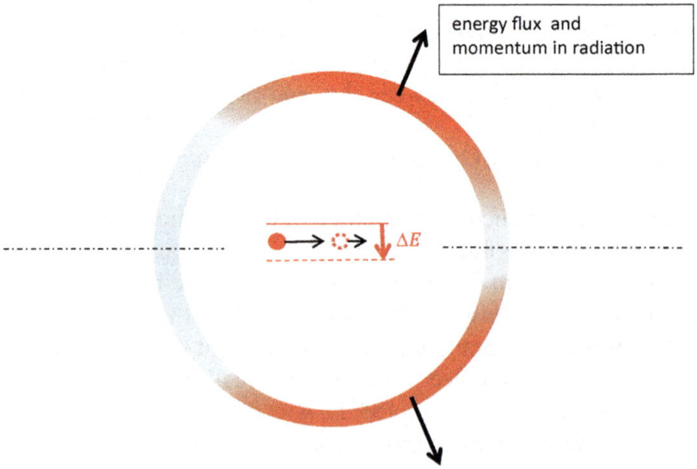

Fig. 12.14 Asymmetry in radiation of decelerating charge

forces like e.g. interactions with the more massive nuclei of the atoms of the rod. Together, these forces cause that the momentum (i.e. the product of mass and velocity) of the charge decreases.

In the far field we will find a flow of energy inside a shell-shaped region. The outer surface of the shell corresponds—taking into account the retardation time—with the start of the deceleration, and the inner surface with the end of the deceleration. Outside the shell the electromagnetic field is zero. The radius of the shell grows with the speed of light. Summing over the entire volume within the shell, the total energy in it must be equal to the radiation energy lost by the charge.

As was discussed, the field inside the shell also represents the transportation of momentum, and this momentum in the field is everywhere pointing radially outwards. Now, the energy flux due to a decelerating charge is asymmetric, as symbolised in Fig. 12.14. The energy is radiated predominantly in sideways direction, as shown by the shades in the red colouring. These sideward "lobes" of the energy radiation are slightly tilted in the same direction as the velocity of the charge. It is this slight fore-aft asymmetry of the field that causes an asymmetry in the local field momenta too (see the proportionality given above between energy flux and momentum density). If we sum all the volume elements inside the shell we will find that, due to the asymmetry, there is a total momentum in the radiation pointing to the right. At first sight, we would think that this overall momentum in the field must be equal to the momentum lost by the charge during its deceleration. This is not true. In actual fact, it will be seen that a role is also played by the momentum of the more massive parts of the antenna rod. This shows up when we consider the total balances of energy and momentum.

12.10 Conservation of Both Energy and Momentum

As already stated, at first sight the process described seems simple and clear: the charge loses energy and momentum, and both are found back in the radiation field. Actually, it is not so simple at all. The problem is, that there is a definite ratio $(momentum)/(energy)$ in the field, whereas this same ratio is *different* for a single electron in the rod, if this electron is considered apart from its surroundings. Therefore, if it is required that the energy in the field equals the energy lost by the charge, then it is not possible that also the momentum in the field equals the momentum lost by the charge. And the other way round too, if we require that the momentum in the field equals the lost momentum of the charge then the energy balance is not correct. If the electron is considered on its own, the conservation of energy and the conservation of momentum cannot be satisfied at the same time.

The conclusion is that an isolated charge cannot simply decelerate under the emission of radiation. It is something well-known of course: an electron freely flying through space will not spontaneously radiate to shed some of its energy.

The radiation of an electron in an antenna rod is possible only thanks to the fact that there is some interaction between the electron and its surroundings. And in reverse, such an interaction must also exist in a receiver antenna. To quote from the textbook by Alonso and Finn (vol.III, Chap. 1.5): "… in the case of an electron bound to either an atom, a molecule, or a solid, the energy and momentum absorbed are shared both by the electron and the atom, the molecule, or the solid lattice to which the electron is coupled. In such circumstances it is always possible to split both energy and momentum in the correct proportion so that both quantities are conserved."

One could state this conclusion even more generally as: there is always a "third actor" needed, apart from the radiation field and the electron, to enable interaction between radiation and an electron. In the next chapter, Chap. 13, an experiment in more dimensions is mentioned, the famous "Compton scattering experiment", which shows up the same. In Fig. 13.6 this experiment is shown in principle, and it is seen that *scattered radiation* here serves as the "third actor".

Returning to the one-dimensional situation of Fig. 12.14, the question that arises is: how can we understand radiation by a charge in a potential box, when the charge jumps from one of the allowed energy levels to a lower one? Apart from the restriction of one-dimensionality, there is no extra force acting when the charge is flying somewhere in between the walls of the box and sheds its energy in the form of radiation.

Bohr had to make a postulate of such an event, stating that it is one of the many things that can happen in the atomic world, whereas they seem to be forbidden by the macroscopic laws of nature. This is fine within the scope of the usual quantum theory, but considering our own droplet of charge we need a better explanation. Fortunately, in the case of the droplet we can identify a "third actor". This is the additional energy reservoir in the form of the variable stretching of the droplet. It was pointed out earlier that energy transfer from this reservoir to the translational energy of the droplet indeed takes place during a radiating condition. In fact, this process of energy transfer was shown to be rather essential, or else no radiation would be possible. The "third actor" thus plays an active role, and must be taken into account when determining the energy and momentum balance of a radiating droplet of charge.

A quantitative analysis of all the balances has not been done yet so that the above given conclusions are still tentative. Nonetheless, *qualitatively* a few more observations are possible. One of them, described now, may be found rather debatable. Another observation, related in the next section, in contrast lends strong support to the described process because it explains an experimentally established phenomenon, viz. the extreme directivity of laser beams.

It should be realised that the "third actor" in the case of a radiating droplet of charge has limited capabilities. The extra freedom here consists of a transfer of *energy* from or into an extra energy reservoir, but no internal exchange of *momentum* within the droplet is possible by the action of this "third actor".

The momentum in the radiation is thus not affected, only the energy can be modified such that the correct balances are restored and the required ratio $(energy)/(momentum)$ in the field has been achieved. The energy poured into the field in this way is *more* than just the kinetic energy lost by the droplet. Would that mean that another droplet, inside another well, could absorb more kinetic energy from the field than the lost kinetic energy of the sending well? No, it does not. One should realise that in the case of a receiver well the same internal processes will occur, be it in the reverse direction. There may be more energy present in the radiation between the sender and receiver, this will not be noticeable in terms of *kinetic* energy lost by the sender and absorbed by the receiver. To avoid confusion, one could call the transferred *kinetic* energy the "perceived" energy transfer, to discern it from the actual energy in the field.

To go to the extremes of speculation: remembering the equivalence of energy and mass, could the difference between actual and perceived energy in radiation have something to do with "dark matter", the mass that is missing in cosmological models?

12.11 Another Restriction Due to Conservation of Momentum: Directivity of Energy Transfers

It would seem that radiation by a *droplet* in a potential well is not restricted in the severe way as applies to a "marble" which does not have an extra degree of freedom. However, there are still some restrictions left, the situation is not fully comparable with a radiating electron inside an antenna rod. In this latter case, the electron may exchange both momentum as well as energy with the surrounding material of the antenna. Such complete freedom of the kind of interactions is not permitted to a droplet in a potential box.

The extra freedom in the case of the droplet consists of a transfer of energy from or into an extra energy buffer, but no internal exchange of momentum is possible with this "third actor" within the droplet. If the droplet in a potential box loses momentum during the radiation process (due to radiation resistance), this momentum will be found *unmodified* in the radiation field. The correct ratio $(energy)/(momentum)$ in the field has been achieved by an injection of extra energy. The momentum itself is unchanged, despite the existence of a "third actor". Let us now consider what the consequences are of this remaining restriction (by "restriction" is meant: compared with the radiation by an antenna rod).

Figure 12.15 shows two potential wells, one which emits energy and momentum, the other (on the right) absorbing all of it, due to an upward "quantum jump".

The upward jump might have been stimulated by the field of the emitting well. This possibility is further considered in the next chapter. For now we just assume that there are both a radiation source and an absorber in the field, and that they have

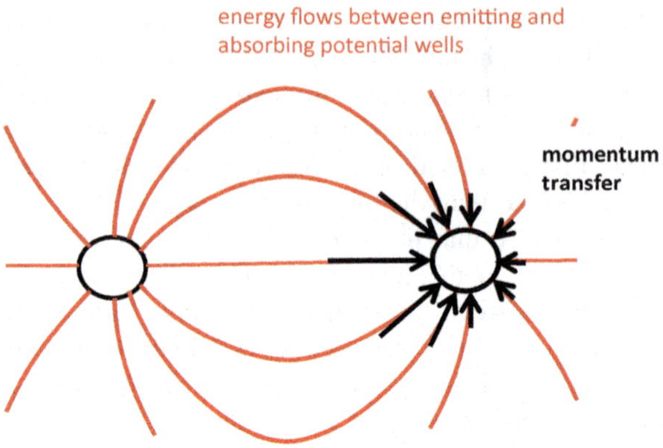

Fig. 12.15 The net force on the receiver is directed along the line of sight from the emitter to the receiver

an equal strength. The absorber must be described by *in*coming electromagnetic waves, as a sort of mirror of the emitting well on the left. The matter of incoming waves has been dealt with earlier in this chapter. If the absorbing well would exist in isolation, incoming waves from infinity would violate causality, as has been discussed. Such a problem does not exist here, where a *combination* of two identical wells is considered, one spewing out energy, the other absorbing it. The pattern of flux lines in the field of such a combination configuration is sketched in Fig. 12.15.

The momentum in the field is closely connected with the flux of energy. The black arrows symbolise how the total momentum in the field is transferred to the absorbing well. The resulting momentum transfer from the left well to the right has the same direction as the line of sight between the wells.

But the restriction is more severe than just momentum transfer along the line of sight. Recall that the momentum in the field equals both in quantity as well as in direction the momentum lost by the emitting droplet of charge. The overall momentum in the field is thus parallel to the motion of this droplet within the well. The same kind of restriction applies to the absorbing well. When the event of emission and absorption has been completed, no energy flux will be found in the field, and neither any momentum will be found in the field. In fact, the field is after the event no longer there, the space around the two wells has become void. In contrast to the energy, the momentum has a definite direction. In order to achieve a complete transfer of momentum the motion of the droplets inside the two wells must have been aligned (strictly for readers who are familiar with the laws of mechanics: the moment of momentum puts this further restriction on the event). It all comes down to the conclusion that the wells *themselves* must have been *aligned along the line of sight* between the two interacting wells.

The last conclusion is graphically illustrated in the following figure, Fig. 12.16. Only interaction between the emitting well and well B is allowed. This is the consequence of the requirement that not only energy should be conserved, but also

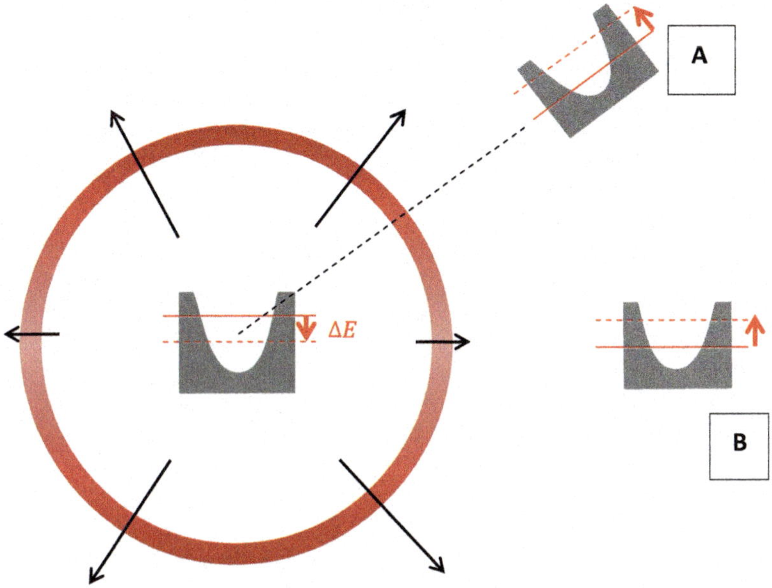

Fig. 12.16 The overall momentum balance can only be correct for the potential well B

momentum. Interactions of this type are thus highly directional, much more so than the directionality of a common antenna. This conclusion is fully in agreement with usual quantum theory (see e.g. the subject of "production of a cloud chamber track" in Schiff: "Quantum mechanics"). This large amount of directivity is also the reason why laser beams can be pencil sharp.

Chapter 13
On Bohr's Radiation "Out of Nothing", and On Photons, the "Particles With a Wavelength"

13.1 The Frequency of Emitted Light: Bohr's Stroke of Genius

In Chap. 10 (Sect. 10.2), the visibility of a droplet of charge was discussed. If the droplet was so tiny that it contained just the unit charge and thus became "electron-like", most of the time it would be invisible. In a stationary state, for example in a potential box, it would race back and forth between the walls of the box, and every pass from wall to wall would be identical to any other pass. This is what was defined to be a "stationary state". A stationary state therefore also implies that the energy of the droplet would remain unchanged from pass to pass. There is no loss of energy by radiation possible, and no radiation. Therefore there is no means to "see" the droplet.

It was already sketched in Fig. 1.4 what Bohr assumed that could happen to get a glimpse of an electron, in the case of his atom where electrons are orbiting the nucleus. To save the reader an inconvenient search through the chapters the sketch is here reproduced in a slightly modified form, as Fig. 13.1. Needless to say, the sketch is completely out of proportion. Perhaps you remember the description of a nucleus as a "fly in a cathedral"! The broken circles just symbolically indicate the stationary states that are allowed (that is, "allowed" in view of the energy quantisation hypothesised by Bohr). As long as an electron stays in such a stationary state and thus keeps to its assigned orbit, it will be invisible. But sometimes a sudden jump is possible towards a lower orbit, an orbit representing a smaller energy. The electron after the jump possesses less energy, and must have shed energy in the form of electromagnetic radiation. There is, so to speak, a flash of light, and this is what makes the electron visible. This way to describe the event probably gives the wrong suggestion: we do not "see" the electron in the sense that it can be inspected momentarily, we only get a signal that something has happened inside the atom resulting in a lower energy of the total constellation, the assembly of the nucleus and electron. The rest of the picture as sketched in Fig. 13.1, for instance the

T. van Holten, *The Atomic World Spooky? It Ain't Necessarily So!*
DOI 10.2991/978-94-6239-234-2_13

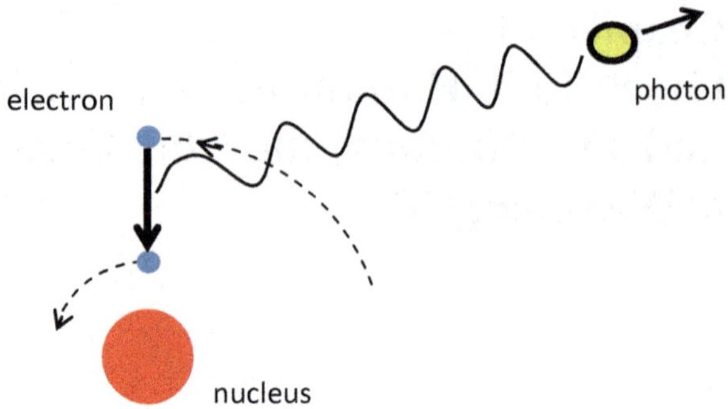

Fig. 13.1 Jump of electron from higher to lower orbit, releasing a photon

geometry of this assembly, is really a mental picture, a constructed visualisation by Bohr which would appear to be in agreement with the only thing that can be observed, the flash of light.

Bohr did not concern himself with what happens to the electron *during* the jump, and how the jumping itself would look like. His view was: at one instant the electron is still in the higher energy level, and the next instant it suddenly makes its appearance in the lower level. Bohr, for the purpose he pursued, did not need any assumption about how the transition *process* would evolve. He had made a few conjectures from which the energies of the different stationary orbits could be calculated, and went on with another rather wild assumption: he declared that the colour of the light flash (in other words: its frequency) would be determined by how much energy the electron loses during its sudden jump. The more energy present in the light flash, the higher the frequency. To phrase it differently: the larger the jump of the electron, the more the colour of the emitted light would be shifted from red to blue (if the emission is in the visible range of frequencies at all, it could also be in the IR (infra-red) region, or in the UV-band or still further removed from what the eye can detect).

This last assumption by Bohr, viz. that the frequency depends on the energy bookkeeping of the electron, is odd—although not more than the other assumptions on which his whole story rested—because one would expect something different. In the human world, i.e. on the macro scale, we always see that the frequency of radiation is bound up with the *motion* of electrons. If for instance electrons are oscillating back and forth in the metal of an antenna, and if they do that with a certain rhythm (impressed on them by the electrical circuits of the radio installation), the frequency of the emitted radio waves is the same as the frequency of the motion of the electrons. The oscillation found in the waves is similar to the oscillation of the electrons, and this is logical: the electromagnetic field is "rooted" in the electron. If the electron moves, the field moves with it. If the electron is oscillating, the field will oscillate too. And it does so in the same rhythm as the electron's motion. At least, that is so in the macro world. Therefore, if an electron is orbiting around a nucleus at a certain rpm, i.e. if it revolves with a certain circular frequency, one would expect that any radiation coming from it would have that

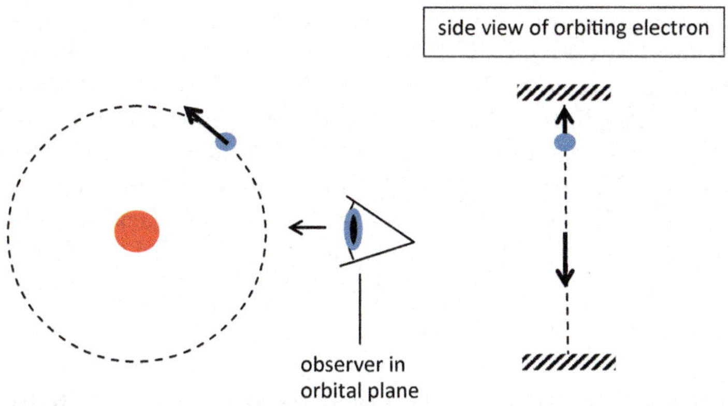

Fig. 13.2 For an observer in the orbital plane the circular motion looks like the translational motion in a potential well

same frequency. This is even more clear if one would look at the revolving electron from a position standing in the orbital plane. It is sketched in Fig. 13.2: looking sideways at the orbiting electron a one-dimensional motion would be seen as if the electron were moving in a one-dimensional potential well. The back and forth motion then seen has a definite, clearly defined frequency. One would expect that any radiation coming from the electron would have this same frequency.

Of course, during a transition from one orbit to another, the situation is more complicated because each one of the orbits would be associated with its own characteristic frequency. In that case perhaps the average of the two orbital frequencies would be expected to be present in the radiation, or some other value that is anyway related to the *motion* of the electron.

In contrast, according to Bohr's hypothesis the frequency found in the radiated light flash is not at all related to the motion of the electron. It is related to the energy difference of the two orbits. And the relation proposed by him is—mathematically speaking—a simple one, but physically rather mysterious:

$$(\textit{frequency of radiation}) \div (\textit{energy difference})$$

This conjecture by Bohr was very successful. In combination with his assumptions about the quantisation of the energy levels in an atom, he could now explain and even correctly quantify a part of the light spectrum of hydrogen. In one masterly stroke, all of his bold assumptions and visualisations appeared to be justified! The above shown proportionality reads, using a proportionality constant to write it in the form of an equation instead of a proportionality:

$$(\textit{frequency of emitted light}) = \frac{(\textit{energy difference})}{h}$$

The constant of proportionality occurring in this formula is $1/h$, miraculously there is again Planck's constant which has the habit to stick up its head almost everywhere in the formulae of quantum mechanics. Why I said above that the formula is "mathematically simple" but "physically mysterious" is, that the oscillations found in the electromagnetic waves (i.e. the frequency of the emitted light) do not seem to be "rooted" in something concrete: no part of the atom is oscillating in this frequency. How successful the theory may be, the question remains where the radiation might come from, or: where are the "roots" of the oscillating electromagnetic field? It can certainly not be explained by Maxwell's laws!

13.2 Bohr's Frequency as Derived by the Droplet Theory, and Planck's Formula

The above stated formula, specifying the relation between on the one hand the frequency of the emitted light and on the other the energy difference of the stationary states between which the electron jumps, was conceived by Bohr with the specific purpose to gain insight in the workings of a three-dimensional atom. However, the same relation is valid too for one-dimensional potential wells, such as we consider in this book. We have seen that in potential wells there are also definite stationary states, leading to quantisation of the energy of droplets inside a well. Let us revert again to the simplest case, the potential box. The droplet may not have any arbitrary velocity, otherwise the situation would not be stationary: there can only be a perfect repetition of the back and forth motion for certain discrete values of the velocity, and there is a tendency to drive the droplet towards these particular values of the velocity.

Now consider a jump from one state to another state with a lower energy. To begin with, take the state with the higher energy and accordingly the higher—average—velocity of the droplet, and call this velocity v_{high}. According to Chap. 8 (Sect. 8.2) the pulsation frequency that is associated with this velocity is given by the expression

$$f_{high} = f_0\left(1 + k \cdot {v_{high}}^2\right)$$

Recall that the symbol f stood for "frequency", f_{high} thus meaning the pulsation frequency at speed, and f_0 the frequency when the velocity is zero, the so-called "zero-speed frequency". The formula expresses that the actual frequency during the motion is higher than this "zero-speed frequency". To be more precise, what is here called the high velocity v_{high} is the *average* velocity around which fast fluctuations take place, fluctuations of the velocity induced by the fast pulsation. The fluctuations of the velocity have the same frequency f_{high} as given by the above given formula for the pulsation. If the droplet would come in the situation that it could radiate (it must then have left the chaotic state, and must be in certain favourable positions within the well, see Chap. 10), the radiation would at the first instant have the frequency f_{high}.

The situation in the finally reached lower energy state to which the electron jumps down is described similarly by the expression

$$f_{low} = f_0\left(1 + k \cdot {v_{low}}^2\right)$$

Like Bohr, for now we will not concern ourselves with the transition *process*. We just content ourselves with the two expressions given above which describe the situation before the jump and after the transition to the lower energy level is completed. The near electromagnetic field of the droplet before the jump consisted of—almost—standing waves with the frequency f_{high}. After the process of energy shedding has been completed we again have a near field of standing waves, now with the frequency f_{low}.

If we compare the situation with the standing waves in a string such as seen in Fig. 12.10, the transition from f_{high} to f_{low} would be equivalent to a sudden adjustment of the tension in the string. As shown in Fig. 13.3, the strings of e.g. a violin are tensioned by tuning pegs. Turning these carefully, the correct ground tone of the strings may be set.

When using the bow to let the string vibrate, and when at the same time the tuning peg is suddenly turned, the tone changes. In physical terminology, the tension in the string is suddenly changed, and the string is going to vibrate with a different frequency. There is a brief moment where both tones sound together, the original vibration extinguishing, and the final vibration coming up and taking over.

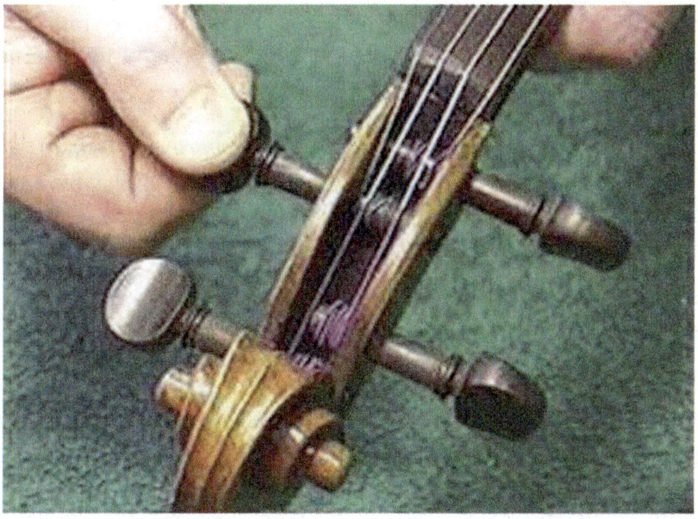

Fig. 13.3 The tuning pegs of a violin, by which the tension in the strings is adjusted, and the correct ground tone of the strings is obtained

Similarly, we may expect that in the near field of a pulsating droplet there will be a mix of the two frequencies f_{high} and f_{low} during a brief moment.

The two frequencies are different, but are not very far apart from each other. The result is—you feel it coming—interference leading to a beat phenomenon. These beat phenomena were mentioned so often in this book, that they will have become familiar to the reader by now and hopefully need no further explanation. In the present case the beat frequency equals, as always in the case of beats, the difference frequency:

$$f_{beats} = f_{high} - f_{low}$$

Combining this result with the expressions that were written above for the actual values of f_{high} and f_{low}, a proportionality relation is found:

$$f_{beats} \div ({v_{high}}^2 - {v_{low}}^2)$$

And here we have almost arrived at Bohr's hypothesis: velocity squared is proportional to the kinetic energy. In a potential box kinetic energy represents the total energy of the droplet, so that this last formula says:

$$(\textit{frequency of beats in the field}) \div (\textit{energy difference})$$

Now, recall the remarks that have been made in the preceding chapter about the observability of the electromagnetic field. The beats have a so much smaller frequency than the frequencies f_{high} and f_{low} of the pulsation itself, that they are inside the "observability window". In radio language: if the antenna signal (after amplifying and rectifying it) would be fed to a loudspeaker, one cannot hear the carrier wave, but one does hear the audio signal carried by it. The *observable* frequency of the radiation due to a transition, i.e. the colour of the emitted light we are able to see, is thus

$$(\textit{observable frequency of radiation}) \div (\textit{energy difference})$$

and this is exactly the same result as was hypothesised by Bohr.

The next, obvious question: is the fully quantified expression by Bohr

$$(\textit{frequency of emitted light}) = \frac{(\textit{energy difference})}{h}$$

also shown up by the droplet theory? It is!

To show this, we must make a side step, and recall another subject dealt with in an earlier chapter. In Chap. 8 we began everything with De Broglie's formula, giving a relation between the velocity of a particle and the wave length of the accompanying matter waves:

$$(\textit{velocity of particle}) \div \frac{1}{(\textit{length of matter waves})}$$

or sticking in some constants to change the proportionality into an equation:

$$(\textit{velocity of particle}) = \frac{h}{(\textit{mass}) \times (\textit{length of matter waves})}$$
$$(\textit{De Broglie's formula for matter waves})$$

From the droplet theory came a similar relation:

$$(\textit{average velocity of droplet}) = \frac{K}{(\textit{average mass}) \times (\textit{wavelength of beats})}$$
$$(\textit{relation from droplet theory})$$

and finally, after having quantified the proportionality factor K, it was found that "De Broglie" was applicable to our droplet of charge just as well, because it appeared that K has practically the same value as Planck's constant h:

$$K \approx h$$

Now returning to the case of the radio transmissions by a droplet of charge changing one stationary state for another, the result from droplet theory was given above:

$$(\textit{observable frequency of radiation}) \div (\textit{energy difference})$$

Again, we can quantify this relation, and it is then found from the droplet theory:

$$(\textit{observable frequency of radiation}) = \frac{(\textit{energy difference})}{K} \quad (\textit{droplet theory})$$

Since $K \approx h$ this latter result from droplet theory is the same as Bohr's theory, which as earlier mentioned resulted in

$$(\textit{frequency of emitted light}) = \frac{(\textit{energy difference})}{h} \quad (\textit{Bohr's theory})$$

The quantitative results from the two theories are the same. But there is a difference in the interpretation. We had noted that in Bohr's theory there is nothing in his atom model that vibrates in the same frequency as the emitted light: the oscillations in the radiation are not rooted in something substantial. What we now have found in the droplet theory is, that the oscillations in the electromagnetic field do not come "out of the blue". The oscillations are definitely "rooted" in something substantial, viz. in the fast speed fluctuations of the droplet of charge, induced by its pulsation. When the average velocity drops down or jumps up to another value, what we do observe during a brief moment is the beat phenomenon in the electromagnetic field, due to the mix of initial and end conditions.

Finally, how much energy in total is present in the travelling "energy burst"? This is an easy one to answer: the total energy in the burst must be equal to the energy lost by the droplet when it jumped to a lower energy level (note that we here take "energy in the field" to be the *perceived* energy, as it was defined in the last part of Sect. 12.10). The total energy equals what was earlier called $(energy\,difference)$ or ΔE, the energy difference between the two stationary levels involved. Recalling the formula

$$(observable\,frequency\,of\,radiation) = \frac{(energy\,difference)}{K} \qquad (droplet\,theory)$$

which was valid for a droplet in a potential box, we thus find for the "travelling energy burst" emitted by the droplet of charge (just invert this last expression, and substitute E_{burst} for ΔE):

$$(total\,energy\,in\,burst) = K \times (observable\,frequency\,of\,radiation)$$
$$(droplet\,theory)$$

One of the properties of photons, as found in the combined theories by Planck and Einstein (the theories that laid the very first foundations of quantum theory, as mentioned in Chap. 1) was:

$$(energy\,of\,emitted\,photon) = h \times (frequency\,of\,emitted\,light)$$

or, in the form of the famous formula by Planck:

$$E = h \cdot f$$

Clearly, the view offered on radiation processes by the droplet theory is not in conflict with the usual quantum mechanical view (recall that $K \approx h$). One should be careful though. What the above given formulae state is, that the process of *shedding* energy—no matter whether we consider a droplet of charge or an electron in usual quantum theory—is similar (even quantitatively). But so far we have not covered the process of *further transportation* of the energy. This matter will be dealt with in the next section. The process of *shedding* in both cases (droplet theory and usual quantum mechanics) takes the form of an energy burst, and in both cases we can assign the same frequency to these bursts, a frequency that is related to the energy difference between two stationary states. The energy of the stationary states has been shown to be the same in the droplet theory as in quantum mechanics, and in both cases we thus find a total energy in the burst conforming to Bohr's expression.

The results derived above might give the feeling that everything is unstandable now (at least: about the process of shedding energy by a droplet). Alas, there are still many riddles left. It should be realised that we considered just initial and end conditions, the process in between was not yet touched on. And for a good reason: the mathematics in the appendix do give a few clues concerning the causal chain of events during a quantum jump, but a complete picture cannot yet be built up. In later chapters more aspects of the quantum jump in a potential well will be mentioned, in a speculative way. What we would like to understand, for instance, is: how does the droplet "know" when exactly to stop radiating, so that the newly obtained energy level corresponds with one of the allowed levels? Of course, the energy level obtained after the radiation need not exactly be one of the quantisation levels, a rough approximation is sufficient. The subsequent bounces against the walls will later fine-tune the motion so that it becomes exactly stationary (see Chap. 9). But even such a rough approximation of the amount of radiation needed for a correct jump seems mysterious. There are some clues in the mathematical derivation to explain it, but these are no more than hints at present. The clues are interesting enough, because they give a better understanding of what in usual quantum theory is called the "superposition of states", one of the famous magical features of the quantum world (think of "Schrödinger's cat", and if you are not familiar with this cat, read the story in Chap. 15). In view of the still speculative character, it has been decided to postpone for now a further discussion. In a later chapter (Chap. 15) several items with a rather more speculative character are collected, and we will there come back to the "superposition of states".

13.3 A Mental Picture of the "Photon" Possible?

The remainder of the present chapter will be devoted to rounding up the above, and to show how we may form a mental picture of one of the most intriguing phenomena of quantum theory: the photon. More generally stated: we now come to the question how the energy is further transported, after having been released by an electron or by a droplet of charge.

After the work by Maxwell on electromagnetic waves it all seemed so clear and simple: light appeared to be just another type of electromagnetic wave. The only thing that discerned light from other forms of radiation such as heat- or Röntgen radiation is, that our human eyes are sensitive for it. This view of "light" explained the outcome of many experiments showing it to be a wave phenomenon, and after Maxwell the question "a wave of what" was solved. Or so it seemed.

But then later, rather disturbing experiments spoiled the game, and the pair of Planck and Einstein told the world that there was only one explanation possible for these new experiments: light consists of a stream of separate *particles*. The particles of light were baptised "photons", and most people are now used to them and talk about photons as if they were the commonest things in life. But of course, they are not. The view on light as a stream of photons in its turn cannot explain the multitude of earlier, very convincing experiments that had proved light to be a wave phenomenon. Nowadays, most people have become used to the idea, conceived by physicists as a refuge, that light sometimes has to be considered as a particle and at other times as a wave (the so-called particle/wave duality). Perhaps the idea is even loved by mystically inclined people who also believe in werewolves behaving most of the time like ordinary humans, except during full moon when they turn into repulsive creatures. One may laugh about these people, but in essence physicists have similar beliefs when it comes to the question what light fundamentally is. De Broglie made it even worse, when advocating the—very successful—idea that this same dichotomy applies to *all* elementary particles, because every one of them can display wave-like behaviour. In the case of light our eyes tell us that light has a certain colour, which is according to the wave-view the same as saying that the light waves have a certain frequency and wavelength. The wave/particle duality then leads to the rather incomprehensible view of photons being "particles with a wavelength". It really is a contradictio in terminis, not mitigated by the fact that physicists can perform extremely reliable calculations based on the duality paradigm. The pragmatic approach in physics is "shut up, and calculate", but for the non-physicist the matter remains worrying.

Even more worrying is the illustration Feynman gives how the theory about photons must be visualised. This interpretation of photon theory may be found in his book "QED, the strange theory of light and matter". One of his sketches is reproduced in Fig. 13.4. Here it is assumed that a photon leaves a source of light (the point S), and arrives at a receiver R to deposit its energy there.

One might have had the primitive thought that a photon behaves like a sort bullet, flying straight from S to R. The theory about photons tells an entirely different thing. One must imagine that the photon can follow a multitude of different paths, as sketched, and not at all straight paths. And what is more: to visualise the theory, it is necessary to imagine that the photon actually follows *all* these possible trajectories before it hits the receiver R. It is as if the photon has split itself in an infinite number of fragments, so that it can simultaneously "sniff out" all these possibilities. Finally, the photon reassembles and materialises again in the form of a "particle-like thing" when hitting the receiver. It is a strange picture, but this is one way to explain the observed behaviour of light, at least if we adhere to the idea of light as a granular phenomenon which transmits energy from S to R in definite packets.

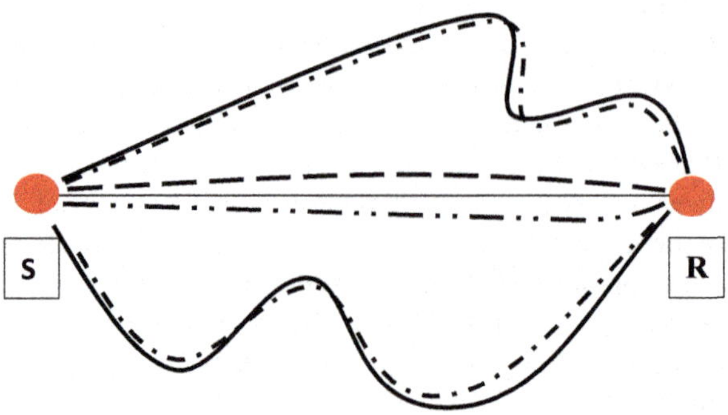

Fig. 13.4 Sketch by Feynman (from the book "QED, the strange theory of light and matter") symbolising the many paths "sniffed" by a photon on its journey from source S to receiver R

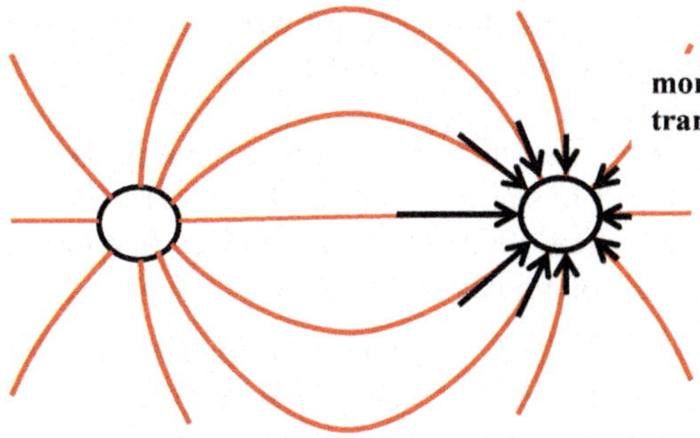

Fig. 13.5 (12.16 repeated): Energy and momentum transfer by waves. To be compared with the transfer by particles (photons) visualised in Fig. 13.4

The view of light as continuous waves leads to a picture which is not drastically different from Feynman's picture. It was already sketched in Fig. 12.15, and is repeated below as Fig. 13.5, so that it can directly be compared with Fig. 13.4. Especially if it realised that the pattern of Fig. 13.5 was stylised too and may in reality be more fickle, the figures are basically similar. The essential difference is, that the wave view does not explain why the energy transmitted would be quantised, i.e. why not any arbitrary amount of energy can be transported from S to R, and why it cannot be a stream of energy, continuous in time.

Actually, what until now has been derived using the droplet theory agrees with the visualisations of the photon theory in more aspects than shown in Figs. 13.4 and 13.5. Let us sum up what already has been found, during our systematic search for what the consequences are of the working hypothesis of a "droplet of charge":

1. The droplet of charge can under exceptional circumstances enter a radiating state. If it is moving inside a potential well, it then falls from one of the allowed energy levels to a lower level.
2. Something that is found from quantifications done in the formula appendix: the time it takes to shed energy by the droplet is very small compared with the time to complete a full pass between the walls of the potential well. On the time-scale of a macroscopic observer the radiation thus gives the impression of a brief flash of light.
3. At sufficiently large distances from the source of radiation (i.e. in the so-called "far field") the "flash" is transported with the speed of light by a radiation field that itself is *un*observable, the frequency of the constituting parts (the electrostatic and magnetic fields) is too high.
4. However, as far as the associated *energy flux* is concerned, this is observable because it contains the definite positive product of the field components and—after the time averaging act of observing it—has the observable, low frequency of beats (in fact, this beat frequency is equal to the Bohr frequency). The travelling "energy burst" is *not* supported by a "visible" electromagnetic field, it appears as energy travelling on its own, in isolation.

5. The total amount of energy in such a burst is exactly the same as that of a photon. The relation between energy and the observable frequency is the same, and agrees with Planck's formula $E = h \cdot f$ (i.e. the energy in the burst is proportional to the observed frequency, with Planck's constant as the proportionality factor).
6. Despite the fact that the field representing the energy burst, after its emission, spreads out in space the interaction it can have with receivers is strictly limited in its direction. So much limited in fact, that the emission is not capable to affect more than one receiver at a time.

So far what already was known to the attentive reader (except item 2, which is a new addition to the list). This kind of picture does remind one strongly of the process of transmitting energy by a photon, which too can be described as an isolated "packet of energy" travelling with the speed of light, and endowed with a certain frequency (colour). Before having seen Fig. 13.4 one might have thought that there is still an essential difference. One would have thought that a photon is particle-like, whereas the radiation pattern in Fig. 13.4 is spread out in space. Feynman's picture in Fig. 13.4 has rubbed out even this difference.

What remains as a—seemingly crucial—difference is found on the receiving side, where the energy is absorbed. In the picture of Fig. 13.4 concerning the history of a photon, the photon "materialises" at the receiver end, and *all* of the energy is absorbed in one single stroke. The wave view, in contrast, permits to imagine that energy is absorbed from the field in arbitrarily small or large portions. This is a fundamental difference. At least, it would be a fundamental difference, if these statements were consistently and generally true. But they are not. In the next section an experiment is described (Compton's scattering experiment), which shows that photons sometimes leave only a part of their total energy with a receiver (the receiver must then be a free electron, instead of an electron bound to a potential well). On the other hand, in the same section it will be described that most of the time a receiver in a continuous field of waves cannot do anything else than absorb in one single stroke all of the available energy (the receiver must then be a droplet in a potential well). The so-called "fundamental" difference thus becomes dependent on the circumstances. The line dividing the particle-view and the wave-view is not sharp, it becomes a bit blurred. This is perhaps not so surprising: after all there are good reasons to speak about the particle/wave *duality* of photons.

Everything boils down to the following question: can the droplet theory describe the available experimental data just as good as the theory based on photons? The droplet theory is based on continuous electromagnetic waves, and for this reason alone it *seems* to be contradictory to the views of usual quantum theory. But the droplet theory leads to very peculiar characteristics of this so-called continuous field, making it quite different from macroscopic radio transmissions. Might it be that it is an *alternative* visualisation, equally useable as the particle view to describe measurable events?

Crucial in the last sentence is the "*useability to describe measurable events*". If we accept this as a criterion, we have shifted the question to a phenomenological investigation: if two types of visualisation seem to be entirely different, but if they are able to describe the same observable phenomena, then it does not make sense to quarrel about which visualisation is the better one. But of course, we must make sure that the two types of view are phenomenologically indeed equivalent. This will be the subject of the next few sections. It goes without saying that the experiment which brought Einstein to his particle view of light (the photoelectric effect) deserves special attention.

13.4 On the Receiving Side, the Photon-View Has a Few More Mysteries in Store

After having discussed the process of releasing a photon and the process of transporting the energy through space, it is now time to give some thoughts to the receiving end. We can begin the discussion by offering a few more observations to think over, and to add to the confusion. First of all, one should realise that a "free" photon cannot be observed. One can only infer that a photon must have been flying around, because at some other place it has transferred its energy to a measuring device. This device can for instance take the form of another potential well, where the energy of an electron suddenly has been increased so that it now finds itself in a higher stationary state. At the same time, the photon ceases to exist. A "free" photon cannot be "seen" directly, any attempt to observe it destroys the original photon. It is only through its *interactions* that we can conclude that there must have existed a packet of energy travelling somewhere between the place and time of creating the photon and the place and time of its destruction.

This is not to say that always a complete destruction of the entire photon will happen. A famous experiment by Compton shows that the "collision" of a photon with a free electron (see Fig. 13.6) may take the form of a deflection of the photon's path, resulting in the disappearance of the *original* photon and a metamorphosis into another photon having another colour and therefore another energy. It is only upon measuring this "transformed" photon that the final destruction is completed. It shows that the observation (i.e. the "measurement") of a photon can be done in

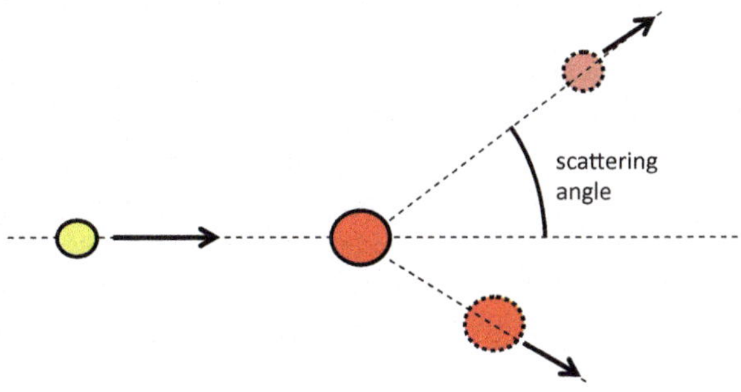

Fig. 13.6 Compton scattering. Photon hits free electron, and transfers energy and momentum. The resulting velocities of the electron and the scattered photon are such that total energy and momentum are preserved. As a consequence, the colour of the photon is changed, and depends on the scattering angle

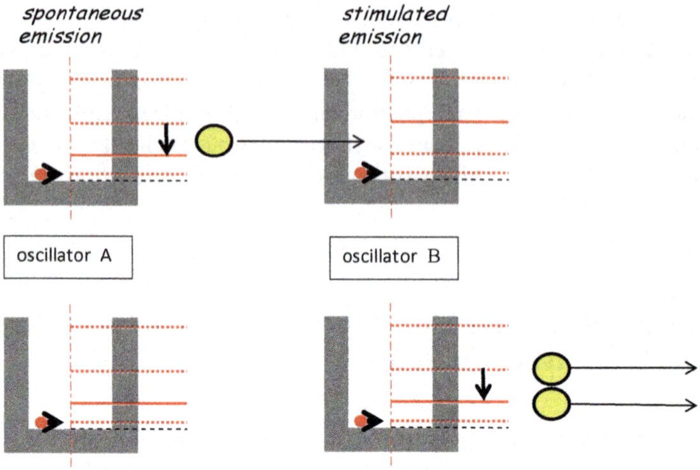

Fig. 13.7 Stimulated emission. Photon emitted by *oscillator A* hits second *oscillator B*. If energy of photon is correct, *oscillator B* is stimulated to emit second photon

more than one step. Even so, we always see the *results* of *interactions*, and it is a fundamental matter that we never directly see a "free photon".

Another phenomenon may be mentioned that can change our primitive view on a photon as just "a small marble packed with energy and having a colour" which flies around with the speed of light and during interactions can transfer part or all of its energy. It is the phenomenon of stimulated emission, playing an essential role in lasers. It is symbolically sketched in Fig. 13.7. In the upper part of the figure two oscillators are sketched, A and B. Oscillator A emits a photon after a spontaneous jump of an electron to a lower energy level, and it is assumed that the photon hits a second, identical oscillator B. The energy of the photon corresponds exactly with one (but only one!) of the energy steps between the quantised levels of B.

The lower part of Fig. 13.7 shows what may be the result: an electron in oscillator B may then be stimulated to jump to the corresponding lower level too. The final result is that two identical photons fly away. This is what is called *stimulated emission*.

At a first glance, it seems a logical process without any mysteries. But at second thought, this sequence of events is not so self-evident at all. The first photon at the end of the process has not been changed, so we may ask how precisely it interacted with the second oscillator. How did the second oscillator "experience" the hit by the photon? The photon did not transfer net energy or momentum to the oscillator. One might think that perhaps the photon acted as a catalyst, first giving part or all of its energy to the oscillator, and afterwards getting it back again. But such an intermediate step would not be possible, just because of the quantisation of the energy levels in oscillator B, which does not allow such a temporary energy absorption: the difference of energy with the next higher level does not fit the energy of the photon.

Then, as an escape from this problem, one might think that the stimulation goes through the "colour" of the photon. One might then imagine that the frequency that in some mysterious way resides in the energy packet is able to cause a sort of resonance in the oscillator B, which triggers the process. But there are two objections to such a view. The first one is, that the starting up of an oscillation of some part of B, even if it is a resonance, still requires an input of energy from outside. The "outside" is in this case the photon that hits B. And we have just seen that an energy transfer—in part or in whole—from this photon is not in agreement with the restrictions set by the quantisation of the energy levels in the oscillator. A second objection is, that resonance would imply that "something" in oscillator B is going to vibrate in the Bohr frequency. But, just as in the case of the emission of the photon by A, there is no part of the system that vibrates in this frequency (remember that we here take the point of view of quantum theory). Earlier it was formulated as: "the frequency of the emitted electromagnetic energy is not *rooted* in something substantial". At the receiving end, the immitted energy neither would have roots in anything that is going to co-vibrate with the photon.

All this is a warning that our first, intuitive hunch about stimulated emission may be too primitive. There must be processes involved in the stimulation that are not so easily understandable as we might have thought at a first superficial glance.

Similarly, *absorption* of energy from a photon is more problematic than a superficial glance would suggest. It will be called here—somewhat unusually, but explained later—as "stimulated" absorption. It is the situation sketched in Fig. 13.8.

In this case an oscillator emits a photon, which is absorbed by a second oscillator so that an electron is raised to a higher energy level. The energy of the photon is equal to that required for a jump of the electron to the higher level. The energy balance is thus simple, and would not give rise to any further questions. But here again, we may ask what it is exactly that *initiates* the jump in the second oscillator. Is it a resonance effect that excites the electron in sympathy with the photon's vibration frequency? Once again, there is nothing substantial to be found in the oscillator that possesses this frequency! So, how would resonance be possible?

Now, we could say, that only the final result counts and that we leave the intermediate processes to the realm of mysteries of quantum mechanics, not amenable to our human intuition. After all, there are so many anti-intuitive

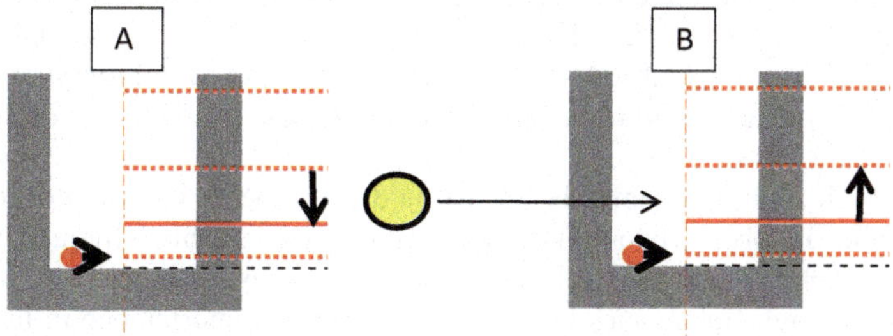

Fig. 13.8 "Stimulated" absorption of energy from a photon by a second oscillator

phenomena in quantum mechanics that cannot be "explained" by our macroscopic ideas about causality. This is the approach Bohr adopted when proposing his model of atoms. He did not concern himself with what happens *during* the jumps of electrons between the stationary states, he just analysed the before/after situation (with great success).

Nevertheless, sometimes quantum theory seems to contradict itself. Einstein, when he set up the theory of stimulated emission, noted on theoretical grounds that stimulated emission and the absorption of a photon are similar processes. They are just the inverse of each other, as Einstein's quantifications show. The word "stimulated", which is always used in the context of photon multiplication as in Fig. 13.7, does suggest a causal process. The visualisation evoked by using the term "stimulated" is more than just a cool, detached observation that the final energy balances come out right. "Stimulation" is the expression of a feeling that there is a chain of causal events causing the release of the second photon, even if it is not clear how it happens.

If the process of photon absorption is indeed comparable with stimulated emission, except for the fact that a photon is absorbed instead of released, we must suspect that the causality of the events taking place is the same. This is the reason why in the present text the liberty has been taken to systematically use the adjective "stimulated" in connection with photon absorption too, although the term is not usual. It is done to emphasise that the conceptual problems noted in the above with respect to stimulated emission similarly apply to "stimulated" absorption of photons.

13.5 A Tentative Visualisation of Emission and Immission According to Droplet Theory

Now consider the processes sketched in the Figs. 13.7 and 13.8 from the point of view of the droplet theory. The causality of these processes can then be restored. First, it should be remembered that in droplet theory a droplet of charge has its energy stored in two different "reservoirs", the energy stored in its elongation and in the pulsation thereof (and coupled to it: the velocity fluctuations caused by the pulsation), and secondly the energy stored in the average velocity of translation. It is this second reservoir that is subject to energy quantisation in a potential well. The first reservoir is not quantised, and can absorb or release arbitrary amounts of energy, either interacting with the "translation reservoir" or with outside sources of energy.

Secondly, it should be remembered that in droplet theory emission or immission of radiation involves two—ultra-high and unobservable—frequencies. The observable "colour" assigned to these emissions is just a *metric,* i.e. a scale to measure these high frequencies by way of their mutual interference in the form of beats.

The incoming radiation which in Fig. 13.7 is "experienced" by a droplet contains a frequency that exactly corresponds with the droplet's pulsation frequency. One can thus imagine that resonance may occur. This resonance would involve energy transfer between the radiation and the droplet, but does not immediately affect the energy quantisation in the other reservoir, i.e. in the translation energy. In other words, the quantisation does not block the interchange of energy between the droplet and the incoming radiation. If the droplet "is ready" to jump to another level, there is a sudden readjustment in the division of energy between the two internal reservoirs. This view certainly does not explain the entire process, not at all. What it does show is, that in droplet theory there might be less severe problems than in quantum mechanics when one tries to imagine a sort of resonance which "tickles" a droplet into stimulated emission. The crux is again the fact that in droplet theory the radiation is rooted in something concrete, and that there exists a second energy reservoir, which both fall outside the scope of the usual quantum theory.

The same reasoning is applicable to the process of "stimulated" absorption (Fig. 13.8). Both in the case of emission as well as absorption the radiation contains exactly the frequency that can cause resonance. On top of that, there is something material that is able to resonate. The stirring up of the resonance requires an energy exchange between the radiation and the droplet, but this is not blocked by any quantisation requirements: it is not a matter of "all or nothing", as it is in the usual quantum theory. The droplet theory thus presents less difficulties than quantum theory to explain the initialisation of an interaction between a field and a particle.

13.6 The Photoelectric Effect

Looking back at the history of quantum theory, our discussion until now goes as far as Planck's first ideas about quantisation of energy. He showed that the only tenable assumption about packets of energy sent out by oscillators is that they have a definite, accurately prescribed amount of energy conforming to the above mentioned expression $\Delta E = h \times f_{light}$ (with ΔE a discrete jump of energy in the oscillator). Oscillators send out (or receive) radiation in lumps, was the conclusion he reached. But mind: this only says something about the properties of oscillators, it does not say anything about the process between the emission and reception. Planck's conclusion, together with the formula he derived for the size of the energy lumps, were later considered to be the very beginning of quantum theory, for which he acquired the name "father of quantum mechanics".

But in a later addition to quantum theory by Einstein, it was concluded that also the transportation process between oscillators is intrinsically lumpy. Photons, on arriving at their target, *always* take the form of energy packets conforming to the formula $E_{photon} = h \times f_{light}$. It is an intrinsic property of light, and does not depend on the way a photon came into existence. There could be other, not quantised, ways to make photons than by energy jumps inside atoms, but that does not matter: the

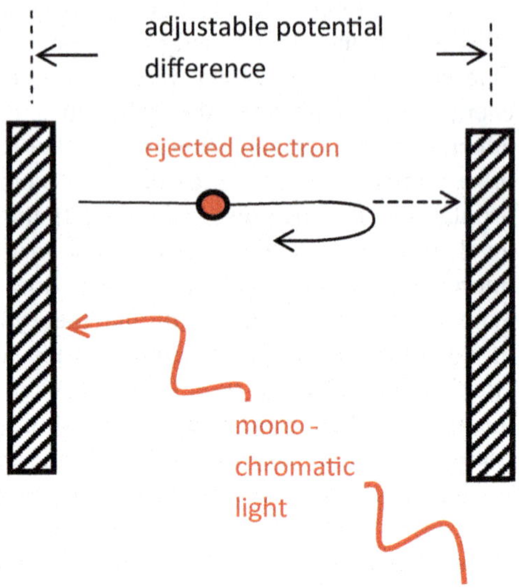

Fig. 13.9 Photoelectric effect

formula is always true. Light always comes in granular form with energy and frequency coupled to each other as in Planck's formula. Einstein's addition thus went further than Planck's conclusions, it said something about the process in between the actual emission and immission of radiation.

How did Einstein reach this conclusion? It was the result of an explanation of the photoelectric effect (see Fig. 13.9). We must therefore next consider this effect, both from the point of view of quantum mechanics and that of droplet theory.

Figure 13.9 shows two metal plates, kept at a different electric potential. The voltage difference can be varied at will. First make the difference zero, and then throw light of one specific colour on one of the plates (so-called monochromatic light). What one sees is that electrons are excited by the light, and acquire such a high velocity that they are able to fly out of the slab of metal. The escaped electrons can reach the other plate, so that an electrical current runs between the plates and can be measured.

It is easily measured how large the kinetic energy of the released electrons is. For this purpose a braking potential is applied between the plates, and at a certain voltage difference the electrons are no longer able to reach the second plate. The "stopping voltage" is a measure for the kinetic energy of the freed electrons. It now appears that the kinetic energy of the electrons depends *only* on the colour of the light. Red light needs a smaller braking potential to stop the electrons than, for instance, blue light (see Fig. 13.10).

Intuitively one would perhaps expect that the intensity of the light would have an influence on how large the voltage difference should be to stop the electrons. Surprisingly, it has no influence at all. If we would permit the electrons to reach the second plate, the intensity of the light does influence the current running between the plates. The higher the intensity, the larger the current. But intensity does not

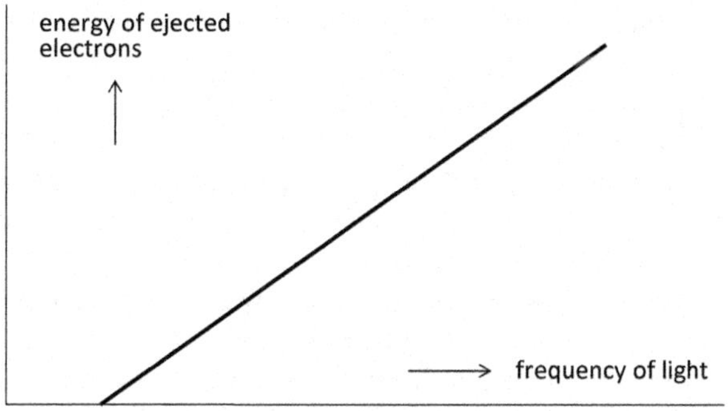

Fig. 13.10 Photoelectric effect. *Thick black line* shows measurement results when monochromatic light of different colours is thrown on the cell

affect the value of the stopping potential. In other words, the intensity of the light has no influence on the kinetic energy of the freed electrons. The kinetic energy is only influenced by the colour of the light thrown on the left plate.

Einstein explained these observations by assuming that the light consists of a stream of particles (photons), each of them having a fixed amount of energy, the amount of energy being determined by the colour (Planck's relation between colour and energy). Each photon can free one electron from the metal plate, by transferring its energy to the electron. The kinetic energy of the freely flying electron is thus determined solely by the colour of the light, in agreement with the experiment. Increasing the intensity of the light means that there are more photons (each of them still having the same energy), freeing more electrons. But each electron separately still has the same kinetic energy. The stopping potential is therefore only affected by the colour of the light. Light intensity influences just the amount of electrons freed, and is thus measured as an increase of the electrical current.

From this Einstein concluded that the photons themselves are particles with their energy determined by their colour, according to the relation $E_{photon} = h \times f_{light}$. Quantitative experiments confirmed his views on the character of photons and verified his quantitative theory on the photoelectric effect.

13.7 The Model of the Metal Plate in the Photoelectric Cell

It should be noted that the event sketched in Fig. 13.9, i.e. the ejection of a single electron after the hit by a single photon, implies that the metal plate will experience a recoil force. In Chap. 12 it was mentioned that the ratio $(energy)/(momentum)$ is different for a free electron and a photon. In case the energy is transferred completely from the photon onto the electron, there must consequently be a discrepancy between the respective momenta. In order to maintain the balance of momenta, it is therefore necessary that the metal plate plays an active role.

This is not really surprising, because one might compare the situation with the electrons in an antenna rod. They can absorb energy from radiation thanks to the fact that there is a certain momentum exchange between these electrons and the fixed atoms of the metal of the rod. They may share momentum in the required proportion, so that the conservation laws governing both energy and momentum are not violated.

It is the active role of the fixed atoms in the metal plate of a photoelectric cell, i.e. the provision of the necessary momentum, which explains the recoil force that must be present in the process of Fig. 13.9.

Keeping this in mind, the following—very simplified—model can be made of the plate. Certain electrons in the plate (the electrons in the conduction band) have the freedom to move from one side of the plate to the other, although they are not entirely free from any interaction with the fixed atoms. We could imagine this as a potential well as sketched in Fig. 13.11. Semi-free electrons can travel the distance between the sides of the plate, the width of the well taken as the thickness of the plate. Roughly, the well has a rectangular shape, with perturbations along the bottom in the form of corrugations. These perturbations of the potential symbolise the influence of the fixed atoms in the metal. The walls at the end are not very high, thus permitting tunneling of the electrons out of the plate if the energy of the electrons is sufficiently high.

Due to the macroscopic width of the well (the width equals the thickness of the plate), the quantisation of energy is practically a continuum of energy levels. One might alternatively interpret this as an absence of quantisation: the semi-free electrons may have any (average) velocity. However, there are non-negligible interactions possible between these electrons and the fixed atoms of the slab, as symbolised by the "corrugated" potential.

One may furthermore assume that the (semi)free electrons moving in the plate will have non-zero, arbitrary values of their energy. This can be attributed to thermal excitation of the electrons. An extreme example of this type of excitation is the cathode in old fashioned radio valves. Such a cathode was heated, and as a result electrons in it obtained so much energy that they were able to escape from the metal. A cloud of free electrons formed within the tube, which subsequently could be manipulated by variations of the electric field between the cathode and anode.

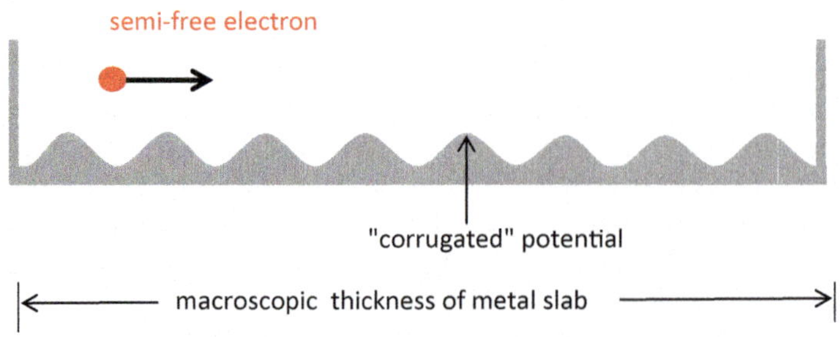

Fig. 13.11 Model of plate in photoelectric cell: almost rectangular potential well, with "corrugation" to represent the interaction with fixed atoms

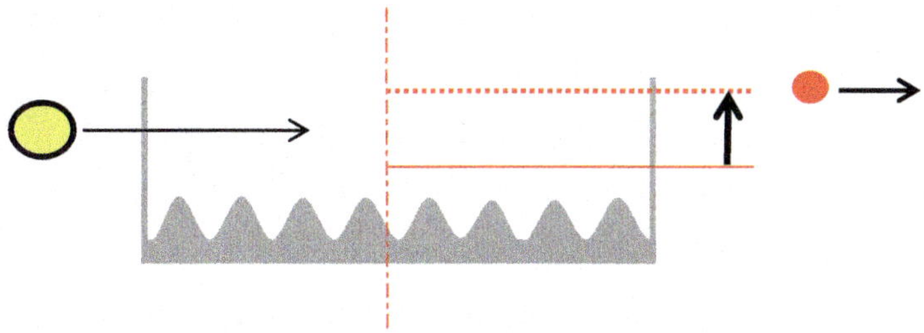

Fig. 13.12 Photoelectric effect from the point of view of photons

The photoelectric effect can be pictured as shown in Fig. 13.12, a picture that is not much different from Fig. 13.8 concerning the "stimulated" absorption of energy by a potential box.

There are two differences. In Fig. 13.12 the rectangular well does allow tunneling of energetic electrons out of the well. In Fig. 13.8 we assumed a potential box with infinitely high "walls", so that an electron cannot escape.

The more important difference between the two Figs. 13.8 and 13.12 is found in the red dotted energy level which after a hit with a photon is obtained by an electron. In the case of the small potential box of Fig. 13.8 there is a quantisation of the allowed levels. The level of the red dotted line is not arbitrary, and therefore the energy difference between the drawn red line and the dotted red line has a strictly specified value. This must correspond almost exactly with the energy of the photon, otherwise no energy transfer is possible. "Almost", because the energy width of the quantisation does allow a slight mismatch between the energy of the photon and the energy difference between the allowed levels.

On the other hand, the wide potential well shown in Fig. 13.12 is not quantised. The only thing that now can determine how far an electron's energy is lifted towards a higher level is the energy of the photon itself. The photoelectric experiments show that the electrons so excited all experience the same addition to their energy, on top of their thermal energy. The fact that there is an addition to their energy by a fixed amount must thus be attributed to the fact that the photon itself contains a certain, fixed amount of energy. This is what Einstein led to his conclusion that the photons themselves are quantised.

13.8 Photoelectric Effect from the Point of View of the Droplet Theory

The model in this case is shown in Fig. 13.13. The difference with Fig. 13.12 is that energy is now supplied to the potential well by a field of electromagnetic waves, instead of by a photon. We have seen that the total energy in this field is the same as the energy in a photon, but it is now spread out in space. It means that the energy

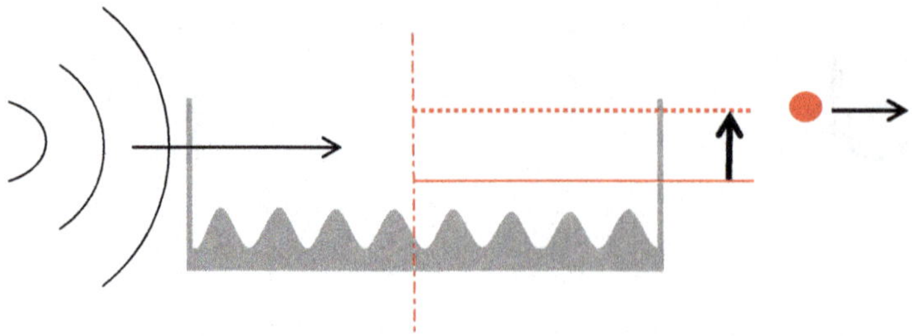

Fig. 13.13 Photoelectric effect from the point of view of droplet theory

addition to the well is not lumped, but will come in arbitrarily small portions. One would think that in this case there is nothing that fixes the distance between the drawn red line and the dotted red line symbolising the energy jump by an irradiated electron. The energy of an electron tunneling out of the well would not be fixed, and could have any value. At first sight, the process sketched in Fig. 13.13 is not able to explain the observed quantisation of the energy of escaping electrons. However, this first impression is false.

There is one property of the incoming radiation that has a fixed value: the frequency content. The frequencies in the radiation of course do not depend on how small or large the portions of transferred energy are. Below, it will be shown that this fixed quantity still determines the amount of energy that will be absorbed by a droplet.

Before proceeding, a limitation of the analysis of the process must be pointed out. Ideally, one would like to describe the process of energy absorption in every detail. From the first hit of the cell by radiation until the final tunnelling of excited droplets a causal chain of events must exist, and one would like to see how the process develops from link to link. At the present state of the theory this is not possible, a complete picture of such a sequence has not been built up yet. A complete description would have to consider all the interactions between the incoming radiation, the self generated radiation of the droplet (in particular the radiation that gives rise to radiation resistance), flows of Schott energy, the two kinds of vibration of the droplet (pulsation and hopping and their mutual phase difference), and the other energy reservoirs in the droplet (mean translational energy and potential energy in the stretching). All these interactions are variables during the proces. It should not surprise that this process is by far too complex to describe in detail. Even the simpler transition of one tone to another in an overblown trumpet is hard to describe in detail. The usual mathematical description of the trumpet neither attempts to give a detailed description of what are exactly the forces working on the particles of air, and how they consequently change their mode of vibration. Rathermore a less detailed, more overall description is usually found to be more convenient. What will be attempted in the case of the droplets of charge in the photoelectric cell is similar, it likewise ignores all the details and tries to give an overall, more global picture.

The latter choice is in line with earlier chapters, e.g. Chap. 10 (concerning the visibility of a droplet inside a well) where just the *initialisation* of radiation was considered. In the present chapter, Sect. 13.2 (Bohr's and Planck's frequency formulae) and 13.5 (stimulated emission and immission) the results were derived by "*before/after*" considerations. Later, in Chap. 15 an attempt will be done to clarify one of the aspects of energy jumps in potential wells (Sect. 15.3 on the superposition of states), but this neither covers the whole of the transition and, on top of that, is for good reasons included in the chapter "full of speculations". Therefore, an "overall approach" will be chosen too in the case of the present problem concerning the photo-electric effect, and no attempt will be made to delve in detail into the separate parts of the entire process.

The radiation impinging on the cell will be described here as an AM (amplitude modulated) high frequency signal. A droplet on the sender side has jumped from one energy level to a lower state, and what is found in the ensuing radiation is a mix of two, slightly different ultra-high frequencies. An alternative description is: the sum is an ultra-high frequency carrier wave containing low frequency beats. Going back to the source, the emitting droplet, we have the relation

$$(beat\,frequency) = (energy\,jump)/K$$

In the receiver (the photoelectric cell symbolised by Fig. 13.13) one may assume that there are a sufficient number of semi-free droplets in the conduction band which are potential receivers. The potential receiver droplets must have a pulsation frequency near the frequency of the carrier wave of the radiation. As in an AM radio receiver, radiation is then absorbed because of resonance effects. It was pointed out earlier (Sect. 12.12), that this interaction between the transmitter and receiver is directionally extremely selective. We may therefore assume that in practice not more than one (if any) of the potentially resonating droplets is actually able to act as a proper receiver. If it so happens that there is more than one droplet affected by the radiation, this would be an extremely rare occurrence, the chance that no one of the droplets is a good candidate is much larger. In the latter case the radiation just passes by without any interaction with the cell.

If there happens to be a proper receiver droplet, the pulsation and the translational hops of this droplet are greatly magnified by the resonance with the radiation, again like the excitation of the circuits in a radio receiver that has been tuned in on a particular broadcasting station. The excited pulsation will show the same beat frequency as was present in the radiation. This is equivalent with the audio signal picked up in AM receivers.

The processes of emitting energy and absorbing are similar, they are the inverse of each other. Therefore, the above given relation between beat frequency and energy difference is valid and now works the other way round:

$$(energy\,jump\,in\,receiver) = K \times (beat\,frequency)$$

It must here be stressed once again that we consider just the before/after situation. The statement above about the energy jump concerns only what the *final* outcome of the process must be. If it would not be so, there would have been a violation of the conservation laws: the equations of motion for a droplet are in agreement with the conservation laws, whereas these equations specify the relation between frequency and energy. The *final* outcome is therefore fixed, but of course nothing has been revealed about the intermediate processes. It may well be that the energy jump (which by the way concerns only the average translational energy of the droplet) involved borrowing energy from the potential energy within the droplet for a start. It would also be conceivable that the final energy balance within the field is not immediately accomplished, but takes a longer time. Think of the situation where the sender is a part of a distant star, whereas the receiver is the eye of an observer. Problems like these have all been bypassed by the before/after consideration, in the same way as they were bypassed when we derived the Bohr frequency of a sender.

A remark of a more philosophical nature may be made about the relation between frequency and energy that both in the sender as well as in the receiver determines the final outcome. The " = " sign in the relation $(energy\,jump) = K \times (beat\,frequency)$ does not indicate a causal relation. It does not say that the beat in the pulsation *causes* an increase of the translational velocity. Neither the other way round: the frequency of the pulsation is not a consequence of the droplet's velocity. Rather, the frequency and the translational energy are both metrics of the same, more fundamental physical phenomenon. One could compare it with the relation $E = mc^2$ where both the energy and mass are metrics of a more fundamental property of matter. Einstein's formula is not causal and does not "explain" why the loss of mass during, for instance, a nuclear fission process will lead to high velocities of the resulting fragments. It just states how the *final* balance is made up. It also bypasses the intermediate processes, but is perfectly usable for a before/after analysis. This may be a justification for our own approach, chosen to analyse the photoelectric cell.

Anyway, the functioning of a photoelectric cell now can be compared with the process going on in the case of "stimulated" absorption (Sect. 13.5). The higher energy level to which the receiving droplet is raised is in both cases fixed. In the narrow potential box of Sect. 13.5 the level is fixed by the quantisation of stationary states, *as well as* by the "colour" of the radiation which must exactly fit this quantisation. In the case of the macroscopically wide potential well of Fig. 13.13 the higher energy level is only fixed by the "colour" of the radiation. But it is nevertheless fixed, in accordance with the formula $(raise\,of\,energy\,level) \div (colour\,of\,received\,light)$.

Another difference between "stimulated" absorption and the photoelectric cell is of course, that the excited electrons (or, in our case: droplets) may tunnel out of the cell (or have been raised to such a high energy level that they have become completely free). This is not really an essential difference, as far as Einstein's arguments are concerned.

There is still the disturbing aspect about the interaction between radiation fields and charges which was already briefly mentioned. It applies to the descriptions of the photoelectric effect, both as given by quantum theory and by the droplet theory. When we think of the light coming from a distant star, for instance millions of light years distant from the earth, the energy in the light waves, spreading through space, becomes almost infinitely diluted before the waves are interacting with a receiver. It is difficult to imagine that such a faint wave still is able to trigger the described process in the receiver's eye. A first thought would be that the photon view does not suffer from this conceptual problem. But that is only if we imagine a photon as a sort of bullet, which it is not. The same problem is met when we try to imagine a photon as sketched in Fig. 13.4. It is also spread out in space, and only materialises at the point of reception. The difficulties then are analogous to those of the droplet theory. As this book does not attempt to dispute or improve the results of usual quantum mechanical theory (after all the experimental evidence to support quantum theory is abundant), the matter will here be noted as something that still is a mystery, but not as something that gives rise to contradictions between the droplet theory and quantum mechanics.

In conclusion, the photoelectric effect may be described by droplet theory just as well as it is described by the usual quantum mechanical concepts. *Phenomenologically* the two types of visualisation are not in conflict with each other. However, this is a preliminary conclusion based on the few examples of interaction between charges and radiation that have been considered until now. Examples do not provide a definite proof that the conclusions are generally valid. For now we must be careful, and we may *not* declare the model of a pulsating droplet (still having the status of a working hypothesis, and not more than that) to be always tenable. The working hypothesis helps to form *visualisations* or *mental pictures*, that is all one may say at present.

Chapter 14
Summing Up the Successes and the Remaining Mysteries

14.1 Two Subjects in This Chapter: Successes and Mysteries (as yet?)

This chapter sums up the visualisations of quantum phenomena, made possible by the "droplet theory". We have taken a deformable droplet of charge as the model of an electron. What is meant by "droplet theory" is the systematic (and quantitative) derivation of the consequences of this working hypothesis. The purpose of the present chapter is, to go through the results obtained so far. The list of strange phenomena mentioned in Chap. 2 will be ticked off, as well as a few additions to the list made in Chap. 3. For these phenomena the possible visualisations are listed in the present chapter, i.e. the visualisations in terms of classical physics as obtained so far. The format chosen for this chapter is, to present the results primarily in pictorial form, with comments to help read these pictorials, but without extensive explanations of their background. Many pictures are copied from the previous, more technical chapters, and for the underlying arguments the reader is referred to these previous chapters.

Many readers will have jumped from Chap. 2 directly to the present chapter, thus—wisely—skipping the rather technical chapters in between. Inevitably, these readers will then have to take the visualisations just for granted, but of course they can page back to the relevant technical chapter if they find some results particularly difficult to accept. They may then form their own opinion about the strength of the underlying arguments. The whole of this book might be viewed as a wedding cake. On top (the present chapter) is the final wedding between classical physics and quantum theory. The middle layers of the cake (the preceding technical chapters) concern the different aspects why the bride and groom have fallen in love with each other, and there are also metal parts (the mathematical appendix) supporting the cake. The readers, as wedding guests, are invited to cut pieces from the cake from any arbitrary layer they like, and as large a wedge of the cake as they like. Other readers, presumably a minority, may have

T. van Holten, *The Atomic World Spooky? It Ain't Necessarily So!*
DOI 10.2991/978-94-6239-234-2_14

read systematically through all the previous chapters. They will find that the present chapter is no more than a brief summary. At least, this is true for the first part of the chapter. They may find new things in Sect. 14.22, titled "remaining riddles", where discussions will be found about what is missing in the theory so far.

The point is, that our wedding cake is a bit malicious. What may be unusual for an actual wedding is, that on the top level (the present chapter) also an indication is given of potential irritations and frictions between the two partners. As marriages go, such potential irritations may in future be smoothed over, or in the worst case may later escalate and lead to a divorce. Both scenario's are possible, but one cannot foresee what will be the future. Readers are warned about the fact that both scenario's might materialise!

Leaving the metaphore and becoming concrete, the full list of Chap. 2 will be taken as the guiding line for the present chapter, though not necessarily in the same order as in this Chap. 2. A few subjects mentioned in Chap. 3 are also mixed in. The summary cannot be anything else—naturally—than a compilation of the results from the droplet theory *as it stands now*. Therefore, do not expect that *all* the strange quantum phenomena can already be supplemented with a mental picture based on the classical laws of physics. The present chapter points out where no compatibility between classical physics and quantum theory has been found yet.

14.2 The Culprit of It All: Pulsation

How is it possible that in quantum mechanics many phenomena are "mysterious and anti-intuitive", whereas in this book these same phenomena appear to be almost "natural" and "self-evident"? What is the essential difference that causes the wide gap between the two points of view? The answer is: all the differences can be traced back to one thing, and one thing only: it is the working hypothesis that an electron is viewed as a tiny, *deformable* droplet of charge instead of a singularity without internal structure. The droplet not only has a translation motion, it also continuously deforms in shape.

The simplest possible model has been assumed with this property of being able to deform. Just as a first attempt to explore what could be the consequences of deformability, a *one-dimensional* model has been made (Fig. 14.1) and has been subjected to further investigation.

The term "one-dimensional model" refers to the fact that the centre of the droplet moves along a straight path in one direction, and that it is allowed to stretch and contract only in this same direction. However, the blob of charge does have thickness, and in this respect it is *not* one-dimensional. The electromagnetic field induced by it is not one-dimensional either, it spreads around the droplet in all directions.

The freedom of this model to stretch and contract in length is called the *pulsation*. Pulsation is the heart of the "droplet theory". It cannot occur when an

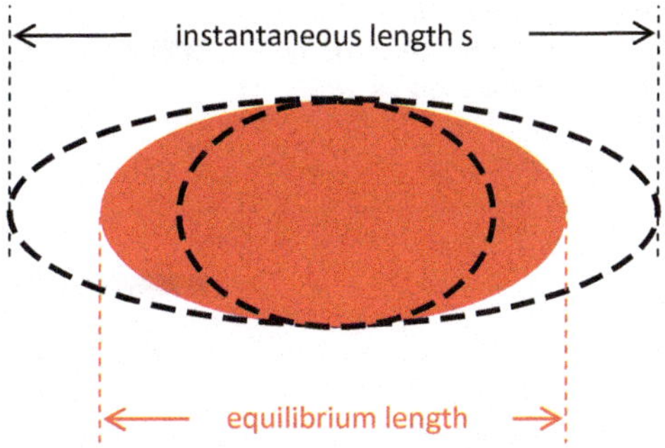

Fig. 14.1 The one-dimensional model investigated

electron is just a singularity as in normal quantum theory. Pulsation explains why the views of droplet theory and quantum mechanics are so different.

Pulsation is governed by two opposite forces on the droplet, as illustrated in Fig. 14.2. First, there is the tendency to explode under the influence of the electrostatic repulsion between all the parts of the droplet. To obtain a consistent model, a surface-tension-like effect is added as an anti-explosion device. At first sight such an anti-explosion device is a rather artificial "deus ex machina" addition, but in fact this must be an inherent part of the model, or else the working hypothesis would have to be abandoned right from the start, even before embarking on any further analysis. That the force opposing the explosion tendency takes the particular form of something like surface tension has been shown (in one of the technical chapters) to be unavoidable, and—curiously—is not wholly incompatible with established quantum mechanics views.

An equilibrium length of the droplet exists, where the explosion tendency and the squeezing by the surface tension are equally large, but of course opposite in direction. It is the shape indicated in Fig. 14.2 by the broken red line.

The droplet is subject to all sorts of electromagnetic forces, both self-generated (so-called self forces) as well as influences from outside. If the droplet is shortened by some of these influences, as shown symbolically in Fig. 14.2, it will try to spring

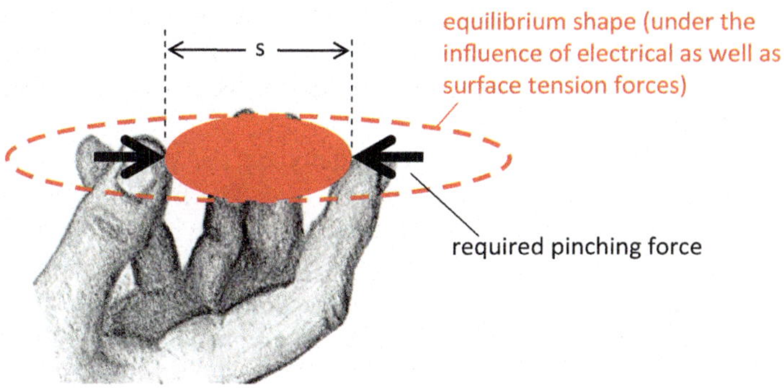

Fig. 14.2 The droplet must forcibly be pinched to make it shorter than its equilibrium length

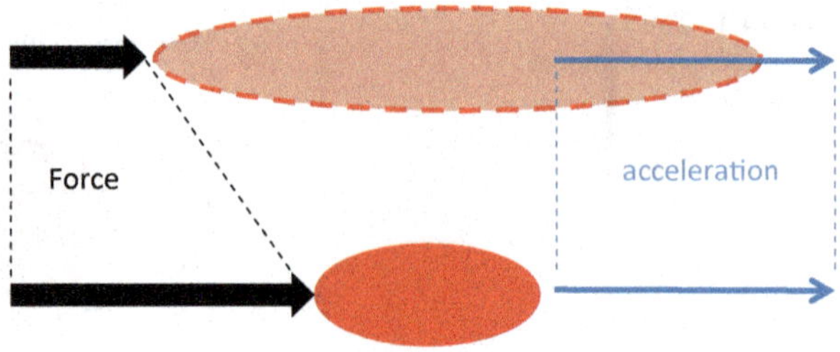

Fig. 14.3 The (electromagnetic) mass of the droplet depends on its elongation

back to its equilibrium shape. Coupled to another self force, the "squeezing inertia", the spring effect then leads to almost continuous oscillations in length: the periodic pulsation symbolised by Fig. 14.1.

The pulsation is the key to everything else. In Fig. 14.3 it is sketched that the mass of the droplet depends on its elongation. If the charge becomes diluted because the droplet is being stretched, the electromagnetic mass of the droplet is small (relatively speaking). If the droplet is being compressed to the small shape in the lower part of Fig. 14.3, its mass is relatively much larger.

A continuous oscillation of the length, i.e. the periodic pulsation, thus leads to the curious effect that the droplet becomes a body with a mass that is going up and down in an almost endless repetition. In the usual theory of mechanics it would be a rather strange object, with its fluctuating mass.

14.3 The "Sack-Race" of the Droplet; Influence of Average Speed

An effect of the fluctuating mass is shown in Fig. 14.4: the velocity of the droplet fluctuates around an average value. In the upper part of Fig. 14.4 this is symbolically shown, and the resulting kind of motion is called "the sack-race" of a moving droplet.

Considering the actual values, the frequency of the pulsation is so high (in the order of 10^{26} pulsation cycles per second), that the separate "hops" are much smaller than the length of the droplet, so much so that a drawing to scale is impossible to make. The frequency of the pulsation is too high to measure, and cannot be observed directly.

There is an influence of the average velocity on the pulsation frequency, as symbolically sketched in the lower part of Fig. 14.4. The frequency increases with speed. Therefore, if we take the horizontal scale in the drawing as the *time*-scale, we see more pulsation cycles in every second. On the other hand, considering the horizontal distances in the figure as a *space*-scale, the picture is different. The increased velocity dominates over the increased frequency, resulting in a larger spacing between the separate "hops".

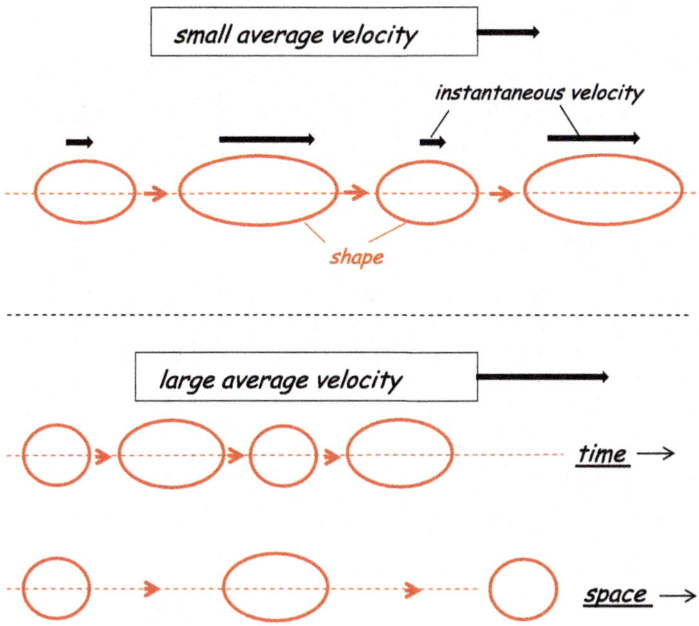

Fig. 14.4 Pulsation causing velocity fluctuations ofthe droplet. Influence of average velocity

14.4 Chaotic Pulsation: A "Drunkard's Sack Race"

Figure 14.5 shows the influence of the amplitude of the pulsations. The figure can also be read as the influence of the energy present in the *fluctuations* of the droplet's elongation and speed (to be discerned from the energy represented by the *average* elongation and *average* speed). The energy in the fluctuations can go up and down in a random way. The total energy is preserved, so that these variations go at the cost of the average elongation (which represents potential energy).

For large amplitudes, the pulsations have a chaotic character, as sketched in the upper part of Fig. 14.5. The jumps in the instantaneous *velocity* show the same chaotic behaviour, resulting in what in earlier chapters is called a "drunkard's sack race". In very exceptional cases the pulsation amplitude may sometimes fall to such a low value that the situation occurs as sketched in the lower part of Fig. 14.5. For small amplitudes the motion "degenerates" into an ordered sinusoidal pulsation. The accompanying speed fluctuations "degenerate" as well, and we obtain a "regular sack race".

The motion of the droplet randomly changes between the modes sketched. We have thus a shift up and down through the pictorials of Fig. 14.5. The change from a chaotic mode to the "degenerated" sinusoidal motion occurs very seldom, the most usual mode is the chaotic one.

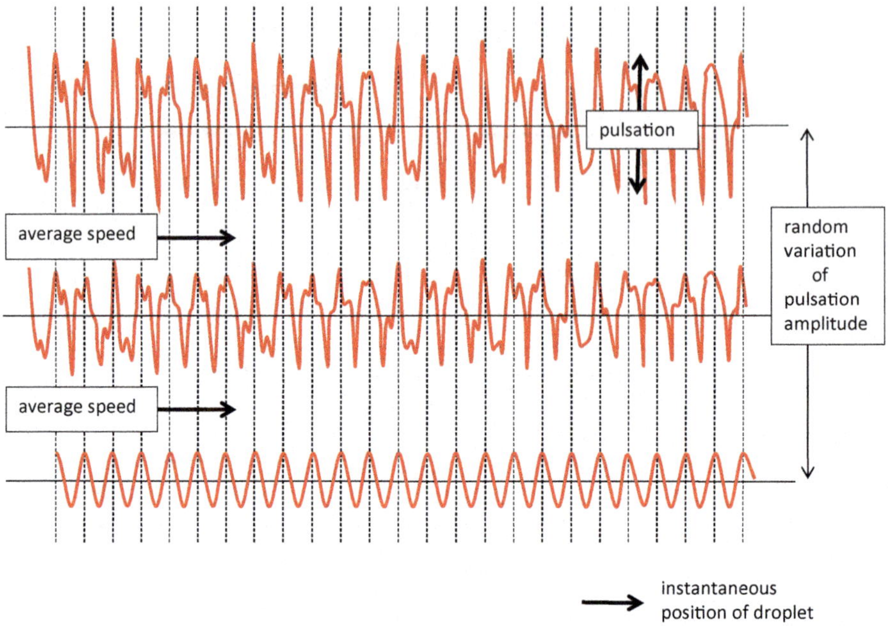

Fig. 14.5 Influence of random variations of pulsation amplitude

14.5 Bohr's Radio Silence

As long as the droplet is in the chaotic pulsation mode, there can be no radiation. Although there is a wild fluctuation of the droplet's velocity, the chaotic character of the motion blocks the possibility to radiate. At first sight this is in conflict with the predictions of the classical theories about radiation: the wilder the velocity fluctuations, the more intense the radiation would be. In chaotic motion, however, the fluctuations become so abrupt that we have surpassed a critical threshold. The sharp spasms of the velocity have now become a barrier, preventing radiation. A physical view how this may be explained was already written down in Chap. 1, and was further elaborated in later chapters. It was a long story, and it will not be repeated here. However, one may also look at it from a different angle than in these previous chapters. Thinking about it, one must conclude that there is bound to exist at some point a certain cut-off threshhold for radiation. Consider the almost razor-sharp peaks of the velocity in the upper sketch of Fig. 14.5. The acceleration of the droplet in these points would be tremendous. According to electromagnetic theory, the intensity of the radiation emitted to infinity is proportional to the acceleration squared: $(radiated\,energy) \div (acceleration)^2$. Therefore, there must be conditions where this relation between radiation and acceleration can no longer hold. The amount of radiated energy would far exceed the amount of energy which is actually available in the accelerating charge and its field. When the motion of the droplet is too wild, the relation $(radiated\,energy) \div (acceleration)^2$ is thus bound to break down. This is just a heuristic argument, and it does not tell at which point to expect such a break down. The earlier given description (e.g. in Chap. 1) how to understand the blocking of radiation by a chaotic motion was more causal and did provide some quantitative information, which the heuristic consideration does not.

14.6 The Break of Radio Silence at Random Times

At times the motion of the droplet can "degenerate" into the regular mode shown in the lower part of Fig. 14.5, though it is a rare occurrence (in the previous text is was compared with a heat wave in mid winter, not entirely impossible but improbable).

The blockage of radiation by chaotic motion is then lifted. This, however, does not mean that radiation will actually take place, because there are more locks on radiation than just the chaotic motion. The regular, sinusoidal pulsation of Fig. 14.5 is a *potential* radiating state. Actual radiation cannot materialise, unless other conditions are favourable as well.

The "free flight" of a droplet with a constant average velocity is such a case where momentarily a potentially radiating state may exist, but where never *all* the necessary conditions for actual radiation can be satisfied. In this respect the behaviour of a droplet of charge is comparable with the "free flight" of an electron, which cannot spontaneously lose energy by radiation, unless there is interaction (collisions for instance) with other particles.

The other barriers preventing radiation can be unlocked by the special conditions which prevail when a charge is speeding back and forth inside a potential well. We will come back to it in a later section of this chapter. The removal of the first lock, i.e. the degeneration of the chaotic pulsation, is a random event. This explains why radiation, if it occurs, starts at unpredictable instants of time.

14.7 The Conditions Inside a Potential Well

The word "potential well" stands for an external electric field by which the droplet is forced to move back and forth between certain limits. The situation is—macroscopically—comparable with a marble rolling in a bowl. The bowl (and the potential well) may have all sorts of shape, the two shapes that are most convenient for our quantitative analyses being shown in Fig. 14.6. The pictorial on the left in the figure symbolises the so-called "parabolic well", and the combination with a charge is called the "harmonic oscillator". On the right side in Fig. 14.6 the so-called potential *box* is symbolised. The latter is an idealisation of an electron

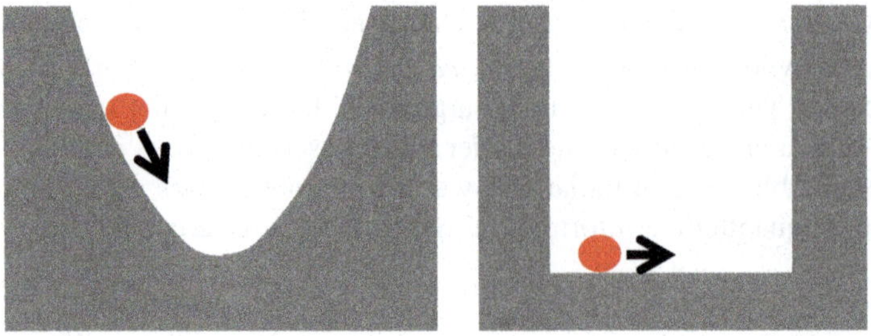

Fig. 14.6 Schematics of different types of potential well (*left* the parabolic well, *right* the box), based on the analogy with the marble inside a bowl

moving freely inside a slab of conducting material, but being bounced back when it tries to escape through the surface of the slab.

In the case of the deformable droplet moving in such a potential well, the velocity of the marble in Fig. 14.6 symbolises the *average* speed of the droplet, on which the velocity fluctuations due to the pulsation are superimposed.

In the parabolic well, the droplet has its highest velocity in the middle. Therefore, the pulsation will have its highest frequency at this point. On the other hand, in the high points where the droplet reverses its direction of motion, the velocity is momentarily zero. Here the pulsation frequency has a lower value, the "zero-speed frequency".

In the potential *box* we see a droplet moving back and forth with a constant velocity in between the walls, the motion interrupted by brief moments of a zero speed.

The next few sections will be concerned with the *box*, before we return to the more general type of potential well such as exemplified by the parabolic well.

14.8 Beats in the Amplitude of the Pulsation: Comparable with De Broglie's Matter Waves

Inside a potential *box* the average velocity of the droplet is constant (not considering the direction of the motion), except when it is interrupted by brief "zero-speed" moments during the hits with the walls. The pulsation frequency becomes a mix of the "frequency at speed" with the "zero-speed frequency". This mixing results in a beat: a slow variation of the pulsation amplitude, as sketched in Fig. 14.7.

In this figure the individual pulsations are not shown, in view of their—unobservably—high frequency. The pulsation, drawn vertically in the diagram, therefore shows up as a reddish area. What does fall within the window of observability is the variation of the pulsation amplitude. This variation, the beat phenomenon, is caused

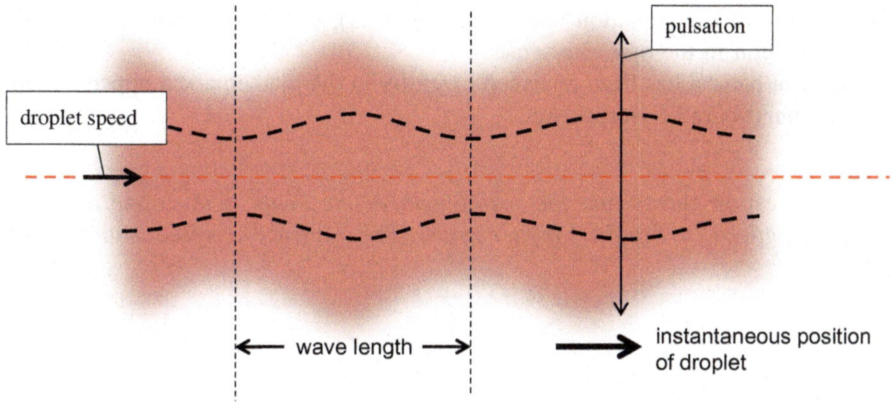

Fig. 14.7 Beats in the pulsation, taking into account the unobservable individual pulsations

by the interference between the two frequencies occurring simultaneously in the pulsation. The outline of the reddish area is most often not sharp, due to the chaotic character of the pulsation. It is only during the time intervals when the chaotic pulsation "degenerates", that a sharp boundary of the reddish area can be observed. In Fig. 14.7 the black dashed lines symbolise how the faint boundaries then shrink into sharply defined curves.

The wavelength of the beats is defined as the distance travelled by the droplet between the consecutive beats. Figure 14.8 shows the influence of the velocity of the droplet on the beat pattern. A doubled speed results in a shorter wavelength, that has been halved compared with Fig. 14.7. This is due to the higher pulsation frequency, and the increased difference with the zero-speed frequency which is mixed with it. The beats then have a higher frequency too, so much higher that it dominates over the direct effect of the speed on the wave length. Comparing with

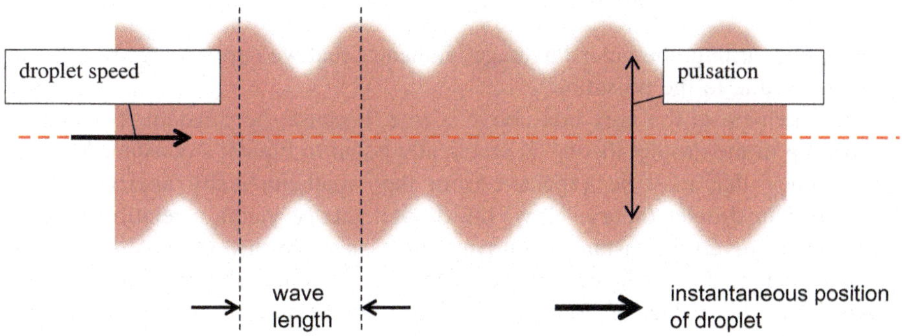

Fig. 14.8 Beats in the pulsation. Comparing with Fig. 14.7, the influence of the droplet's speed is shown

the response of the pulsation itself (see Fig. 14.4), the beats thus respond to the speed variation in the opposite sense.

The beat pattern has been analysed carefully in the mathematical appendix, and what is found is rather surprising.

Quantification shows that *the wavelength of the beats is the same as the wavelength of De Broglie's matter waves*. The influence of speed is identical too.

14.9 Planck's Constant Found to Be Expressible in Terms of Classical Constants

The last, boxed sentence has yet another implication one may not realise immediately when reading the above, but important enough to write it down explicitly. According to De Broglie's theory in usual quantum mechanics, the relation between the velocity of a particle and the wavelength of a matter wave is:

$$(velocity\ of\ particle) = \frac{1}{m}\frac{h}{(wavelength\ of\ matter\ wave)} \qquad (De\ Broglie)$$

where the symbol m stands for the mass of the particle, and h is a universal constant, named Planck's Constant.

In the droplet theory a similar expression is found:

$$(average\ velocity\ of\ droplet) = \\ = \frac{1}{m_{average}}\frac{K}{(wavelength\ of\ beat)}(droplet\ theory)$$

where, as a warning, the index "average" indicates that we here average out the fast fluctuations due to the pulsation.

The statement in the grey box above is true, because the constant K which has cropped up in the droplet theory, is practically equal to Planck's constant h, within the accuracy that may be expected from the—mathematically approximated—derivations in droplet theory. Now, the droplet theory yields a relation between K and other universal constants occurring in classical physics, such as the value of the unit charge, the speed of light, etc. This also implies that we have found a formula for the value of Planck's constant, expressed in terms of classical constants!

14.10 Quantisation of Energy in a Potential Box

A stationary state is defined as a situation where the droplet, travelling back and forth between the walls of the box, exactly repeats the beat pattern during each cycle, no matter whether it is going to the right or to the left, nor depending on the number of cycles it has gone through. In other words, in a stationary state the beat pattern is constant, it does not depend on time.

This is only possible if the droplet reaches the walls at exactly the moments when the beats have a maximum or minimum amplitude. The different possibilities that a stationary state can arise are shown in Fig. 14.9. The wavelength of the beats

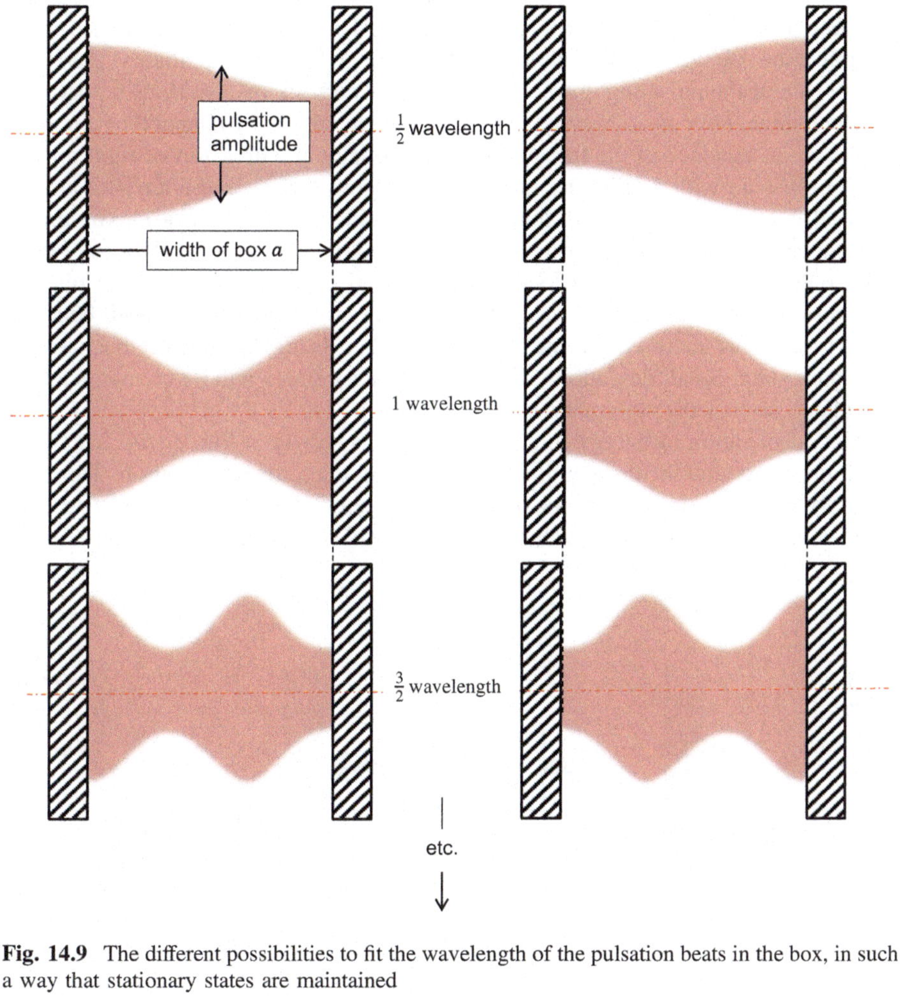

Fig. 14.9 The different possibilities to fit the wavelength of the pulsation beats in the box, in such a way that stationary states are maintained

must fit in the width of the box, and this is only possible if the width of the box equals $\frac{1}{2}, 1, \frac{3}{2}, 2, \ldots$. wavelengths of the beats.

This is exactly the same criterion for stationary states as is found in usual quantum mechanics, where it is found that the width of the box must equal $\frac{1}{2}, 1, \frac{3}{2}, 2, \ldots$. wavelengths of the matterwaves.

Each value of the wavelength requires its own value of the droplet velocity, as illustrated in Figs. 14.7 and 14.8. The message conveyed by Fig. 14.9 is therefore, that only certain values of the droplet's velocity are allowed, any arbitrary speed most of the time would not lead to a stationary state. The kinetic energy (which in a potential box is the same as the total energy) is thus *quantised*, it cannot have any arbitrary value (for stationary states).

This conclusion must be coupled to another consideration. The stationary state is a *preferred* state, it is not just an arbitrary mode of motion which is equally probable as any other—*non*-stationary—motion. The reason for this is once again the pulsation of the droplet, which everywhere in the droplet theory plays the central role. If the pulsating droplet is being bounced back by a wall, its velocity (in absolute value, not considering its direction) in general will be changed by the wall. This is a consequence of the fact that we are not dealing with a sort of rigid billiard ball, but instead with a vibrating object. The velocity before and after a hit against the wall is changed in general, and it only remains unchanged if a maximum (or minimum) in the beat pattern coincides with the collision. Once in a stationary state, the droplet will thus continue to move at the same average velocity, despite the bounces against the walls. If a stationary state momentarily is perturbed, the velocity changes incurred by the hits will tend to bring the droplet back to the stationary state. Small deviations from the stationary states will be counteracted by this effect, so that the stationary state is stable.

The quantisation of energy in a potential box is shown in Fig. 14.10. The usual quantum theory and the droplet theory both predict the same quantisation. This does not only apply to the relative position of the energy levels with respect to each other. Calculating the actual values of the energies, both theories give the same result.

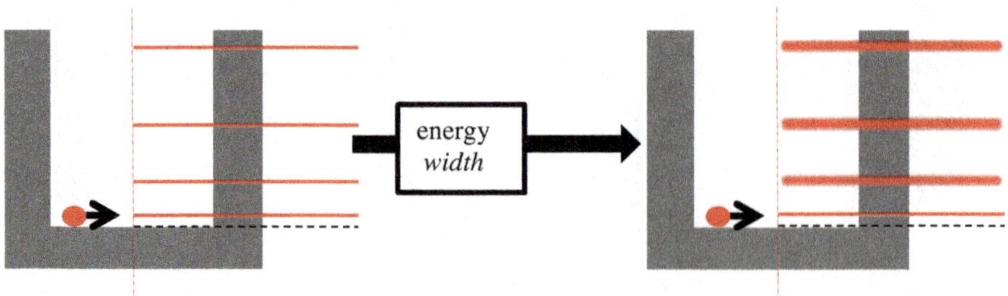

Fig. 14.10 *Left* quantisation of the energy levels in a potential box, identical in quantum theory and in the droplet theory. On the *right* symbolisation of "energy width" (Sect. 14.12)

14.11 Zero-Point Energy

In Fig. 14.10 the black dashed line symbolises the—zero energy—datum level. What strikes the eye is that the lowest possible energy of the droplet in the box, according to the quantisation, is somewhat higher than this datum level. The lowest possible energy is *not* zero! It is called the "zero-point energy".

In usual quantum mechanics this is attributed to the uncertainty relations of Heisenberg: according to this theorem it is not possible that both the position and the velocity of a particle *at the same time* are known with 100 % certainty. A particle lying at rest at a given, well-defined place, is therefore a fundamental impossibility according to the uncertainty theorem. To be more precise, the theorem says that a particle completely at rest would have an infinitely large uncertainty in position. Now, it is known that the particle must be somewhere between the walls of the box, which comes down to a position uncertainty less than infinite. Hence, there must be some uncertainty in the velocity, which therefore cannot be exactly zero all the time. In conclusion, the particle inside a potential well always maintains some "wriggling" motion. This lowest level of the energy coincides with the lowest red line in Fig. 14.10.

A similar theorem is lacking (yet) in the droplet theory. In droplet theory, we must be more cautious with the conclusions from the diagram of Fig. 14.10. We have here, as an initial assumption, taken a droplet which is already moving within the box, with a non-zero velocity. For such a droplet the theory says that no *stationary* state is possible which has a smaller energy than the lowest red line in Fig. 14.10. This is something else than saying that a droplet completely at rest is an impossibility.

However, it was seen that a particle in a *non*-stationary state will tend to change its motion such that a stationary state is achieved. If we assume that perturbations always exist, a droplet at a complete standstill in a known position will not remain for long in this state. Once in motion, it will tend to go to the nearest stationary level, which is the lowest quantum level. The droplet theory thus predicts that *in practice* a droplet completely at rest will hardly ever be found inside a potential well. This still is different from the uncertainty principle, which says that it is *fundamentally* impossible.

14.12 Energy Width

The term "energy width" describes that in actual fact there is a—small—amount of uncertainty in the energies of the stationary states. In a figure like 14.10 this is indicated by a certain vagueness or "width" of the horizontal lines symbolising the allowed energy levels in the box (right hand side of the figure).

In quantum mechanics the energy width is once more attributed to a fundamental uncertainty. It is again Heisenberg's uncertainty principle, although another part of the principle than mentioned above which involved position-momentum uncertainty. In the present section so-called "energy-time uncertainty" is involved. It causes the "blurring" of the quantised energy levels. If an electron is in one of the higher stationary energy states, it will sooner or later fall back to a lower energy state. The higher stationary state thus has a finite life-time, although it is uncertain how long the actual life-time will be. According to the "energy-time uncertainty relation" the uncertainty in the life-time will be reflected in an uncertainty of the precise energy level. This is in quantum theory the cause of the energy width.

The same kind of energy uncertainty is found in the droplet theory, but it is here caused by the unpredictable character of the chaotic motion. In Fig. 14.9 the beat patterns were drawn with somewhat vague contours, to symbolise the effect of the chaotic motion. Due to this vagueness, fitting the beats in the potential box cannot be exact, and a small spread of allowable wave lengths is the result. This in turn translates itself into a small spread of the associated energy values. Mind however, that this applies to the energy in the translational motion: during the chaotic motion the *total* energy is still conserved, the fluctuations of the droplet's velocity are going at the cost of energy in the elongation.

14.13 The Droplet Is "Invisible" in Certain Places of the Box (Corresponding to "Zero Position Probability" in Quantum Mechanics)

In a stationary state the energy of the droplet is constant, and no radiation is emitted (it is blocked by the chaotic mode of motion). Without radiation there are no means to know where the droplet is, since it cannot be observed.

As symbolised in Fig. 14.5, at unpredictable times a "degeneration" into a more regular motion may occur, and one of the locks on radiation has then been removed. The droplet is now in a *potentially* radiating state. The blurred beat patterns inside the box at the same time have shrunk into well-defined outlines, such as shown on the left hand side of Fig. 14.11. If the droplet, passing some position inside the box, suddenly finds itself executing a regular "sack race" instead of an irregular "drunkard's sack race", the amplitude of the pulsation at that moment is given by the local value of the contour shown in Fig. 14.11.

Now, there appears to be a second lock on radiation: if the droplet happens to be at or near the blue circles in Fig. 14.11, it still is unable to radiate. It will move on with the same energy as it had before, and soon will revert to the "drunkard's sack race". The opportunity to radiate has been missed, and even the first lock is closed again.

The reason why certain places in the box are more conducive than other places to initiate actual radiation (so that the droplet can be observed), is quite complicated to explain (see Chap. 10). Some feeling why this might be so can be gotten when we

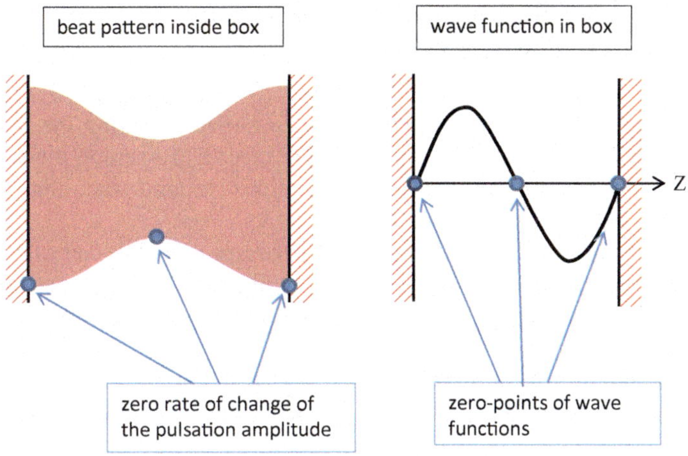

Fig. 14.11 Relation between the beat patterns of droplet theory and the wave functions of quantum mechanics

consider the *slope* of the reddish contours in Fig. 14.11. Near the blue circles the slope is zero or very small. Here the conditions are comparable with those of a "free-flying" droplet having a constant pulsation amplitude. A "free-flying"droplet will not spontaneously radiate, just like an electron that is following a straight path and does not interact with other particles.

It can be shown that the conditions to radiate are more favourable, the larger the slope of the reddish contour. All this corresponds with the results from quantum theory: in the same positions where droplet theory predicts conditions of no-radiation, quantum theory predicts that the chance to observe the particle-in-box is zero. This is shown in the right hand side of Fig. 14.11. The zero-points of the wave functions of quantum mechanics indicate a zero chance to "see" the electron at this place, and these points coincide with the positions predicted by the droplet theory where radiation cannot be started up.

The interpretation of the zero-points is sometimes a matter of debate in usual quantum mechanics: sometimes the zero chance to observe an electron in a certain place is translated as "zero probability that the electron passes this place". This leads to the enigma whether an electron can or cannot step over from one part of the potential well to another, when these parts are separated by a zero-point. The interpretation according to the droplet theory is just, that the droplet may move unrestricted within the potential well, but at some places is unable to radiate and then cannot "be seen".

14.14 When Radiating, the Emission Is in the Form of Isolated Energy Flashes

Once the droplet has seized the opportunity to radiate, opening the first lock by going over to a civilised pace (the regular sack race) and opening the second lock by avoiding the "forbidden" zero-slope regions around the blue circles in Fig. 14.11, it sheds energy at a fast rate.

What is "fast" in this respect depends on the perspective. We can calculate that a typical time interval taken by the energy shedding spans many thousands of pulsation cycles. On the time-scale of the individual pulsations the shedding of energy is not fast. However, if we take the time-scale of the bouncing cycles of the droplet in the box, the process of energy shedding is very fast. In other words, the process is already completed when the droplet has travelled only a tiny fraction of the total width of the box.

The latter time-scale is the more relevant for a macroscopic onlooker. For such an observer it will appear as if the energy is radiated in a "flash-like" way. It is symbolised in Fig. 14.12: due to the short radiation time, the radiation will be found in a thin "shell", expanding outwards with the velocity of light. Outside this shell no radiation travels through the field.

Inside the shell, the radiation is not uniformly distributed either. The shading of the red colour in Fig. 14.12 suggests that there are certain directions in which the intensity of the radiation is larger than in other directions.

The *total* amount of energy in the shell, which will eventually be carried off to infinity, obviously must be equal to the amount of energy lost by the droplet.

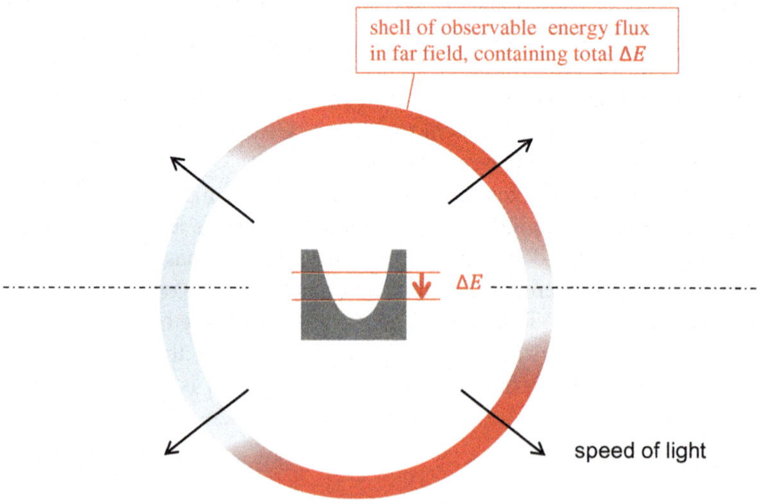

Fig. 14.12 Energy released by a droplet in a potential well, travelling as an isolated energy-"burst" through space

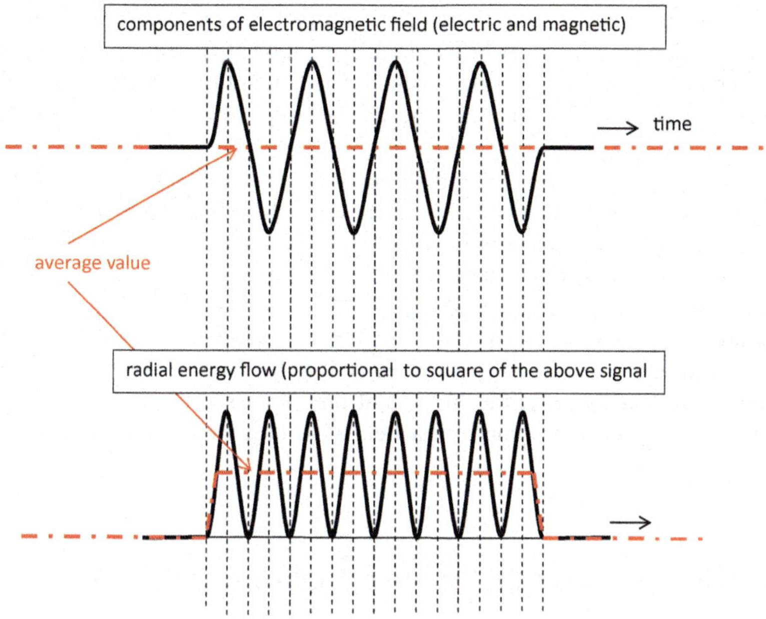

Fig. 14.13 Electric and magnetic field strength in the far field, both fluctuating around a zero average value with the frequency of the pulsation of the droplet. The energy flux in the far field is proportional to the product of the electric and magnetic field, and shows a variation around a non-zero average

In the caption of Fig. 14.12 the thin shell of radiation is described as an "isolated energy burst". What is meant here is that the burst of electromagnetic energy displays rather extraordinary characteristics. It is not at all comparable to the field of a macroscopic antenna. Figure 14.13 shows how the transmission by a macro antenna is experienced by a macroscopic receiver. It is assumed—just to show the principle—that the carrierwave is transmitted during a short time interval, without any modulation, as if it were a dash in a Morse-message. The field consists of a fluctuating electric component, as well as a magnetic component, both fluctuating with the same frequency. Both these components vary around a zero value. The energy flow in the field is shown in the lower part of Fig. 14.13. Its value is equal to the square of the oscillations in the upper part of Fig. 14.13. The energy flow is seen to have a doubled frequency, and it does *not* have a zero average. The red dash-dot signal, representing the average of the energy flow, is the same type of signal as would be detected by a radio receiver after rectifying the antenna signal and averaging it. It is the Morse-dash that can be put on a rolling paper tape by a Morse-writer.

If the transmission is coming from a pulsating droplet, we should remember that the ultra-high frequency of the transmission is outside the "observation window". An observation of the signal will always produce a time-average of the signal. This is inherent in the basic assumptions of the droplet theory. The droplet was conceived as the smeared out image of a singularity subject to "zitter", i.e. the droplet theory inherently deals only with time-averaged quantities in case of ultra-high frequencies. An attempt to measure the components of the electromagnetic field thus leads to a null-result.

However, the energy flow in the field does not give such a null-result when it is measured. Again, a measurement will come down to observing a time-average of the real signal. The time-average is not zero in the case of the energy flow. What can be observed is now more than a null-result, it is the signal shown by the red average line in the lower part of Fig. 14.13.

Returning to the earlier figure Fig. 14.12, we conclude that inside the radiation shell only a flow of electromagnetic *energy* can be "seen", whereas a complete electromagnetic *field* is invisible. In a point that is passed by the expanding shell one experiences a flash-like energy packet, without a supporting electromagnetic field!

14.15 The Colour of the Emitted Light Flashes

Until this point something has been left outside the discussion, which now has to be considered. If the droplet is moving in a potential well and if it has shed some of its energy by radiating, it will finally settle into *another* stationary state, with a lower energy. The radiation process has a definite beginning and end, both with very specific properties.

To make more clear how we may deduce some global characteristics of the transition, an analogy will be used. Take a trumpet without valves (Fig. 14.14). This instrument can produce just a limited series of tones, e.g. playing the "last post" is possible using only these tones. The wavelengths associated with these different tones, determined by the embouchure, all fit exactly in the length of the trumpet tube.

Fig. 14.14 Trumpet without valves ("natural" or "Baroque" trumpet), only able to sound a limited series of tones

When the ground tone is being played the particles of air inside the tube of the trumpet oscillate, *mainly* at the frequency of the ground tone, but higher frequencies are also superimposed. These higher frequencies have a smaller amplitude, but are still important for the timbre of the sound. If the next overtone is being played, all these frequencies are still present in the vibration of the air particles, but the amplitude of the first overtone has become dominant. It is quite difficult to describe exactly how any individual particle of air changes its motion *during* the transition from one tone to another. A global view on the process is, that we expect for a brief moment a mix of the two modes of vibrations: the vibrations of the ground tone gradually extinguishing, and giving way to the new vibration mode where another tone gradually becomes dominant. Although the word "gradually" is being used here, the transition is actually so fast that the listener perceives it as a sudden jump of the tone. Nevertheless, on a smaller time-scale than relevant and observable for the human ear, there will be a brief moment where the two modes exist together at the same time: the transition cannot be a discontinuous process. Sometimes one can hear the transition of the tones as a short "thudding" sound.

One can have a similar visualisation of what happens when a droplet changes its motion between two of the energy levels shown in Fig. 14.10. Let us assume that the energy drops to a lower level, which is comparable with the return of the trumpet from an overtone back to the ground tone. In the case of the droplet we similarly expect that during a brief moment the two vibration modes overlap each other, and "sound" together. Figure 14.15 shows on the left the situation of the

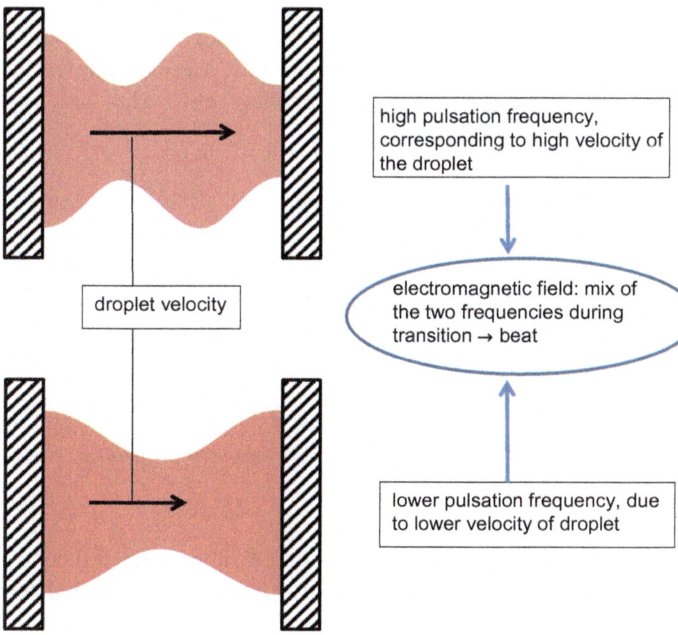

Fig. 14.15 Before and after transition from one stationary state to another

droplet before and after the flip from one stationary state to another. The lower energy state corresponds to a reduced velocity of the droplet, and therefore it entails a smaller pulsation frequency.

The electromagnetic field induced by the pulsation (and by the corresponding "hops" in the "sack-race" of the droplet) has after the transition a slightly lower frequency than before the transition. When, during the transition, both situations overlap each other for a brief moment, the electromagnetic field surrounding the droplet contains a mix of the two frequencies, giving rise to a beat phenomenon. The beat in the electromagnetic field has a very much lower frequency than the original overlapping signals, in fact the beat frequency is the difference frequency of the two original signals. The beat frequency has thus descended to the observable region, in contrast to the pulsation frequencies themselves.

Note that the beat phenomenon which is observable in the field during a transition is not the same as the beats in the pulsation amplitudes symbolised by the reddish wavy contours on the left side of Fig. 14.15. The latter are beats too, but correspond to the mix of the pulsation frequency with the "zero-speed pulsation frequency". The transition beats in the field, in contrast, are the result of mixing the pulsation frequencies belonging to the speed of the droplet before and after the transition.

According to the droplet theory the beat frequency found in the electromagnetic field depends on the energy difference of the two states involved. The theory predicts the following relation:

$$\textit{beat frequency} = (\textit{energy difference before and after transition})/K$$

The frequency of the beats is—relatively—low, and is within the observability window. This is in contrast to the ultra-high frequency of each of the two oscillations that mix together, they are both unobservable. The beats thus determine the colour (frequency) of the *perceived* light phenomenon. The last formula may therefore be rephrased as follows:

$$\begin{aligned} &(\textit{frequency of perceived light}) = \\ &\quad = (\textit{energy difference before and after transition})/K \end{aligned}$$

So far the result obtained by the droplet theory.

Now comparing with quantum theory, according to Bohr's theorem the colour of the emitted light is related to the energy difference of the two stationary states between which the electron jumps:

$$(\textit{frequency of emitted light}) = (\textit{energy difference of stationary states})/h \quad (\textit{Bohr})$$

Remembering that the proportionality constant K in the droplet theory practically equals Planck's constant h in quantum theory, it can be concluded that the two theories are in good agreement with each other.

The droplet theory solves one of the riddles in Bohr's theory. An oscillation in the electromagnetic field must according to Maxwell's equations always be accompanied by a vibration of a charge, with the same frequency. In Bohr's theory, however, the colour (or frequency) of the light emitted by an atom is not related at all to a corresponding vibration of any electrical charge. It is related to an energy difference, and nothing in the atom vibrates with the same frequency as the vibrations in the field. The riddle is: where do the oscillations in the field come from? What are the "roots" of the electromagnetic radiation?

This riddle has now been solved by the droplet theory: the field indeed has firm roots in the vibrations of the droplet, i.e. in the pulsation. However, the result that is finally observable is only the beat between two diferent frequencies of the pulsation. This beat frequency is indeed nowhere found in the kinematics of the droplet, but its roots are.

14.16 The Energy-Colour Relation in the Emitted Light Flashes and in Photons

The law of conservation of energy requires that the energy found in the field must be equal to the energy lost by the droplet by jumping between the two stationary states:

$$(\textit{energy poured into field}) = (\textit{energy difference of stationary states})$$

Combining this with the results of Sect. 14.15, the total energy in the light flash of the droplet theory is therefore related to its perceived colour:

$$(\textit{total energy in the light flash}) = K \times (\textit{frequency of emitted light})$$

It is the same relation as, according to Planck's and Einstein's theory, is valid for photons:

$$(\textit{energy of photon}) = h \times (\textit{light frequency})$$

14.17 The Interactions of the Shell of Energy Do not Differ from the Effects of a Photon

Already there appeared to be many similarities between the shell of energy depicted in Fig. 14.12 and the photons of usual quantum theory. Both have the appearance of flash-like bursts, both have the character of energy packets without an observable supporting electromagnetic field, traveling with the speed of light. They have the same energy content, and the same colour. The difference is in their geometry: one

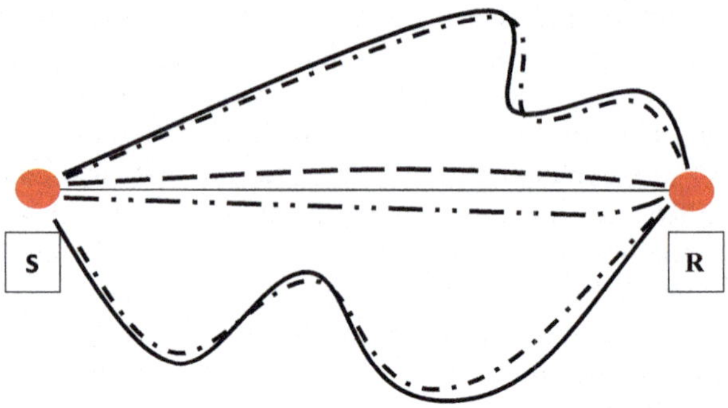

Fig. 14.16 Sketch by Feynman (from the book "QED, the strange theory of light and matter") symbolising the many paths "sniffed" by a photon on its journey from source S to receiver R

of them is spread-out, the other is point-like. This may seem a gross, and essential difference.

Actually, the visualisation of a photon as a point-like particle, as a sort of bullet shot from one place and hitting a target, is far too simplistic. One should realise that neither the energy shell of droplet theory nor a photon can be observed without interactions taking place. A photon can only betray its presence when some kind of measurement is done, which in turn affects the photon. Before an interaction has taken place, one can only build up a mental picture, a theoretical construction which tries to explain the observable interactions. Some idea about the "appearance" of a photon is given by Feynman (in the book "QED, the strange theory of light and matter"). One of his sketches is reproduced in Fig. 14.16. It is a picture visualising the essence of the energy transfer by photons, such as described by the QED-theory (quantum electrodynamics theory). Here it is assumed that a photon leaves a source of light (the point S), and arrives at a receiver R to deposit its energy there.

A photon does not behave at all like a bullet, flying straight from S to R. The theory about photons tells an entirely different thing. One must imagine that the photon can follow a multitude of different paths, as sketched, and not at all straight paths. And what is more: to visualise the theory, it is necessary to imagine that the photon actually follows *all* these possible trajectories before it hits the receiver R. It is as if the photon has split itself in an infinite number of fragments, so that it can simultaneously "sniff out" all these possibilities. Finally, the photon reassembles and materialises again in the form of a "particle-like thing" when hitting the receiver. It is a strange picture, but it is necessary to explain the observed behaviour of light.

This visualisation should be compared with Fig. 14.17, which sketches the transfer of energy by a spread-out electromagnetic field. The field is assumed to be caused by a potential well, in which a droplet of charge has jumped down to a lower energy level, so that it emits a flash of light. Not too far from the emitter the field looks almost as it is sketched in Fig. 14.12 for an unhindered emission towards infinity. But this is only in the close vicinity. Further out, the pattern will be deformed by the presence of a receiver. If the receiver absorbs just as much energy as has been emitted, then the pattern of the energy flow is—roughly—the one as

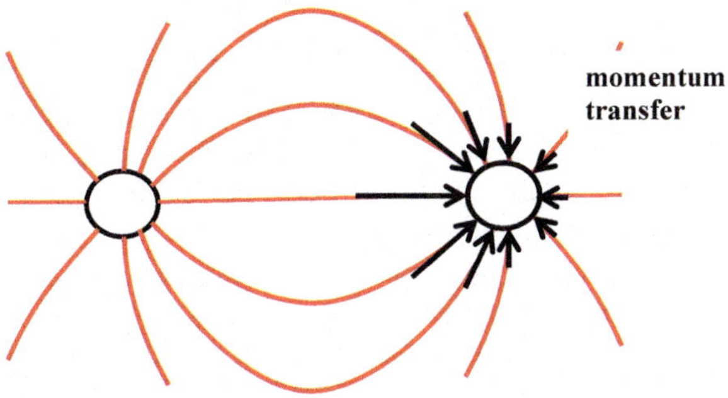

Fig. 14.17 Energy and momentum transfer by waves

sketched in Fig. 14.17. It is a picture that is not drastically different from the one sketched in Fig. 14.16.

Before bothering about the consequences of the similarity between Figs. 14.16 and 14.17, let us first give some attention to the process going on at the receiver end. The receiver will initially, at the very beginning of the whole process, be hit by some of the radiation, and experiences a certain "colour" of the radiation. This colour corresponds with the relatively low beat frequency in the field, which is caused by the interference between two ultra-high frequency signals (shown schematically In Fig. 14.15). The colour, or beat frequency, was given by the expresssion

$$\begin{aligned} &\mathit{beat\,frequency}(\mathit{colour}) = \\ &\quad = \mathit{difference\ between\ pulsation\ frequencies\ before\ and\ after\ transition} \\ &\quad = (\mathit{energy\ difference\ before\ and\ after\ transition})/K \end{aligned}$$

A number of different kinds of receivers have been considered in Chap. 13. In all cases we see that the receiving droplet, initially having its own pulsation and "hopping" frequency, comes in resonance with the ultra-high frequency in the field. One may compare it with an AM (amplitude modulated) radio transmission, where the receiver circuits are tuned to the frequency of the carrier wave. In the receiver—before rectifying—currents run representing the high frequency signal modulated by the audio signal. Likewise, a droplet in resonance with the external field will show a pulsation which has a variable amplitude. The variable amplitude has a relatively low frequency, the same frequency as the beats in the incoming field.

Resonance of a vibrating object is always a way to rapidly absorb energy, by an almost limitless amount. But in this case there is an upper limit to the amount of energy that will be absorbed. The upper limit is here set by the "colour" of the field from which the energy is being tapped. To see this, one has to realise that the emission and absorption of energy are similar processes, one is the inverse of the other. The same relation between frequency and energy may be derived for the two processes. To find the upper limit of the energy absorption one therefore only has to invert the sequence of equalities given above:

$$(\textit{upward energy jump in receiver}) = K \times (\textit{beat frequency})$$

In other words, the "colour" of the field determines the—quickly reached—upper limit of the energy that will be taken up by the receiver. The result is, that the emission and absorption of energy are equally large, which leads to the picture of energy flows which is roughly sketched in Fig. 14.17.

As far as the *result* and *character* of the interactions is concerned, there is hardly any difference between the droplet theory and the photonic view. The study of the photoelectric effect (see Chap. 13) led Einstein to the conclusion that light always consists of separate "particles of light", each carrying a "lump of energy", the size of the lump being fixed by the "colour of the photon". The droplet theory can also describe the photoelectric effect. The extra degree of freedom in a droplet of charge, as opposed to the view of an electron as a singularity without internal structure, is the key. One wonders whether the differences between the theories are not more a matter of using a different terminology and different visualisations to describe the same things.

14.18 Energy Quantisation in Parabolic Potential Wells

Until now the potential *box* was used to illustrate the droplet theory. As an example of another type of potential well than the special case of a box, the parabolic well will be considered. This is to make sure that the above presented results, which seem to conform quite well with established quantum theory, are not coincidents applicable only to the potential box.

In a parabolic well the velocity of the droplet (mind: the *average* velocity, ignoring the ripples in the velocity due to the pulsation) varies gradually from zero to a maximum value in the middle. The "local" wavelength of the beats (as defined in the technical chapters) is also continuously variable, with the shortest wavelength in the middle. This is shown in Fig. 14.18, where a medium high energy level is assumed for clarity. Once again, there can only be a stationary state, if the beat pattern has maxima (or minima) in the points where the droplet reverses its velocity. This condition is rather restrictive, it does not allow any arbitrary value of the maximum velocity of the droplet in the lowest point of the well. Only discrete values of this maximum velocity can lead to a stationary state. Hence, there is again quantisation of the energy of the droplet.

The energy quantisation found by the droplet theory is schematically shown in Fig. 14.19. It is in quantitative agreement with the quantisation found in quantum theory. Note that the allowed energy levels and their distribution are quite different from those found for a potential box.

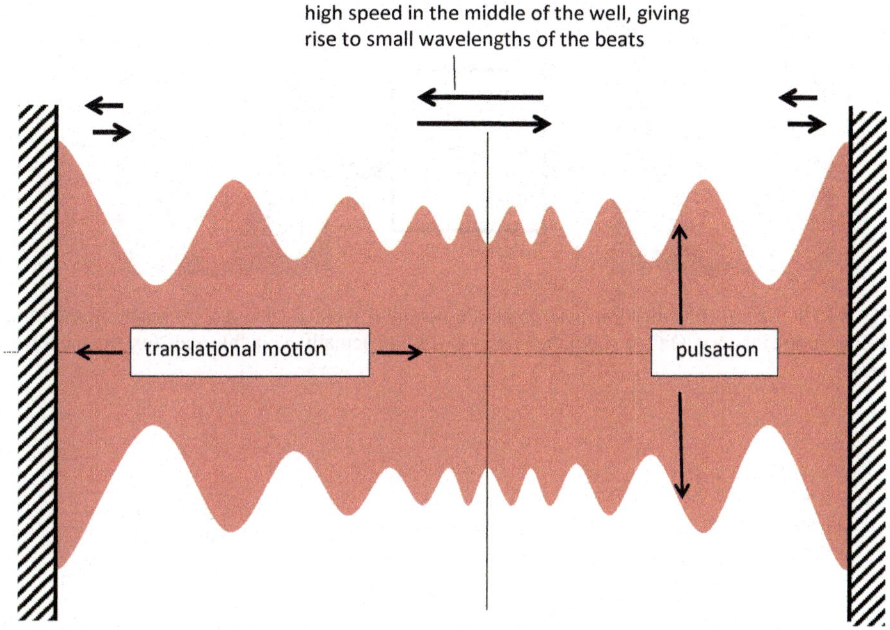

Fig. 14.18 Schematic of pulsation pattern with beats inside a parabolic potential well (intermediate high energy level)

In this case too the quantisation levels are a bit "blurred", due to the chaotic character of the pulsation. The phenomenon was earlier called the "energy width" of the quantisation, occurring in both the droplet theory as well as in usual quantum theory. In the latter case it was attributed to Heisenberg's energy-time uncertainty. In the droplet theory the "blurring" of the levels results from the chaotic character of the "drunkard's sack race" of the droplet between the walls of the well.

14.19 Violation of the Classical Limits of Motion in a Well

Figure 14.20 shows how the energy width explains how it is possible that the droplet sometimes *seems* to be transgressing the limits set to its motion by classical energy considerations. If we would limit our attention just to the *nominal* quantisation of the energy (i.e. the quantisation in the left hand side of Fig. 14.19), the droplet would not be able to travel any further than the red line in Fig. 14.20. Outside this boundary the energy needed to climb higher up the wall would be larger than the energy the droplet actually possesses. However, recognising that there is a spread of energy around the nominal value, we see that the droplet sometimes can indeed come a little bit higher, whilst at other times it will not be able to reach the limits corresponding to its nominal energy. There is thus a spread in the limits of the droplet's motion, without any violation of the law of energy

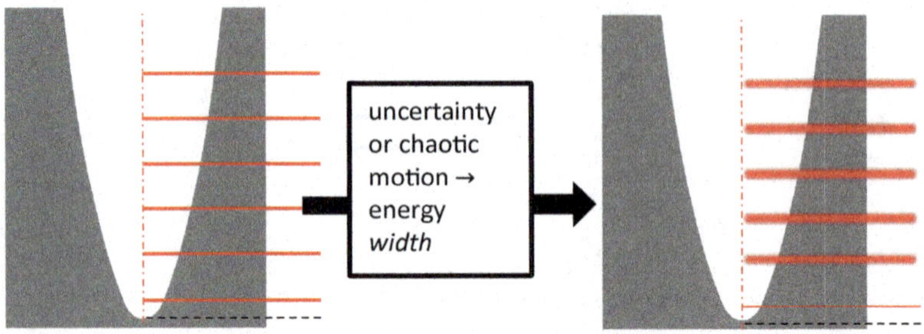

Fig. 14.19 Energy quantisation in a parabolic potential well, as derived by quantum theory as well as droplet theory. On the *right*: the energy levels are actually a bit "blurred" (except the lowest one)

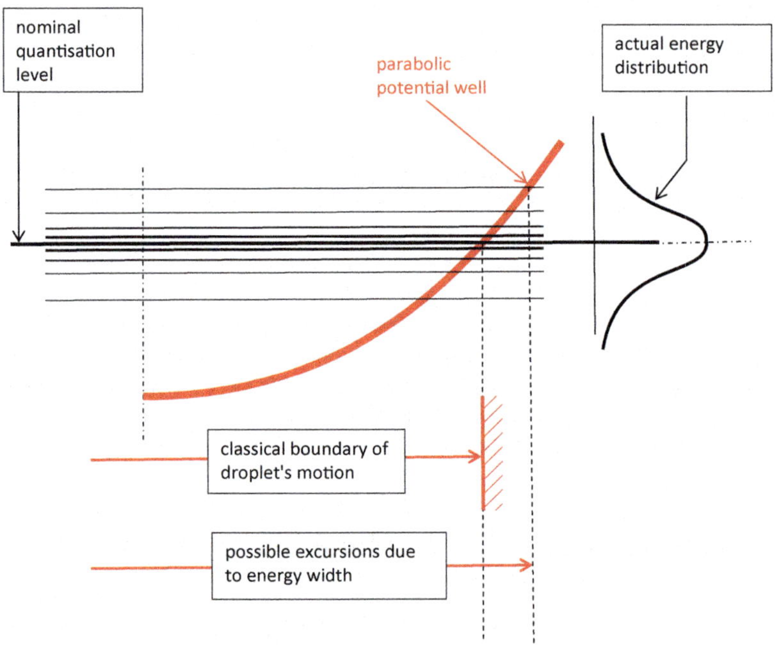

Fig. 14.20 Due to energy width (spread of energy around the nominal quantisation level), there is some vagueness in the position where the droplet reverses its motion

conservation: the chaotic energy fluctuations, seen in the form of energy width, are caused by energy transfers between translation and the variable elongation of the droplet. At any moment the law of conservation of the *total* energy in the droplet is respected.

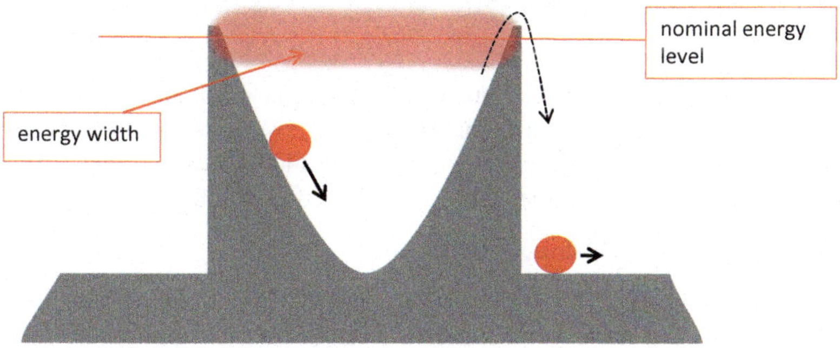

Fig. 14.21 Capped parabolic well, and tunnelling due to energy width

14.20 Tunnelling Out of the Well

Until now the parabolic well was assumed to have infinitely high "walls", so that the droplet can never escape from it, even though there is some spread in the positions where the droplet reverses it direction of motion. If the well is capped, as shown in Fig. 14.21, the droplet can escape completely under favourable circumstances. The nominal energy level would need to be close to the highest point of the well, and the instantaneous energy in the translation motion would have to deviate rather much to the high side. As symbolised by Fig. 14.21, the droplet may then "fall over the edge", and may thus escape. This is equivalent to the tunnelling effect of quantum mechanics.

14.21 General Wells: The Beat Patterns Satisfy Schrödinger's Equation

After the examples of the potential box and the parabolic well, one begins to suspect that more general types of potential well will show the same agreement between quantum mechanics and the droplet theory. This can be made plausible, since it appears that the beat pattern of the pulsations satisfies Schrödinger's equation, i.e. it satisfies the same mathematical equation that determines the wave functions of quantum theory. The beat patterns and the wave functions thus have a close kinship, although they are different members of the same family.

In fact, in the special case of the potential box Schrödinger's equation reduces to a simpler equation, and the kinship then becomes more evident: the slope of the beat pattern corresponds to the shape of the wave functions. This was already illustrated in Fig. 14.11. For other kinds of wells the relationship is less easy to see by a superficial glance. Their relation is then accomplished through the full, unabridged Schrödinger equation.

14.22 Remaining Riddles

What is here meant by "riddles" are the phenomena of quantum mechanics for which as yet no visualisations in terms of classical laws of nature have been found, or where the predictions by the droplet theory would—at the present state of this theory-seem to be in disagreement with established quantum theory.

1. **Spin and Pauli's exclusion principle**.

Conspicuously lacking still is the phenomenon of "spin" of electrons. Originally physicists had in the back of their minds a visualisation resembling the rotation of bodies. However, one is often warned in textbooks *not* to compare "spin" directly with a rotation of the electron. This picture has been abandoned, and within the scope of our working hypothesis of deformable droplets of charge it is neither useful to search for actual rotations.

It is not so surprising that the phenomenon of spin did not crop up yet in the previous text: it does not play a—prominent—role in the usual analysis of potential wells. Nonetheless, spin does play a very important role in other aspects of quantum theory. For instance, in three-dimensional atoms one assumes that on each energy level (or rather: in each of the quantised states) at most two electrons may be found, one having "spin up" and the other "spin down". This means that the electrons in the atom are *all* different from each other: they occupy different energy levels (or rather: states) and have different spins. This is formalised in "Pauli's exclusion principle", which states that all the electrons *should* be different in these respects, it is never possible to find any identical electrons (again: identical in these respects). It forms an important restriction to the jumps electrons are allowed to make between two different quantisation levels. Take for instance the situation of Fig. 14.22, and let us assume that the lower level drawn in this figure corresponds with the lowest energy in an atom. If there would already be two electrons in this orbit, one with "spin up", the other with "spin down", there would not be room anymore for a third electron. Pauli's exclusion principle entails that in such a case the jump down from a higher energy level is forbidden, and never will happen. The radiation is blocked, because the lower states are already fully occupied.

In the one-dimensional potential wells considered in this book the same exclusion principle is applicable. Before proceeding any further, it is first necessary

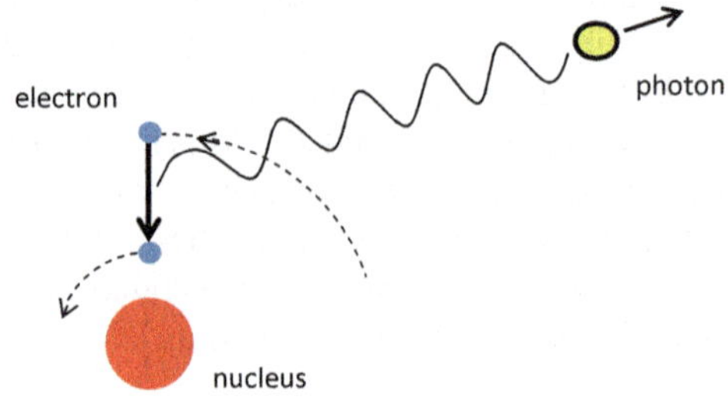

Fig. 14.22 Jump of electron from higher to lower orbit, releasing a photon

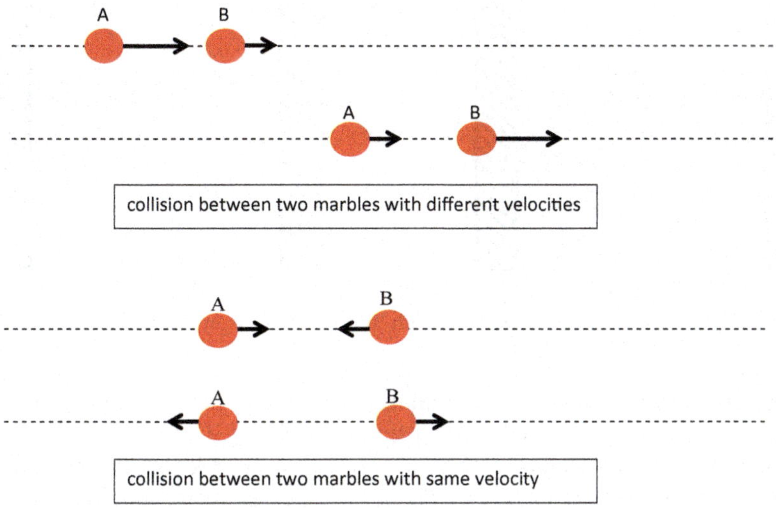

Fig. 14.23 If marble A cannot be discerned from B, the collisions look as if the marbles have gone through each other without a hindrance

to build up a picture about a *one-dimensional* potential well containing more than one droplet. This is a new situation that did not occur earlier in the text. Is it possible at all, considering that the droplets will then sometimes meet each other? It is, as shown in Fig. 14.23. As a first step, the charges are assumed to be rigid marbles.

In the upper part of Fig. 14.23 marble A tries to overtake marble B, which has a lower speed. What happens is a collision, leaving A with a smaller speed and B travelling faster. However, we are dealing with identical marbles (apart from their different speeds), and we do not have any means to identify which one is marble A and which is B. Then the event has the appearance as if the marbles have gone through each other. It looks as if the faster marble just continues its journey, without being bothered by the presence of another one.

A similar optical illusion happens when the marbles have the same speed, as shown in the lower part of Fig. 14.23. There is no question now of overtaking, but if marble B has been reflected by a wall the two will somewhere meet in a head-on collision. Again, if there is no possibility to discern A from B, then the end result looks as if the two marbles have passed through each other.

The conclusion is, that one may have more than one marble inside a potential well, and they will behave in a way as if they were there just on their own, not hindered by the presence of other marbles. Each marble may have a different velocity from the others, but that does not change the picture.

There is in one respect an exception to this general statement. If we are dealing with marbles which both have the same energy, i.e. if a head-on collision occurs as in the lower part of Fig. 14.23, then exactly at the moment of the collision the

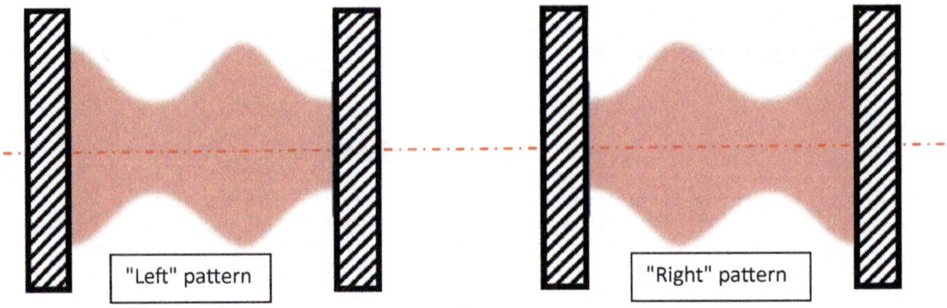

Fig. 14.24 Example of beat patterns, both with 3/2 wavelengths in the width of a potential box

marbles come to a stop before continuing their journey. In a minute you will see that this fact may explain Pauli's exclusion principle.

In fact, we are not dealing with rigid marbles in our droplet theory, and things will be different from the picture of Fig. 14.23. Let us, for simplicity, assume that there are two droplets inside a potential box, occupying the same energy level. The associated pulsation frequency will be the same for both droplets, as well as the wave length of their respective beat pattern. In order to maintain a stationary state of motion, the beat pattern must have a zero slope right at the walls of the box (derived in Chap. 9). This always leads to two possibilities, as shown in Fig. 14.9. Figure 14.24 above gives the example where there are 3/2 wavelengths fitted between the walls of the potential box. The two possible beat configurations are marked as "Left" and "Right" patterns, which is of course just an arbitrary label without any physical meaning.

If there are two droplets in this box, occupying the same energy level, they will repeatedly suffer head-on collisions. As has been shown, the motion is then *almost* as if the two droplets pass each other unhindered, but not *quite*. At arbitrary points within the box their motion is interrupted by a sudden stop, after which they continue their motion. Such a momentary stop is comparable with what happens during their collisions with the walls of the box. As is the case with the stops at the walls, the stationary motion will not be destroyed by these interruptions, on condition that their *combined* beat patterns have a zero slope. Why their combined patterns? In case of a collision with a wall the droplet is the only thing that has a hopping motion, the wall itself is static. In order to maintain symmetry of the motion before and after the collision—thus maintaining the stationary condition—the sack race itself should be symmetric w.r.t. the point of collision with the wall, which also implies a maximum or minimum of the amplitude of the hopping. When two droplets collide, they both perform their own sack race. The strength of the recoil is then determined by the sum of the individual hops. Therefore, if it is required that a droplet has the same absolute velocity before and after the collision, the *sum* of the two beat patterns must be symmetric w.r.t. the collision point.

The points where collisions can take place are completely randomly placed in the box. What may be concluded is, that the stationary state of both the droples can be maintained, *on condition* that the sum of their respective beat patterns everywhere in the box has the same, constant value.

It is clear that this condition for maintaining a stationary state may be satisfied only, if one of the droplets has a "Left" beat pattern, whereas the other one has a "Right" beat pattern. The sum of their beat patterns is then constant, and has zero slope everywhere within the box. In case both the droplets have a "Left" pattern, the condition for stationarity cannot be satisfied. Of course, neither is it possible to have two droplets within the same box with the "Right" signature. A further consequence is, that it is impossible to add a third droplet without violating the condition for a stationary state.

On the other hand this kind of problem, arising from the instantaneous stops, does not occur in the case of two droplets with *different* speeds (different energy levels). If the droplets have a different energy level, the encounters are of the type shown in the upper part of Fig. 14.23, and do not involve sharp interruptions of their speed.

The behaviour of droplets as sketched, is practically the behaviour as prescribed by Pauli's exclusion principle. One might, if wished, use the described processes as a mental picture of the exclusion principle, to have it in the back of one's mind as a visualisation. Whether the results from the droplet theory are really *identical* with Pauli's principle is another matter. In Pauli's principle the "spin up" and "spin down" properties of an electron play the same role as the "Left beat pattern" and the "Right beat pattern" in the droplet theory. However, one may not equate them to each other. The beat patterns only occur within potential wells, and do not materialise outside wells. In the case of free-flying droplets the beat patterns are just *potentially* there, and will arise only with the help of an external field having the zero-speed frequency. On the other hand, spin is an intrinsic characteristic of electrons, no matter whether free-flying or bound. So: plenty of mysteries left! It was the reason to take up this subject in the "riddles" part of the present chapter.

2. **Heisenberg's uncertainty principle.**

This well-known principle was not completely absent in the foregoing text. It was mentioned in relation with the subjects of zero-point energy (Sect. 14.11), energy width (Sect. 14.12), and tunnelling (Sects. 14.19 and 14.20). In all these cases we met situations where the instantaneous energy of an electron can differ a little from its nominal value. Random fluctuations of the energy may occur. In other words: there is an amount of *uncertainty* in the energy, which explains the name of the principle. What we are really talking about is a violation of the law of energy conservation, despite the euphemism "uncertainty". But some sort of consolation is also given by Heisenberg. The larger the violation, the shorter the time that it can persist. In reverse, the longer the violations continue, the less serious the violation is in terms of energy. The random fluctuations of energy are said to happen because "energy can be borrowed from the uncertainty", which is again a euphemism.

In the droplet theory random fluctuations of the translation energy do occur as well.

The basic cause here is the chaotic motion of the droplet, and the fact that there are two energy reservoirs in the droplet (translation energy and the energy in the variable elongation). The chaotic motion entails that the translation energy may fluctuate quite randomly, "borrowing" the energy needed for the speed excursions from the second reservoir. Statistics then lead—qualitatively—to the same relation as the one mentioned above: very large excursions can be expected to occur only during very short time intervals. Less severe excursions are less rare and may exist during longer spans of time. It is just as in the case of extreme weather conditions, a well-known example of a chaotic process. In droplet theory a process of "borrowing energy" takes place as well, but no violation of the conservation laws is needed to explain it.

What however is lacking at present in the droplet theory is, that the scale of the uncertainty has not been fixed yet. In Heisenberg's uncertainty relation the scale is implied by his well-known formula

$$(uncertainty\, in\, knowledge\, about\, energy) \geq \frac{h}{(uncertainty\, in\, the\, time\, domain)}$$

Remarkably, the proportionality constant in this expression is Planck's constant again!

Qualitatively, the formula says the same as above about the relation between time and how large the energy fluctuation is, but it also quantifies a minimum level of the uncertainties. The product of the uncertainty in energy and the time uncertainty can, according to the formula, never be smaller than the value of Planck's constant. Planck's constant is small, but certainly not completely zero, meaning that there is *always* an amount of uncertainty. On an atomic scale this obviously has much more importance than in macroscopic life, which is dominated by—much larger—uncertainties due to errors of observation and measurement.

Another thing that is lacking in the present droplet theory is a second expression by Heisenberg, relating uncertainties in position and momentum of a particle:

$$(uncertainty\, in\, the\, knowledge\, about\, the\, position\, of\, a\, particle) \geq$$
$$\geq \frac{h}{(uncertainty\, about\, its\, momentum)}$$

We can thus never know with absolute accuracy both the position as well as the momentum of a particle. The scale of this fundamental uncertainty is again given by Heisenberg, as seen in the above given formula.

In the droplet theory we could, with some willingness, derive this second uncertainty from the energy-time uncertainty (divide by the average velocity of a droplet), but what again is lacking is the scale factor. This matter is certainly something worth to investigate more deeply in future.

Chapter 15
A Chapter Full of Speculations: Bohr's Atom, Schrödinger's Cat, "Spooky" Interactions, and the Double-Slit

15.1 The Most Fascinating Magic Phenomena of Quantum Mechanics

It may well be that you feel a bit disappointed at this point. You probably had heard about the utter magic of quantum theory, where particles may be in different places at the same time, where cats (Schrödinger's cat) may be dead and alive at the same time, and where particles have a telepathic knowledge of the fate of remote particles. It could be that your interest in quantum theory has arisen in the first place out of curiosity to hear an explanation about this kind of "big magic". Therefore, possibly you opened this book hoping that everything would be unveiled, and that you could finish this book with a sigh of relief, now being able to see through the tricks of the illusionist that nature is.

Until now, having read through so many pages of this book, you have only been presented with rather dry, technical stories. There may have been a little magic in certain phenomena, and this may have been removed, but the really *big* magic, the things that are the most astonishing have not been dealt with yet. It is the task of this chapter to give you a feeling for the really big magic.

It was unavoidable that we started with the "small fish", one has to build up gradually from basics to the more complex. But there is another reason why the big things were postponed: until now everything could be based on a solid foundation of mathematics. The one thing that appeared out of the blue was the idea to adopt the droplet of charge as a model for the electron, instead of the usual point-like singularity. It was just a working hypothesis, to find out what would be the consequences of such a model. But the rest, i.e. determining these consequences, was a matter of conscientiously working out the mathematics, and interpreting the results in physical terms. The author hopes that this task has been performed correctly, without too severe flaws.

When we now come to the really big stuff, the high-school magic tricks of nature, the way to proceed must be slightly different, unfortunately. Everything that

T. van Holten, *The Atomic World Spooky? It Ain't Necessarily So!*
DOI 10.2991/978-94-6239-234-2_15

could be given a complete, solid mathematical foundation was summarised in the preceding Chap. 14. The remainder of this book rests on less solid ground. That is not to say that the following stories are completely conjectural. In fact, in the mathematical derivations many hints are found on which speculations can be based. But they are just hints, and only vaguely point the way towards possible mental pictures about the big magic. The full road to satisfactory explanations has not been traveled yet to the end (in the mathematical sense). The author could have chosen to leave these hints hidden in the formula appendix, because an account of them including some necessary conjectures to make it into full stories would make him very vulnerable to criticism and attacks. On the other hand, it would be a pity not to mention the definite hints found in the formula appendix. Anyway, as stated on the first page of this book, *all* of the stories in this book will be found controversial by quantum mechanics experts. So let us go on with the heresies! But it cannot be repeated too often: the whole of the present chapter is highly speculative. The overall picture rests on pieces of hard information wrung from Maxwell's theory, but they are just fragmentary pieces. Like a bridge resting on firm piers and pillars, but with the main superstructure connecting the piers still somewhat vaguely seen in the mist.

15.2 Bohr's Atom Model

Bohr's atom model is three-dimensional, whereas the theory developed in this book is strictly one-dimensional. Extrapolation to Bohr's atom of the mental pictures in this book is therefore conjectural by definition. However, it is not very daring to try such an extrapolation and it is quite plausible that the working hypothesis of a deformable droplet of charge orbiting a nucleus could give new insights.

Bohr had to introduce a number of postulates to "explain" the workings of his atom. This was necessary to circumvent clashes with classical physical laws. One of his first postulates was, that electrons in his model would have to maintain "radio silence" most of the time, which seems to contradict the classical electromagnetic theory. Now, we found in the one-dimensional droplet theory that there is not necessarily a contradiction, if we take into account the chaotic motion of the droplet. It is not a big step to extrapolate this conclusion to Bohr's electrons orbiting a nucleus in three dimensions.

Similarly, why stationary orbits can exist only for certain, quantised energy levels may be shifted from the realm of postulates to the realm of strictly reasoned consequences, if we have the courage to extrapolate the findings in one-dimensional potential wells to three-dimensional orbits. The stationary orbits are then those trajectories where we find closed beat patterns, the pulsation cycles of the droplet repeating themselves indefinitely and unchanged, no matter how often the droplet has gone round. This is not really a new thought: already Schrödinger in the old days sketched a picture where the matter waves accompanying an electron will show stationary patterns, only possible for certain quantised values of the electron's velocity in its orbit (a consequence of De Broglie's formula). There are differences between Schrödinger's mental picture and the theory developed in this book. First, in Schrödinger's view the orbiting electrons are not really particles at all, but must be viewed as pure wave-like things all the time. And secondly, these matter waves were mysterious phenomena, whereas the droplet theory identifies them with very concrete physical quantities, i.e. the beat patterns of vibrating droplets.

Bohr's model was unable to explain *all* the spectral data produced by the multitude of available experiments, although the initial successes of his model were striking. In hindsight, the most logical conclusion at that point would have been, that this model of orbiting electrons around a nucleus could not be entirely wrong, although something was still missing. This was *not* the conclusion arrived at by the contemporary physicists. They supplanted Bohr's visualisation of the atom by an entirely new mental picture, and threw away the planet-like view. It has been pictured in Chap. 1 (Fig. 1.9) what came in its place: just a probability distribution, a picture of the regions where electrons may dwell in the space around the nucleus. There are large forbidden regions where electrons are not allowed to come, and consequently one could no longer imagine that electrons would be able to complete full orbits like planets.

This later view, called the modern quantum theory, was the result of Schrödinger's formulation of the laws of quantum mechanics. It was clearly a large improvement, since it could predict correctly *all* of the experimental evidence. Schrödinger's equations, however, did away entirely with the simple view of electrons orbiting in closed paths. It was a completely new approach which did not bring a stepwise improvement of Bohr's model, but implied the message that the old model had to be radically abandoned.

Extrapolation of the one-dimensional droplet theory to the three-dimensional atom could tell a different story. It could merge the theories of Bohr and Schrödinger, and blend them together into one single consistent picture.

First of all it should be remarked that the droplet theory is not in disagreement with Schrödinger's equation. On the contrary, it has been made plausible that Schrödinger's equations are fully applicable, and govern the motions of droplets of charge. Now, in a one-dimensional potential well where a droplet is travelling the full width of the well between the bounces against the walls, we found places where the droplet will never be seen. In and around these so-called "zero-points" the conditions for radiating cannot be satisfied, and for that reason there is never a way to observe the droplets when they are passing these stretches of the well. Concerning the location of these zero-points, the droplet theory gives exactly the same results—quantitatively correct—as the predictions by modern quantum theory. But the interpretation is more down to earth than many explanations about the significance of these zero-points found in textbooks. The droplet traverses the whole width of the well, without being hindered by the zero-points, but it is sometimes visible and at other, fixed places in the well, *in*visible.

Extrapolating towards the atom, it would mean that we can still adhere to Bohr's picture of closed orbits, and nevertheless obtain the same picture as predicted by the modern quantum theory (Fig. 1.9). Once again, this is a speculation: the extrapolation is not yet founded on a complete three-dimensional mathematical analysis. But it is an attractive thought that perhaps the old and modern quantum theory could be merged in this way. It would also explain how it was possible that the original, old model could provide a lot of good answers.

15.3 Superposition of States (The Magic of Schrödinger's Cat), or Superposition of Frequencies (A Common Phenomenon)?

Bohr left the question unanswered what happens to an electron *during* a jump from one energy level to another. He did not need this information, and was able to predict the colour of the emitted light just by considering the before/after situations. In books you will find a lot of speculation about the whereabouts of an electron *during* the jump. Some people say that the electron is not there at all: it disappears from one of the states, and reappears in another state, with nothing in between. Other people find this hard to believe, because physics (almost?) never confronts us with complete discontinuities.

If one ponders about the consequences of this latter belief, viz. that the electron must be somewhere and must be doing something during the jump, we meet grave conceptual difficulties. We then must assume that there is a sequence of events, beginning with the starting-up of radiation which commences the shedding of energy. At the end of the process we arrive at the final energy level, where the radiation stops. The problem is: right from the start the radiation had a certain colour, and this colour is related to the energy difference between the two states. So: the electron "must have had thoughts", right from the start, about where its jump is going to end in future and what energy level will then have been reached. How does the electron "know" what are the allowed energy levels? Does it have a table with allowed levels, and does it decide which of all the possibilities it will choose?

Schrödinger's equations may provide a partial answer to these questions. It is the time-dependent version of the equations, which predicts how quantum systems evolve in time. In the case at hand, Schrödinger's equation describes the process of jumping in a way somewhat similar to the pictures given in this book about violin strings (Fig. 13.3), or the overblowing of a Baroque trumpet (Fig. 14.14). In these latter cases it was argued that one tone fades away and another takes over, with a brief moment of overlap in between. The overlap of the tones is most often not noticed by the listener, but sometimes a "duh"sound between the two tones is heard, a sign that there is a short beat between them. A solution of Schrödinger's time-dependent equation similarly shows that one of the states extinguishes, and another state rises to dominance. The states are both there all the time, but only one of them is "active" whilst the other is dormant, and they change roles during the transition.

This may be nice mathematics, but the physical picture is rather problematic. What is meant by "state" in this equation, is expressed as an energy level. When we visualise that the two states are present all the time, but change from dormant to active (or the other way round), we are implicitly imagining an electron that has two energy levels. Or, if we think of a potential box where the energy is only present in the form of kinetic energy, we must imagine that the electron possesses two velocities, superimposed on each other. It is a brain twister how to imagine such a thing, but this is the consequence of assuming that the electron contains both states already, even before the actual jump.

It will be no surprise that this "superposition of states", as it is called, has given rise to heated debates. How must the mathematical theory be interpreted physically? Schrödinger himself pointed out how absurd the above sketched physical interpretation of his own formulae is, by giving an example in the macroscopic world. This is the famous "cat of Schrödinger", where one of the possible states is "being alive ", and the other is "being dead". No doubt a transition from the first state to the second state is possible. But how to imagine that both states are originally superimposed, in other words can we have a cat that is both dead and alive at the same time?

The author of this book is aware that his summary of the extensive debate about these matters is too simplistic. In reality the discussions raging between Bohr ("advocating the Copenhagen interpretation"), his opponent Einstein and others like Schrödinger were more subtle than summarised in the above. If the reader wants to have a fuller account, he/she can be referred to almost any of the many popular-scientific books about quantum mechanics, where the "Copenhagen interpretation" is usually explained in much detail.

Let it suffice to say that the superposition of states, occurring in the mathematical expressions, is difficult to interpret physically, and is one of the famous pieces of magic inherent in quantum theory.

Now the droplet theory. What is the form the phenomenon takes here? As was often the case in this book, the interpretation according to droplet theory is rather down to earth, and strips the magic from the phenomenon. It is almost a disillusionment, taking the thrill out of the whole matter. A pity, really! What we meet in the droplet theory, as explained in the text below, is just a superposition of *frequencies*, not of states. And the superposition of frequencies is a common thing, found in any musical instrument which vibrates with a ground tone on which overtones are superimposed to give the instrument its specific timbre.

To explain this view, have a look at Fig. 15.1. What has been drawn there is a general property of beats formed by the mixing of two high frequencies. It is shown how the shape of the beats depends on the amplitudes of the two vibrations. In the uppermost part of the figure a sinusoidal vibration is shown, without mixing effects.

Next, we assume that there is a beginning of interference, because a second vibration is mixed in with a slightly different frequency. The second vibration still has a small amplitude, and forms no more than a small disturbance of the original vibration. A beat pattern has become clear in this second line of the figure, with a practically sinusoidal shape. However, the beat does not have an *exact* sine-shape.

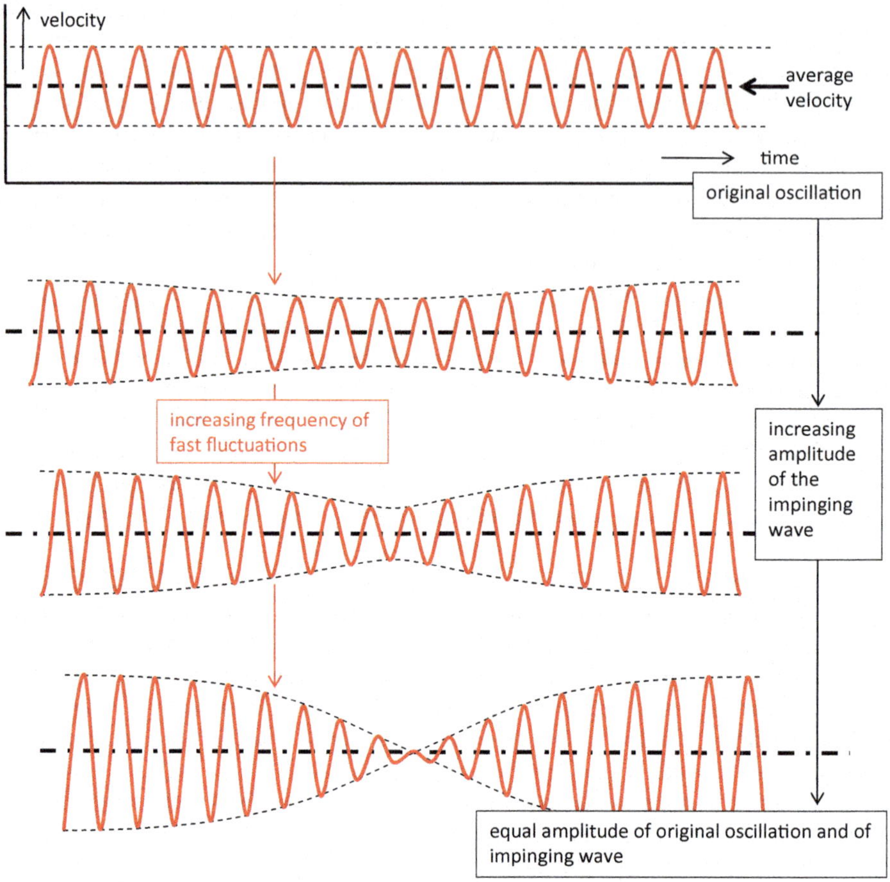

Fig. 15.1 Velocity fluctuations of droplet under the influence of impinging electromagnetic wave with slightly higher, but constant, frequency. From *top* to *bottom*: increasing amplitude of impinging wave

There are overtones (and possibly sub-harmonics) buried in it, although on this second line this is hardly or not at all visible. That there must be overtones in the beat shape becomes clear when we gradually increase the amplitude of the second vibration which is mixed in. The third and fourth line of Fig. 15.1 show an increasing departure from a pure sine-shape, indicating an ever growing importance of the overtones. So far a general property of beats. A tentative picture of the transition of a droplet from one energy level to another can now be built up as follows. For clarity, consider again a droplet inside a potential *box*. If the droplet were completely free, the high frequency "sack race" of the first line in Fig. 15.1 would be seen. Actually, the droplet is not completely free, and bounces at regular intervals against the walls of the box, where it momentarily stops and then during brief moments has a slightly lower pulsation frequency, the so-called "zero-speed frequency". The mathematics show that the result is a mix of the two frequencies during the entire journey of the droplet from wall to wall: the second line of Fig. 15.1 is representative for the resulting "sack race".

Until now we were mostly interested in the length of the beat pattern and—rather superficially—described the shape of the beat as sinusoidal. Actually, a lot more overtones and sub-harmonics are present. When energy is lost by radiation, a consequence will be that some of the energy of the original vibration (the uppermost line in Fig. 15.1) is being shed, and the amplitude of this particular ingredient of the mix becomes smaller. The ratio of amplitudes changes, we descend to the lower pictures of Fig. 15.1 and it is conceivable that one of the overtones becomes dominant, instead of the original beat frequency. This new dominant frequency has a different wavelength, which must in the end again fit between the walls of the box. The described process may thus be the prelude to a jump towards another stationary state.

It is not very clear yet how exactly the entire transition between two energy levels can be imagined. This is the reason why the present subject has been included in the "chapter of speculations". Something that is clear and founded on firm analysis, is the fact that in every state occupied by the droplet, the "seeds" for other possible states are buried. This is not different from usual quantum mechanics. However, these "seeds" in the droplet theory take the form of superimposed *frequencies*, instead of a superposition of *states*. Such a superposition of frequencies does not present any conceptual problems, it is a common phenomenon in physics. Whereas a superposition of states, as has been mentioned earlier, poses grave conceptual problems, bordering on magic.

15.4 Telepathic ("Spooky") Contact Between Widely Separated Particles (Or: Superluminal Signalling?)

And now follows the most spectacular (and the most speculative) part of the chapter. In quantum physics so-called *entangled* systems occur. They are systems consisting of more than one particle, where the separate particles have a close kinship, for instance two photons produced by one and the same event. Even if the separate parts fly apart, they keep their "remembrance" of a shared past. In some mysterious way they can still quickly interact with each other, as if by telepathy, their communication taking place at "superluminal" speeds. "Superluminal" means: faster than the speed of light, something that is considered impossible in physics. The "entangled" system apparently has a very strange, inexplicable behaviour. If we catch one of them and do something to it (for instance: change the direction of its spin), the other one "feels" it and it too can respond to our intervention, practically immediately.

Einstein did not believe that such interactions are possible, although they are consequences of well-established quantum theory. He called them "spooky" actions at a distance, and rather used these theoretical predictions to underline that the theory as it stands cannot yet be complete: if these "spooky" interactions are indeed logical, inescapable consequences of the theory, then something must be wrong with the theory, according to Einstein.

The puzzling thing is, that experiments repeated in various parts of the world and in several different forms seem to confirm that such "spooky" interactions can happen in actual fact. The matter is still a subject of some debate, but it seems that "spooky", immediate influences at a large distance cannot be ruled out absolutely, at least if we are dealing with entangled systems.

That two particles, at different positions and with nothing than a void space between them can interact with each other is strange, but not the strangest aspect in the above: we are used to gravity and electromagnetic effects, where there is action at a distance. Gravity, with an unknown velocity of propagation estimated between the speed of light and infinite, is too small a force to bother about when dealing with elementary particles. Electromagnetic actions at a distance are weird when you think deeper about them, but we are used to them. This is therefore a possible candidate for the spooky interaction, were it not that electromagnetic interactions are never superluminal. If we would invent other messenger fields, the same objection would still exist, because Einstein's speed limit is thought to be universal.

It so happens that the mathematics of the droplet theory, as developed in the formula appendix, *does* cover the above described phenomenon of immediate influences! It does so despite the fact that the usual electromagnetic retardation effects due to the finite velocity of electromagnetic waves are faithfully taken into account in the theory. This would seem to be a paradox. Below, it will be attempted to describe in physical terms how it works. But first, it should be stressed that special conditions (very special indeed) must prevail, otherwise the mathematics will show up the usual retardation (at the normal speed of light) between two interacting objects, and nothing peculiar would be found in the solution of the equations.

The special condition is, that the two objects must share the same "near field". In other words they must be parts of the same configuration. Until now we considered a deforming droplet, whose separate elements could have different velocities as if they were living their own life, although they are subject to some coherence. It was natural to define a near field encompassing the whole droplet, with all its elements. Now imagine that such a droplet would break up in two parts, in the way a droplet of water can sometimes break up (see Fig. 15.2). Obviously, the droplet must contain at least twice the unit charge before breaking up. This could still be considered as just a special kind of deformation, and for the purpose of analysing the electromagnetic field there would be no compelling reason to treat the near field differently from that of a droplet where the connections between all the elements are still intact. This is what was meant above by two objects sharing the same near field. By the way, in the mathematics of course a sharper criterion is found to distinguish between objects sharing their near field and objects that must be analysed as if they were completely separate.

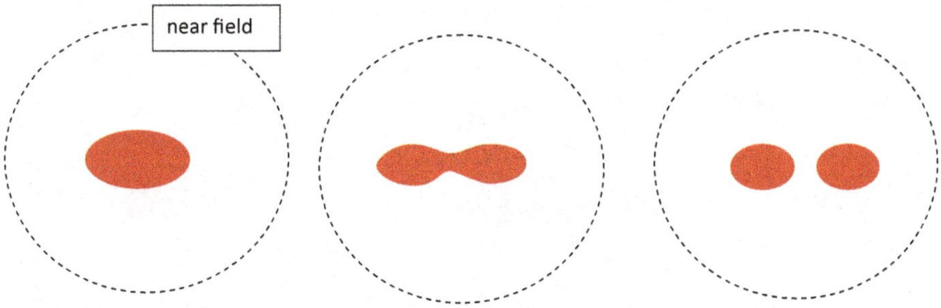

Fig. 15.2 The concept of "two droplets sharing the same near field", viewed as an extreme type of deformation

As you may remember, in Chap. 12 a clear distinction was made between the near field, immediately surrounding the wriggling and jerking charge, and the more orderly structured "far field". In the far field a description in terms of *running* waves was most applicable and useful. In the near field in contrast, one should forget the picture of running waves, a description in terms of—almost—standing waves is more applicable. We also met there the strange flows of "Schott-energy", occurring in the near field. "Schott-energy" acts as a buffer of energy, and the flows of "Schott-energy" supply or drain energy at points and at times where and when a lack or surplus of energy exists. The action of Schott-energy has the appearance of immediate interventions, not bothering about distances either. The concept of Schott-energy, how strange it may be, is an unavoidable concept if one wants to reconcile Lorentz's self forces with Larmor's theory about the final radiation escaping to infinity.

To make it more clear that we cannot apply our instinctive feelings about running waves to the near field an example may be given. Figure 15.3 considers a droplet of charge, on which suddenly an external force acts, which causes an acceleration. The figure shows what is expected if we would adhere to the view that the near field also may be described by running waves, like the far field. The final configuration of the near field then would take some time to come into existence. Now, the near field is the origin of the self forces, or rather: we calculate how large the self forces are by taking into account the fine details of the electromagnetic field which directly surrounds (and pervades) the elements of the droplet. If the view of Fig. 15.3 were correct, the consequence would be that the electromagnetic mass would *gradually* obtain its final value.

If the near field would indeed need some time to adapt itself fully to the new value of the acceleration, the "acceleration resisting self force", i.e. the inertia, would show a delay. The acceleration already has increased, and the self force resisting it is hastening after it. It would mean that the electromagnetic mass is not a constant. As a further consequence, this would give to electromagnetic mass a different character than "normal" mechanical mass. The total mass, electromagnetic and normal mass taken together, would also be variable, and would depend on the

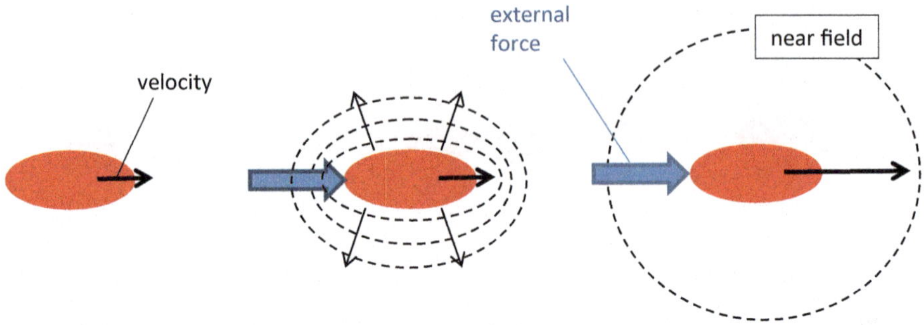

Fig. 15.3 Droplet undergoing sudden acceleration. The influence of suddenly changed boundary conditions would, if the near field consisted of running waves, imply a slowly increasing electromagnetic mass

jerk of a body. This is something that has never been observed, and actually it would undermine the whole of dynamics theory if it were true. We must reject the idea that the near field needs some time to completely adapt itself to changes of the acceleration. More generally, what mathematicians call "a change of the boundary conditions" works out *immediately* in the configuration of the near field and in the value of the self forces.

Incidentally, the self force of "radiation resistance" is a different story. Why it does show delays has another cause, which can be traced back in the mathematics, for which the reader who really wants to know is referred to the appendix.

How can this anti-intuitive behaviour of the near field be explained in physical terms? A possible "explanation" may be found in the Schott-energy. We have just recalled the phenomenon of Schott energy in its role to even out the discrepancies between Lorentz's and Larmor's theories. Schott-energy can in other circumstances, like the acceleration of the droplet above, also act as an instantaneous "stop gap", supplying energy in the near field where it is needed, and depleting the field energy at other places. It *looks* as if the Schott energy is flowing around without caring about the speed limit set by the velocity of light, but this is not necessarily the correct picture. One could also have as an interpretation of its behaviour, that energy is *locally* taken up or dumped from or into a sort of "back ground" electromagnetic field. Afterwards, flows of energy, respecting the speed limit, may take place within that field to return to an equilibrium situation. An example of such a process is given by Feynman in his famous lectures.

A sketch of the problem treated by Feynman is given in Fig. 15.4. It concerns the charging of a capacitor, consisting of two oppositely placed circular plates. An external power supply drives electrons into the upper plate, and sucks them away from the lower plate. Work is being done by this external source, which is found back in the form of potential energy of the electrons. If the external driving voltage is taken away, the potential energy is released again, and the capacitor discharges.

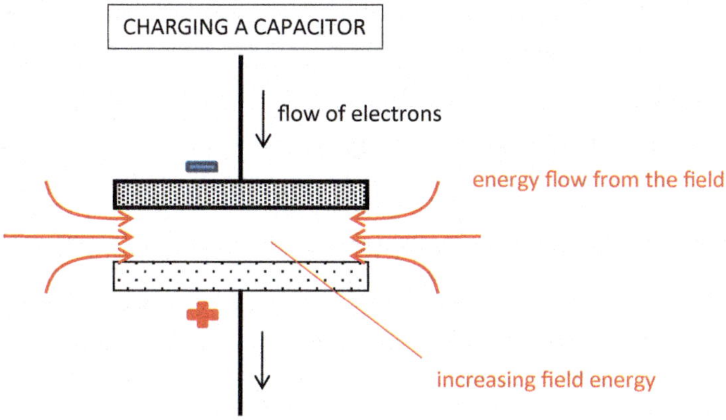

Fig. 15.4 The increasing energy of a capacitor charging up is not the result of work done by the electrons, but is taken from the external field

But there is also a parallel energy accumulation in the space between the plates. Here the strength of the electric field increases during the charging up. This implies that the energy in the field between the plates increases. Where does this energy come from? The configuration is so simple that a calculation is easily done, and Feynman shows that this energy enters through the open sides of the capacitor. The energy is supplied by the external field! It is not supplied by the external power source, as our first thought would have been. No matter how fast the capacitor is being charged, the energy can be taken up from the field, without any delay.

The particular distribution of electrons in space, "enforced" by the presence of the power source and the capacitor, is thus accompanied by a redistribution of energy in the field: a higher density of energy in between the plates, at the cost of a lower energy density elsewhere, but such that the total energy in the field is being conserved. Of course, the process of charging is a *dynamic* process. In the first instance, the energy between the plates is drawn locally from the parts of the field immediately adjacent to the plates. Then, this depletion of energy is "felt" by parts of the field further away, and they will transport some energy towards the inner regions, etc. The extent of space where the energy redistribution has taken place grows with the speed of light. This gradual process taking place afterwards, does not hinder an immediate suppletion of energy between the plates, since this is *locally* provided by the adjacent field, on the spot, which does not directly require extensive flows of energy over large distances.

Likewise, one can imagine a similar process in the case of Fig. 15.3. The changed boundary conditions require an immediate adaptation of the entire near field, otherwise one would be confronted with impossible things like a "variable mass" during accelerations. The immediate adaptation of the near field configuration might be imagined to come about by purely local energy exchanges with the "back ground" electromagnetic field. It is only afterwards, that redistributions of energy take place with the "normal" speed of energy flows.

To return to the configuration consisting of two different charges sharing the same near field, have a look again at Fig. 15.2. This concept was clarified by the assumption that the configuration of two droplets is just an extreme form of deformation. The two droplets are then just two elements of the same charge distribution. The near field has the property to respond immediately to changes of the boundary conditions. If one of the parts changes its shape or its motion, this results in an immediate interaction with the other parts within the same near field. Could it be that this phenomenon is the same as an "entangled " system in quantum mechanics? It is just a conjecture, of course.

The above given physical description is an example of the type of speculation which was earlier compared with a bridge based on solid piers, but for the rest having somewhat vague contours in the mist. It may be helpful for some of the more interested readers to point out what precisely are the hints given by the mathematical analysis. Therefore, a few words to explain the mathematical basis, only for readers who are really curious. In Chap. 5 a short, qualitative description was given of the mathematical procedure to determine the electromagnetic field surrounding a charge in motion, whilst deforming in shape. The procedure had the abbreviated name "max-technique". The basis of the method is schematically presented in Fig. 15.5.

On physical grounds it is assumed that the far field is not much affected by very fine details of the instantaneous charge configuration. In an earlier chapter it was compared with the gravitational field of the Earth, as it influences the orbit of the Moon. Small aberrations of the gravitational field caused e.g. by mountain ranges

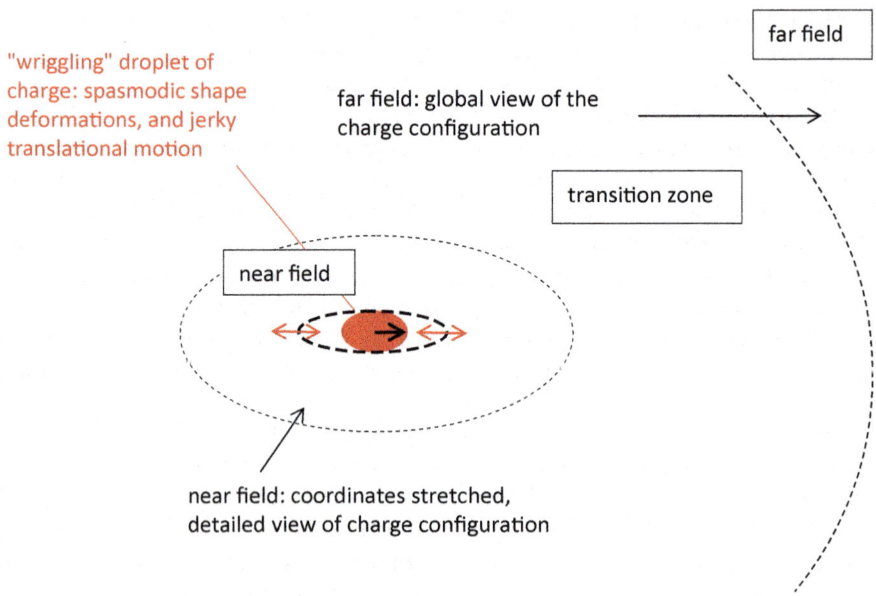

Fig. 15.5 Basic approach of the "max-technique"

will not appreciably affect the Moon's trajectory, and may be mostly neglected at these large distances. On the other hand, small details of the Earth's mass distribution can be important on or close to the surface, and to describe these details properly we need a finer coordinate mesh in the near field of the Earth. Likewise, in the near field of the "wriggling" charge the coordinates must be stretched, so that all the details of the charge configuration and its motion can be taken into account.

Another consequence of the division into a near- and far field is, that retardation effects work out in different ways. At large distances, we will find an electromagnetic field having a character as sketched in the earlier Chap. 12, see Fig. 12.2. In the far field one can characterise the field as a mix of running waves with different wavelengths. Running waves have a distinct direction of motion, and there is a distinct time interval between passing two different points.

The near field has a quite different character. It is not fruitful to think about the near field as "a bunch of running and interfering waves". It was mentioned earlier, that the character of the near field is primarily "a bunch of standing waves", although there is the secondary effect that radiation energy is seeping out of the near field to feed the energy transportation to infinity by the running waves of the far field. In acoustics one could compare it with a concert hall, outside which the music can be heard faintly, as it leaks through the walls and subsequently propagates further. Inside the music hall the sound of the orchestra reflects against walls, ceilings, chairs, people etc. These reflections represent retardations due to the finite velocity of sound, but they are not perceived as real echo's, they just lead to a change of timbre. The sound field inside the hall is perceived as a field of standing waves, not as a field of running waves interfering with each other.

The "max-technique" to calculate the electromagnetic field of a "wriggling" droplet of charge does exactly make this distinction, in a very natural way. The user of the method, without explicitly aiming at it, will find that the method automatically directs him/her to another equation in the near field than the usual wave equation. It is an equation which mimics the effect of Schott-energy by additional—artificial—energy sources, distributed all over the near field. Instead of *flows* of energy, the near field equation describes how everywhere in the field additional sources and sinks of energy occur, just what was earlier called the *local* supply and depletion as needed (Fig. 15.6).

In this way something is represented similar to the perceived field of standing waves within the concert hall. The effects of retardation due to the finite velocity of light are thus taken into account, but in a way which is closer to the actual appearance of the field than a description in terms of running waves.

What is even more important for the present discussion: the so obtained near field describes *immediate* interactions between the different elements of a charge configuration, because it is not the solution of a wave equation.

The distribution of extra energy sources thus mimics the effects of retardation and of flows of Schott-energy in the near field, without the problem that self forces such as electromagnetic inertia (mass) would become variable.

A separate text box is provided for the mathematically curious readers.

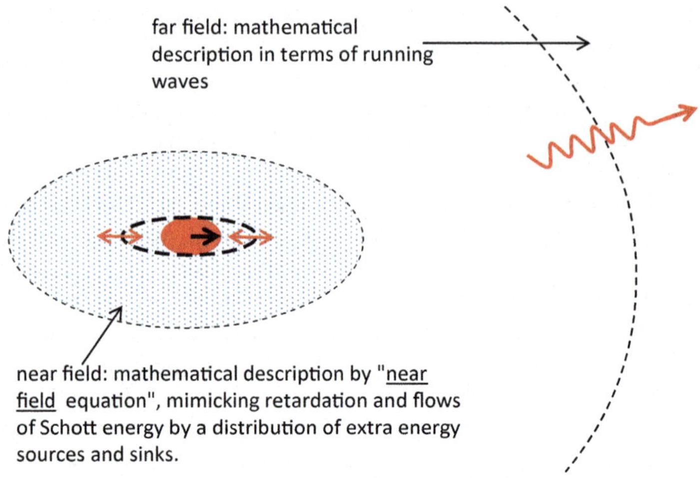

Fig. 15.6 The mathematical consequences of the basic assumptions shown in Fig. 15.4

Text Box: The Asymptotic Equations of the Near Field

The scalar potential $\Phi(x, y, z, t)$ of the electromagnetic field, as well as the Cartesian components of the vector potential $\underline{A}(x, y, z, t)$ all four have to satisfy the wave equation (outside the region where charges are situated):

$$\nabla^2\Phi = \frac{1}{c^2}\frac{\partial^2\Phi}{\partial t^2}$$

This form of the field equation is most appropriate to the "far field", leading to the solution of running waves and showing clear retardation effects. Typical wavelengths are of the order $O(\lambda)$.

The near field is the region close to the charge configuration, where typical distances are of order $O(a)$, with a/λ small, by definition. If, using an asymptotic procedure, successive approximations are made, the wave equation in the near field takes the form:

$$\nabla^2\Phi_n = \frac{1}{c^2}\frac{\partial^2\Phi_{n-2}}{\partial t^2}$$

where the indices indicate the order of the asymptotic approximation. The wave equation is thus reduced to a Poisson equation in the region close to the charge configuration. This is independent of the asymptotic order, it is purely a consequence of the definition of "near field" and of using an asymptotic ordering scheme to group together terms of equal orders of magnitude.

The effects of retardation are taken into account to the required accuracy by the presence of the r.h.s. of the Poisson equation, which represents a distribution of sources and sinks. The solution of the Poisson equation is only *parametrically* dependent on time: the asymptotic sequence is *not* a time-sequence.

15.5 One Particle, at the Same Time in Two Different Places

This consequence of the quantum theory also belongs to the repertoire of the great illusionist, i.e. nature as it behaves at quantum scales. In fact, we have already met this trick of nature in the first parts of Chap. 1, where the double-slit experiment was discussed. Particles, like electrons, were shot one by one through the arrangement, see Fig. 15.7.

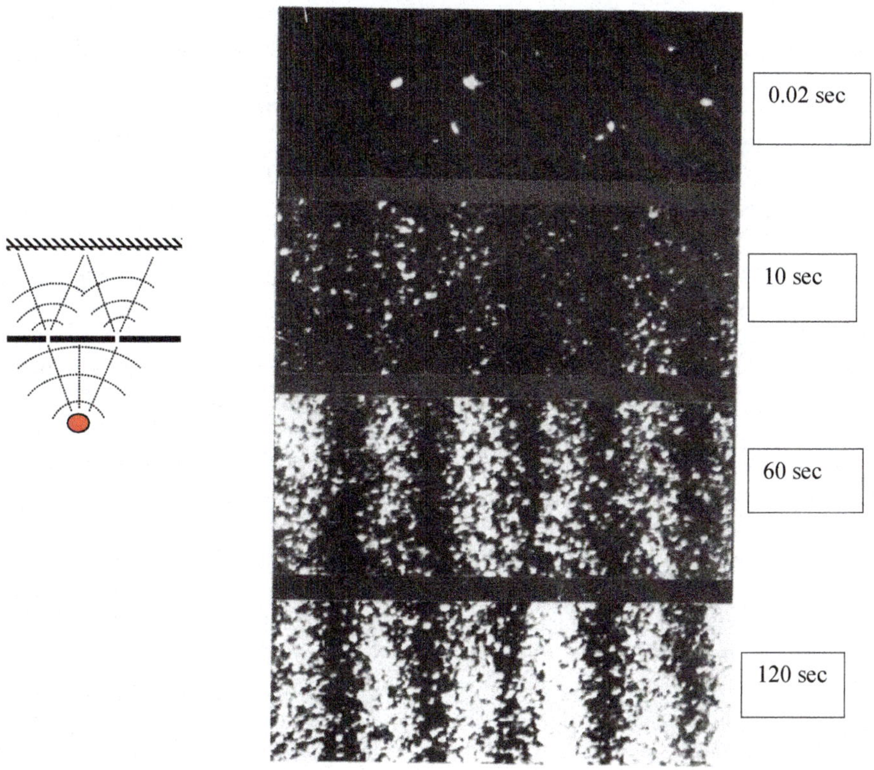

Fig. 15.7 Electrons are shot at a photographic plate through a double-slit arrangement. The plate is developed after different exposure times

The nice interference pattern disappears when one of the slits is closed and when the electrons can only pass through one single slit. The conclusion is, that electrons seem to "know" whether one or two slits are open, even if we think they can pass just a single one. The latter view is apparently wrong, and we are compelled to believe that something is happening as was shown in Fig. 13.4. This figure showed how we must imagine that photons are travelling through space: they seem to "sniff out" any path that is possible, to converge in the end to a single point where the photon exposes its presence. In a similar way we are compelled to assume that a particle like an electron can divide itself, to go through *both* of the slits of the experiment. In quantum mechanics there is nothing weird about such a view, since particles are subject to particle/wave duality. The electron, before hitting the photographic plate, behaves in this view like a wave so that it in this guise can pass through both the slits.

In the droplet theory another view has been adopted. Here we have a droplet of charge which cannot morph at will into a wave. We have assumed that a particle is a real particle as we know particles in macroscopic life. We did not model these particles as rigid marbles. The model was slightly more intricate, with pulsations and accompanied by a—mostly unobservable—electromagnetic field. Can the results of the double-slit experiment be explained using this model? Already in Chap. 1 it was announced that we regrettably cannot know the answer for certain, since the double-slit experiment is two-dimensional, whereas the droplet theory developed as yet is strictly one-dimensional.

But a few speculations are allowed in this chapter. The first basic question is, why a droplet which goes through a slit may show a kinked path, so that the trajectories of electrons passing the slit in sequence can fan out in different directions. The answer to this question may be found in a combination of two earlier given figures, Figs. 12.12 and 12.14.

The droplet we are talking about is in free flight (another feature that strictly speaking is not covered by the present theory), and does not radiate. The latter means that no energy is transported from the electron to infinity. It does *not* mean that there is no radiation field at all, as is shown in Fig. 12.12. This schematic depicts the field of standing waves, a field that accompanies the droplet. There are "nodes" in this field, or "barriers" through which there is no flow of energy. Now, if we put obstructions close to the droplet, the inner part of the field will obviously be altered. This in turn must unavoidably alter the self forces on the droplet.

Figure 15.8 very schematically shows what may be expected. The droplets passing the narrow slit can do so at different lateral distances from the walls of the slit, they do not necessarily fly exactly along the centre line. A droplet which follows a path away from the centre line will experience an asymmetric distortion of its accompanying field.

Figure 12.14 is now recalled to see what one of the consequences of an asymmetric field distortion might be. Figure 12.14 is drawn for a situation where the field is symmetric w.r.t. the line of flight. This is really the only symmetry found in the field, other symmetries are absent as symbolised by the colour shading. The asymmetries explain the presence of momentum in the field. It was argued that the

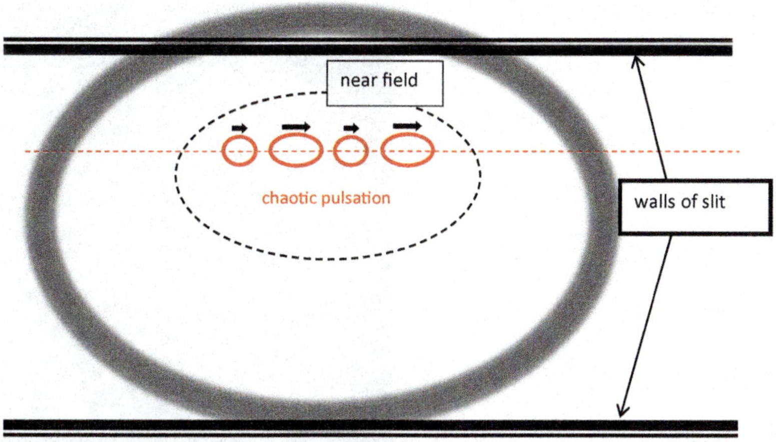

Fig. 15.8 The walls of the slit deform the droplet's field asymmetrically, when the droplet passes not exactly on the centre line

momentum due to this field pattern is directed along the line of flight, to the right. In the case of Fig. 12.14 an additional asymmetry w.r.t. the line of flight would clearly entail a momentum which no longer is parallel to the line of flight. This gives a first clue about the possible consequences of an asymmetrically distorted field. However, Fig. 12.14 concerned a droplet that was forced to slow down, e.g. by an external force, so that these results are not directly applicable to Fig. 15.8.

We must go back to the basic relations quoted in Chap. 12, concerning the relation between the flows of energy and the momentum in an electromagnetic field. Without bothering about the details (they have not yet been worked out with mathematical rigour), it is to be expected that the asymmetry of the energy flows sketched in Fig. 15.8 will bring along an asymmetry of momentum as well, i.e. a lateral component of the momentum which drives the droplet away from its straight course.

So far the visualisation of the effect of one single slit, and the fanning out of the possible trajectories of a droplet. How strong the deviation from the original flight direction will be, depends on how far from the centre line the droplet finds itself when it begins its passage through the slit. Every directional change is possible (within a certain limiting angle), so that a stream of charged droplets, impinging on a suitable screen, will show a continuous pattern of scintillations with the highest density of hits right opposite the slit.

In case a second slit has been opened, the electromagnetic waves will go through both slits, although the droplet can still only pass through one of them. In the electromagnetic field accompanying the droplet one will then find interference effects. This is not different from the water waves that were shown in Fig. 1.6, repeated here as Fig. 15.9.

The basic frequency found in the field is equal to the—unobservable—pulsation frequency of the droplet, but the amplitude of these very high frequency waves is

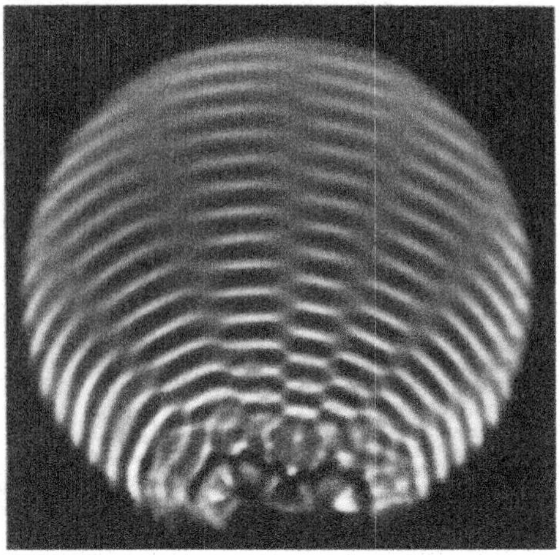

Fig. 15.9 Photograph of the double-slit experiment in water

now modulated. This amplitude modulation (like AM radiowaves), is what is seen in Fig. 15.9 in the form of the nearly radial stripes. The droplet, traversing this interference field will experience a field strength which is sometimes almost zero, and at other times may excite strongly its pulsation. The amplitude of its pulsation thus becomes variable, depending on the place where the droplet happens to be.

Now, recall that in the one-dimensional theory of a droplet, we met a somewhat similar situation. In a one-dimensional well, we had beats in the amplitude of the pulsation, which had another cause (they were caused by mixing two different pulsation frequencies), but with the same effect of alternating large and small pulsation amplitudes, depending on the place of the droplet within the well. The result was, that there were places in the well conducive to radiation, and other places where radiation could never be started up, so that the droplet would never be observed in these points.

In the three-dimensional atom, the conjecture was that the electrons can orbit the nucleus in closed trajectories, but would not be visible in certain regions due to this same effect. This would give rise to the patterns of visibility as shown in the beautiful figure reproduced in Chap. 1, Fig. 1.9.

The speculation is, that something similar will be the case in the double-slit experiment: the droplet travels unhindered from one of the slits to the photographic plate, and can do so along any of the fanning paths, but will meet places where it will always remain invisible. The pattern of potential visibility/invisibility corresponds with the interference pattern of the waves through the two slits. Finally, the potential visibility of the droplet is actually shown up by the hit with the photographic plate. On the plate bands will appear corresponding with the interference pattern of the electromagnetic field accompanying the droplet. Remember that the ultra-high frequency field itself is unobservable, so that only the *consequences* of the low frequency interference effects are visible, which thus would explain the behaviour of electrons in the double-slit experiment.

15.6 Emergent Quantum Mechanics and the Bell Test

This whole book might be considered to be a contribution to the branch of physics called "emergent quantum mechanics". This relatively new branch deals with classical laws of nature conspiring in such a way that quantum-like behaviour emerges. Such conspiracies are found under very special circumstances only, remember for instance the bouncing fluid droplets mimicking quantum-like behaviour (textbox at the end of Chap. 1). Apart from the stories in the present book, more examples of emergent quantum mechanics than only the bouncing droplets have been found, as will be seen below (although all these examples are *analogues*, not quantitative replications of quantum mechanics as in this book). All of them have in common that the "rules and dogmas" of quantum mechanics are *not* invoked, whilst nonetheless similar behaviour is found as predicted by quantum mechanics. One of the intriguing questions is, whether such "emerging" phenomena coincidentally imitate a few selected results of quantum mechanics, or whether they can be "complete" analogues covering *all* the predictions by quantum theory. If it is suspected that a particular example of emergent quantum mechanics is a complete analogy (or even a replication as in this book), then it must cover all the quantum phenomena in every aspect and without exceptions, including the most "weird" aspects. Only then one could begin to think about a conceivable reconciliation between quantum theory and classical physics.

Undoubtedly the most extreme predictions by the "real" quantum theory are the "spooky actions at a distance", i.e. the instantaneous influence a particle can exert on another particle even if it is in the remote distance. Section 15.4 of the present chapter already dealt with the subject, and concentrated on how the phenomenon might be viewed in the theory of the droplet of charge. The expression "telepathic contact" was used, to emphasise how strange it is that widely separated particles seem to "be aware"—be it under exceptional circumstances—of the fate of sister particles, and actually respond to it when the outside world operates on a sister. Remember that in the quantum world any measurement influences the properties of a quantum particle, so that a measurement on one member influences the other. If the distance between the two is large, such an influence can only be described by a sort of superluminal or even instant signalling. This in turn is incompatible with Einstein's speed limit stating that nothing can surpass the speed of light. There is clearly a paradox, which has become known under the name "EPR-paradox", an acronym for the authors Einstein-Podolsky-Rosen who first pointed it out in 1935.

In established quantum theory such systems of "sister particles" are called *entangled* systems. Entanglement means that particles (or systems) once intimately interacted with each other before they went off to live their own life. However, it only *seems* that they are living their own individual lifes after the separation. Quantum theory tells that in the case of entanglement it is impossible to completely dissect a whole constellation and treat it as the sum of separate parts. Instead, these parts still must be considered as components of one and the same configuration (in jargon: the system is still described by one wave function, which function can mathematically not be separated completely to get a sum of different wave functions). Quantum theory therefore might be called to be "holistic". Holistic means "wholeness" of the world. The world cannot be divided into small, independent components that individually interact only *locally* with neighbouring components. Instead, in the holistic view the world should be considered as a whole, as one big entity, resulting in *non-local* interactions. Quantum theory implies such a holistic view, at least when we are dealing with entangled systems.

Problems arise if we expect that one part of an entangled system should have the same properties as a "normal" particle. In particular, an individual, "normal" particle can only display *local* interactions, meaning that a particle exclusively interacts with its immediate surroundings (or at most with surroundings within the reach of electromagnetic forces). If one part of an entangled system is mistaken for an individual particle, then we would conclude that the locality of its interactions is violated. We then do not understand how it can immediately respond to something that happens elsewhere, possibly at large distances, and we will then call it a "mysterious *non*-local event". One could thus argue that *non*-local interactions are the result of a wrong identification: mistakenly, individuality has been assigned to the particle, whereas it in actual fact is just one part of a larger system.

A possible comparison is the ventriloquist. Very young children will not really be impressed by the performance. What they perceive is just one object: what we, as adults, consider as a group consisting of a human and a doll is one object to them, it is like their father just wearing another kind of clothing, somewhat bulging and more colourful than usual. And the funny speech is something infants hear all the time when adults are addressing them. Therefore: nothing to be surprised about! Children at a later stage in their development recognise a doll and have learnt that dolls and people belong to different categories of objects. Therefore, the group is then seen as an assembly of separate, individual objects. Problems now arise, because everybody knows that dolls do not speak. Confusion is the result! It is caused by the—natural but incorrect—separation into different objects, and by expecting the usual behaviour of these separated objects. The amusing thing is of course, that the holistic view of the infant is the correct one in this particular metaphore, whilst older children are deceived by their better knowledge of the real world.

This is the kind of story told by quantum mechanics about entangled systems and non-local interactions, and it sounds reasonably enough. Nonetheless, the holistic view is counter-intuitive, as so many chapters in quantum mechanics are. Think for instance about an event in our macroscopic world such as a billiard ball hitting another one. The ball is locally slightly indented and elements of the ball directly contacting the other ball can only, by internal shock waves, transmit the forces which cause the subsequent accelerations of the other elements. A "holistic billiard ball", i.e. one that immediately as a whole responds to the collision (without any internal mechanisms) may often be a usable simplified model but it does not correspond with the real physics. At least not with what we humans experience as real physics.

Now, Einstein offered another possible view on the phenomenon of entanglement, much closer to our intuition. Einstein supported the idea that in nature nothing else than local interactions exist, he did not believe in non-locality. He long ago (in the same EPR-paper of 1935) pointed out a different explanation for the closely correlated behaviour of entangled particles, an explanation that does not require the concept of holistic behaviour. If we consider the two parts of an entangled system to be completely and truly separated after their common history, so that they are capable of local interactions only, we could still have "holistic-like" behaviour. His idea was: let us assume that something is travelling together with the particles that contains a program prescribing how the particles should behave when they encounter measuring equipment. This "something" would be a sort of DNA contained by the particles, and there is some similarity in the DNA of both particles because of their common origin. Such DNA has never (or: not yet) been discovered, but it could be one of the hidden variables lying underneath quantum theory. He thus left the possibility open that the present quantum theory is not complete. We could imagine that the programming in the DNA is such that it is just *mimicking* holistic behaviour, despite the fact that the interactions with the measuring equipment are actually purely local. Quantum theory in its present form is in this view wrong, or rather: incomplete. The mathematical inseparability of entangled wave functions is then just *apparent*, and can be solved by assuming a different model of the situation. One is just fooled into thinking that "spooky" interactions exist because there are manco's in the theory. So far the warning by Einstein to avoid easy and premature conclusions, or to fall into the trap of "false negative" experimental outcomes. Actually it was more than a warning, Einstein hoped that by this argument the "spooky actions at a distance" could be disposed of as being rubbish. He was opposed by other people, for instance by Bohr, who considered quantum theory as a complete theory, so that everything that ever can be known about the world is covered by the present theory. The crux is, that Einstein believed that the real world can only show up local interactions, whereas his opponents believed in the existence of non-local interactions. If an experimental test could be devised such that it without any doubt shows a counter example of undeniable non-locality, then the matter could be resolved.

Once again we can clarify this discussion by thinking about the ventriloquist. It may surprise an observant onlooker that the doll and the man are never speaking at the same time, they never have arguments shouting through each other. This can be explained in two ways. The holistic view is, that there is just one brain and one voice mechanism, but different organs through which the sounds can be transmitted. The other explanation, applicable to the view that the doll and man are separate, individual objects is that both behave politely, in accordance with social rules to keep your mouth shut when the other is speaking. The latter would correspond to Einstein's view of an ingrained but hidden program underlying the interactions between the two. Something similar applies to the mouth movements and facial expressions of the doll, which never occur when both hands of the man are visible. This could be the result of either some sort of agreement arranged beforehand, or alternatively of the fact that the man and doll are not really separate systems.

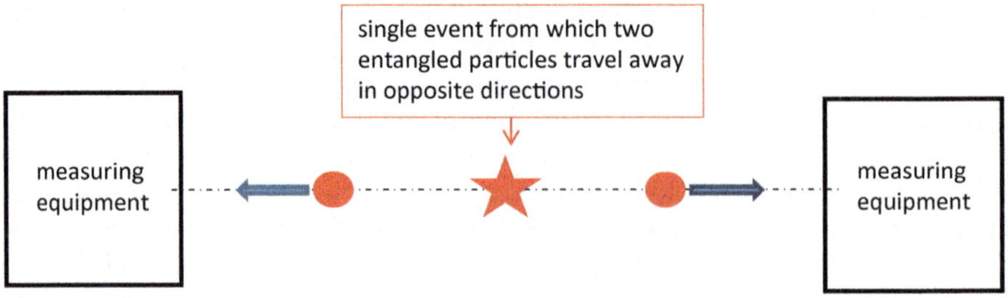

Fig. 15.10 General arrangement of the Bell test

Long after these discussions between Einstein and Bohr a test has indeed be conceived to find out whether a system really is holistic (in other words: whether it truly displays non-local characteristics). The test was devised by the physicist J.S. Bell (at the time working at the CERN centre of nuclear research), and it can be applied both to theoretical models as well as to actual experiments. A very brief and superficial impression of what this test entails will be given below, ignoring all the rather subtle details (Fig. 15.10).

It is supposed that two particles emerge from one and the same event, and that they fly away in opposite directions. Think for instance of an electron and a positron formed by the same event, and leaving their birth place in opposite directions. Whether they really may be considered as individual, separate particles is the crucial question: they are entangled because of their shared history, and must therefore perhaps be considered as two parts of the same system. Even so it is possible to let them trigger measuring equipment in two different, unconnected places.

The measurements on the left and on the right will be correlated to a certain extent, because of the close kinship of the two particles. For instance, if two particles or photons are formed by a spin-less process, then we will always measure spin-up on the left and spin-down on the right, or the other way round. This is comparable with the alternative speaking and maintaining silence of a doll and a ventriloquist. As explained, these measurement results—although correlated—do not yet give a clue whether we are dealing with a holistic system or with a system containing hidden variables.

Bell showed that taking into account more types of correlation and considering them together would give the possibility to discern a holistic system from a system with hidden variables, where the latter has strictly purely local interactions with the measurement equipment. In the metaphore of the ventriloquist we would then look for instance at the sequence of speaking, as well as at the correlation between the arm position and the speaking, the correlation between the mouth movements and the arm position, etc.

A test to determine such correlations will obviously require that the experiment is done on many pairs of particles: a sign of similarity between the measurement results could be a coincidence if just one pair of particles has been shot through the arrangement. Therefore, the Bell test necessarily takes the form of a statistical criterion.

Bell analysed two different scenarios, the first one based on the assumption that there are no holistic properties at play. Each particle has its own influence on the measuring equipment, by purely local interactions. A second scenario assumed that there are indeed non-local interactions, as if they would still belong to the same holistic system. The first scenario (the local interaction one) corresponds to the view classical physics (and Einstein) would have on the process. The second one (non-locality inherent in a holistic system) is the view according to established quantum mechanics (and Bohr). As explained, in both scenario's a statistical analysis of the results would show up a non-negligible correlation between the left and the right measurements: after all, the particles possess some sort of "family relation" due to their shared past. However, Bell showed that the *amount* of correlation would be larger in the holistic scenario than in the scenario with only local interactions. And the fortunate thing is that the amount of correlation between the two scenario's would differ so much that it would easily show up, and thus would be testable. Like litmus paper in a chemical test, Bell's test will betray any trace of non-local effects by an increased amount of correlation.

Many experiments have been performed to reveal if indeed non-local (holistic) effects occur in nature. The experiments are difficult, and require much ingenuity, so that it took some time before Bell's theory (1964) could be applied. Pioneering experimental work was done by Freedman and Clauser (1972) of the University of California, which confirmed the holistic predictions by quantum mechanics about non-local effects. Perhaps the most famous are the even more elaborate experiments by Aspect (1982) of the Paris University, which avoided the possible "loophole" that the measuring equipment itself could have some sort of secret communication with each other, through electromagnetic signals. To exclude such a "loophole" in an actual experiment one has to make sure that the *type* of measurements is continually changed in an unpredictable way. For instance, the spin orientation is sometimes inverted by one of the measurements, and in other cases it is not. The distance between the left and right measurement equipment must be chosen so large that the information about such changes cannot be transferred by "normal" electromagnetic transmissions within the available time. In other words, the left side equipment cannot be warned in time about a change in the type of measurement on the right. The experiment by Aspect and his team really was very ingenious!

Recently (2015) tests were done at the Delft University of Technology by Hanson and his team claiming to have "closed *all* the loopholes", in other words excluding any thinkable form of escape from the conclusion that non-local interactions do exist. However, the discussion still does not seem to be closed completely (Vervoort in arXiv 2016, see the chapter "references").

The conclusion from all the mentioned experiments so far invariably is: non-local interactions do exist! It leaves us with the problem how to view the consequences, viz. how to interpret the holistic behaviour. Is the assumption of superluminal signals between entangled particles unavoidable? And if so, would this not be a violation of one of the other fundamental ideas in physics, viz. the speed limit set by the velocity of light? A violation of the speed limit would be a catastrophe, necessitating the revision of large parts of physical theory. Emergent quantum theory might help to sort out our confused thoughts. If there would exist an example where classical laws of nature lead to holistic-like behaviour, this could serve as a crowbar to lift some tiles hiding the solutions to all these questions. What we would ideally like to find is that nature is no more than a superb illusionist, seemingly not caring about fundamental laws of nature, but in fact of course just as well obeying these laws.

However, very often it is thought that analogues based on classical physics will *never* be able to survive the Bell test. Classical physics would describe only local interactions, or interactions at a distance taking place with the speed of light. Instantaneous interactions or interactions with superluminal speed between two distant objects do not have a place in classical physics, and are typical for—and are strictly reserved to—the realm of quantum mechanics, it is thought. If this general statement about classical physics describing merely local interactions were indeed true, the consequence would be that emergent quantum mechanics can never be more than a *cripple* analogy of the real quantum world, since the latter must sometimes be considered as holistic. However, the interesting thing is that counter examples can be given where classical systems—under special circumstances—indeed display holistic-like properties. Such a counter example will be described below. What will be done in the present chapter is to consider in more detail what it exactly is that can give holistic-like properties to a classical system, and then check from this point of view the contents of Sect. 15.4 about spooky interactions in the droplet theory.

Brady and Anderson (the same scientists as mentioned in the textbox about bouncing droplets, but in an inverted order of first- and co-author) found a fluid analogue which stands up to the Bell test and indicates non-local effects. And what is most important: it can be traced back what is the origin of the non-locality effect. Their publication (2015), entitled "Maxwell's fluid model of magnetism", is interesting for historians of science as well because—surprisingly—their fluid model has a lot of commonality with mental pictures Maxwell had in his mind when he developed the Maxwell-equations!

To give a first idea of this fluid model, have a look at Fig. 15.11. The actual model of Brady and Anderson is slightly different, but Fig. 15.11 may serve as a first step to explain it. What is shown is a demonstration of a vortex in water (San Francisco Exploratorium), formed by tangential injection of water at the top, at a rate which is controlled by the girl on the right. It is the well known snake-like shape which in the bathtub can frighten young children. In the photograph it is seen that in this particular instance there are waves running along the length of the vortex core, shown up by the patches of light. It is one of the many manifestations of so-called "inertial waves" in fluids. Other examples of inertial waves are found in rotating atmospheric and oceanic flows. Internal waves in these rotating masses of fluid are driven by Coriolis forces, in contrast to surface waves which are driven by gravity. The picture of Fig. 15.11 shows that these inertial waves in that particular

Fig. 15.11 Inertial waves on a bathtub vortex. Milton Van Dyke: An album of Fluid Motion

case have the visual appearance of a sort of beads travelling along the "vortex snake".

There is a remarkable similarity with some sketches by Faraday and Maxwell how they visualised an electromagnetic field (Fig. 15.12) by an analogy with fluid flow. The lower sketch by Maxwell shows a circulating electrical current accompanied by a magnetic field line (the arrow), a sort of solenoidal arrangement very similar to vortex flow, where the vortex core is the analogue of the magnetic force line. Mathematically speaking, this is a correct fluid analogue of the electromagnetic field. Maxwell even went so far that he viewed the attraction force between the poles of a magnet as the analogue of the low pressure near the core of a vortex which exerts, as in Fig. 15.11, opposite pressure forces on the water surface and on the bottom surface of the cylinder.

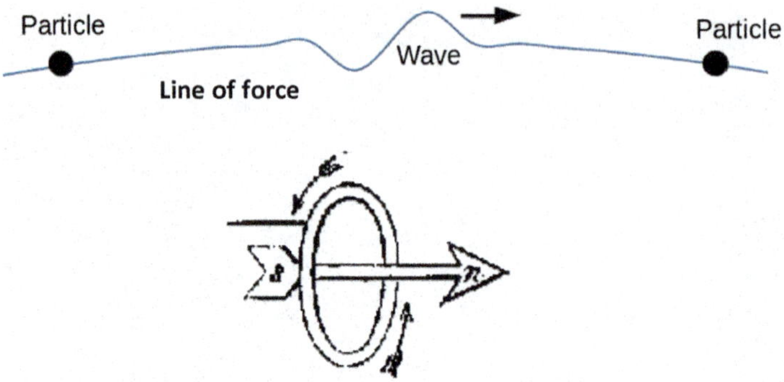

Fig. 15.12 Sketches found in publications by Faraday and Maxwell, how they visualised the propagation of light in an electromagnetic field

Just as remarkable is the upper sketch in Fig. 15.12, drawn by Faraday, showing how he imagined light waves: these would be disturbances propagating along such a magnetic force line. Combining the visualisations by Faraday and Maxwell, we almost arrive at the picture of Fig. 15.11, where the inertial waves running along the vortex core would be analogues of light waves.

In view of the later theory of light, following from the complete equations of Maxwell, these are of course merely beautiful—although cripple—mental pictures, but with considerable historical interest because it shows how the imagination and physical intuition of scientists often helps to develop theory. Unfortunately, the final theories are mostly presented in the textbooks as if they were the outcome of logical and rational, almost mathematically exact thinking, which is often far from the truth.

To return to the fluid model of Brady and Anderson, Fig. 15.11 just served as a preliminary step to explain their model. They investigated whether a disturbance (or "wave packet" in their terminology) moving along a vortex core could be made into a more complete model of a photon. For this purpose such a wave packet must show polarisation, because this is one of the important properties of light. It is also desirable to have polarisation in the model for another reason, because it would then be possible to subject the model to the Bell test, which is almost always performed using the spin of particles or the polarisation of photons.

Brady and Anderson found an example of inertial waves in a rotating fluid showing definite directionality (like the polarisation of light waves), in contrast to the wave packets of the simple vortex with their circular symmetry. The core of such an extended vortex model is shown in Fig. 15.13.

It is not just a theoretical model of a possible flow pattern, what is shown on the right hand side is based on actual experimental results by D.H. Kelley and coworkers, and has been realised in the laboratory (see the chapter "References"). One can thus have vortex flows with a clear "polarisation" in the azimuthal direction, i.e. in a section through the core.

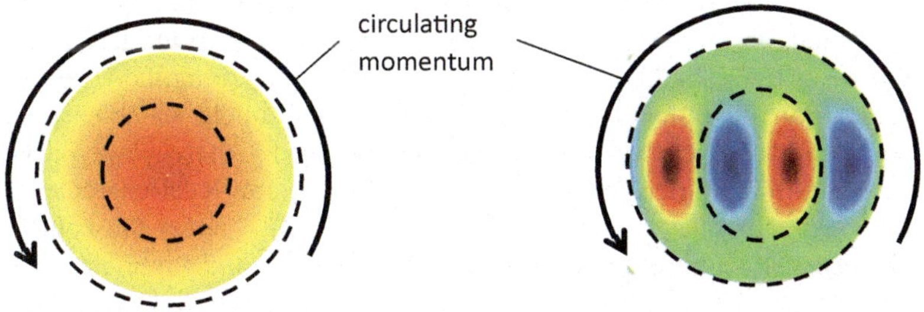

Fig. 15.13 Two possible configurations of a vortex core. *Colours* indicate the pressure variations. On the *left* the simple vortex, on the *right* an extended vortex model with directionality. *Right hand* picture based on experiments by D.H. Kelley e.a. (see Chap. "References")

This kind of vortex flow was modelled in mathematical form, so that it enabled Brady and Anderson to study what would happen if wave packets (now having polarisation) are moving along the vortex line. In particular, they studied a situation similar to that in Fig. 15.10, where a Bell test is performed. It was assumed that two wave packets are sent away from the mid point in opposite directions, and that the test would then be performed. The conclusion was striking: the Bell test would indicate that the fluid analogue of photons (i.e. the polarised wave packets moving along the vortex) is an almost perfect analogue. The experiment with the inertial waves running along such an modified vortex would give results completely in accordance with the predictions of quantum mechanics! The correlations found by the Bell test indicate *non*-local behaviour of the wave packets in the fluid! For the historian too this is an exciting conclusion: the physical intuition of Faraday and Maxwell had set them on the right track when they pondered about the phenomenon of "light".

For people interested in emergent quantum mechanics it also is exciting news: this fluid analogue of photons proves that, given an appropriate model, the classical laws of nature indeed can conspire such that quantum behaviour emerges, even quantum behaviour in its most mysterious form. Of course it is just an analogue, but it now enables us to inspect how this kind of mysterious behaviour can arise. In the fluid analogue based on a vortex flow, this is not difficult to see. The wave packets here form a part of a larger system, there is—so to speak—a form of long-range order impressed by the vortex flow. Often there are a lot of misunderstandings when vortex flows are described. One then talks (and thinks) about this kind of flow as if there is a line of singularities (the "trunk" in Fig. 15.11) which would *cause* the rest of the fluid to rotate. In (too) many textbooks this kind of wrong mental picture is conveyed by saying that the velocities in the field are *induced* by the vortex line. The Biot-Savart law in both fluid mechanics as well as in electromagnetism is likewise wrongly interpreted in the same causal sense, whereas it actually is no more than a quantification of how the velocities are distributed in the *entire* field. "Vortex" is the name of the entire field, it is nonsense to consider the central line of singularities as an individually existing thing. One should not talk about the vortex field of a singularity, but instead about the singularity of a vortex field.

Disturbances in this field, like the wave packets sliding along the vortex line, do not have individuality either, they are intrinsical parts of a field which extends to far distances, in principle out to infinity. This explains why these wave packets are subject to long-range order, and how they can display non-local characteristics.

These conclusions have much relevance for Sect. 15.4 about the behaviour of droplets of *charge* sharing the same near field. The first and foremost conclusion from the Brady-Anderson theory is, that systems purely described by classical laws *can* stand up to the Bell test, and can thus display holistic-like behaviour. The droplet theory of the present book likewise is not *fundamentally* barred from being a "complete" replication of quantum mechanics, including the holistic properties of the real quantum mechanics. The classical laws do not fundamentally stand in the way.

The second conclusion is, that such holistic-like behaviour in classical systems requires special circumstances. It is possible only when there is some form of long-range order present in the system. It was argued in Sect. 15.4 that in that particular case of near field sharing instantaneous actions at a distance may be expected. In other words: something is expected that can be interpreted as a non-local influence on the two separated particles. Can we identify here too the presence of a long-range ordering? Yes, we can easily pinpoint the actor which causes that the two droplets behave as if they still belonged to one and the same larger system. The culprit is the far field, which is identical for two separate particles sharing the same near field. The mathematical "max"-analysis on which everything in this book is based shows that self forces on charges can be calculated from the near field. But the near field in turn depends to a large extent on the structure of the far field. This is so since the latter "injects" into the near field additional terms that would not be present if we would limit our view strictly to the immediate surroundings of the charges. Finally, from the point of view of the far field the two separate objects in the near field are "seen" as one singularity. It can thus be concluded that the "spooky actions at a distance" identified in Sect. 15.4 are in agreement with the findings in the Brady-Anderson model.

Chapter 16
Conclusions: The Direction Towards Einstein's "Hidden Variable"?

What are the conclusions—if any—to be drawn from the collection of mental pictures presented in this book? This chapter is a reflection on the essence of what has been achieved so far.

Of course, as has been stressed on a large number of occasions, strictly speaking no more has been achieved than presenting a picture book. A collection of mental pictures, or—if you prefer: visualisations—of quantum phenomena, hopefully less mysterious than quantum theory "naturel". The intention was emphatically *not*, to present some half-baked analogies, in the form of comparisons—unavoidably imperfect—with better known phenomena in the macroscopic world. You will often find these in popular-scientific books. The comparisons do not have the intention to "explain" quantum phenomena, but to bring these phenomena closer to physical intuition and make them appear less weird. Such analogies are useful ("find this mysterious? Then think of……, which is a little bit similar and is an effect that you are used to"), but they always have limitations. Of necessity such comparisons must be "cripple" and they cannot be anything else, as long as it is accepted that there are two different physics, one only applicable to the macroscopic world and the other describing the atomic world.

In contrast, the mental pictures in the present book were based on a particular model of the electron, and on a systematic mathematical analysis of the consequences of this working hypothesis. Obviously, the character of this exercise is essentially different from attempts to think up arbitrary metaphores. Nonetheless, the ultimate aim of it all was not different, it was no more than to arrive at a better understanding of quantum phenomena, i.e. to bring them closer to the intuition. What is meant by "understanding" is, just as in the approach by metaphores, to have a picture in terms of classical physical laws, so that one may obtain a more intuitive grasp on the stories of quantum theory.

But this officially stated aim of the book might be an easy trick to forego criticisms. In case people find mathematical errors or flaws in the interpretation of the formulae (and such shortcomings will certainly be there), the author now has an easy escape by saying "the basis under picture no….. in Chap. 14 may have fallen

T. van Holten, *The Atomic World Spooky? It Ain't Necessarily So!*
DOI 10.2991/978-94-6239-234-2_16

away, but nevertheless it is a helpful picture (like the metaphores in other popular-scientific books)". However, this would not be entirely honest. The *suggestion* given by the approach chosen in this book is definitely that an alternative theory *might* be feasible concerning quantum mechanic phenomena. Of course, the book does not have the pretention that it already can present a complete alternative theory. And of course, it cannot be ruled out that this alternative crumbles to nothing upon continued attempts to make it a full fledged theory. Nevertheless, at the present state of affairs, it cannot be denied that there is a strong suggestion in the book that a possible alternative for the usual quantum theory is thinkable. To push things to the extreme: *what if* the droplet theory in the end (after a lot more work has been done on it, and no doubt a lot of corrections have been made to it) would appear to be *not* all nonsense? *What if* some grains of truth were found in the model of electrons as vibrating droplets instead of singularities? What would then become the view on the usual quantum mechanics?

Quantum theory as it stands is an almost perfect theory, although it "feels strange". Without doubt quantum theory is an excellent, extremely accurate tool to quantify what happens in the atomic world, and it is in complete agreement with all the available experiments. Therefore, one may better ask whether the view on classical mechanics would change if it would appear that it too can give an insight into the atomic world. No longer would it be justified to look upon quantum theory and classical physics as two separate, sometimes conflicting, branches of physics.

Remarkably, quantum theory itself does *not* lead to the notion that there are two entirely different kinds of physics. It is almost never mentioned—at least not in the popular-scientific books—that Newton's dynamic law $F = m \cdot a$ can actually be derived from Schrödinger's equations. This is called Ehrenfest's theorem. The name of Ehrenfest is too seldom mentioned in the popular-scientific literature, which is a shame in the author's opinion. Let alone that his important theorem gets the attention it deserves.

Furthermore, there is the better known "correspondence principle"—something Bohr often used to advantage—which says that in the limit of large energies all the quantum phenomena should be in agreement with the classical laws of physics, despite the conflicts on the smallest scales.

On the basis of Ehrenfest's theorem and the correspondence principle one would be tempted to look upon the classical laws as a special case of the more general quantum theory. There would be a ranking of the theories: quantum theory on top, and a sub-class of it would be classical physics, the latter valid only under special circumstances.

This kind of ranking would then be similar to the relation between the special theory of relativity and Newton's law of motion. Special relativity theory under particular circumstances narrows down to Newton's dynamical law $F = m \cdot a$. Einstein's formulation is the more general, but for small velocities (relative to the velocity of light) Newton's law—with a constant value of the mass—is recovered from it. The hierarchy of the theories is clear in this case.

On the other hand, the hierarchy in the case of quantum theory versus classical physics is not so clear. It may be so that classical laws may be derived from quantum theory, the classical theory has more to offer in some respects. Classical physics is not restricted to probabilistic predictions, for instance. It does predict the path of celestial bodies with large accuracy, instead of being limited to vague predictions, e.g. about the general region of space where the Moon can probably be found.

Furthermore, the classical laws imply strict conservation of energy, whereas quantum mechanics does not. The energy-time uncertainty, and the expression that sometimes "energy can be borrowed from the uncertainty" is a disguised way to concede that the theory admits violations of the law of energy conservation.

Of more fundamental importance is, that the classical theory is strictly causal, which quantum theory is often not. In classical physics every effect is considered to have a definite cause, and mostly this cause can be pinpointed very precisely. Effects without any cause, i.e. just the unpredictable results of a sort of lottery, do not occur in the classical domain. On the other hand, they do in quantum physics. It is one of the things that Einstein found so irritating (and really repulsive and unacceptable) in quantum theory. Certainly, a ranking of the two theories where the lower ranked theory (classical physics) derives everything from the higher ranked one is not tenable. Where would the causality of classical physics come from, if classical physics is just a "sub-class"?

At this point the droplet theory of this book might have something to offer (if it were proven correct). It is an attempt to describe quantum phenomena not only in terms of classical physics, but also—intimately related to this—as causal sequences of events. Certainly not everywhere these attempts have yet been successful: this is an area that has to be improved by future efforts, and on which much of the validity of the theory will depend. However, it is inherent in a "what if" story that the validity of the theory is assumed as an a priori. So, if the droplet theory in future will be able to describe all the known quantum effects in terms of classical laws, it is inherent that it can construct *causal* sequences of events for everything. The causality in quantum phenomena would then have been restored by the droplet theory. Seemingly non-causal effects, events in quantum mechanics that would appear to be determined by pure chance, can then be given a causal explanation by the droplet theory. Examples of this have already been given in many places of this book.

The relative ranking of quantum theory and classical physics would then no longer be hierarchical, but they would instead both have a shared place at the top of the ranking. The mutual relation of the "two separate physics" would then be comparable to the relation thermodynamics and kinetic gas theory have to each other. Both these theories are valid, one of them (thermodynamics) excellent for making technical sums, the other (kinetic gas theory) more useful to show that gases conform to general dynamics theory. As a shortcut thermodynamics introduces quantities meaningful for large conglomerates of molecules (like pressure, or entropy) that do not have a meaning for single molecules. However, working upwards from the motions of single molecules towards large conglomerates, kinetic gas theory reveals the relation of quantities like pressure or entropy with general dynamics. The two theories are complementary, and certainly not mutually exclusive.

Likewise, quantum theory and classical physics might be complementary to each other (from macroscopic down to atomic scales), instead of mutually excluding at the smallest scales. Is this not what Einstein would have wanted to see when he called for a search of the "hidden variable" in quantum theory? The classical theory would then be able to bring causality back to the quantum world. No longer would one have to live with the idea that "Der lieber Herrgott würfelt" (to paraphrase a bit). In practice it would not have any consequence, unfortunately, because causality does not imply predictability. The chaotic motion of droplets, for instance, leads to poor predictability in practice, but this kind of motion is in principle the result of strictly causal processes. The kind of unification of quantum theory and classical theory that would have been achieved perhaps does not have more than a philosophical value.

The more concrete advantage obtained in case the "what if" scenario would come true is perhaps found in bringing science to the general public. It does not help (except some people's ego's), when physicists say that the atomic world is mysterious and can only be understood by plunging into the mathematical descriptions.

If we have valid mental pictures in terms of everyday physics, for the interested layman this is a real advantage. Because we humans live in a macroscopic world, where we are used to the classical laws of nature, we naturally like to see everything "explained" in terms of everyday physics. It is closer to our intuition, and we feel more comfortable with it. We would even tend, from this point of view, to rank classical theory higher than the "unheimisch" feeling quantum theory. We would like to see that quantum theory is the derived—and thus lower ranked—one, so that we can claim that ours is the "reality".

This would certainly help to popularise physics although, examining it more carefully, such a feeling is not rational. An important part of the set of classical laws consists of Maxwell's equations. They describe interactions between objects that are not in contact with each other, objects separated by a void space in between them. They influence each other through nothing tangible. Although the artificial concept of "fields" seems to help, this is mainly semantics, after Einstein showed that something like "ether" as the carrier of electromagnetic waves is not a meaningful idea. Therefore, are such electromagnetic interactions not more "spooky" than the view of photons flying between objects (even if the interpretation as a sort of bullets is wrong)?

The author prefers to leave it open questions, and is content with having compiled a picture book. Hopefully, the pictures in it at least do have an esthetic value.

Appendix A
General

A.1 Configuration and Notations

Figure A.1.1 shows the configuration studied as well as the most important notations used to define the geometry. The figure is no indication of the actual relative proportions. The sketched droplet of charge moving along the Z-axis would in the case of an electron correspond to a time-averaged view of a "zittering" electron. The dimensions of the-in this way observed- charge distribution are very much smaller than the width of the potential well in which it is moving, although the sketch A.1.1 suggests differently. Actually, the scale of the time averaged "blob" of charge and the scale of the potential well are of entirely different orders of magnitude.

The XYZ-axis system (Fig. A.1.1) is an inertial system. The position of the potential well is fixed w.r.t. this system. The potential well is in Fig. A.1.1 symbolized by a barrier, not necessarily to be interpreted as a hard barrier (as in a potential box), it may be "soft" as in a parabolic well.

The "droplet of charge" moves in Z-direction w.r.t. the inertial system XYZ, but remains centred around the Z-axis. The velocity of its elements is purely in Z-direction (this is what is meant by "one-dimensional" assumption; the electromagnetic field associated with the droplet is nevertheless 3-D). The charge distribution is assumed to be elongated, so much so that "slender body theory" may be applied to analyse the electromagnetic field around it (i.e. $r/s \ll 1$). Furthermore, the charge distribution is taken to be axi-symmetric w.r.t. the Z-axis, and also symmetric w.r.t. a transverse plane through the mid-point of the distribution (the equator plane).

The centre of mass and centre of charge, defined in the usual way, both coincide with this mid-point thanks to these symmetries. The translation of the droplet is defined as the motion in Z-direction of this mid-point, and is thus described by the Z-coordinate $z_m(t)$ of the mid-point, with velocity $v_m(t) = \dot{z}_m$, and acceleration, jerk etc. defined likewise.

The length of the droplet is denoted by $s(t)$. This is the only extra degree of freedom taken into account next to the translation, and is called the "pulsation". The pulsation velocity is defined as $V_{puls}(t) = \dot{s}$, and pulsation acceleration, pulsation jerk etc. are defined likewise.

T. van Holten, *The Atomic World Spooky? It Ain't Necessarily So!*
DOI 10.2991/978-94-6239-234-2

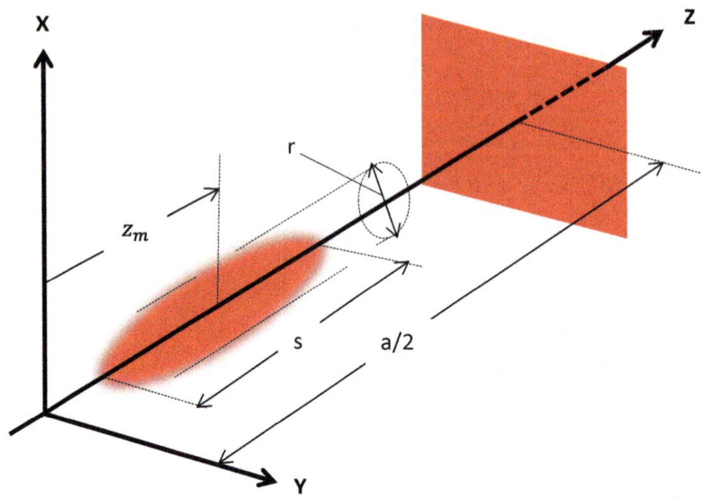

Fig. A.1.1 Notations

The sketch in Fig. A.1.1 depicts the most general distribution of charge density σ, variable throughout the droplet. As sketched, the droplet might rather be called a "cloud" of charge. However, in order to simplify the mathematical analysis, a constant charge density has been assumed. This implies that the boundaries of the droplet are sharp, hence the indication "droplet of charge". Essentially, an "incompressible" droplet has been assumed in this way, implying that a change in length s is always accompanied by a change in thickness r. Strictly speaking, the lateral velocities of the charge elements cannot be zero if $\dot{s}$ is non-zero. However, these lateral velocities are antisymmetric w.r.t. the Z-axis, and therefore have a small influence on the electromagnetic field, except at distances of the order r. Most importantly, the assumption of a slender configuration means that the lateral velocities have an order $O(r/s)$ with respect to the longitudinal velocity $\dot{s}$. In the present first order analysis, the influence of the lateral velocities on the electromagnetic field will be neglected. Or, stated the other way round: we assume the slenderness so large that to the presently desired accuracy the influence of the lateral velocities may be neglected. Under this assumption the velocities that are relevant for the electromagnetic field are only the velocities in Z-direction, which is the same as the assumption of one-dimensionality.

The Z-velocities of the elements of the droplet are assumed to depend only on their lengthwise position in the droplet, the velocity thus does not depend on their lateral position.

The pulsation is taken as a quasi-static lengthening or shortening of the droplet. This means that, looking in a reference frame moving with the droplet, with its origin fixed in the mid-point, during an elongation the displacement of the elements of the droplet is proportional to their distance from the mid-point.

A specific element of the droplet is characterised by its lateral position w.r.t. the Z-axis (its x- and y-coordinate) and by its *relative* position w.r.t. the midpoint of the droplet: the ratio $\frac{z_{elem}(t)-z_m(t)}{s/2(t)}$ is constant for a given charge element, despite the variable elongation $s(t)$ of the droplet.

This leads to the expression for the velocity of an element:

$$\begin{aligned} v_{element}(t) &= v_m(t) + \frac{z_{elem}(t) - z_m(t)}{s/2(t)} \left(\frac{\dot{s}}{2}\right) \\ &= v_m(t) + const.\dot{s}(t) \quad (Lagrangian\, description) \end{aligned} \tag{A.1.1}$$

The terminology "Lagrangian description" refers to the fact that we here follow a specific element, and specify its velocity as a function of time, its Z-position not explicitly occurring in the specification of velocity (as clearly seen in the second line of (A.1.1) where the constant is just the specification of which element is considered).

A given charge element with an amount of charge dq is considered. If the complete droplet is built up from a given, fixed number of elements, each of these elements must have a constant charge dq.

A Lagrangian type of analysis results in a time history $z_{elem}(t)$ for a specific element. The Lagrangian description will be used in Chaps. C.1–C.5, where the electromagnetic field of a single, moving point charge is constructed. The value of the potentials of the whole droplet in a point (x, y, z) at time t should then be determined by summing the contribution of all the point charges. The integration should thus extend over all the possible values of $\frac{z_{elem}(t)-z_m(t)}{s/2(t)}$, i.e. from $\frac{z_{elem}(t)-z_m(t)}{s/2(t)} = -1\, to\, 1$ if the Lagrangian type of description is used.

The integration over all the elements of the droplet is, however, more conveniently performed by taking the Eulerian point of view to describe the field. In this case we again determine the field induced in a certain point (x, y, z) of the inertial system (this in itself is not different from the Lagrangian type of analysis), but now do it as it is "induced" by an amount of charge that happens to pass the position (x_0, y_0, z_0) in the inertial system. The amount of charge passing this fixed position is $\sigma(x_0, y_0, z_0).dx_0.dy_0.dz_0$ and belongs at every instant of time to a different element of the droplet.

The velocity field in the Eulerian description is denoted as $v(x_0, y_0, z_0, t)$. Again it is the velocity of an element of charge that happens to pass at time t the point (x_0, y_0, z_0) and, therefore, it is at different instants of time the velocity of different elements of charge.

In the case of the assumed deformation modes of the complete droplet (only translation and pulsation in Z-direction), and taking into account that the velocities do not depend on the lateral coordinates, the velocity field is thus described by (using a Galileian, non-relativistic summing of velocities):

$$v(z_0, t) = v_m(t) + \frac{z_0 - z_m(t)}{s/2(t)} (\dot{s}/2) \quad (Eulerian\, description) \tag{A.1.2}$$

with the coordinate z_0 fixed. As long as the point (x_0, y_0, z_0) lies inside the boundaries of the droplet, the charge density (which in the most general case would be $\sigma(x_0, y_0, z_0, t)$) is, due to the assumption of constant density, independent of the position and time. When we move along with a given charge element, its change of density in time is the socalled substantial differential of the density, denoted as $D\sigma/Dt$, i.e. the sum of the local and the convective change of density:

$$\frac{D\sigma}{Dt} = \frac{\partial\sigma(z_0, t)}{\partial t} + v(z_0, t).\frac{\partial\sigma(z_0, t)}{\partial z_0} \quad (\textit{Eulerian description}) \tag{A.1.3}$$

From the fact that the density in the field does not depend on either position or time, it thus follows that

$$\frac{D\sigma}{Dt} = 0 \quad (\textit{moving with a specific charge element}) \tag{A.1.4}$$

Moving along with a given charge element, the total charge $\sigma(x_0, y_0, z_0).dx_0.dy_0.dz_0$ of such an element is therefore constant. We may identify such an element with the infinitesimally small point charges occurring in the analysis based on the Lagrangian description.

The acceleration of a given charge element is in the Eulerian description given by the substantial differential of the velocity, denoted as Dv/Dt. It is the sum of the local and the convective change of velocity with time:

$$\frac{Dv}{Dt}(z_0, t) = \frac{\partial v(z_0, t)}{\partial t} + v(z_0, t).\frac{\partial v(z_0, t)}{\partial z_0} \quad (\textit{Eulerian description}) \tag{A.1.5}$$

Substituting the assumed form (A.1.2) of the velocity field $v(z_0, t)$ yields

$$\begin{aligned}\frac{Dv}{Dt} &= \dot{v}_m + \frac{z_0 - z_m}{s/2}.\ddot{s}/2 - \frac{v_m}{s/2}.\dot{s}/2 - \frac{z_0 - z_m}{s/2}.\left(\frac{\dot{s}}{s}\right)^2 + \left(v_m + \frac{z_0 - z_m}{s/2}\dot{s}/2\right).\left(\frac{\dot{s}}{s}\right) \\ &= \dot{v}_m + \frac{z_0 - z_m}{s/2}.\ddot{s}/2\end{aligned} \tag{A.1.6}$$

which corresponds with the acceleration found by differentiating the Lagrangian form of the velocity.

In a similar way the jerk of a given charge element is, when the Eulerian description is used:

$$\frac{D^2v}{Dt^2}(z_0, t) = \frac{\partial}{\partial t}\left(\frac{Dv}{Dt}\right) + v(z_0, t).\frac{\partial}{\partial z_0}\left(\frac{Dv}{Dt}\right) \quad (\textit{Eulerian description}) \tag{A.1.7}$$

A.2 Field Equations to Be Satisfied

The electromagnetic field, specified by the electric field strength $\underline{E}(x, y, z, t)$ and the magnetic field strength $\underline{B}(x, y, z, t)$ is not one-dimensional, even though the velocity $v(z, t)$ of the charge elements $\sigma(x, y, z, t).dx.dy.dz$ has been taken to be in Z-direction only. We have to use the complete Maxwell equations.

Maxwell's equations in vacuum read (following the notations used in Feynman's Lectures on Physics, except the charge density which is here denoted by σ):

$$\begin{aligned} div\underline{E} &= \frac{\sigma}{\varepsilon_0} \\ rot\underline{E} &= -\frac{\partial \underline{B}}{\partial t} \\ div\underline{B} &= 0 \\ c^2.rot\underline{B} &= \frac{\underline{j}}{\varepsilon_0} + \frac{\partial \underline{E}}{\partial t} \end{aligned} \tag{A.2.1}$$

where according to the force law

$$\underline{F} = q(\underline{E} + \underline{v} \times \underline{B}) \tag{A.2.2}$$

the electric field strength $\underline{E}$ is the force on a *static* unit charge placed in the electro-magnetic field, and the magnetic field strength $\underline{B}$ is defined by the additional force on the unit charge when it is moving with velocity $\underline{v}$, which additional force equals $\underline{v} \times \underline{B}$.

In Maxwell's equations σ equals the (volume) charge density distribution $[C/m^3]$, and $\underline{j}$ is the current density distribution in the field $[C/(s.m^2]$.

In the following the boundary value problem for the present problem will be formulated in terms of the potentials of the field. The (magnetic) vector potential $\underline{A}$ is defined by its relation to the magnetic field strength:

$$\underline{B} = rot\underline{A} \tag{A.2.3}$$

The scalar (electric) potential Φ is defined by the relation

$$\underline{E} = -grad\Phi - \frac{\partial \underline{A}}{\partial t} \tag{A.2.4}$$

These two equations relate the potentials to the measurable quantities $\underline{E}$ and $\underline{B}$. However, in order to avoid any ambiguity, another relation between the potentials should also be satisfied (the gauge condition):

$$div\underline{A} = -\frac{1}{c^2}\frac{\partial \Phi}{\partial t} \tag{A.2.5}$$

The field equations satisfied by the potentials are:

$$\begin{aligned} \nabla^2 \Phi - \frac{1}{c^2}\frac{\partial^2 \Phi}{\partial t^2} &= -\frac{\sigma}{\varepsilon_0} \\ \nabla^2 \underline{A} - \frac{1}{c^2}\frac{\partial^2 \underline{A}}{\partial t^2} &= -\frac{1}{c^2}\frac{\underline{j}}{\varepsilon_0} \end{aligned} \tag{A.2.6}$$

where the symbol $\nabla^2 \Phi$ stands for the scalar Lagrangian $\nabla^2 \Phi = div(grad\Phi)$ which, written out in Cartesian coordinates, is $\nabla^2 \Phi = \frac{\partial^2 \Phi}{\partial x^2} + \frac{\partial^2 \Phi}{\partial y^2} + \frac{\partial^2 \Phi}{\partial z^2}$.

Caution should be exercised when this same symbol ("nabla squared") operates on a vector. In this case the vector Lagrangian is defined as $\nabla^2 \underline{A} = grad(div\underline{A}) - rot(rot\underline{A})$. Only in Cartesian coordinates (and only then) this equals $\nabla^2 \underline{A} = \underline{i}\nabla^2 A_x + \underline{j}\nabla^2 A_y + \underline{k}\nabla^2 A_z$.

The solution of the field equations for the potentials will always have the structure

$$\begin{aligned} \Phi(1,t) &= \frac{1}{4\pi\varepsilon_0}\int \frac{\sigma(2, t - r_{12}/c)}{r_{12}} dV_2 \\ \underline{A}(1,t) &= \frac{1}{c^2}\frac{1}{4\pi\varepsilon_0}\int \frac{\underline{j}(2, t - r_{12}/c)}{r_{12}} dV_2 \end{aligned} \tag{A.2.7}$$

where the potentials are determined in the point 1 at time t, and the charge density as well as the current density "inducing" the potentials are located in the point 2 at an earlier (retarded) time $t - r_{12}/c$, with r_{12} the distance between the points 1 and 2. The integrals have to be taken over the whole volume where the charges and currents are located.

A.3 The Boundary Value Problem for the Potentials

The model of the slender axi-symmetric charge distribution, moving and deforming in Z-direction only, implies that the potentials will be axi-symmetric around the Z-axis. In addition, since the charge velocity is parallel to the Z-axis, according to the above given integral representation of the field the vector potential $\underline{A}$ can only have a non-zero component A_z in Z-direction, with $A_x = A_y = 0$.

The potentials thus will be the axi-symmetric solution of the equations

$$\nabla^2 \Phi - \frac{1}{c^2}\frac{\partial^2 \Phi}{\partial t^2} = -\frac{1}{\varepsilon_0}\sigma(x, y, z, t) \tag{A.3.1}$$

$$\nabla^2 A_z - \frac{1}{c^2}\frac{\partial^2 A_z}{\partial t^2} = -\frac{1}{c^2}\frac{1}{\varepsilon_0}\sigma(x,y,z,t).v(z,t) \tag{A.3.2}$$

Outside the boundaries of the droplet of charge the potential $\Phi(x,y,z)$ and the component $A_z(x,y,z)$ of the vector potential will everywhere satisfy the homogeneous wave equation:

$$\begin{aligned} \nabla^2\Phi - \frac{1}{c^2}\frac{\partial^2\Phi}{\partial t^2} &= 0 \\ \nabla^2 A_z - \frac{1}{c^2}\frac{\partial^2 A_z}{\partial t^2} &= 0 \end{aligned} \tag{A.3.3}$$

Furthermore there is a boundary condition at infinity, where both the potentials are required to vanish:

$$\begin{aligned} \Phi \to 0 \quad for \quad \sqrt{x^2+y^2+z^2} \to \infty \\ A_z \to 0 \quad for \quad \sqrt{x^2+y^2+z^2} \to \infty \end{aligned} \tag{A.3.4}$$

Appendix B
Matched Asymptotic Expansion Procedure (Principles)

B.1 Asymptotic Approximation of the Near Field

The Eulerian description shows clearly some of the geometric characteristics of the potential fields. Close to the droplet, there are two different characteristic length scales in the electromagnetic field, a small scale of the order of the "thickness" r of the droplet, and a larger scale of the order of the "length" s. Consider first the most general case where, apart from the assumption of a slender distribution $(r/s << 1)$, no other assumptions are made yet concerning the shape of the distribution (no symmetries assumed). In general the length scale of the droplet may have any value as long as it is smaller than the width of the potential well, the shape may thus be varying from a point charge to the other extreme of a charge filling the entire potential well like a sloshing fluid in a slender bath tub.

The situation is schematically shown in Fig. B.1.1. The charge density sketched in this figure is the *linear* density (charge per unit length in Z-direction), where by "charge" is meant the charge in infinitesimally thin discs perpendicular to the Z-axis.

The typical length scale that concerns us first of all is the scale in Z-direction, which is at most of the order of a, i.e. the distance between the "barriers" of the potential well. In order to find an approximate solution to the boundary value problem, we introduce the following physical assumption: *in the close vicinity of the moving charge the spatial variations of the electric potential have at most a characteristic length scale of the order of a.*

Evidently, this assumption about the field can be valid only in the *near field*, i.e. the field at distances from the origin of the order a. Further out, away from the origin, this assumption is not tenable.

In the near field the largest possible characteristic scale is thus equal to a. Evidently, the assumption of slenderness of the droplet also implies that in a direction perpendicular to the Z-axis (in the following indicated as the "radial direction") the characteristic length scale is smaller. The difference in these two length scales close to the droplet will be used later, when "slender body theory" is applied to simplify the field expressions.

What concerns us for this moment is that there is a third length scale, which is relevant at large distances from the droplet. This is the length scale of the

T. van Holten, *The Atomic World Spooky? It Ain't Necessarily So!*
DOI 10.2991/978-94-6239-234-2

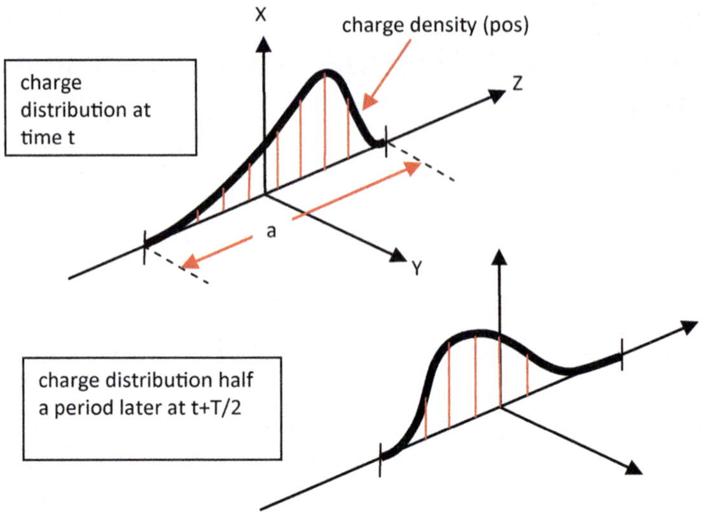

Fig. B.1.1 The amount of charge in lateral slices through the droplet, variable with position and time (the extreme case of a "sloshing bath tub" is drawn)

electromagnetic waves induced by the variable charge distribution. A typical wave length λ will be determined by the *time* scale of the shape variations of the charge distribution.

We assume that the time variations in the field have a characteristic scale of the order T, that is the average time the droplet needs to complete a "cycle" (although the motion is not necessarily harmonic). Rewritten in terms of the characteristic coordinates x/r, y/r, z/a and t/T the wave equation for the electric potential reads:

$$\frac{\partial^2\Phi}{\partial(x/r)^2}(a/r)^2+\frac{\partial^2\Phi}{\partial(y/r)^2}(a/r)^2+\frac{\partial^2\Phi}{\partial(z/a)^2}=\frac{a^2}{c^2T^2}\frac{\partial^2\Phi}{\partial(t/T)^2}=(a/\lambda)^2\frac{\partial^2\Phi}{\partial(t/T)^2} \tag{B.1.1}$$

where $\lambda = cT$ is a characteristic wavelength in the electromagnetic field induced by the motion of the charge distribution. On account of the physical assumptions mentioned above, the partial derivatives in (B.1.1) are all of the same order of magnitude.

It follows immediately from (B.1.1) that the electric potential satisfies a Laplace equation $\nabla^2\Phi = 0$ in case the ratio a/λ is very small $(a/\lambda \to 0)$. The parameter a/λ will be used as perturbation parameter in an asymptotic expansion. An equivalent expansion parameter, which will be used as well in the following developments is the parameter v/c, a typical velocity of the charge elements compared with the speed of light. The equivalence between the expansion parameters a/λ and v/c may be seen as follows:

$$v/c = \frac{1}{c}\frac{dz}{dt} = \frac{d(z/a)}{d(t/T)}\frac{a}{\lambda} \tag{B.1.2}$$

where the ratio $\frac{d(z/a)}{d(t/T)}$, expressed in terms of the characteristic coordinates, is of order unity.

The zero'th order approximation of the near field in which the wave equation (implying retardation effects) reduces to a Laplace equation (implying no retardations) will be called the *quasi-static* electric potential of the charge. Considering the equivalence of the expansion parameters a/λ and v/c this is understandable.

One may refine the analysis by writing the near field in the following form:

$$\Phi(x,y,z,t) = \Phi_0 + (a/\lambda)\Phi_1 + (a/\lambda)^2\Phi_2 + (a/\lambda)^3\Phi_3 + \ldots \quad for \quad (a/\lambda) \to 0 \tag{B.1.3}$$

This is an asymptotic expression, in which the first term is the quasi-static field, whereas the other terms describe the way in which the field becomes quasi-static if the ratio (a/λ) becomes smaller and smaller. Substituting the assumed type of solution (B.1.3) into (B.1.1):

$$\begin{aligned}&\left(\frac{\partial^2}{\partial(x/r)^2}\left(\frac{a}{r}\right)^2 + \ldots\right)\left[\Phi_0 + (a/\lambda)\Phi_1 + (a/\lambda)^2\Phi_2 + \ldots\right] \\ &= (a/\lambda)^2\frac{\partial^2}{\partial(t/T)^2}\left[\Phi_0 + (a/\lambda)\Phi_1 + (a/\lambda)^2\Phi_2 + \ldots\right]\end{aligned} \tag{B.1.4}$$

If Φ_0 has been determined such that it satisfies $\nabla^2\Phi_0 = 0$ this equation is reduced to:

$$\left(\frac{\partial^2}{\partial(x/r)^2}\left(\frac{a}{r}\right)^2 \ldots\right)[\Phi_1 + (a/\lambda)\Phi_2 + \ldots] = (a/\lambda)\frac{\partial^2}{\partial(t/T)^2}\left[\Phi_0 + (a/\lambda)\Phi_1 + (a/\lambda)^2\Phi_2 + \ldots\right] \tag{B.1.5}$$

On taking once again the limit $(a/\lambda) \to 0$ it follows that the field Φ_1 also has to satisfy the Laplace equation. This process of repeatedly taking the limit $(a/\lambda) \to 0$ yields the following result: even when terms of order $O(a/\lambda)$ are included, the field still satisfies the quasi-static Laplace equation

$$\nabla^2\Phi = 0 \quad up\ to \quad O(a/\lambda) \tag{B.1.6}$$

In the next higher order approximations, where Φ_2 and Φ_3 are also included, the near field should satisfy a Poisson equation:

$$\nabla^2 \Phi = \frac{1}{c^2}\frac{\partial^2 \Phi_0}{\partial t^2} \quad \textit{up to} \quad O(a/\lambda)^3 \tag{B.1.7}$$

where $\Phi_0(x, y, z, t)$ is the quasi-static field determined by (B.1.6).

All this concerns the *field equations* to be satisfied in the near field. The *boundary conditions* to be satisfied by the near field are the boundary conditions mentioned above for the complete problem (Chap. A.3), *except* the conditions at infinity, since the near field approximation is not valid far from the origin. The set of boundary conditions is thus incomplete, and there will be some amount of indeterminacy in the near field solution.

The near field will consist of the particular solution of the Poisson equation (B.1.7), plus solutions of the Laplace equation. Evidently, these additional terms—solutions of the Laplace equation $\nabla^2\Phi = 0$—must at least include the quasi-static field since the quasi-static field sees to it that the boundary conditions are satisfied, i.e. it has the required singularities in the region of the droplet. However, there might also be terms present which do satisfy Eq. (B.1.6) but do not vanish at large distances. Such terms cannot be ruled out a priori (because of the missing conditions at infinity), but neither can they be determined by the available boundary conditions. In fact, it will appear later that such terms indeed do arise, and even play an essential role in the analysis.

B.2 Determining the Far Field

In order to remove the indeterminacy, an additional requirement will be used. This requirement, the socalled *matching condition*, can be applied if another asymptotic approximation has been determined which is valid in the *far field*. The far field is the region of space which is more remote from the origin, i.e. the region at distances of the order λ and larger, up to infinity. See the sketch in Fig. B.2.1.

Here a different approximation can be made. The field equation cannot be simplified, and is the full wave equation:

$$\nabla^2 \Phi = \frac{1}{c^2}\frac{\partial^2 \Phi}{\partial t^2} \tag{B.2.1}$$

In contrast, in the far field the charge configuration can be simplified.

The asymptotic limit $a/\lambda \to 0$ implies, on the scale of the far field, that the dimension a shrinks to zero. Consequently, the far field becomes the electromagnetic field of a set of multipoles in the origin. As will be seen the far field is, like the

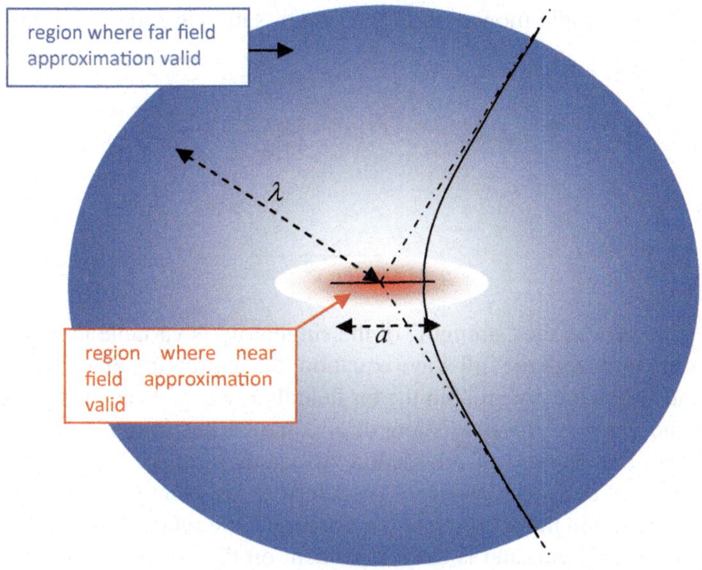

Fig. B.2.1 Sketch of the definition of near- and far field. The drawn line is a hyperbola through a part of the droplet, approaching in the far field a radius emanating from the origin

near field, neither completely determined, since it misses the "inner" boundary conditions. The number as well as the type of poles representing the far field are thus a priori unknown. Again, they can be determined by the *matching* condition.

The far field will be expressed in terms of spherical coordinates, as sketched in Fig. B.2.2.

Fig. B.2.2 Spherical coordinates used in the far field

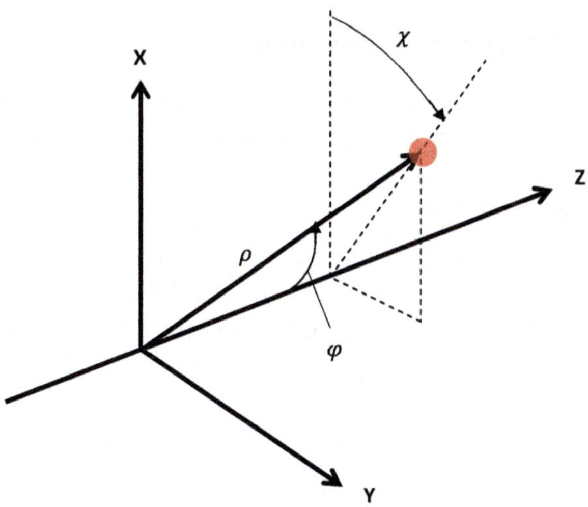

In general, a variable monopole in the origin, satisfying the wave equation will have the field

$$\Phi_{monopole}(x,y,z,t) = -\frac{1}{4\pi\varepsilon_0}\frac{q(t-\rho/c)}{\rho} \tag{B.2.2}$$

together with

$$\rho = \sqrt{x^2+y^2+z^2} \tag{B.2.3}$$

In this general case the "strength" of the singularity is variable in time and, since we are dealing now with the full wave equation, the retardation effect has to be taken into account. When the potential in the far field $\Phi(x,y,z,t)$ is calculated, one should base it on the value of the pole strength at some earlier time $t-\Delta t$, the socalled retarded time. The amount of retardation Δt equals the time taken by a signal to travel the distance between the source (which from the point of view in the far field has a fixed position in the origin) and the observer. The retardation equals $\Delta t = \rho/c$ with c the speed of light, and is only dependent on the position of the field point.

In the case at hand the total charge of the droplet is constant, and the monopole field will reduce to its static form:

$$\Phi_{monopole}(x,y,z) = -\frac{1}{4\pi\varepsilon_0}\frac{q}{\rho} \quad (\text{constant q}) \tag{B.2.4}$$

A dipole in the origin with variable dipole-moment in z-direction $m(t)$ is accompanied by the retarded field

$$\Phi_{z-dipole}(x,y,z,t) = -\frac{1}{4\pi\varepsilon_0}\frac{\partial}{\partial z}\left[\frac{m(t-\rho/c)}{\rho}\right] \tag{B.2.5}$$

Performing the differentiation:

$$\Phi_{z-dipole}(x,y,z,t) = \frac{1}{4\pi\varepsilon_0}\left[\frac{1}{c}\frac{\dot{m}(t-\rho/c)}{\rho^2}z + \frac{m(t-\rho/c)}{\rho^3}z\right] \tag{B.2.6}$$

together with

$$\rho = \sqrt{x^2+y^2+z^2} \tag{B.2.3}$$

Alternatively, completely expressed in terms of polar coordinates:

$$\Phi_{z-dipole}(\rho,\varphi,t) = \frac{1}{4\pi\varepsilon_0}\left[\frac{1}{c}\frac{\dot{m}(t-\rho/c)}{\rho} + \frac{m(t-\rho/c)}{\rho^2}\right].cos\varphi \qquad \text{(B.2.7)}$$

The latter form shows more clearly than the first form (B.2.6) how fast the different components of the potential fall off with distance.

The fields of other multipoles are similarly denoted, for instance

$$\Phi_{xz-quadrupole} = -\frac{1}{4\pi\varepsilon_0}\frac{\partial^2}{\partial x\partial z}\left[\frac{n(t-\rho/c)}{\rho}\right] \qquad \text{(B.2.8)}$$

and these fields may be written out in the same way as shown for the dipole.

One of the things needed later is, how these multipole fields behave close to the origin, i.e. how they behave for small retardation times ρ/c. One may use a Taylor expansion to find this behaviour. This will be called the "inner expansion of the far field". Taking as an example again the z-dipole, the dipole moment is expanded as

$$m(t-\rho/c) = m(t) - \frac{\rho}{c}\dot{m}(t) + \frac{1}{2}\left(\frac{\rho}{c}\right)^2\ddot{m}(t) - \frac{1}{6}\left(\frac{\rho}{c}\right)^3\dddot{m}(t)\ldots\ldots \qquad \text{(B.2.9)}$$

whereas

$$\dot{m}(t-\rho/c) = \dot{m}(t) - \frac{\rho}{c}\ddot{m}(t) + \frac{1}{2}\left(\frac{\rho}{c}\right)^2\dddot{m}(t) + \ldots\ldots \qquad \text{(B.2.10)}$$

Substituting these expansions into

$$\Phi_{z-dipole}(\rho,\varphi,t) = \frac{1}{4\pi\varepsilon_0}\left[\frac{1}{c}\frac{\dot{m}(t-\rho/c)}{\rho} + \frac{m(t-\rho/c)}{\rho^2}\right].cos\varphi \qquad \text{(B.2.7)}$$

gives

$$\Phi_{z-dipole}(\rho,\varphi,t) \overset{inner\ expansion}{\longrightarrow} \frac{1}{4\pi\varepsilon_0}\left[\frac{m(t)}{\rho^2} - \frac{1}{2}\frac{\ddot{m}(t)}{c^2} + \frac{1}{3}\rho\frac{\dddot{m}(t)}{c^3} + \ldots\right]cos\varphi \qquad \text{(B.2.11)}$$

It is seen that the leading term in the inner expansion is a quasi-static dipole field. The other terms do not have the structure of multipoles, in fact they are not singular.

Such inner expansions of the multipole far field will be seen to be needed for the purpose of matching the near- to the far field. The matching condition will next be considered.

B.3 Construction of a Composite Field

The final goal is to find an approximation for the entire field, covering the whole region from the origin to infinity. It is possible to construct such a field, using the near- and far field as building blocks. This is called the *composite field*, having the property that the near- and far field are smoothly embedded in it.

The composite field is formed by summing the near- and far field, and subtracting a "correction field", called the *common part*:

$$\Phi_{composite} = \Phi_{near} + \Phi_{far} - \Phi_{common} \tag{B.3.1}$$

The common field has the task to remove any "double countings". For instance, when considering a point which is close to the origin, the potential there should be just the near field as determined by the Poisson equation (B.1.7). Close to the origin the common part should cancel the far field from the expression for the composite field. Similarly, the common part should cancel the near field when we consider points far from the origin.

The common part is a field $\Phi_{common}(x, y, z, t)$ with properties as sketched in the following Fig. B.3.1.

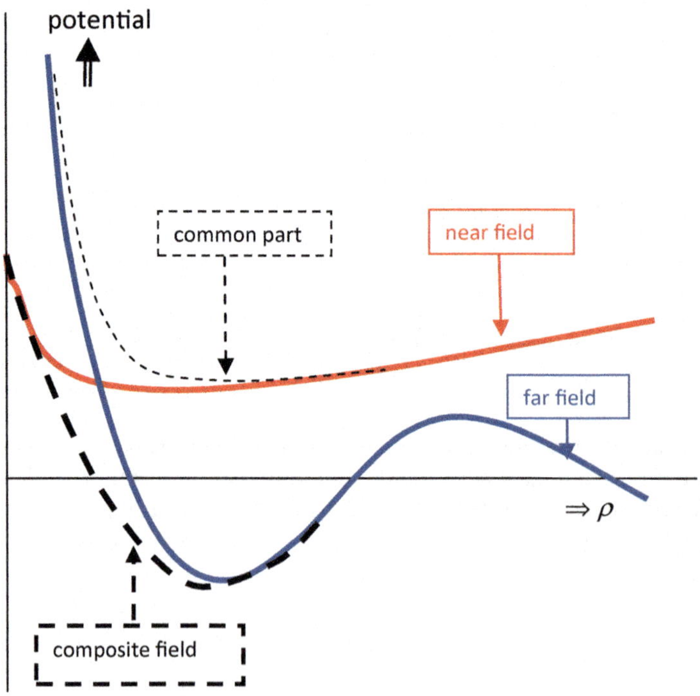

Fig. B.3.1 Example of the value of the potential along a hyperbola (see Fig. B.2.1) through the droplet

In this figure, showing the value of the potential (at a given instant of time) along a hyperbola as sketched in Fig. B.2.1, the far field is an oscillating function. This is due to the fact that at any distance ρ from the origin the potential is associated with the strength of the singularity in the origin at an earlier time. The spatial oscillations of Φ_{far} thus reflect the time variations of the singularity. The far field potential goes to zero at an infinitely large distance. Near the origin it is singular.

The near field reflects the fact that the field in this region is not singular. As explained earlier, the near field may contain terms not vanishing at infinity. Even terms continuing to grow at large distances may occur in the near field, which is the situation sketched.

The field called the "common part" is chosen such that it smoothly blends both with the near- as well as with the far field. At small distances from the charge configuration the common part behaves like the far field, whilst at large distances it approaches the near field.

Looking back at the structure of the composite field, this means that close to the origin the far field is cancelled by the common part, so that only the near field remains in this region. Conversely, at large distances the near field is eliminated by the common part, so that the composite field there behaves like the far field.

Summarizing, the common field Φ_{common} behaves like the near field Φ_{near} at distances ρ of the order of a typical wavelength (this is called the outer expansion of the near field):

$$\Phi_{common} \sim (\Phi_{near})_{\rho \to order\ \lambda} \tag{B.3.2}$$

and it behaves like the far field at distances of the order a (the earlier introduced inner expansion of the far field):

$$\Phi_{common} \sim (\Phi_{far})_{\rho \to order\ a} \tag{B.3.3}$$

where the symbol " $\sim$ " means identity up to a certain specified order of accuracy.

B.4 The Matching Condition

Combining (B.3.2) and (B.3.3) yields the matching condition:

$$(\Phi_{near})_{\rho \to O(\lambda)} \sim (\Phi_{far})_{\rho \to O(a)} \tag{B.4.1}$$

The matching condition can alternatively be formulated in terms of the characteristic coordinates (ρ/a for the near field and ρ/λ for the far field). Taking the asymptotic limit $a/\lambda \to 0$ then gives the condition in the form used later:

$$\lim_{\rho/a\to\infty} \Phi_{near} \sim \lim_{\rho/\lambda\to 0} \Phi_{far} \quad (a/\lambda \to 0) \tag{B.4.2}$$

The crucial assumption in the above is that a potential field Φ_{common} can indeed be found which satisfies (B.3.2) and (B.3.3) simultaneously. The assumption is equivalent to the assumption that Φ_{near} and Φ_{far} can satisfy (B.4.1). Now, Φ_{near} and thus the expression $(\Phi_{near})_{\rho\to order\ \lambda}$ satisfy the field equations (B.1.6) and (B.1.7). But these equations are also satisfied by the expression $(\Phi_{far})_{\rho\to order\ a}$ because the physical assumptions leading to (B.1.6) and (B.1.7) apply just as well to a charge moving on a *vanishingly* small interval $(a \to 0)$. Therefore, $(\Phi_{near})_{\rho\to order\ \lambda}$ and $(\Phi_{far})_{\rho\to order\ a}$ satisfy the same partial differential equation, and it is always possible to apply condition (B.4.1).

B.5 Remark on "Instantaneous" Interactions Within the Near Field

The field equations describing the near field depend on time in a *parametric* way only. If the charge configuration is time-dependent, retardation effects are taken into account by solving the Poisson equation instead of a Laplace equation. The Poisson equation itself, however, is only parametrically dependent on time. This means that different parts of the charge configuration have an *immediate* influence on each other, despite the fact that retardation effects are included. This is true only if these different parts may be considered to have the same near field.

If for instance the configuration consists of two point charges, and if these point charges belong to the same system, the near field is defined as a region of space comprising both these particles. Interactions between these two parts of the same system are then immediate, without any delays, as long as we take care that these interactions are consistent with the Poisson equation. If, on the other hand, the two particles cannot be considered to be parts of the same system, for each of them the complete composite field has to be determined separately. Any interaction between them then involves the far field as well, and will show up explicit retardation effects.

This interesting property, inherent in the matched asymptotic analysis, remains true even if we would extend the asymptotic approximations to higher orders of accuracy. Higher order approximations will always take the form of a Poisson equation to be solved, be it with more complex right hand sides than derived above.

Appendix C
The Potential Fields of a Moving Point Charge

C.1 Near Field of a Moving Point Charge (Scalar Potential)

The complete field of the moving and deforming droplet of charge will be built up by summing the contributions of its elements. We will therefore first construct the potentials of a moving point charge dq (see for the notations Fig. C.1.1).

It is important to realise that we here use the Lagrangian description, where a given charge element is followed. The position of the charge element dq is given by $(x_0, y_0, z_0(t))$, and its velocity is $v_{element}(t) = \dot{z}_0(t)$.

A field point is given by the coordinates (x, y, z). The value of the potentials in such a field point will be denoted by the lower case symbols $\phi(x, y, z, t)$ and $a_z(x, y, z, t)$ in order to emphasize that they are induced by a small element dq, instead of by the complete droplet q where the potentials are denoted as $\Phi(x, y, z, t)$ and $A_z(x, y, z, t)$.

In the near field the quasi-static potentials are given by

$$(\phi_{near})_{qu.static} = \frac{dq}{4\pi\varepsilon_0}\frac{1}{r} \quad \text{with } r = \sqrt{(x - x_0)^2 + (y - y_0)^2 + (z - z_0)^2} \tag{C.1.1}$$

$$(a_{z_{near}})_{qu.static} = \frac{v}{c^2}(\phi_{near})_{qu.static} \tag{C.1.2}$$

Because the charge is moving the potentials in the field point are time dependent. However, if we would take the substantial derivative of the field (field point co-moving with the charge) the electric potential would be constant:

$$\frac{D(\phi_{near})_{qu.static}}{Dt} = \frac{\partial(\phi_{near})_{qu.static}}{\partial t} + v\frac{\partial(\phi_{near})_{qu.static}}{\partial z} = 0 \tag{C.1.3}$$

which is also confirmed by substituting the above given expression (C.1.1) for the electrostatic field into the definition of a substantial derivative. Note that this does not apply to the magnetic potential, since its source strength $dq.v$ is variable on account of the variable velocity $v(t)$.

T. van Holten, *The Atomic World Spooky? It Ain't Necessarily So!*
DOI 10.2991/978-94-6239-234-2

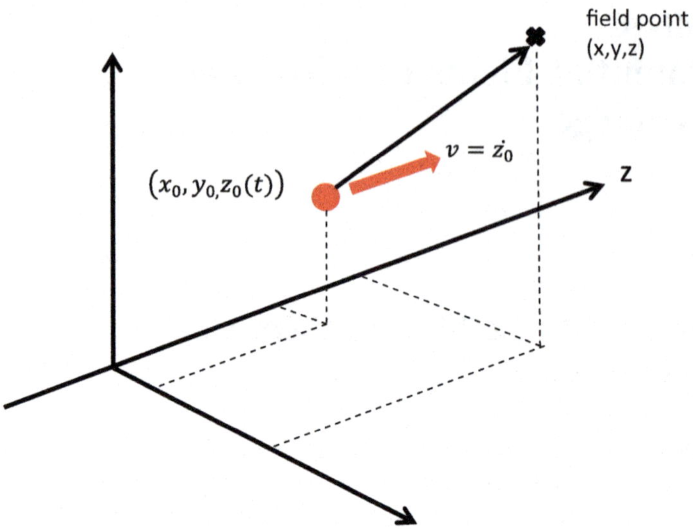

Fig. C.1.1 Moving point charge (Lagrangian description)

The next asymptotic approximation for the near field, where the effects of retardation are taken into account, is obtained by solving the Poisson equation

$$\nabla^2 \phi_{near} = \frac{1}{c^2} \frac{\partial^2 (\phi_{near})_{qu.static}}{\partial t^2} \tag{C.1.4}$$

The right hand side of this equation is written, using Eq. (C.1.3):

$$\begin{aligned} \frac{1}{c^2} \frac{\partial^2 (\phi_{near})_{qu.static}}{\partial t^2} &= \frac{1}{c^2} \frac{\partial}{\partial t} \left[-v \frac{\partial (\phi_{near})_{qu.static}}{\partial z} \right] \\ &= \frac{1}{c^2} \left[-\dot{v} \frac{\partial (\phi_{near})_{qu.static}}{\partial z} - v \frac{\partial}{\partial z} \left(\frac{\partial (\phi_{near})_{qu.static}}{\partial t} \right) \right] \end{aligned} \tag{C.1.5}$$

In order to find a particular solution of this Poisson equation use can be made of the following general property.

If we have a Poisson equation

$$\nabla^2 \Phi = \frac{\partial \Psi}{\partial z} \tag{C.1.6}$$

then, on condition that Ψ satisfies a Laplace equation

$$\nabla^2 \Psi = 0 \tag{C.1.7}$$

a particular solution of the Poisson equation is:

$$\Phi_{partic} = \frac{z}{2}\Psi \tag{C.1.8}$$

which may be checked by substitution.

Thanks to the fact that the right hand side (C.1.5) of the Poisson equation (C.1.4) has been written in the form of z-derivatives, and since $\frac{\partial(\phi_{near})_{qu.static}}{\partial z}$ as well as $\frac{\partial(\phi_{near})_{qu.static}}{\partial t}$ both satisfy the Laplace equation, we may immediately write down a particular solution of the Poisson equation:

$$(\phi_{near})_{partic} = -\frac{z}{2}\left[\frac{\dot{v}}{c^2}(\phi_{near})_{qu.static} + \frac{v}{c^2}\frac{\partial(\phi_{near})_{qu.static}}{\partial t}\right] \tag{C.1.9}$$

or, using Eq. (C.1.3) again:

$$(\phi_{near})_{partic} = -\frac{z}{2}\left[\frac{\dot{v}}{c^2}(\phi_{near})_{qu.static} - \frac{v^2}{c^2}\frac{\partial(\phi_{near})_{qu.static}}{\partial z}\right] \tag{C.1.10}$$

The complete solution of the Poisson equation (C.1.4) consists of the particular solution, and solutions of the Laplace equation. In order to represent the singularity, we therefore have to add the field $(\phi_{near})_{qu.static}$. Another contribution to the complete solution may be a non-singular field $\Delta\phi$, satisfying the Laplace equation, and satisfying at infinity the matching conditions. The complete solution for the near field, accurate to the order $(a/\lambda)^3$, thus becomes:

$$\phi_{near} = (\phi_{near})_{qu.static} - \frac{z}{2}\left[\frac{\dot{v}}{c^2}(\phi_{near})_{qu.static} - \frac{v^2}{c^2}\frac{\partial(\phi_{near})_{qu.static}}{\partial z}\right] + \Delta\phi \tag{C.1.11}$$

with $(\phi_{near})_{qu.static} = \frac{dq}{4\pi\varepsilon_0}\frac{1}{r}$ and $r = \sqrt{(x-x_0)^2 + (y-y_0)^2 + (z-z_0)^2}$

C.2 The Outer Expansion of the Near Field

In order to apply the matching condition with the far field the behaviour of the near field at large distances from the origin $(r \to O(\lambda))$ should be investigated. For this purpose again the spherical coordinate system of Fig. B.2.2 with its origin in $x = y = z = 0$ will be used.

The transformation between the Cartesian system used until now and the spherical system is

$$
\begin{aligned}
x &= \rho.\sin\varphi.\cos\chi \\
y &= \rho.\sin\varphi.\sin\chi \\
z &= \rho.\cos\varphi
\end{aligned}
\tag{C.2.1}
$$

The distance r between the charge element and a field point is then given by

$$
\begin{aligned}
r &= \sqrt{(x-x_0)^2+(y-y_0)^2+(z-z_0)^2} \\
&= \rho.\left[1+\frac{x_0^2+y_0^2+z_0^2}{\rho^2}-2\frac{x_0}{\rho}\sin\varphi.\cos\chi-2\frac{y_0}{\rho}\sin\varphi.\sin\chi-2\frac{z_0}{\rho}\cos\varphi\right]^{1/2}
\end{aligned}
\tag{C.2.2}
$$

The function $1/r$ can for large ρ thus be expanded into the Taylor series:

$$
\begin{aligned}
\frac{1}{r}=\frac{1}{\rho}\Bigg[&1+\frac{1}{2}\left(\frac{x_0}{\rho}\right)^2.(-1+3.\sin^2\varphi.\cos^2\chi)+\frac{1}{2}\left(\frac{y_0}{\rho}\right)^2.(-1+3.\sin^2\varphi.\sin^2\chi) \\
&+\frac{1}{2}\left(\frac{z_0}{\rho}\right)^2.(-1+3.\cos^2\varphi)+\frac{x_0}{\rho}\sin\varphi.\cos\chi+\frac{y_0}{\rho}\sin\varphi.\sin\chi+\frac{z_0}{\rho}\cos\varphi \\
&+3\frac{x_0}{\rho}\frac{y_0}{\rho}\sin^2\varphi.\sin\chi.\cos\chi+3\frac{x_0}{\rho}\frac{z_0}{\rho}\sin\varphi.\cos\varphi.\cos\chi+3\frac{y_0}{\rho}\frac{z_0}{\rho}\sin\varphi.\cos\varphi\sin\chi+\ldots\ldots\Bigg]
\end{aligned}
\tag{C.2.3}
$$

When investigating the behaviour of the near field for $\rho \to O(\lambda)$, keeping only terms up to $O(a/\lambda)^3$ and neglecting higher orders, we see that the above given series expansion may be truncated after the last term that has been written out explicitly (thus truncating the terms in the formula symbolised by the dots "……..").

The remaining expansion result can be written in the alternative form:

$$
\begin{aligned}
\frac{1}{r}\overset{\rho\to O(\lambda)}{\longrightarrow}&\frac{1}{\rho}+\frac{1}{2}\frac{\partial^2}{\partial x^2}\left(\frac{x_0^2}{\rho}\right)+\frac{1}{2}\frac{\partial^2}{\partial y^2}\left(\frac{y_0^2}{\rho}\right)+\frac{1}{2}\frac{\partial^2}{\partial z^2}\left(\frac{z_0^2}{\rho}\right) \\
&-\frac{\partial}{\partial x}\left(\frac{x_0}{\rho}\right)-\frac{\partial}{\partial y}\left(\frac{y_0}{\rho}\right)-\frac{\partial}{\partial z}\left(\frac{z_0}{\rho}\right)+\frac{\partial^2}{\partial x.\partial y}\left(\frac{x_0.y_0}{\rho}\right)+\frac{\partial^2}{\partial x.\partial z}\left(\frac{x_0.z_0}{\rho}\right)+\frac{\partial^2}{\partial y.\partial z}\left(\frac{y_0.z_0}{\rho}\right) \\
&\text{up to } O(a/\lambda)^3
\end{aligned}
\tag{C.2.4}
$$

The outer expansion of the quasi-static near field is therefore:

$$
(\phi_{near})_{qu.static}\overset{outer\ expansion}{\longrightarrow}\frac{dq}{4\pi\epsilon_0}\frac{1}{r}\bigg|_{\rho\to O(\lambda)}
\tag{C.2.5}
$$

with $\frac{1}{r}\Big|_{\rho\to O(\lambda)}$ given by the above expression (C.2.4).

It is important to realise that the near field, as well as its outer expansion, is the solution of field equations where the time enters just *parametrically* (which does *not* mean that retardation effects are neglected; retardation effects are taken into account by solving the Poisson equation instead of a Laplace equation for the potentials).

One should therefore read the above equation as

$$\frac{1}{r}(x,y,z;t)\Bigg|_{\rho\to O(\lambda)} = \ldots + \frac{1}{2}\frac{\partial^2}{\partial z^2}\left(\frac{z_0^2(t)}{\rho(x,y,z)}\right) + \ldots etc. \tag{C.2.6}$$

where the position $z_0(t)$ of the point charge is taken at the same instant of time as the time in the function $1/r(x,y,z;t)$. For this reason the strength $z_0^2(t)$ of the pole can either be written inside the spatial differentiation $\partial^2/\partial z^2$ or in front of it, without any further consequences.

In the case of the far field, which satisfies a wave equation, this is different. The outer expansion of the near field shows that matching with the far field will require the far field to consist of a set of multipoles, ranging from a simple monopole representing the constant charge, to three types of dipoles and six different quadrupoles. All these poles in the far field are subject to explicit retardation effects, so that e.g. the far field pole corresponding to $\frac{\partial^2}{\partial z^2}\left(\frac{z_0^2(t)}{\rho}\right)$ will have the field $\frac{\partial^2}{\partial z^2}\left(\frac{z_0^2(t-\rho/c)}{\rho}\right)$.

Before writing out the far field, we must first complete the outer expansion of the near field.

The outer expansion of the particular solution (C.1.10) of the Poisson equation (C.1.4) is:

$$\begin{aligned}(\phi_{near})_{partic} &= -\frac{z}{2}\left[\frac{\dot{v}}{c^2}(\phi_{near})_{qu.static} - \frac{v^2}{c^2}\frac{\partial(\phi_{near})_{qu.static}}{\partial z}\right]^{\overrightarrow{outer\ expansion}} \\ &\quad -\frac{1}{2}\frac{\dot{v}}{c^2}\frac{dq}{4\pi\varepsilon_0}z\left[\frac{1}{\rho} - \frac{\partial}{\partial x}\left(\frac{x_0}{\rho}\right) - \frac{\partial}{\partial y}\left(\frac{y_0}{\rho}\right) - \frac{\partial}{\partial z}\left(\frac{z_0}{\rho}\right)\right] + \frac{1}{2}\frac{v^2}{c^2}\frac{dq}{4\pi\varepsilon_0}z\frac{\partial}{\partial z}\left(\frac{1}{\rho}\right) \\ &= -\frac{1}{2}\frac{\dot{v}}{c^2}\frac{dq}{4\pi\varepsilon_0}cos\varphi\left[1 - \frac{x_0}{\rho}sin\varphi.cos\chi - \frac{y_0}{\rho}sin\varphi.sin\chi - \frac{z_0}{\rho}cos\varphi\right] + -\frac{1}{2}\frac{v^2}{c^2}\frac{dq}{4\pi\varepsilon_0}\frac{cos^2\varphi}{\rho}\end{aligned} \tag{C.2.7}$$

The reason why this expansion has been truncated at an earlier point than the expansion of the quasi-static field is, that in the coefficients the terms $\frac{\dot{v}}{c^2}$ and $\frac{v^2}{c^2}$ occur. Because of the equivalence of the expansion parameters v/c and a/λ both these coefficients are of order $O(a/\lambda)^2$. From the second form of the outer expansion it is seen that the smallest terms occurring in this expression are therefore of order $O(a/\lambda)^3$, and no need to continue the expansion exists.

C.3 The Far Field and Its Inner Expansion

The far field must, in order to match with the near field, consist of the same types of multipoles as the outer expansion of the near field. However, the far field satisfies the complete wave equation, so that these far field multipoles take into account the retardation effect. An exception is the monopole, representing the constant total charge, which is not time dependent and is therefore not subject to retardation effects. The same applies to the poles having a pole strength which depends solely on x_0 or y_0.

Thus, considering Eq. (C.2.4), the far field is given by:

$$\begin{aligned}\phi_{far}\cdot\frac{4\pi\varepsilon_0}{dq} &= \frac{1}{\rho}+\frac{1}{2}\frac{\partial^2}{\partial x^2}\left(\frac{x_0^2}{\rho}\right)+\frac{1}{2}\frac{\partial^2}{\partial y^2}\left(\frac{y_0^2}{\rho}\right)+\frac{1}{2}\frac{\partial^2}{\partial z^2}\left(\frac{z_0^2(t-\rho/c)}{\rho}\right)\\ &-\frac{\partial}{\partial x}\left(\frac{x_0}{\rho}\right)-\frac{\partial}{\partial y}\left(\frac{y_0}{\rho}\right)-\frac{\partial}{\partial z}\left(\frac{z_0(t-\rho/c)}{\rho}\right)+\frac{\partial^2}{\partial x.\partial y}\left(\frac{x_0.y_0}{\rho}\right)+\frac{\partial^2}{\partial x.\partial z}\left(\frac{x_0.z_0(t-\rho/c)}{\rho}\right)\\ &+\frac{\partial^2}{\partial y.\partial z}\left(\frac{y_0.z_0(t-\rho/c)}{\rho}\right)\text{up to } O(a/\lambda)^3\end{aligned}\tag{C.3.1}$$

For the purpose of matching the inner expansion of the far field is needed, determined by a series expansion for small values of the retardation time ρ/c. The general procedure has already been explained in Chap. B.2, where the variable dipole was used as an example:

$$\begin{aligned}(\phi_{far})_{z-dipole}\frac{4\pi\varepsilon_0}{dq} &= -\frac{\partial}{\partial z}\left[\frac{z_0(t-\rho/c)}{\rho}\right]\\ &= \left[\frac{1}{c}\frac{v(t-\rho/c)}{\rho}+\frac{z_0(t-\rho/c)}{\rho^2}\right].cos\varphi \xrightarrow{\text{inner expansion}}\\ &\xrightarrow{\text{inner expansion}}\left[\frac{z_0(t)}{\rho^2}-\frac{1}{2}\frac{\dot v(t)}{c^2}+\frac{1}{3}\rho\frac{\ddot v(t)}{c^3}+\ldots\right]cos\varphi\\ &= -\frac{\partial}{\partial z}\left[\frac{z_0(t)}{\rho}\right]-\frac{1}{2}\frac{z}{\rho}\frac{\dot v(t)}{c^2}+\frac{1}{3}z\frac{\ddot v(t)}{c^3}+\ldots\end{aligned}\tag{C.3.2}$$

Where to truncate the expansion is in this case, where $\rho\to O(a)$, solely determined by the order of the coefficients. In view of the equivalence between the expansion parameters (a/λ) and (v/c) it is seen that the smallest term in the above expression is $O(a/\lambda)^3$. A further expansion is not needed.

Similarly, the other time-dependent poles in the far field are written out and expanded for small retardation times. In the following for the sake of efficiency the time functions $x_0.z_0(t-\rho/c)$, $y_0.z_0(t-\rho/c)$ *and* $z_0^2(t-\rho/c)$ are symbolized by the general notation $g(t-\rho/c)$.

$$(\phi_{far})_{xz-quadrupole}.\frac{4\pi\varepsilon_0}{dq} = \frac{\partial^2}{\partial x.\partial z}\left[\frac{g(t-\rho/c)}{\rho}\right] = -z.\frac{\partial}{\partial x}\left[\frac{g}{\rho^3}+\frac{1}{\rho^2}\frac{g'}{c}\right]$$
$$= \frac{xz}{\rho^5}\left[3.g(t-\rho/c)+3.\left(\frac{\rho}{c}\right).g'(t-\rho/c)+\left(\frac{\rho}{c}\right)^2.g''(t-\rho/c)\right] \xrightarrow{inner\ expansion}$$
$$\rightarrow 3\frac{xz}{\rho^5}\left[g(t)-\left(\frac{\rho}{c}\right).g'(t)+\frac{1}{2}\left(\frac{\rho}{c}\right)^2.g''(t)-\frac{1}{6}\left(\frac{\rho}{c}\right)^3.g'''(t)\right]$$
$$+3\frac{xz}{\rho^5}\left(\frac{\rho}{c}\right).\left[g'(t)-\left(\frac{\rho}{c}\right).g''(t)+\frac{1}{2}\left(\frac{\rho}{c}\right)^2.g'''(t)\right]$$
$$+\frac{xz}{\rho^5}\left(\frac{\rho}{c}\right)^2.\left[g''(t)-\left(\frac{\rho}{c}\right).g'''(t)\right] = \frac{xz}{\rho^5}\left[3.g(t)-\frac{1}{2}\left(\frac{\rho}{c}\right)^2.g''(t)\right]$$
$$= 3.g(t)\frac{xz}{\rho^5}-\frac{1}{2}\frac{g''(t)}{c^2}\frac{xz}{\rho^3}$$
$$= \frac{\partial^2}{\partial x.\partial z}\left(\frac{g(t)}{\rho}\right)+\frac{1}{2}.z.\frac{\partial}{\partial x}\left(\frac{g''(t)/c^2}{\rho}\right) \tag{C.3.3}$$

Substituting $g = x_0.z_0$:

$$(\phi_{far})_{xz-quadrupole}.\frac{4\pi\varepsilon_0}{dq} = \frac{\partial^2}{\partial x.\partial z}\left[\frac{x_0.z_0(t-\rho/c)}{\rho}\right] \xrightarrow{inner\ expansion} \frac{\partial^2}{\partial x.\partial z}\left[\frac{x_0.z_0(t)}{\rho}\right]+\frac{z}{2}\frac{\dot{v}}{c^2}\frac{\partial}{\partial x}\left[\frac{x_0}{\rho}\right] \tag{C.3.4}$$

A cyclic change of coordinates shows:

$$(\phi_{far})_{yz-quadrupole}.\frac{4\pi\varepsilon_0}{dq} = \frac{\partial^2}{\partial y.\partial z}\left[\frac{y_0.z_0(t-\rho/c)}{\rho}\right] \xrightarrow{inner\ expansion} \frac{\partial^2}{\partial y.\partial z}\left[\frac{y_0.z_0(t)}{\rho}\right]+\frac{z}{2}\frac{\dot{v}}{c^2}\frac{\partial}{\partial y}\left[\frac{y_0}{\rho}\right] \tag{C.3.5}$$

In the case of the quadrupole $(\phi_{far})_{zz-quadrupole}\frac{4\pi\varepsilon_0}{dq} = \frac{1}{2}\frac{\partial^2}{\partial z^2}\left[\frac{z_0^2(t-\rho/c)}{\rho}\right]$ there is an additional term compared with the other quadrupoles (note the factor 1/2 multiplying this term in the complete far field (C.3.1)):

$$2.(\phi_{far})_{zz-quadrupole}.\frac{4\pi\varepsilon_0}{dq} = -z.\frac{\partial}{\partial z}\left[\frac{g}{\rho^3}+\frac{1}{\rho^2}\frac{g'}{c}\right]-\left[\frac{g}{\rho^3}+\frac{1}{\rho^2}\frac{g'}{c}\right] \rightarrow$$
$$\xrightarrow{inner\ expansion}$$
$$\rightarrow \frac{z^2}{\rho^5}\left[3.g(t)-\frac{1}{2}\left(\frac{\rho}{c}\right)^2.g''(t)\right]-\frac{1}{\rho^3}\left[g(t)-\frac{1}{2}\left(\frac{\rho}{c}\right)^2.g''(t)+\frac{1}{3}\left(\frac{\rho}{c}\right)^3.g'''(t)\right]$$
$$= \frac{\partial^2}{\partial z^2}\left[\frac{g(t)}{\rho}\right]+\frac{1}{2}.z.\frac{\partial}{\partial z}\left[\frac{g''(t)/c^2}{\rho}\right]+\frac{1}{2}\left[\frac{g''(t)/c^2}{\rho}\right]-\frac{1}{3}\frac{g'''(t)}{c^3} \tag{C.3.6}$$

After substitution of $g = z_0^2$:

$$(\phi_{far})_{zz-quadrupole}.\frac{4\pi\varepsilon_0}{dq} \rightarrow \frac{1}{2}\frac{\partial^2}{\partial z^2}\left[\frac{z_0^2(t)}{\rho}\right] + \frac{z}{2}\frac{\partial}{\partial z}\left[\frac{(v^2/c^2 + z_0.\dot{v}/c^2)}{\rho}\right] + \frac{1}{2}\left[\frac{v^2/c^2 + z_0.\dot{v}/c^2}{\rho}\right] - \frac{1}{3}\frac{3v.\dot{v} + z_0.\ddot{v}}{c^3} \tag{C.3.7}$$

C.4 Matching the Near- and Far Field; Complete Expression for the Near Field (Scalar Potential)

Matching the near and far field requires

$$(outer\,expansion\,of\;\phi_{near}) = (inner\,expansion\,of\;\phi_{far})$$

Now summarising:

$$\begin{aligned}\left(outer\,expansion\,of\;\phi_{near}.\frac{4\pi\varepsilon_0}{dq}\right) &= (quasi - static\,poles).\frac{4\pi\varepsilon_0}{dq}\\ &+ -\frac{1}{2}\frac{\dot{v}}{c^2}z\left[\frac{1}{\rho} - \frac{\partial}{\partial x}\left(\frac{x_0}{\rho}\right) - \frac{\partial}{\partial y}\left(\frac{y_0}{\rho}\right) - \frac{\partial}{\partial z}\left(\frac{z_0}{\rho}\right)\right] + \frac{1}{2}\frac{v^2}{c^2}z\frac{\partial}{\partial z}\left(\frac{1}{\rho}\right)\\ &+ \left(outer\,expansion\,of\;\Delta\phi.\frac{4\pi\varepsilon_0}{dq}\right)\end{aligned} \tag{C.4.1}$$

$$\begin{aligned}\left(inner\,expansion\,of\;\phi_{far}.\frac{4\pi\varepsilon_0}{dq}\right) &= (quasi - static\,poles)\\ &+ -\frac{1}{2}\frac{z}{\rho}\frac{\dot{v}(t)}{c^2} + \frac{1}{3}z\frac{\ddot{v}(t)}{c^3} + \frac{z}{2}\frac{\dot{v}}{c^2}\frac{\partial}{\partial x}\left[\frac{x_0}{\rho}\right] + \frac{z}{2}\frac{\dot{v}}{c^2}\frac{\partial}{\partial y}\left[\frac{y_0}{\rho}\right]\\ &+ \frac{z}{2}\frac{\partial}{\partial z}\left[\frac{(v^2/c^2 + z_0.\dot{v}/c^2)}{\rho}\right] + \frac{1}{2}\left[\frac{v^2/c^2 + z_0.\dot{v}/c^2}{\rho}\right] - \frac{1}{3}\frac{3v.\dot{v} + z_0.\ddot{v}}{c^3}\end{aligned} \tag{C.4.2}$$

The quasi-static poles are not written out, because they automatically satisfy the matching condition on account of the choice of poles in the far field (the number and type of these poles being the same as in the outer expansion of the near field).

The matching condition yields the result

$$(\Delta\phi)\frac{4\pi\varepsilon_0}{dq} \xrightarrow{outer\,expansion} \frac{1}{3}\frac{\ddot{v}(t)}{c^3}.z + \frac{1}{2}\frac{v^2/c^2 + z_0.\dot{v}/c^2}{\rho} - \frac{1}{3}\frac{3v.\dot{v} + z_0.\ddot{v}}{c^3} \tag{C.4.3}$$

so that in the near field

$$(\Delta\phi)_{near}\frac{4\pi\varepsilon_0}{dq}=\frac{1}{3}\frac{\ddot{v}(t)}{c^3}.z+\frac{1}{2}\frac{v^2/c^2+z_0.\dot{v}/c^2}{r}-\frac{1}{3}\frac{3v.\dot{v}+z_0.\ddot{v}}{c^3} \tag{C.4.4}$$

We may add, to enhance the symmetry of the final expression, another term of irrelevant order (because its outer expansion is of a negligible asymptotic order):

$$-\frac{1}{2}\frac{v^2}{c^2}.z_0.\frac{\partial}{\partial z}\left(\frac{1}{r}\right)\xrightarrow{\text{outer expansion}}O(a/\lambda)^4 \tag{C.4.5}$$

which finally results in the following expression for the near field:

$$(\phi_{near})=\frac{dq}{4\pi\varepsilon_0}\left[\left(1+\frac{1}{2}\frac{v^2}{c^2}\right).\frac{1}{r}+\frac{1}{2}\frac{v^2}{c^2}(z-z_0)\frac{\partial}{\partial z}\left(\frac{1}{r}\right)-\frac{1}{2}\frac{\dot{v}}{c^2}\frac{z-z_0}{r}+\frac{1}{3}\frac{\ddot{v}}{c^3}(z-z_0)-\frac{v.\dot{v}}{c^3}\right] \tag{C.4.6}$$

C.5 The Vector Potential of a Moving Point Charge

The quasi-static near field is

$$(a_{z_{near}})_{qu.static}=\frac{dq.v}{c^2 4\pi\varepsilon_0}\frac{1}{r}=\frac{v}{c^2}(\phi_{near})_{qu.static}. \tag{C.5.1}$$

Retardation effects are given by the particular solution of the Poisson equation:

$$\begin{aligned}\nabla^2(a_{z_{near}})_{retard}&=\frac{1}{c^2}\frac{\partial^2(a_{z_{near}})_{qu.static}}{\partial t^2}\\&=\frac{\ddot{v}}{c^4}(\phi_{near})_{qu.static}-3\frac{v.\dot{v}}{c^4}\frac{\partial(\phi_{near})_{qu.static}}{\partial z}+\frac{v^3}{c^4}\frac{\partial^2(\phi_{near})_{qu.static}}{\partial z^2}\end{aligned} \tag{C.5.2}$$

In the second line use has been made of Eq. (C.1.3).

The particular solution is (making use again of the fact that several terms in the r.h.s. are z-derivatives):

$$(a_{z_{near}})_{partic} = \frac{\ddot{v}}{c^4}\frac{r}{2}\frac{dq}{4\pi\varepsilon_0} - 3\frac{v.\dot{v}}{c^4}\frac{z}{2}(\phi_{near})_{qu.static} + \frac{v^3}{c^4}\frac{z}{2}\frac{\partial(\phi_{near})_{qu.static}}{\partial z} \tag{C.5.3}$$

To obtain the complete solution, we have to add the quasi-static field, as well as additional terms $\Delta(a_{znear})$ satisfying the Laplace equation and injected by the matching process.

When determining the outer expansion up to $O(a/\lambda)^3$, it should be taken into account that the magnetic potential

$$(a_{z_{near}})_{qu.static} = \frac{v}{c^2}(\phi_{near})_{qu.static} \tag{C.5.4}$$

has a smaller asymptotic order than the electric potential, due to the extra factor v/c. As a consequence the outer expansions contain fewer terms than in the case of the electric field. In particular, we will find only mono- and dipoles in the outer expansion. A slight complication is, that the monopole is time dependent now.

The outer expansions are:

$$(a_{z_{near}})_{qu.static}\cdot\frac{4\pi\varepsilon_0}{dq} = \frac{v}{c^2}\frac{1}{r} \rightarrow$$

$$\xrightarrow{\text{outer expansion}}$$

$$\rightarrow \frac{v}{c^2}\frac{1}{\rho}\left[1 + \frac{x_0}{\rho}\sin\varphi.\cos\chi + \frac{y_0}{\rho}\sin\varphi.\sin\chi + \frac{z_0}{\rho}\cos\varphi\right] \tag{C.5.5}$$

$$= \frac{v}{c^2}\left[\frac{1}{\rho} - \frac{\partial}{\partial x}\left(\frac{x_0}{\rho}\right) - \frac{\partial}{\partial y}\left(\frac{y_0}{\rho}\right) - \frac{\partial}{\partial z}\left(\frac{z_0}{\rho}\right)\right]$$

summed with the expansion of the particular solution of the Poisson equation

$$(a_{z_{near}})_{partic}\frac{4\pi\varepsilon_0}{dq} = \frac{\ddot{v}}{c^4}\frac{r}{2} - 3\frac{v.\dot{v}}{c^4}\frac{z}{2}\frac{1}{r} + \frac{v^3}{c^4}\frac{z}{2}\frac{\partial}{\partial z}\left(\frac{1}{r}\right) \rightarrow$$

$$\xrightarrow{\text{outer expansion}}$$

$$\rightarrow \frac{\ddot{v}}{c^4}\frac{\rho}{2}\left[1 - \frac{x_0}{\rho}\sin\varphi.\cos\chi - \frac{y_0}{\rho}\sin\varphi.\sin\chi - \frac{z_0}{\rho}\cos\varphi\right] - \frac{3}{2}\frac{v.\dot{v}}{c^4}\frac{z}{\rho} + O(a/\lambda)^4$$

$$= \frac{1}{2}\frac{\ddot{v}}{c^4}\left[\rho - \frac{x.x_0}{\rho} - \frac{y.y_0}{\rho} - \frac{z.z_0}{\rho}\right] - \frac{3}{2}\frac{v.\dot{v}}{c^4}\frac{z}{\rho} \tag{C.5.6}$$

The outer expansion of the quasi-static near field gives the number and types of poles to be included in the far field. These are a monopole and three dipoles. They are all poles with a time dependent strength. The monopole in the far field is:

$$\frac{(a_{z_{far}})_{monopole}}{dq/(c^2.4\pi\varepsilon_0)} = \frac{v(t-\rho/c)}{\rho} \rightarrow \xrightarrow{\text{inner expansion}} \rightarrow \frac{v(t)}{\rho} - \frac{\dot{v}}{c} + \frac{1}{2}\frac{\ddot{v}}{c^2}.\rho \tag{C.5.7}$$

The three dipoles in the far field are:

$$\frac{(a_{z_{far}})_{dipole}}{dq/(c^2.4\pi\varepsilon_0)} = \left[-\frac{\partial}{\partial x}\left(\frac{x_0.v(t-\rho/c)}{\rho}\right) - \frac{\partial}{\partial y}\left(\frac{y_0.v(t-\rho/c)}{\rho}\right) - \frac{\partial}{\partial z}\left(\frac{z_0.v(t-\rho/c)}{\rho}\right)\right] \rightarrow \xrightarrow{\text{inner expansion}} \rightarrow -\frac{\partial}{\partial x}\left(\frac{x_0.v(t)}{\rho}\right) - \frac{1}{2}\frac{x_0.\ddot{v}(t)}{c^2}\frac{x}{\rho} - \frac{\partial}{\partial y}\left(\frac{y_0.v(t)}{\rho}\right) - \frac{1}{2}\frac{y_0.\ddot{v}(t)}{c^2}\frac{y}{\rho} - \frac{\partial}{\partial z}\left(\frac{z_0.v(t)}{\rho}\right) - \frac{1}{2}\frac{3v.\dot{v} + z_0.\ddot{v}(t)}{c^2}\frac{z}{\rho} \tag{C.5.8}$$

Matching is achieved by adding to the near field the term

$$\frac{\Delta(a_{z_{near}})}{dq/(4\pi\varepsilon_0)} = -\frac{\dot{v}}{c^3} \tag{C.5.9}$$

On top of that, we may add terms of fourth order, which are again intended to enhance the symmetry of the final expression for the complete near field:

$$\frac{(a_{z_{near}})}{dq/(4\pi\varepsilon_0)} = \frac{v}{c^2}\frac{1}{r} + \frac{\ddot{v}}{c^4}\frac{r}{2} - 3\frac{v.\dot{v}}{c^4}\frac{z-z_0}{2}\frac{1}{r} + \frac{1}{2}\frac{v^3}{c^4}(z-z_0)\frac{\partial}{\partial z}\left(\frac{1}{r}\right) - \frac{\dot{v}}{c^3} \tag{C.5.10}$$

An alternative expression is:

$$(a_{z_{near}}) = \left[\frac{v}{c^2}(\phi_{near})_{qu.static} + \frac{\ddot{v}}{c^4}\frac{r}{2}\frac{dq}{4\pi\varepsilon_0} - \frac{3}{2}\frac{v.\dot{v}}{c^4}(z-z_0)(\phi_{near})_{qu.static} + \frac{1}{2}\frac{v^3}{c^4}(z-z_0)\frac{\partial(\phi_{near})_{qu.static}}{\partial z} - \frac{\dot{v}}{c^3}\frac{dq}{4\pi\varepsilon_0}\right] \tag{C.5.11}$$

Appendix D
Self Forces on a Droplet of Charge (Translation Direction)

D.1 Force Exerted by Moving Point Charge on Another Unit Charge

The derived expressions will later be used to determine the self forces on a droplet of charge in Z-direction. The contributions by all the charge elements forming the droplet will then be integrated to find the total force.

Using the general relations

$$\underline{E} = -grad\Phi - \frac{\partial \underline{A}}{\partial t}$$
$$\underline{B} = rot\underline{A}$$
$$\underline{F} = q(\underline{E} + \underline{v} \times \underline{B})$$

and considering that the velocities are only in Z-direction (i.e. no contribution by $\underline{v} \times \underline{B}$), we see that the Z-force on a unit charge induced by the element dq is

$$f_z = -\frac{\partial \phi}{\partial z} - \frac{\partial a_z}{\partial t} \tag{D.1.1}$$

If this expression is used to determine the self forces on a droplet of charge, it is not necessary to use the composite fields, we can substitute the near fields instead. To avoid that terms of an irrelevant asymptotic order of magnitude are included, the terms in Eq. (C.5.10) for $a_{z_{near}}$ where c^4 is in the denominator are neglected. Furthermore, using the relation

$$\frac{\partial}{\partial t}\left(\frac{1}{r}\right) = -v.\frac{\partial}{\partial z}\left(\frac{1}{r}\right) \tag{D.1.2}$$

we find

T. van Holten, *The Atomic World Spooky? It Ain't Necessarily So!*
DOI 10.2991/978-94-6239-234-2

$$
\begin{aligned}
(f_z)_{near} &= -\frac{\partial(\phi_{near})}{\partial z} - \frac{\partial(a_{znear})}{\partial t} = -\frac{dq}{4\pi\varepsilon_0}\left[\left(1+\frac{v^2}{c^2}\right)\frac{\partial}{\partial z}\left(\frac{1}{r}\right)\right. \\
&\left. + \frac{1}{2}\frac{v^2}{c^2}(z-z_0)\frac{\partial^2}{\partial z^2}\left(\frac{1}{r}\right) - \frac{1}{2}\frac{\dot{v}}{c^2}(z-z_0)\frac{\partial}{\partial z}\left(\frac{1}{r}\right) - \frac{1}{2}\frac{\dot{v}}{c^2}\frac{1}{r} + \frac{1}{3}\frac{\ddot{v}}{c^3}\right] \\
&- \frac{dq}{4\pi\varepsilon_0}\left[-\frac{v^2}{c^2}\frac{\partial}{\partial z}\left(\frac{1}{r}\right) + \frac{\dot{v}}{c^2}\frac{1}{r} - \frac{\ddot{v}}{c^3}\right]
\end{aligned}
\tag{D.1.3}
$$

D.2 Slender Body Approximation

Thanks to the assumption of a slender shape of the droplet $(t/s \ll 1)$ a considerable simplification is possible in the near field. We have here (in this case the thickness of the droplet is denoted as t to avoid confusion with the distance r):

$$(x-x_0) = O(t), \quad (y-y_0) = O(t), \quad (z-z_0) = O(s) \tag{D.2.1}$$

from which:

$$
\begin{aligned}
\frac{1}{r} &= \frac{1}{\sqrt{(x-x_0)^2 + (y-y_0)^2 + (z-z_0)^2}} \\
&= \frac{1}{\sqrt{\left(\frac{x-x_0}{t}\right)^2 .t^2 + \left(\frac{y-y_0}{t}\right)^2 .t^2 + \left(\frac{z-z_0}{s}\right)^2 .s^2}} \\
&= \frac{1}{s}\frac{1}{\left|\frac{z-z_0}{s}\right|}[1 + O(t/s)^2]
\end{aligned}
\tag{D.2.2}
$$

and

$$
\begin{aligned}
&(z-z_0)\frac{\partial}{\partial z}\left(\frac{1}{r}\right) \\
&= -\frac{(z-z_0)^2}{\left[(x-x_0)^2 + (y-y_0)^2 + (z-z_0)^2\right]^{3/2}} \\
&= -\frac{1}{s}\frac{1}{\left|\frac{z-z_0}{s}\right|.}\left[1 + O(t/s)^2\right]
\end{aligned}
\tag{D.2.3}
$$

If in slender body approximation terms of order $O(t/s)^2$ are neglected, we see that to this accuracy we may equate:

$$(z - z_0)\frac{\partial}{\partial z}\left(\frac{1}{r}\right) \approx -\frac{1}{r} \tag{D.2.4}$$

A similar consideration leads to

$$(z - z_0)\frac{\partial^2}{\partial z^2}\left(\frac{1}{r}\right) \approx -2\frac{\partial}{\partial z}\left(\frac{1}{r}\right) \tag{D.2.5}$$

The force on a unit charge in the near field can, using the slender body approximation, be written in the simple form:

$$(f_z)_{near} \approx -\frac{dq}{4\pi\varepsilon_0}\left[\left(1 - \frac{v^2}{c^2}\right)\frac{\partial}{\partial z}\left(\frac{1}{r}\right) + \frac{\dot{v}}{c^2}\left(\frac{1}{r}\right) - \frac{2}{3}\frac{\ddot{v}}{c^3}\right] \tag{D.2.6}$$

In view of the developments in the next chapter, it is important to emphasize that this expression relates to a *Lagrangian* description of the moving charge. We have considered a given point charge (with a constant total amount of charge dq), and have followed its history in time. Its position, velocity, acceleration etc. were all described as functions of time: $z_o(t)$, $v(t)$, $\dot{v}(t)$, ...*etc*.. However, since the expression (D.2.6) contains only the instantaneous derivatives of $z_o(t)$ and does not really depend on the history before the time t, we may interpret (D.2.6) also as the force on another unit charge when the point charge passes the point z_o and has, when passing this point, the instantaneous velocity v, the instantaneous acceleration $\dot{v}$, etc. This kind of interpretation makes a switch to an Eulerian description relatively easy.

D.3 Switch from Lagrangian to Eulerian Description

A switch to an Eulerian description is easily done. What the above equation expresses is the force in a field point (x, y, z) due to a charge dq which at time t passes the point (x_0, y_0, z_0), and has, when passing this point, a velocity v, an acceleration $\dot{v}$, and a jerk $\ddot{v}$.

In the Eulerian description the amount of charge which is passing a small volume around this point, at time t, is $\sigma(x_0, y_0, z_0, t)\, dx_0.dy_0.dz_0$, where $\sigma(x_0, y_0, z_0, t)$ is the *field* of charge densities. The velocity of this amount of charge is given by the velocity *field* $v(x_0, y_0, z_0, t)$. (Compare with the Lagrangian description above, where the symbol $v(t)$ stood for the velocity of a *given* charge, and is thus a function of time only).

Simplifying model assumptions, already stated in Chap. A.1, were that the velocity field does not depend on the x_0- and y_0-coordinates: $v = v(z_0, t)$. Furthermore, the charge density was taken to be constant, in time as well as in space.

If we follow a *particular* charge element which happens to pass the point (x_0, y_0, z_0) at time t we see that it experiences a change of charge density per unit of time given by the substantial derivative

$$\frac{D\sigma}{Dt} = \frac{\partial\sigma}{\partial t} + v\frac{\partial\sigma}{\partial z} = 0 \tag{D.3.1}$$

The substantial derivative is zero, because of the assumption $\sigma = constant$. The amount of charge contained in the particular element does not change in time, so that in Eq. (D.2.6), which was derived for a point charge whose charge is conserved, we may replace

$$dq \xrightarrow{\text{Eulerian description}} \sigma.dx_0.dy_0.dz_0 \tag{D.3.2}$$

We also step over from the velocity $v(t)$ of a given charge element to the velocity *field* $v(z_0, t)$. The velocity of the charge element, when it is passing the position (x_0, y_0, z_0) equals $v(z_0, t)$ in the Eulerian description. According to the simplifying model assumptions of Chap. A.1 the velocity field is given by:

$$v(z_0, t) = v_m(t) + \frac{\dot{s}}{s}(t).[z_0 - z_m(t)] \tag{D.3.3}$$

This is the expression that must be substituted for v in Eq. (D.2.6).

Likewise we step over from the acceleration of an individual element $\dot{v}(t)$ to the acceleration field, so that

$$\dot{v}(t) \xrightarrow{\text{Eulerian description}} \frac{Dv(z_0, t)}{Dt} = \frac{\partial v(z_0, t)}{\partial t} + v(z_0, t)\frac{\partial v(z_0, t)}{\partial z_0} \tag{D.3.4}$$

And similarly

$$\ddot{v}(t) \xrightarrow{\text{Eulerian description}} \frac{D^2 v(z_0, t)}{Dt^2} = \left(\frac{\partial}{\partial t} + v(z_0, t)\frac{\partial}{\partial z_0}\right)\frac{Dv(z_0, t)}{Dt} \tag{D.3.5}$$

D.4 Two-Parameter Representation of the Force Field (Eulerian Description)

The total force in Z-direction on a unit charge in the near field becomes, using the Eulerian description:

$$
\begin{aligned}
(f_{z,total})_{near}(x,y,z,t) = -\iiint_{droplet} \frac{\sigma.dx_0.dy_0, dz_0}{4\pi\varepsilon_0}\left[\left(1-\frac{v^2(z_0,t)}{c^2}\right)\frac{\partial}{\partial z}\left(\frac{1}{r}\right)\right. \\
\left. + \frac{Dv/Dt(z_0,t)}{c^2}\left(\frac{1}{r}\right) - \frac{2}{3}\frac{D^2v/Dt^2(z_0,t)}{c^3}\right]
\end{aligned}
\tag{D.4.1}
$$

A simple deformation mode has been assumed, depending only on the two parameters $z_m(t)$ (the position of the mid-point), and $s(t)$ (the instantaneous length). The force on a unit charge in the near field may thus be expressed in terms of these two parameters, by substituting into (D.4.1):

$$
v(z_0,t) = v_m(t) + \frac{\dot{s}}{s}(t).[z_0 - z_m(t)] \tag{D.4.2}
$$

which gives

$$
\begin{aligned}
v^2 &= v_m^2 + 2.v_m.(\dot{s}/s).(z_0 - z_m) + (\dot{s}/s)^2.(z_0 - z_m)^2 \\
\frac{Dv}{Dt} &= \dot{v}_m + (\ddot{s}/s).(z_0 - z_m) \\
\frac{D^2v}{Dt^2} &= \ddot{v}_m + (\dddot{s}/s).(z_0 - z_m)
\end{aligned}
\tag{D.4.3}
$$

A last simplification is achieved, when we assume that the length of the droplet s is at least an order smaller than the width a of the potential well: $O(s/a) = O(a/\lambda)$. A term like $v_m.(\dot{s}/s).(z_0 - z_m) = v_m.\dot{s}.\frac{z_0 - z_m}{s}$ is therefore of the order $O(a/\lambda)^3$ and a term like $(\dot{s}/s)^2.(z_0 - z_m)^2$ is of the order $O(a/\lambda)^4$ and may be neglected. It then follows that, under this last assumption, the simplified substitutions may be used:

$$
\begin{aligned}
v &= v_m + (\dot{s}/s).(z_0 - z_m) \\
v^2 &= v_m^2 + 2.v_m.(\dot{s}/s).(z_0 - z_m) \\
\frac{Dv}{Dt} &= \dot{v}_m + (\ddot{s}/s).(z_0 - z_m) \\
\frac{D^2v}{Dt^2} &= \ddot{v}_m
\end{aligned}
\tag{D.4.4}
$$

D.5 Self Force in Z-Direction

The self force on the droplet as a whole is

$$
\begin{aligned}
F_z(t) &= \iiint_{droplet} \sigma.(f_{z,total})_{near}(x,y,z,t).dx.dy.dz \\
&= -\iiint_{droplet} \sigma.dx.dy.dz \iiint_{droplet} \frac{\sigma.dx_0.dy_0.dz_0}{4\pi\epsilon_0}\left[\left(1-\frac{v^2(z_0,t)}{c^2}\right)\frac{\partial}{\partial z}\left(\frac{1}{r}\right)\right. \\
&\quad \left. + \frac{Dv/Dt(z_0,t)}{c^2}\left(\frac{1}{r}\right) - \frac{2}{3}\frac{D^2v/Dt^2(z_0,t)}{c^3}\right]
\end{aligned}
\quad \text{(D.5.1)}
$$

in which the two-parameter representation (D.4.4) of the velocity-, acceleration- and jerk field may be substituted to obtain the dependence of this force on the translation and pulsation velocity.

To keep the expressions manageable, the integral for $F_z(t)$ will be written as the sum of three contributions, in accordance with the three terms within the square brackets of the integrand.

The first of these three integrals is

$$
F_{z,1} = -\iiint_{droplet} \sigma.dx.dy.dz \iiint_{droplet} \frac{\sigma.dx_0.dy_0.dz_0}{4\pi\epsilon_0}\left(1-\frac{v^2(z_0,t)}{c^2}\right)\frac{\partial}{\partial z}\left(\frac{1}{r}\right) \quad \text{(D.5.2)}
$$

Substituting the two-parameter representation of the velocity field:

$$
\begin{aligned}
F_{z,1}(t) &= -\iiint_{droplet} \sigma.dx.dy.dz \iiint_{droplet} \frac{\sigma}{4\pi\varepsilon_0}\left(1-\frac{v_m^2+2.v_m.(\dot{s}/s).(z_0-z_m)}{c^2}\right)\frac{\partial}{\partial z}\left(\frac{1}{r}\right).dx_0.dy_0,dz_0 \\
&= -\left(1-\frac{v_m^2}{c^2}\right)\iiint_{droplet} \sigma.dx.dy.dz \iiint_{droplet} \frac{\sigma}{4\pi\varepsilon_0}\frac{\partial}{\partial z}\left(\frac{1}{r}\right).dx_0.dy_0,dz_0 \\
&\quad + \frac{2.v_m.(\dot{s}/s).}{c^2}\iiint_{droplet} \sigma.dx.dy.dz \iiint_{droplet} \frac{\sigma}{4\pi\varepsilon_0}(z_0-z_m)\frac{\partial}{\partial z}\left(\frac{1}{r}\right).dx_0.dy_0,dz_0
\end{aligned}
\quad \text{(D.5.3)}
$$

In the first of the integrals in the last line we may replace the differentiation under the integral sign by a differentiation of the integral itself. In general this operation is not allowed in the case of integrals in which the integrand has singular points. However, in Kellogg: "Foundations of potential theory" it is shown that the operation is nevertheless allowed for the type of integral we have here. This property is related to the fact that outside the charge distribution there is apparently no problem, and the field strength is simply given by a differentiation of the electric potential. Now, the potential inside the distribution mathematically has the form of a singular function, but it in fact can be shown to be non-singular (compare the gravity potential outside and inside the earth). On physical grounds one would expect that forces and the potential have the same relation in- as well as outside the distribution. It can indeed be shown—mathematically rigorously—that a differentiation to find the field strength inside the distribution is also allowed, and gives the correct field strength. One has to be cautious, because this property of interchangeability of differentiation and integration does not apply to the second derivative of the singular integrand.

The first integral is thus written as:

$$\begin{aligned}
&-\left(1-\frac{v_m^2}{c^2}\right)\iiint_{droplet}\sigma.dx.dy.dz\iiint_{droplet}\frac{\sigma}{4\pi\varepsilon_0}\frac{\partial}{\partial z}\left(\frac{1}{r}\right).dx_0.dy_0,dz_0\\
&=-\left(1-\frac{v_m^2}{c^2}\right)\iiint_{droplet}\sigma.dx.dy.dz.\frac{\partial}{\partial z}\iiint_{droplet}\frac{\sigma}{4\pi\varepsilon_0}\frac{1}{r}.dx_0.dy_0,dz_0\\
&=-\left(1-\frac{v_m^2}{c^2}\right)\iiint_{droplet}\sigma\frac{\partial\Phi_{qu.static}}{\partial z}.dx.dy.dz.
\end{aligned} \tag{D.5.4}$$

Thanks to the symmetry assumed in the droplet's configuration, the quasi-static potential $\Phi_{qu.static}$ is symmetric in Z-direction w.r.t. the midpoint of the droplet of charge. This means that the integral $\iiint_{droplet}\sigma\frac{\partial\Phi_{qu.static}}{\partial z}.dx.dy.dz$, taken over the whole droplet, will be zero.

The second integral in $F_{z,1}$ can be written in two parts, by substituting

$$z_0 - z_m = -\{(z - z_0) - (z - z_m)\} \tag{D.5.5}$$

which gives

$$\begin{aligned}
I&=\iiint_{droplet}\sigma.dx.dy.dz\iiint_{droplet}\frac{\sigma}{4\pi\varepsilon_0}(z_0-z_m)\frac{\partial}{\partial z}\left(\frac{1}{r}\right).dx_0.dy_0,dz_0\\
&=-\iiint_{droplet}\sigma.dx.dy.dz\iiint_{droplet}\frac{\sigma}{4\pi\varepsilon_0}(z-z_0)\frac{\partial}{\partial z}\left(\frac{1}{r}\right).dx_0.dy_0,dz_0+\\
&\quad+\iiint_{droplet}\sigma.(z-z_m).dx.dy.dz\iiint_{droplet}\frac{\sigma}{4\pi\varepsilon_0}\frac{\partial}{\partial z}\left(\frac{1}{r}\right).dx_0.dy_0,dz_0
\end{aligned} \tag{D.5.6}$$

In the first part of the integral I use can be made again of the slender body approximation $(z-z_0)\frac{\partial}{\partial z}\left(\frac{1}{r}\right)\approx-\frac{1}{r}$ so that this part reads

$$\begin{aligned}
&-\iiint_{droplet}\sigma.dx.dy.dz\iiint_{droplet}\frac{\sigma}{4\pi\varepsilon_0}(z-z_0)\frac{\partial}{\partial z}\left(\frac{1}{r}\right).dx_0.dy_0,dz_0\\
&\approx\iiint_{droplet}\sigma.\Phi_{qu.static}.dx.dy.dz
\end{aligned} \tag{D.5.7}$$

The latter integral may be related to the socalled electrostatic energy U_{es} of the droplet, i.e. the energy needed to collect all the charge elements (supposed to be initially present at infinity) and "compress" them into the confines of the droplet's boundaries. According to classical electrostatic theory the electrostatic energy equals

$$U_{es} = \frac{1}{2} \iiint_{droplet} \sigma.\Phi_{qu.static}.dx.dy.dz \tag{D.5.8}$$

The second part of I may be written differently by first interchanging the integration sequence:

$$\begin{aligned} &\iiint_{droplet} \sigma.(z - z_m).dx.dy.dz \iiint_{droplet} \frac{\sigma}{4\pi\varepsilon_0} \frac{\partial}{\partial z}\left(\frac{1}{r}\right).dx_0.dy_0, dz_0 \\ &= -\iiint_{droplet} \sigma.dx_0.dy_0.dz_0 \iiint_{droplet} \frac{\sigma}{4\pi\varepsilon_0}(z - z_m)\frac{\partial}{\partial z_0}\left(\frac{1}{r}\right).dx.dy.dz \end{aligned} \tag{D.5.9}$$

where use has been made of the equality

$$\frac{\partial}{\partial z_0}\left(\frac{1}{r}\right) = -\frac{\partial}{\partial z}\left(\frac{1}{r}\right) \tag{D.5.10}$$

Now interchanging the notations:

$$x \leftrightarrow x_0, \; y \leftrightarrow y_0, \quad z \leftrightarrow z_0$$

shows that

$$\begin{aligned} &\iiint_{droplet} \sigma.(z - z_m).dx.dy.dz \iiint_{droplet} \frac{\sigma}{4\pi\varepsilon_0} \frac{\partial}{\partial z}\left(\frac{1}{r}\right).dx_0.dy_0, dz_0 \\ &= -\iiint_{droplet} \sigma.dx.dy.dz \iiint_{droplet} \frac{\sigma}{4\pi\varepsilon_0}(z_0 - z_m)\frac{\partial}{\partial z}\left(\frac{1}{r}\right).dx_0.dy_0, dz_0 - I \end{aligned} \tag{D.5.11}$$

We thus obtain from Eq. (D.5.6):

$$I = 2.U_{es} - I \tag{D.5.12}$$

or

$$I = U_{es} \tag{D.5.13}$$

and finally from (D.5.3), (D.5.4) and (D.5.13):

$$F_{z,1}(t) = \frac{2.v_m(t).(\dot{s}/s)(t)}{c^2}.U_{es} \tag{D.5.14}$$

The second part of F_z is according to Eq. (D.5.1) defined as

$$
\begin{aligned}
F_{z,2}(t) &= -\iiint_{droplet} \sigma.dx.dy.dz \iiint_{droplet} \frac{\sigma.dx_0.dy_0.dz_0}{4\pi\epsilon_0} \frac{Dv/Dt}{c^2}\left(\frac{1}{r}\right) \\
&= -\frac{\dot{v}_m}{c^2} \iiint_{droplet} \sigma.dx.dy.dz \iiint_{droplet} \frac{\sigma.dx_0.dy_0.dz_0}{4\pi\epsilon_0}\left(\frac{1}{r}\right) + \\
&\quad -\frac{\ddot{s}/s}{c^2} \iiint_{droplet} \sigma.dx.dy.dz \iiint_{droplet} \frac{\sigma.dx_0.dy_0.dz_0}{4\pi\epsilon_0}(z_0 - z_m)\frac{1}{r}
\end{aligned}
\tag{D.5.15}
$$

The second integral in the r.h.s. vanishes, as may be seen by first interchanging the order of the integration, and then taking into account the assumed symmetry of the droplet. The first integral in the r.h.s. can again be related to the electrostatic energy, from which it follows:

$$
F_{z,2}(t) = -\frac{\dot{v}_m}{c^2}.2U_{es} \tag{D.5.16}
$$

The third part of $F_z(t)$ is defined as

$$
\begin{aligned}
F_{z,3}(t) &= \frac{2}{3} \iiint_{droplet} \sigma.dx.dy.dz \iiint_{droplet} \frac{\sigma.dx_0.dy_0.dz_0}{4\pi\epsilon_0} \frac{D^2v/Dt^2(z_0,t)}{c^3} \\
&= \frac{2}{3}\frac{\ddot{v}_m}{c^3} \iiint_{droplet} \sigma.dx.dy.dz \iiint_{droplet} \frac{\sigma.dx_0.dy_0.dz_0}{4\pi\epsilon_0} = \frac{2}{3}\frac{q^2}{4\pi\epsilon_0}\frac{\ddot{v}_m}{c^3}
\end{aligned}
\tag{D.5.17}
$$

D.6 Equation of Motion for the Translation of the Droplet of Charge

Collecting Eqs. (D.5.14), (D.5.16) and (D.5.17) and substituting them into the dynamic equation

$$
F_z(t) + q.E_{external} = m_{bare}.\dot{v}_m \tag{D.6.1}
$$

(where m_{bare} stands for the mechanical mass of the droplet, and $E_{external}$ is the field strength due to an external potential), we find the dynamic equation for the translation:

$$
\left(m_{bare} + 2\frac{U_{es}}{c^2}\right)\dot{v}_m - \frac{2}{3}\frac{q^2}{4\pi\varepsilon_0}\frac{\ddot{v}_m}{c^3} = 2\,v_m(\dot{s}/s)\,\frac{U_{es}}{c^2} + q.E_{external} \tag{D.6.2}
$$

The coefficient of the acceleration, i.e. the apparent mass, consists of two terms. It appears that the electrostatic energy contributes to the total mass, in agreement with the equivalence of mass and energy according to the special theory of relativity. It is usual to call the addition $2\frac{U_{es}}{c^2}$ to the total mass the *electromagnetic mass* m_{em} of the charge. Hence:

$$m_{em} = 2\,U_{es}/c^2 \tag{D.6.3}$$

The factor 2 in this expression needs some further investigation, since one would expect that the additional mass would equal the mass m_{es} equivalent to the electrostatic energy:

$$m_{es} = U_{es}/c^2 \tag{D.6.4}$$

Actually, it appears from (D.6.3) and (D.6.4) that the addition m_{em} to the mass of the droplet is *twice* the mass m_{es}:

$$m_{em} = 2\,m_{es} \tag{D.6.5}$$

In the main text of this book the matter is discussed in more detail. The explanation was given by A.D. Yaghjian (Relativistic Dynamics of a Charged Sphere, updating the Lorentz-Abraham model. Lectures in Physics 686, second edition by Springer, 2006). He argues that the mass which is being used in the relativity relation must be the measurable mass:

$$m_{measured} = m_{bare} + m_{em} \tag{D.6.6}$$

so that according to the relativity relations the intrinsic energy must be

$$U_{intrinsic} = m_{measured}.c^2 = (m_{bare} + m_{em}).c^2 \tag{D.6.7}$$

If we require that the intrinsic energy should be the same as the energy of formation:

$$U_{intrinsic} = U_{es} \tag{D.6.8}$$

it follows that m_{bare} is negative. In our case of the slender body approximation:

$$m_{bare} = -U_{es}/c^2 = -\frac{1}{2}m_{em}. \tag{D.6.9}$$

In Chap. B.2 of the main text, the implication of a negative bare mass, i.e. a mass *deficit*, is considered further. The mass deficit is there associated with the binding energy of the droplet.

It is important to realise that the above given relations between m_{bare}, m_{es} and m_{em} are applicable only to *measured* conditions, i.e. the conditions for which actual values of electron masses (or muon or tau masses) are tabulated. If these conditions are perturbed by the application of squeezing *forces* (as distinct from the effects of relativity), one may no longer assume these relations to be valid. The effect of a squeezing *force* on the masses m_{bare} and m_{em} is quite different from relativistic effects. An external squeezing force leads to a further compression of the charges, i.e. a shortening of the length s, implying an increase of the electrostatic energy, and therefore also an increase of the electromagnetic mass. There is no such effect of squeezing on the mechanical mass m_{bare}.

In order to emphasize this, the relations given above will from now on be stated like

$$m_{bare} = -\frac{1}{2} m_{em_0} \tag{D.6.10}$$

where m_{em_0} denotes the *unperturbed* electromagnetic mass, i.e. the value of m_{em} without external squeezing forces:

$$m_{em_0} = 2\, U_{es_0}/c^2. \tag{D.6.11}$$

Therefore, the effective mass $\left(m_{bare} + 2\frac{U_{es}}{c^2}\right)$ in the equation of motion is to interpreted as

$$\left(m_{bare} + 2\frac{U_{es}}{c^2}\right) = \left[-\frac{1}{2} m_{em_0} + m_{em}(s)\right] \tag{D.6.12}$$

where the first term between the square brackets is a constant, whilst the second term depends on s, i.e. it depends on the squeezing forces. Only under *unperturbed* conditions the total apparent mass will be equal to $\frac{1}{2} m_{em_0}$.

It will appear later that this interpretation is essential, or else the requirement of conservation of momentum and energy cannot be satisfied.

Using this result, the dynamic equation for the translation of the droplet may be put in the alternative form

$$\left[-\frac{1}{2} m_{em_0} + m_{em}(s)\right] \dot{v}_m - \frac{2}{3}\frac{q^2}{4\pi\epsilon_0}\frac{\ddot{v}_m}{c^3} = m_{em}(s)\, v_m \frac{\dot{s}}{s} + q.E_{external} \tag{D.6.13}$$

We will often add to the r.h.s. of this equation a fictive force $F(t)$. This artifice is used to study special kinds of motion. For instance, Eq. (19.13) will never describe an undamped motion, because of the term with $\ddot{v}_m$. This term is the radiation resistance, which will later (see chapters about the combination of the radiation field with the droplet's dynamics) be shown to be the "recoil" force associated with radiation. The occurrence of radiation implies a loss of energy from the system, so that Eq. (19.13) always describes an unsteady situation. However, it will be seen later (chap. G.5) that the rate of change of energy is slow compared with the time scale of the pulsating motion. It will appear that this feature facilitates an asymptotic approximation to be made of the unsteady equations, where as asymptotic limit the—artificially—undamped system is taken.

The most general translation equation of motion to be used in the remainder of this text is therefore:

$$\left[-\frac{1}{2}m_{em_0}+m_{em}(s)\right]\dot{v}_m-\frac{2}{3}\frac{q^2}{4\pi\epsilon_0}\frac{\ddot{v}_m}{c^3}=m_{em}(s)v_m\frac{\dot{s}}{s}+q.E_{external}+F(t) \quad \text{(D.6.14)}$$

D.7 The Electromagnetic Mass

If it is wished, one may express the electromagnetic mass m_{em} as a dimensional quantity multiplied by a non-dimensional factor which only depends on the non-dimensional shape of the droplet of charge. This is achieved by going back to Eq. (D.5.8):

$$\begin{aligned} m_{em} &= 2\,U_{es}/c^2=\frac{1}{c^2}\iiint_{droplet}\sigma.\Phi_{qu.static}.dx.dy.dz \\ &= \frac{1}{c^2}\iiint_{droplet}\sigma.dx.dy.dz\frac{\sigma\,dx_0\,dy_0\,dz_0}{4\pi\varepsilon_0}\frac{1}{\sqrt{(x-x_0)^2+(y-y_0)^2+(z-z_0)^2}} \\ &= \frac{1}{c^2}\frac{\sigma^2(t^2\,s)^2}{4\pi\varepsilon_0}\iiint_{droplet}d(x/t)d(y/t)d(z/s)\iiint_{droplet}\frac{d(x_0/t)d(y_0/t)d(z_0/s)}{s.\sqrt{\left(\frac{x-x_0}{t}\right)^2\left(\frac{t}{s}\right)^2+\left(\frac{y-y_0}{t}\right)^2\left(\frac{t}{s}\right)^2+\left(\frac{z-z_0}{s}\right)^2}} \end{aligned} \quad \text{(D.7.1)}$$

which may be written as

$$m_{em}=\frac{1}{c^2}\frac{q^2}{4\pi\varepsilon_0}\frac{1}{s}\alpha \quad \text{(D.7.2)}$$

where α is a non-dimensional factor purely depending on the non-dimensional shape of the droplet.

Some remarks must be made about the effects of relativity. First of all it should be realised that all the derivations in this book are strictly non-relativistic (as far as possible, considering the use of Maxwell's equations, implicitly conforming to the theory of special relativity). The inertia due to charge, as represented by the electromagnetic mass, cropped up in the equation of motion (D.6.2) due to an electromagnetic self force $2\frac{U_{es}}{c^2}\,\dot{v}_m$. From the asymptotic point of view, this is an effect of the order $O(v/c)^2=O(a/\lambda)^2$. Now, looking at the expression (D.7.2) one would be tempted to think that m_{em} conforms to the relativistic speed dependency of mass according to relativity theory because of the presence of the length s in the denominator. Within the framework of relativity theory this length would be subject to Lorentz contraction, with the result that the electromagnetic mass would be velocity dependent in exactly the way as predicted by relativity theory. However, it should be realised that the inclusion of Lorentz contraction in (D.7.2) would imply that effects of an order $O(v/c)^4 = O(a/\lambda)^4$ are taken into account, which means that we are then coming outside the validity of the asymptotic approximations. Relativistic effects must thus be excluded. This is the reason why the equation of motion was set up using Newtonian dynamics, which is thus consistent with the approximations accurate up to order $O(a/\lambda)^3$. The dynamic equation gives the effects of s variations due to external forces, but not due to relativistic effects. This should also be kept in mind when we later consider the squeezing dynamics of the droplet of charge.

D.8 Momentum Equation

Using the expression (D.7.2) for m_{em} it is seen that

$$\frac{d}{dt}[m_{em}(s)v_m] = m_{em}(s).\,\dot{v}_m - m_{em}(s)\,v_m\frac{\dot{s}}{s} \tag{D.8.1}$$

Hence, the equation of motion (D.6.14) may be written in the form

$$\frac{d}{dt}\left[\left\{-\frac{1}{2}m_{em_0} + m_{em}(s)\right\}v_m\right] - \frac{2}{3}\frac{q^2}{4\pi\epsilon_0}\frac{\ddot{v}_m}{c^3} = q.E_{external} + F(t) \tag{D.8.2}$$

The term between the curly brackets is the instantaneous effective mass, so that we can also write

$$\frac{dp}{dt} = \frac{2}{3}\frac{q^2}{4\pi\epsilon_0}\frac{\ddot{v}_m}{c^3} + q.E_{external} + F(t) \tag{D.8.3}$$

where p denotes the momentum. The expression shows that Newton's dynamic equation $F_{ext} = dp/dt$ is satisfied.

A consequence is that, in the case of a steady condition (i.e. if the first term in the r.h.s., the socalled "radiation resistance", is cancelled by the artificially added force $F(t)$) and in the absence of external potential gradients, the momentum is constant. However, this does *not* imply that the velocity is constant, since fluctuations of s cause the effective mass to be variable. To have a constant momentum therefore requires the velocity to fluctuate, together with the pulsations of the droplet.

Here we can note an inconsistency which must be clarified later, when radiation effects are treated in more detail. A free droplet, which we know to be in steady motion like a free flying electron, must according to the above show a fluctuation of its velocity. Classical electromagnetic theory then predicts a loss of energy due to radiation (because radiation depends on the accelerations of the particle). It means that a free flying particle gradually would lose its kinetic energy and would finally come a stand still. Apparently, the model of the pulsating droplet can only make sense if an explanation can be given how a non-radiative condition can come about, despite the "ripples" in the velocity.

Another problem occurs if we consider the translation energy of the droplet. If we consider again the free flying droplet in the absence of an external potential and in steady motion, one would expect a constant kinetic energy. This is clearly not the case, since kinetic energy $T_{translation}$ equals

$$T_{translation} = \frac{p^2}{2m} \tag{D.8.4}$$

Under the conditions stated, the momentum p is constant, but the effective mass $\left\{-\frac{1}{2}m_{em_0} + m_{em}(s)\right\}$ fluctuates in accordance with the pulsations. The translational energy therefore fluctuates as well. This cannot be explained by the equation of motion for the translation alone. The energy in the translation can only be variable, if there is an exchange of energy with the pulsation. We will therefore have to check carefully (this will be done later) whether the sum of the pulsation energy and translation energy is indeed constant under the stated conditions, or else the requirement of the conservation of energy is not satisfied. Non-compliance with the law of conservation of energy would indicate an inconsistent model, and must therefore be inspected carefully.

Appendix E
Squeezing Self Forces (Pulsation)

E.1 The Electromagnetic Generalised Squeezing Force

This generalised self force in s-direction (to be indicated as "squeezing self force" and defined as a Lagrangian generalised force) is defined via the work done on the elements of the droplet during a virtual variation of the length s.

The force on an element of the droplet in (x, y, z) at time t is $\sigma\, dx.dy.dx.f_z$. If we keep the force constant and apply a virtual displacement of the element over a distance δz the virtual work done is

$$\delta W = (\sigma\, dx.dy.dx.f_z).\delta z. \tag{E.1.1}$$

The virtual displacement must be in agreement with the kind of deformations allowed in the model of the droplet. Therefore the virtual displacement associated with a virtual variation of the droplet's length is

$$\delta z = \left.\frac{\delta z}{\delta s}\right|_{\left(\frac{z-z_m}{s}\right)=constant} .\delta s = \frac{\partial v}{\partial \dot{s}}\delta s = \frac{z - z_m}{s}\delta s \tag{E.1.2}$$

Finally, the generalised force in s-direction is

$$Q_s = \frac{\delta W}{\delta s} = (\sigma\, dx.dy.dx.f_z).\frac{z - z_m}{s} \tag{E.1.3}$$

Integrated over all charge elements, we find using (D.4.1) the total generalised force on the droplet:

$$\begin{aligned} Q_s(t) &= \iiint_{droplet} \sigma.(f_{z,total})_{near}\frac{z - z_m}{s}.dx.dy.dz \\ &= -\iiint_{droplet} \sigma.\frac{z - z_m}{s}dx.dy.dz \iiint_{droplet} \frac{\sigma.dx_0.dy_0.dz_0}{4\pi\epsilon_0}\left[\left(1 - \frac{v^2(z_0, t)}{c^2}\right)\frac{\partial}{\partial z}\left(\frac{1}{r}\right)\right. \\ &\quad \left. + \frac{Dv/Dt(z_0, t)}{c^2}\left(\frac{1}{r}\right) - \frac{2}{3}\frac{D^2v/Dt^2(z_0, t)}{c^3}\right] \end{aligned} \tag{E.1.4}$$

T. van Holten, *The Atomic World Spooky? It Ain't Necessarily So!*
DOI 10.2991/978-94-6239-234-2

Again, we split up the generalised force into three parts, in accordance with the three terms within the square brackets of the integrand. Furthermore, the relations of Eq. (D.4.4) are substituted. The first contribution to the "squeezing self force" is

$$
\begin{aligned}
Q_{s1} &= -\iiint_{droplet} \sigma \frac{z - z_m}{s}.dx.dy.dz \iiint_{droplet} \frac{\sigma}{4\pi\varepsilon_0}\left(1 - \frac{v^2(z_0, t)}{c^2}\right)\frac{\partial}{\partial z}\left(\frac{1}{r}\right).dx_0.dy_0, dz_0 \\
&= -\left(1 - \frac{v_m^2}{c^2}\right)\iiint_{droplet} \sigma \frac{z - z_m}{s}.dx.dy.dz \iiint_{droplet} \frac{\sigma}{4\pi\varepsilon_0}\frac{\partial}{\partial z}\left(\frac{1}{r}\right).dx_0.dy_0.dz_0 \\
&+ \frac{2.v_m.(\dot{s}/s).}{c^2}\iiint_{droplet} \sigma \frac{z - z_m}{s}.dx.dy.dz \iiint_{droplet} \frac{\sigma}{4\pi\varepsilon_0}(z_0 - z_m)\frac{\partial}{\partial z}\left(\frac{1}{r}\right).dx_0.dy_0, dz_0
\end{aligned}
\tag{E.1.5}
$$

The first of the integrals in the r.h.s. did occur earlier: see Eq. (D.5.11). It gives the contribution $\left(1 - \frac{v_m^2}{c^2}\right)\frac{1}{s}U_{es}$. The second integral is zero, which may be seen as follows. Defining I as

$$
I = \iiint_{droplet} (z - z_m)dx.dy.dz \iiint_{droplet} (z_0 - z_m)\frac{\partial}{\partial z}\left(\frac{1}{r}\right)dx_0.dy_0.dz_0 \tag{E.1.6}
$$

we can reverse the sequence of the integrations, and take into account that $\frac{\partial}{\partial z}\left(\frac{1}{r}\right) = -\frac{\partial}{\partial z_0}\left(\frac{1}{r}\right)$:

$$
I = -\iiint_{droplet} (z_0 - z_m)dx_0.dy_0.dz_0 \quad \iiint_{droplet} (z - z_m)\frac{\partial}{\partial z_0}\left(\frac{1}{r}\right)dx.dy.dz \tag{E.1.7}
$$

Now interchanging the notations: $x \leftrightarrow x_0$, etc. it follows that $I = -I$, or $I = 0$. We thus find:

$$
Q_{s,1} = \left(1 - \frac{v_m^2}{c^2}\right).\frac{U_{es}}{s} \tag{E.1.8}
$$

An alternative form can be given to this result, using the earlier derived expressions

$$
m_{em} = 2\frac{U_{es}}{c^2} \tag{D.6.3}
$$

$$
m_{em} = \frac{1}{c^2}\frac{q^2}{4\pi\varepsilon_0}\frac{1}{s}\alpha \tag{D.7.2}
$$

from which

$$U_{es} = \frac{1}{2}\frac{q^2}{4\pi\varepsilon_0}\frac{1}{s}\alpha \tag{E.1.9}$$

Differentiation yields

$$\frac{\partial U_{es}}{\partial s} = -\frac{U_{es}}{s} \tag{E.1.10}$$

so that we may write

$$Q_{s,1} = -\left(1 - \frac{v_m^2}{c^2}\right).\frac{\partial U_{es}}{\partial s} \tag{E.1.11}$$

The last form makes this contribution to the " squeezing self force" more understandable, because U_{es} is the potential energy stored by bringing together all the charge elements of the droplet. Its derivative must thus be the associated generalised force. The factor multiplying the derivative represents a relativistic effect, explained in the main text.

The second contribution to the "squeezing self force" is

$$\begin{aligned} Q_{s2} &= -\iiint_{droplet} \sigma\frac{z - z_m}{s}.dx.dy.dz \iiint_{droplet} \frac{\sigma}{4\pi\varepsilon_0}\frac{Dv/Dt(x_0, y_0, z_0, t)}{c^2}\frac{1}{r}.dx_0.dy_0, dz_0 \\ &= -\frac{\dot{v}_m}{c^2}\iiint_{droplet} \sigma\frac{z - z_m}{s}.dx.dy.dz \iiint_{droplet} \frac{\sigma}{4\pi\varepsilon_0}\frac{1}{r}.dx_0.dy_0, dz_0 \\ &\quad - \frac{(\ddot{s}/s).}{c^2}\iiint_{droplet} \sigma\frac{z - z_m}{s}.dx.dy.dz \iiint_{droplet} \frac{\sigma}{4\pi\varepsilon_0}(z_0 - z_m)\frac{1}{r}.dx_0.dy_0, dz_0 \end{aligned} \tag{E.1.12}$$

The first of the integrals in the r.h.s. is zero on account of the symmetry of the droplet w.r.t. the midpoint.

The second integral may be interpreted as the electrostatic energy needed to form an asymmetric charge distribution $\mu(z, t) = \sigma\frac{z - z_m}{s}$.

Let the energy of formation of this kind of asymmetric charge distribution be denoted as U_{es}^*, then

$$Q_{s,2} = -\frac{(\ddot{s}/s)}{c^2}2\,s\,U_{es}^* = -2\frac{U_{es}^*}{c^2}.\ddot{s} \tag{E.1.13}$$

This is apparently an inertia term (in the sense of a d'Alembert inertia force), which may be written in terms of a generalised electromagnetic mass m^*_{em} defined as

$$m^*_{em} = 2\frac{U^*_{es}}{c^2} \tag{E.1.14}$$

As before in the case of m_{em}, one may write this generalised mass as a product of dimensional quantities and a non-dimensional factor depending on the shape of the charge distribution:

$$m^*_{em} = \frac{1}{c^2}\frac{q^2}{4\pi\varepsilon_0}\frac{1}{s}\alpha^* \tag{E.1.15}$$

The third contribution to the "squeezing self force" is

$$\begin{aligned} Q_{s3} &= \frac{2}{3}\iiint_{droplet}\sigma\frac{z - z_m}{s}.dx.dy.dz\iiint_{droplet}\frac{\sigma}{4\pi\varepsilon_0}\frac{D^2v/Dt^2(x_0, y_0, z_0, t)}{c^3}.dx_0.dy_0, dz_0 \\ &= \frac{2}{3}\frac{\ddot{v}_m}{c^3}\iiint_{droplet}\sigma\frac{z - z_m}{s}.dx.dy.dz\iiint_{droplet}\frac{\sigma}{4\pi\varepsilon_0}.dx_0.dy_0, dz_0 \end{aligned} \tag{E.1.16}$$

which is zero on account of symmetry considerations:

$$Q_{s,3} = 0 \tag{E.1.17}$$

The complete generalised force in s-direction, as far as it is associated with electromagnetic effects, is:

$$Q_s = -\left(1 - \frac{v_m^2}{c^2}\right).\frac{\partial U_{es}}{\partial s} - m^*_{em}.\ddot{s} + Q_{s,external} \tag{E.1.18}$$

where $Q_{s,external}$ stands for any "squeezing force" (positive in the direction of increasing elongation s) due to external electrical effects.

E.2 Non-electromagnetic Squeezing Forces

In the main text it is argued why—as a first attempt—the "surface tension model" is taken as a working hypothesis, to represent any Poincaré force preventing an explosion of the droplet under the influence of the internal electrostatic repulsion.

The surface tension model implies a generalised force of the general type

$$Q_{s,Poincare} = -\left[qAs + qB + qC\frac{1}{s}\right] \tag{E.2.1}$$

where the minus-sign and the common factor q are used just for convenience, in order to simplify the later found expressions for the constants A, B, C.

Once again, it should be stressed that the theory developed here is non-relativistic (see the remarks made in Chap. D.7). The length s is therefore, to be consistent, *not* velocity dependent as it would be if it were subject to Lorentz contraction. The variations of s can nevertheless be influenced by the velocity of the droplet, but only through the variation of forces on the droplet.

If it is assumed that external potentials do not have a significant contribution to the squeezing, and if a stationary situation is assumed (so that squeezing is neither influenced by velocity nor inertia effects), the generalised force $Q_{s,Poincare}$ and the electromagnetic generalised force $-\frac{\partial U_{es}}{\partial s}$ (stationary situation, so $v_m = 0$) must be zero when summed together:

$$-\frac{1}{2}\frac{q^2}{4\pi\varepsilon_0}\frac{\alpha}{s^2} + \left[q.A.s + q.B + q.C\frac{1}{s}\right] = 0 \tag{E.2.2}$$

which equation determines the equilibrium values of the droplet's length s in a stationary situation. The equation may be written in the form of a cubic algebraic equation:

$$s^3 + \frac{B}{A}s^2 + \frac{C}{A}s - \frac{1}{2}\frac{q^2}{4\pi\varepsilon_0}\alpha\frac{1}{qA} = 0 \tag{E.2.3}$$

It therefore has three solutions. In the main text it is argued that these three equilibrium values of s might be identified with the three "guises" of an electron: the common electron, the muon and the tau particle (again this is just an attempt to see if any useful result can be obtained from the "surface tension" hypothesis). Let us accordingly denote the three possible real solutions from the cubic as $s_e,\ s_\mu,\ and\ s_\tau$. It should thus be possible to bring the cubic in the form

$$(s - s_e).(s - s_\mu).(s - s_\tau) = 0 \tag{E.2.4}$$

or:

$$s^3 - (s_e + s_\mu + s_\tau).s^2 + (s_e.s_\mu + s_\mu.s_\tau + s_\tau.s_e).s - s_e.s_\mu.s_\tau = 0 \tag{E.2.5}$$

resulting in the requirement that the constants satisfy the following relations:

$$\begin{aligned}\frac{B}{A} &= -(s_e + s_\mu + s_\tau) = -s_e\left(1 + \frac{s_\mu}{s_e} + \frac{s_\tau}{s_e}\right)\\ \frac{C}{A} &= (s_e.s_\mu + s_\mu.s_\tau + s_\tau.s_e) = s_e^2.\left(\frac{s_\mu}{s_e} + \frac{s_\mu}{s_e}\frac{s_\tau}{s_e} + \frac{s_\tau}{s_e}\right)\\ \frac{1}{2}\frac{q^2}{4\pi\varepsilon_0}\alpha\frac{1}{qA} &= s_e.s_\mu.s_\tau = s_e^3.\left(\frac{s_\mu}{s_e}\frac{s_\tau}{s_e}\right)\end{aligned} \tag{E.2.6}$$

Now, it was determined that the total mass of the common electron $m_{measurable} = (m_{bare} + m_{em_0})$ equals $\frac{1}{2}m_{em_0}$ (according to Yaghjian's hypothesis of negative m_{bare} representing the mass deficit associated with binding energy).

Therefore, the measurable electron mass is:

$$m_e = \frac{1}{2}\frac{1}{c^2}\frac{q^2}{4\pi\varepsilon_0}\frac{\alpha}{s_e} \tag{E.2.7}$$

likewise, the muon- and tau masses are

$$m_\mu = \frac{1}{2}\frac{1}{c^2}\frac{q^2}{4\,\pi\,\varepsilon_0}\frac{\alpha}{s_\mu} \tag{E.2.8}$$

$$m_\tau = \frac{1}{2}\frac{1}{c^2}\frac{q^2}{4\,\pi\,\varepsilon_0}\frac{\alpha}{s_\tau} \tag{E.2.9}$$

Now denoting by μ and τ the mass ratio's:

$$\mu = \frac{m_\mu}{m_e} = \frac{s_e}{s_\mu} \tag{E.2.10}$$

$$\tau = \frac{m_\tau}{m_e} = \frac{s_e}{s_\tau} \tag{E.2.11}$$

the requirement on the constants A, B, C transforms into:

$$\frac{B}{A} = -s_e.\left(1 + \frac{1}{\mu} + \frac{1}{\tau}\right) \tag{E.2.12}$$

$$\frac{C}{A} = s_e^2.\left(\frac{1}{\mu} + \frac{1}{\tau} + \frac{1}{\mu.\tau}\right) \tag{E.2.13}$$

$$\frac{1}{2}\frac{q^2}{4\pi\varepsilon_0}\frac{\alpha}{qA} = s_e^3.\frac{1}{\mu.\tau} \tag{E.2.14}$$

We can now, having determined the constants in Eq. (E.2.1), write the complete equation of motion in s-direction as

$$m^*_{bare} . \ddot{s} = Q_{s,electromagnetic} + Q_{s,Poincare} \tag{E.2.15}$$

where Eqs. (E.1.18) and (E.2.1) can be substituted in the right hand side. The explicit expression is shown in the next chapter.

Appendix F
The Equations of Motion and Their Solution

F.1 Summary of the Set of Non-linear Equations of Motion

The equation of motion for the translation was derived in Chap. D.6, Eq. (D.6.14):

$$\left[-\frac{1}{2}m_{em_0} + m_{em}(s)\right]\dot{v}_m - \frac{2}{3}\frac{q^2}{4\pi\epsilon_0}\frac{\ddot{v}_m}{c^3} = m_{em}(s)\,v_m\frac{\dot{s}}{s} + q.E_{external} + F(t) \qquad \text{(F.1.1)}$$

where the electromagnetic mass is given by

$$m_{em}(s) = \frac{1}{c^2}\frac{q^2}{4\pi\varepsilon_0}\frac{1}{s}\alpha \qquad \text{(F.1.2)}$$

having an equilibrium value

$$m_{em_0} = m_{em}(s_e) = \frac{1}{c^2}\frac{q^2}{4\pi\varepsilon_0}\frac{1}{s_e}\alpha \qquad \text{(F.1.3)}$$

An alternative formulation is often useful in terms of the momentum, as derived in Chap. D.8:

$$\frac{dp}{dt} = \frac{2}{3}\frac{q^2}{4\pi\epsilon_0}\frac{\ddot{v}_m}{c^3} + q.E_{external} + F(t) \qquad \text{(F.1.4)}$$

where the momentum p is based on the instantaneous value of the mass:

$$p = \left\{-\frac{1}{2}m_{em_0} + m_{em}(s)\right\}v_m = \frac{1}{2}m_{em_0}\left\{2\frac{s_e}{s} - 1\right\}v_m \qquad \text{(F.1.5)}$$

Recall that $\frac{1}{2}m_{em_0}$ is the rest mass of the droplet.

T. van Holten, *The Atomic World Spooky? It Ain't Necessarily So!*
DOI 10.2991/978-94-6239-234-2

Equation (E.2.15) for the pulsation reads after substitution of (E.1.18) and (E.2.1) and using the values of the constants (E.2.12) through (E.2.14):

$$\begin{aligned}&\left(m^*_{bare}+m^*_{em}\right)\ddot{s}+\left(1-\frac{v_m^2}{c^2}\right)\frac{\partial U_{es}}{\partial s}\\&+\frac{1}{2}\frac{q^2}{4\pi\varepsilon_0}\frac{\alpha}{s_e^3}\mu\tau.s\left[1-\frac{s_e}{s}\left\{1+\frac{1}{\mu}+\frac{1}{\tau}\right\}+\left(\frac{s_e}{s}\right)^2\left\{\frac{1}{\mu}+\frac{1}{\tau}+\frac{1}{\mu.\tau}\right\}\right]=Q_{s,external}\end{aligned}\tag{F.1.6}$$

where

$$U_{es}=\frac{1}{2}\frac{q^2}{4\pi\varepsilon_0}\frac{1}{s}\alpha\tag{F.1.7}$$

and

$$m^*_{em}=\frac{1}{c^2}\frac{q^2}{4\pi\varepsilon_0}\frac{1}{s}\alpha^*\tag{F.1.8}$$

F.2 Conservation of Energy

In general, if we have an equation of motion of the form

$$M(t)\ddot{x}+\frac{\partial U(x)}{\partial x}=f(x,t)\tag{F.2.1}$$

with a time dependent mass, then the sum of kinetic and potential energy is

$$E(t)=\frac{1}{2}M(t)\dot{x}^2+U(x)\tag{F.2.2}$$

and the rate of change of the total energy is

$$\frac{dE}{dt}=M\dot{x}\ddot{x}+\frac{1}{2}\dot{M}\dot{x}^2+\frac{\partial U}{\partial x}\dot{x}\tag{F.2.3}$$

Multiplying the equation of motion (F.2.1) by $\dot{x}$ and substituting the result into the expression for $\frac{dE}{dt}$ Eq. (F.2.3) gives

$$\frac{dE}{dt}=f(x,t)\dot{x}+\frac{1}{2}\dot{M}\dot{x}^2\tag{F.2.4}$$

Applying this general result to the equation of motion for the pulsation:

$$\left(m^*_{bare}+m^*_{em}\right)\ddot{s}+\left(1-\frac{v_m^2}{c^2}\right)\frac{\partial U_{es}}{\partial s}$$
$$+\frac{1}{2}\frac{q^2}{4\pi\varepsilon_0}\frac{\alpha}{s_e^3}\mu\tau.s\left[1-\frac{s_e}{s}\left\{1+\frac{1}{\mu}+\frac{1}{\tau}\right\}+\left(\frac{s_e}{s}\right)^2\left\{\frac{1}{\mu}+\frac{1}{\tau}+\frac{1}{\mu.\tau}\right\}\right]=Q_{s,external} \tag{F.2.5}$$

we have as time derivative of the variable mass:

$$\dot{M}=\frac{dm^*_{em}}{dt}=-m^*_{em}\frac{\dot{s}}{s} \tag{F.2.6}$$

All the terms in the equation of motion (F.2.5) that depend only on s can be thought to be derived from a potential funtion. The remaining terms are brought to the r.h.s. and are denoted as $f(t)$:

$$f(t)=\frac{v_m^2}{c^2}\frac{\partial U_{es}}{\partial s}=-\frac{1}{2}v_m^2\,m_{em}\frac{1}{s} \tag{F.2.7}$$

From the general result (F.2.4) it then follows:

$$\frac{dE_{puls}}{dt}=-\frac{1}{2}v_m^2\,m_{em}\frac{\dot{s}}{s}-\frac{1}{2}m^*_{em}\frac{\dot{s}}{s}\dot{s}^2 \tag{F.2.8}$$

On the basis of the assumption in Chap. D.4 that $s/a=O(a/\lambda)$ we may neglect the second term in the r.h.s. as irrelevant within the scope of the present asymptotic theory which is accurate up to and including $O(a/\lambda)^3$. Therefore:

$$\frac{dE_{puls}}{dt}=-\frac{1}{2}v_m^2\,m_{em}\frac{\dot{s}}{s} \tag{F.2.9}$$

The translation equation of motion is:

$$\left[-\frac{1}{2}m_{em_0}+m_{em}(s)\right]\dot{v}_m-\frac{2}{3}\frac{q^2}{4\pi\epsilon_0}\frac{\ddot{v}_m}{c^3}=m_{em}(s)v_m\frac{\dot{s}}{s}+q.E_{external}+F(t) \tag{F.1.1}$$

from which it is derived that here the variation of the mass is:

$$\dot{M}=\frac{dm_{em}}{dt}=-m_{em}\frac{\dot{s}}{s} \tag{F.2.10}$$

and

$$f(t) = \frac{2}{3}\frac{q^2}{4\pi\epsilon_0}\frac{\ddot{v}_m}{c^3} + m_{em}(s)v_m\frac{\dot{s}}{s} + F(t) \tag{F.2.11}$$

resulting in

$$\frac{dE_{transl}}{dt} = \frac{1}{2}v_m^2\, m_{em}\frac{\dot{s}}{s} + \frac{2}{3}\frac{q^2}{4\pi\epsilon_0}\frac{\ddot{v}_m v_m}{c^3} + F(t).v_m \tag{F.2.12}$$

Comparing Eqs. (F.2.9) and (F.2.12) we obtain the final result:

$$\frac{dE_{transl}}{dt} + \frac{dE_{puls}}{dt} = \frac{2}{3}\frac{q^2}{4\pi\epsilon_0}\frac{\ddot{v}_m v_m}{c^3} + F(t).v_m \tag{F.2.13}$$

In the case of a steady motion (i.e. if the term with the translation jerk—which is associated with energy loss by radiation—is cancelled by the force F(t)), this result shows that the total energy is constant. There are fluctuations of the translation energy and pulsation energy, but these represent a mutual energy transfer without affecting the total energy. The model of the droplet of charge thus satisfies the law of conservation of energy.

F.3 Equations of Motion Expressed in Terms of Pulsation Perturbations

Denoting by Δs perturbations w.r.t. the equilibrium value s_e (for now *without* assuming that the perturbations are small):

$$s = s_e\left(1 + \frac{\Delta s}{s_e}\right) \tag{F.3.1}$$

The length s must be positive, so that from physical considerations it is concluded that the condition $\left|\frac{\Delta s}{s_e}\right| < 1$ must be satisfied, and the following Taylor expansions may be made. It is not yet assumed that $\left|\frac{\Delta s}{s_e}\right|$ is always so much smaller than 1 that linearisation would be allowed. We therefore retain non-linear terms for now.

The electromagnetic mass is

$$m_{em} = \frac{1}{c^2}\frac{q^2}{4\pi\varepsilon_0}\frac{1}{s}\alpha = m_{em_0}\left[1 - \frac{\Delta s}{s_e} + \left(\frac{\Delta s}{s_e}\right)^2 + \ldots\right] \tag{F.3.2}$$

The momentum may be expanded like

$$p = \frac{1}{2} m_{em_0} \left\{ 2\frac{s_e}{s} - 1 \right\} v_m = \frac{1}{2} m_{em_0} v_m \left[1 - 2\frac{\Delta s}{s_e} + 2\left(\frac{\Delta s}{s_e}\right)^2 + \ldots \right] \tag{F.3.3}$$

or conversely

$$v_m = \frac{p}{\frac{1}{2} m_{em_0}} \left[1 + 2\frac{\Delta s}{s_e} + 2\left(\frac{\Delta s}{s_e}\right)^2 + \ldots \right] \tag{F.3.4}$$

which must be combined with the momentum equation

$$\frac{dp}{dt} = \frac{2}{3} \frac{q^2}{4\pi\epsilon_0} \frac{\ddot{v}_m}{c^3} + q.E_{external} + F(t) \tag{F.3.4}$$

By analogy with the electromagnetic mass we assume that the generalised mass equals

$$\left(m^*_{bare} + m^*_{em} \right) = \frac{1}{2} m^*_{em_0} \left[1 - 2\frac{\Delta s}{s_e} + 2\left(\frac{\Delta s}{s_e}\right)^2 + \ldots \right] \tag{F.3.5}$$

The term $\frac{\partial U_{es}}{\partial s}$ in Eq. (F.1.6) is expanded as:

$$\frac{\partial U_{es}}{\partial s} = -\frac{1}{2} \frac{q^2}{4\pi\varepsilon_0} \frac{\alpha}{s_e^2} \left[1 - 2\frac{\Delta s}{s_e} + 3\left(\frac{\Delta s}{s_e}\right)^2 + \ldots \right] \tag{F.3.6}$$

and the last term in the l.h.s. of (F.1.6) is expanded as:

$$\begin{aligned} &\frac{1}{2} \frac{q^2}{4\pi\varepsilon_0} \frac{\alpha}{s_e^3} \mu\tau.s \left[1 - \frac{s_e}{s} \left\{ 1 + \frac{1}{\mu} + \frac{1}{\tau} \right\} + \left(\frac{s_e}{s}\right)^2 \left\{ \frac{1}{\mu} + \frac{1}{\tau} + \frac{1}{\mu.\tau} \right\} \right] \\ &\quad = \frac{1}{2} \frac{q^2}{4\pi\varepsilon_0} \frac{\alpha}{s_e^2} \mu\tau \left[\frac{1}{\mu\tau} + \frac{\Delta s}{s_e} \left\{ 1 - \frac{1}{\mu} - \frac{1}{\tau} - \frac{1}{\mu.\tau} \right\} + \left(\frac{\Delta s}{s_e}\right)^2 \left\{ \frac{1}{\mu} + \frac{1}{\tau} + \frac{1}{\mu.\tau} \right\} \right] \end{aligned} \tag{F.3.7}$$

For convenience, in the final expression for the expansion of the entire pulsation equation (F.1.6) the combination of symbols $\frac{q^2}{4\pi\varepsilon_0} \frac{1}{s_e} \alpha$ is replaced by $(m_{em_0} c^2)$ which leads to the following form of the pulsation equation:

$$
\begin{aligned}
& m^*_{em_0}\frac{\ddot{\Delta s}}{s_e} + \frac{(m_{em_0}c^2)}{s_e^2}\,\mu\tau\left\{1-\frac{1}{\mu}-\frac{1}{\tau}+\frac{1}{\mu.\tau}\right\}\frac{\Delta s}{s_e} \\
& \quad -2\frac{(m_{em_0}c^2)}{s_e^2}\left(\frac{v_m}{c}\right)^2\frac{\Delta s}{s_e}+\frac{(m_{em_0}c^2)}{s_e^2}\left(\frac{v_m}{c}\right)^2 \\
& = 2\,m^*_{em_0}\frac{\ddot{\Delta s}}{s_e}\frac{\Delta s}{s_e}-\frac{(m_{em_0}c^2)}{s_e^2}\,\mu\tau\left\{\frac{1}{\mu}+\frac{1}{\tau}-\frac{2}{\mu\tau}\right\}\left(\frac{\Delta s}{s_e}\right)^2+3\frac{(m_{em_0}c^2)}{s_e^2}\left(\frac{v_m}{c}\right)^2\left(\frac{\Delta s}{s_e}\right)^2+\ldots
\end{aligned}
\tag{F.3.8}
$$

where the terms linear in $\frac{\Delta s}{s_e}$ have been collected in the left hand side, and the non-linear terms in the r.h.s.

F.4 The Character of the Non-linearities; Chaotic Motion and Possible "Degeneration" into Harmonic Motion

The set of equations of motion is non-linear, and leads—as confirmed by a preliminary numerical integration—to chaotic motion. Some insight in the character of the non-linearities and the cause of chaotic motion may be obtained by considering the dominant non-linear term in the r.h.s. of Eq. (F.3.8), i.e. $2\,m^*_{em_0}\frac{\ddot{\Delta s}}{s_e}\frac{\Delta s}{s_e}$. Taken together with the first term on the l.h.s., this represents the fact that the mass in the equation is variable. The equation of motion (F.3.8) may thus be written as

$$
m^*_{em_0}\frac{\ddot{\Delta s}}{s_e}\left(1-2\frac{\Delta s}{s_e}\right)+\frac{(m_{em_0}c^2)}{s_e^2}\,\mu\tau\left\{1-\frac{1}{\mu}-\frac{1}{\tau}+\frac{1}{\mu.\tau}\right\}\frac{\Delta s}{s_e}+\ldots \tag{F.4.1}
$$

or, multiplying by the factor $\left(1-2\frac{\Delta s}{s_e}\right)^{-1}=\left(1+2\frac{\Delta s}{s_e}+\ldots\right)$:

$$
m^*_{em_0}\frac{\ddot{\Delta s}}{s_e}+\frac{(m_{em_0}c^2)}{s_e^2}\,\mu\tau\left\{1-\frac{1}{\mu}-\frac{1}{\tau}+\frac{1}{\mu.\tau}\right\}\left(1+2\frac{\Delta s}{s_e}\right)\frac{\Delta s}{s_e}+\ldots \tag{F.4.2}
$$

The variable mass motion is thus equivalent with a variable stiffness motion. For negative perturbations $\frac{\Delta s}{s_e}<0$ the spring in the equation softens, whereas for positive values it stiffens. A complete numerical investigation of the non-linear equation for s would really be necessary, but without such an extensive investigation one may conjecture that the noted chaotic solution (that is: noted by a quick-scan preliminary numerical calculation) is caused by this one-sided spring-softening. An example of the effect of a softening spring is given in Thompson and Stewart: Nonlinear Dynamics and Chaos (Sect. 6.3). A softening spring (in the case illustrated in this book it is of the double-sided type) is seen to lead to chaotic behaviour. The interesting thing is that this chaotic response coexists with a periodic limit cycle which occurs for small amplitudes *at the same control parameter values*. In our case too the chaotic motion "degenerates" into a harmonic motion for small amplitudes of $\frac{\Delta s}{s_e}$. The latter is confirmed by Chap. F.7 where the equations of motion (F.1.4) and (F.1.6) are combined and completely linearised.

A typical *steady-state* chaotic signal is shown in Fig. F.4.1. Although the signal is whimsical, it is not really completely random. One sees a definite rhythm, which means that amongst the wide spread of frequencies occurring in the signal, certain frequencies are prominently present, in contrast to completely random white noise. The term "steadiness" concerns the fact that the relevant parameters in the equation of motion have a constant value. Visual inspection clearly shows for instance that a certain, constant "average amplitude" exists. From a visual inspection, one also suspects that the frequency content, as determined using a moving time window, will be independent of time. In other words, the power spectral density curve determined by using a moving time window would be constant. This must of course be true, if the parameters in the equation of motion do not change in time.

Now, looking back at Eq. (F.3.8), one might consider the r.h.s. of the equation of motion as a pseudoforce, similar to d'Alembert's concept of inertia forces. The l.h. s. of the equation would, without these pseudoforces, represent harmonic motion at a single frequency (see Chap. F.7). The pseudoforces represent the non-linear effects, giving an additional forced excitation of the motion. If the resulting motion is chaotic, this is due to the non-linear effects in the r.h.s. of the equation of motion.

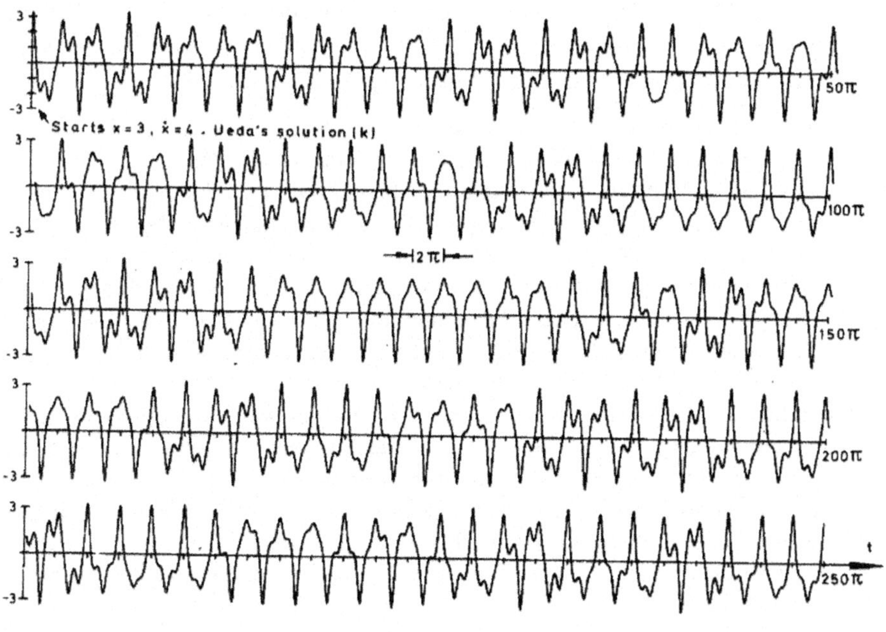

Fig. F.4.1 Typical chaotic signal in a steady-state condition. From: Thompson and Stewart, Non-linear dynamics and Chaos

These pseudoforces are non-conservative, as may be seen by considering for instance the first term in the r.h.s. of Eq. (F.3.8): $2\,m^*_{em_0}\frac{\ddot{\Delta s}}{s_e}\frac{\Delta s}{s_e}$. A change $d\left(\frac{\Delta s}{s_e}\right)$ represents work done by the pseudoforce: $\delta W = 2\,m^*_{em_0}\frac{\ddot{\Delta s}}{s_e}\frac{\Delta s}{s_e}\,d\left(\frac{\Delta s}{s_e}\right)$, so that a change from $\left(\frac{\Delta s}{s_e}\right)_1$ to $\left(\frac{\Delta s}{s_e}\right)_2$ represents the work $W_{1\to 2} = m^*_{em_0}\int_1^2 \frac{\ddot{\Delta s}}{s_e}.d\left(\frac{\Delta s}{s_e}\right)^2$ which is path-dependent. A variation of $\left(\frac{\Delta s}{s_e}\right)$ to some value and back to its original value may thus involve an increase or decrease of the energy present in the elongation mode of the droplet. According to the conservation of the total energy derived in Chap. F.2, this implies an accompanying decrease or increase of the energy in the translation mode. The conclusion is, that during a chaotic motion of the droplet (when the r.h.s. of the equation of motion (F.3.8) is non-zero), the energy in one of the modes (translation or elongation) may accumulate during relatively long time intervals, spanning many periods of the fast oscillations of $\left(\frac{\Delta s}{s_e}\right)$ and v_m. In reverse, long-term depletion of energy in one of the modes is possible too. Because the *total* energy of the droplet is conserved, such a long-term accumulation or depletion of energy in any one of the modes is mirrored by an equally large depletion or accumulation of energy in the other mode.

This is in contrast to the fast exchanges of energy between the two dynamic modes in the case of small amplitudes of $\left(\frac{\Delta s}{s_e}\right)$, when complete linearisation of the equations of motion is allowed and the r.h.s. of Eq. (F.3.8) may be neglected. From Chap. F.7, which considers this linearised case, it will appear that the mutual energy exchanges between the two modes then only have a small period, equal to the period of the fast harmonic variations of $\left(\frac{\Delta s}{s_e}\right)$ and v_m, and long-term accumulations of energy in one of the modes is not possible; the average energy partition stays as it is.

The energy accumulation or depletion in the individual dynamic modes, spanning many fast oscillations, implies that the chaotic pulsation is not strictly steady. Very schematically, the pulsation during the chaotic state may be imagined as shown in Fig. F.4.2. The "amplitude" of the pulsation will during some time intervals be larger than average, and during other periods smaller again. If the amplitude is very small, we see a "degeneration" into a nearly sinusoidal signal, with constant amplitude. The envelope of the pulsations then shrinks into straight lines, as sketched in Fig. F.4.2 by the dashed red lines.

The relation between the mutual energy exchanges and the amplitudes of the pulsation and velocity fluctuations will now be considered in somewhat more detail.

Consider for the sake of simplicity the case of no radiation losses and the absence of an external potential field, a case earlier considered in Chap. D.8 dealing with the momentum equation. From (F.1.4) in this situation it follows that the momentum of the droplet is constant, $p = p_0$. The energy in the translation is purely kinetic energy:

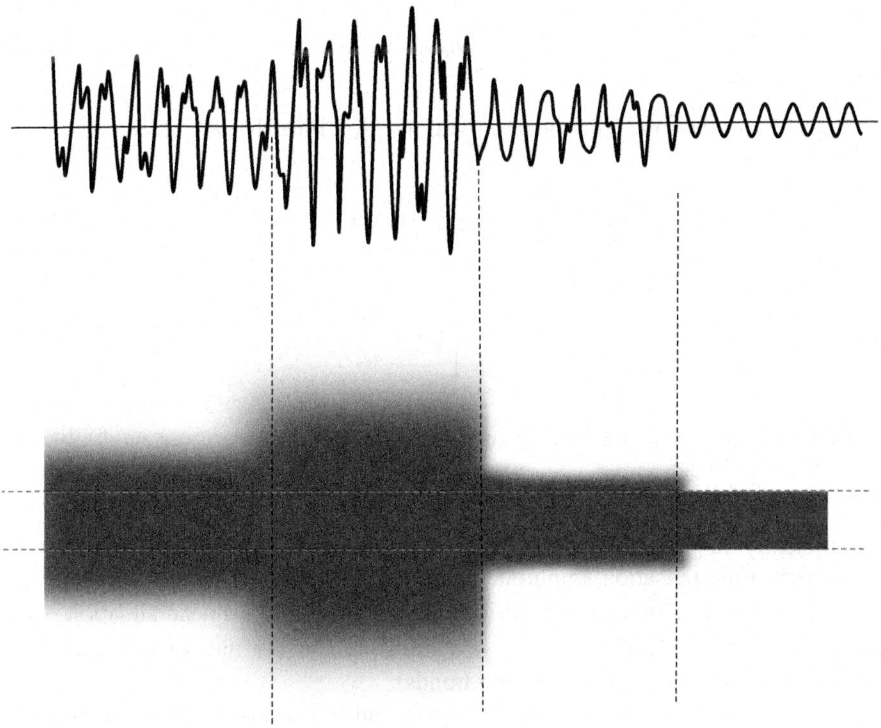

Fig. F.4.2 Picture of the long-term amplitude changes of the pulsation due to spurious changes of the energy partition. *Lower part* shows same picture taking into account the high frequency of the pulsation. *On the right* degeneration into harmonic pulsation

$$E_{transl} = \frac{p_0^2}{2.m(s)} \tag{F.4.3}$$

where the instantaneous mass is given by (see Eq. (F.1.5)):

$$m(s) = -\frac{1}{2}m_{em_0} + m_{em}(s) \tag{F.4.4}$$

and the electromagnetic mass is given by (F.1.2) as:

$$m_{em}(s) = \frac{1}{c^2}\frac{q^2}{4\pi\varepsilon_0}\frac{\alpha}{s} = m_{em_0}\frac{s_e}{s} \tag{F.4.5}$$

with m_{em_0} the equilibrium value (at zero speed) of the electromagnetic mass, corresponding to the unperturbed elongation s_e. From (F.4.4) and (F.4.5) the instantaneous mass follows as:

$$m(s) = \frac{1}{2} m_{em_0} \left(2 \frac{s_e}{s} - 1 \right) \tag{F.4.6}$$

Expanding again in terms of perturbations $\frac{\Delta s}{s_e}$ defined by $s = s_e \left(1 + \frac{\Delta s}{s_e} \right)$:

$$m(s) = \frac{1}{2} m_{em_0} \left(\frac{2}{1 + \frac{\Delta s}{s_e}} - 1 \right) = \frac{1}{2} m_{em_0} \left[1 - 2 \frac{\Delta s}{s_e} + 2 \left(\frac{\Delta s}{s_e} \right)^2 + \ldots \right] \tag{F.4.7}$$

$$E_{transl} = \frac{p_0^2}{2.m(s)} = \frac{p_0^2}{m_{em_0}} \left[1 + 2 \frac{\Delta s}{s_e} + 2 \left(\frac{\Delta s}{s_e} \right)^2 + \ldots \right] \tag{F.4.8}$$

In the chaotic state the non-linear terms like $\left(\frac{\Delta s}{s_e} \right)^2$ etc. are not negligible. We therefore see from (F.4.8) that the average value $\overline{E_{transl}}$ (the average determined over a sufficiently large number of short cycles of the pulsation) is large when the amplitude of the pulsation amplitude is large. One may easily check this by taking $\frac{\Delta s}{s_e}(t) = S.\cos \omega t$ (although we are talking about the non-linear situation where harmonic variations do not occur, this crude model for the actual, much more whimsical oscillations does show the trends).

From the fact that the momentum is constant we conclude that large pulsation amplitudes also imply large velocity fluctuations around an average value:

$$m(s).v_m = p_0 \tag{F.4.9}$$

or

$$v_m = \frac{p_0}{m(s)} = \frac{p_0}{\frac{1}{2} m_{em_0}} \left[1 + 2 \frac{\Delta s}{s_e} + 2 \left(\frac{\Delta s}{s_e} \right)^2 + \ldots \right] \tag{F.4.10}$$

Large amplitudes of $\frac{\Delta s}{s_e}$ therefore correspond to increased velocity fluctuations, as well as an increased value of the average velocity.

The energy which is present in the elongation mode consists of potential energy $(E_{elong})_{pot}$ and kinetic energy $(E_{elong})_{kin}$, unlike the translation energy in the absence of an external potential field. A long-term average (taken over many pulsation cycles) of the kinetic energy in the pulsation $\left(\overline{E_{elong}}\right)_{kin}$ may be assumed to be relatively large when the amplitude of the pulsations $\frac{\Delta s}{s_e}(t)$ is large.

This would, at first sight, seem to contradict the conclusion above that the total energy in the elongation mode must be small, when the amplitudes of the velocity fluctuations and of the pulsation are large. However, the contradiction is removed if we assume that the droplet is "pre-stressed", i.e. that the value of the elongation s_{av} around which the pulsation takes place is unequal to the "zero-speed" equilibrium value s_e. In the later chapter on harmonic pulsations, where a complete quantification is possible, it will be shown that such a pre-stressing is indeed present.

One can now compare the situation with a pre-loaded spring with mass, where oscillations take place around the pre-stressed situation. In this mass-spring system the amplitude of the fast oscillations (and therefore the kinetic energy in the fast oscillation) may have any value, but it is not uniquely related to the *total* energy in the system, since there is a second control parameter in the form of the amount of pre-stress which may be a—slow—variable as well.

In the particular case considered here, viz. the constant-momentum situation, the pre-stressing of the elongation mode can be seen in the second order terms. It should be realised that Eq. (F.3.8) was derived under the condition $\left|\frac{\Delta s}{s_e}\right| < 1$, whereas no restrictions were imposed on the velocity variations $v_m(t)$. In the particular case of constant momentum $p = p_0$, we see from (F.4.10) that $v_m(t)$ in that case is a small fluctuation around an average value, the small variations expressible in terms of $\frac{\Delta s}{s_e}$. Therefore, by substituting (F.4.10) into (F.3.8) we have an uncoupled equation solely in terms of $\frac{\Delta s}{s_e}$, and can for this particular case separate the first-order effects linear in $\frac{\Delta s}{s_e}$ from the higher-order, non-linear effects. Doing this, one finds a pseudoforce acting on the linear system having the value:

$$pseudoforce = \frac{(m_{em_0}.c^2)}{s_e^2}\left(\frac{\Delta s}{s_e}\right)^2\left[2 + 5\frac{1}{c^2}\left(\frac{p_0}{\frac{1}{2}m_{em_0}}\right)^2\right] + \ldots \qquad \text{(F.4.11)}$$

This pseudoforce is always positive (as soon as the non-linear effects become of any importance, and are no longer negligible) and thus causes a "pre-stressing effect" on the elongation of the droplet, in the sense that the droplet is permanently longer than the value it has according to the linear theory. An elongation larger than s_e implies that some of the potential energy (the energy of formation) has been tapped from it. The tapping of energy from the formation energy becomes larger when the amplitude of the pulsation becomes larger (think again of the crude model for the variations $\frac{\Delta s}{s_e}(t) = S.\cos\omega t$ which shows that the average $\overline{\left(\frac{\Delta s}{s_e}\right)^2} \div S^2$). In conclusion: the larger the fluctuations of the pulsation, the smaller the potential energy associated with the average elongation of the droplet. This is exactly what is required by the conservation of the *total* energy of the droplet.

The relation between all the variables is summarised in the following diagram:

Pulsation amplitude	+	−
amplitude of velocity fluctuations	+	−
kinetic energy in translation	+	−
kinetic energy in pulsation	+	−
average potential energy in elongation	−	+
+: larger than average due to long-term energy exchanges in the chaotic state between the two dynamic modes (translation and elongation)		
−: idem, smaller than average		

A transition from chaotic to harmonic motion can now be described as follows. During the chaotic motion during relatively long time intervals the energy in the translation can be larger than a long-term average, at the cost of the potential energy in the elongation mode. This corresponds with large amplitudes of both the velocity fluctuations as well as with large pulsation amplitudes. Non-linear effects will then be large, and the chaotic motion stays chaotic.

On the other hand, if the reverse happens, i.e. if temporarily the energy in the translation becomes smaller than its long-term average, this is accompanied by smaller pulsation amplitudes. If the pulsation amplitude falls below a certain threshold, the character of the motion changes into harmonic motion. This is due to the fact that in the case of small pulsation amplitudes the equations of motion approach linearised equations.

Once such a transition from the chaotic state into a harmonic motion has happened, there will not be a way back. The random, long-term and relatively large energy exchanges between the two dynamic modes, will become "frozen": in the state of harmonic motion the energy exchanges are more like short-term ripples, and large energy transfers cannot accumulate. See the analysis in Chap. F.7 where this has been quantified. The only way to leave the state of harmonic motion is by radiation of energy, as will briefly be described later in the present chapter, and as will be fully analysed in Chap. F.8.

The frequency content of the chaotic signal will also be influenced by the long-term, relatively large energy exchanges between the two dynamic modes. Looking at the schematic of Fig. F.4.2 this is easily seen by visual inspection: small amplitudes of the pulsation give a tendency towards the harmonic respones, where just one discrete frequency is present. This tendency is schematically shown in Fig. F.4.3, in terms of the power spectral density of the signal.

During a fully developed chaotic state there is a broad range of frequencies occurring in $\frac{\Delta s}{s_e}(t)$, and the PSD curve is widely spread out over a large range of frequencies. On the other hand, if for small amplitudes of $\frac{\Delta s}{s_e}(t)$ the motion has "degenerated" into harmonic motion, the SPD diagram will approach a delta function, representing only one value of the frequency.

During the chaotic motion the "amplitude" of $\frac{\Delta s}{s_e}$ shows long-term variations (determined by the long-term energy exchanges between translation and elongation), and associated with this is a long-term broadening and narrowing of the spectral density curve (using a moving time window). All this is going on in a random way as time proceeds, the spectral density is slowly alternating between small and large width ("slow": compared with the time scale of the pulsations).

One may expect that these deformations of the spectral density curve take place around a centre frequency which is the same as the limiting frequency represented by the delta function, although this expectation is speculative as long as no detailed numerical investigation of the non-linear equation of motion has been done. This assumption means that the "rhythm" seen in the chaotic signal is the same as the frequency of the degenerated, harmonic signal. It explains why the later detailed

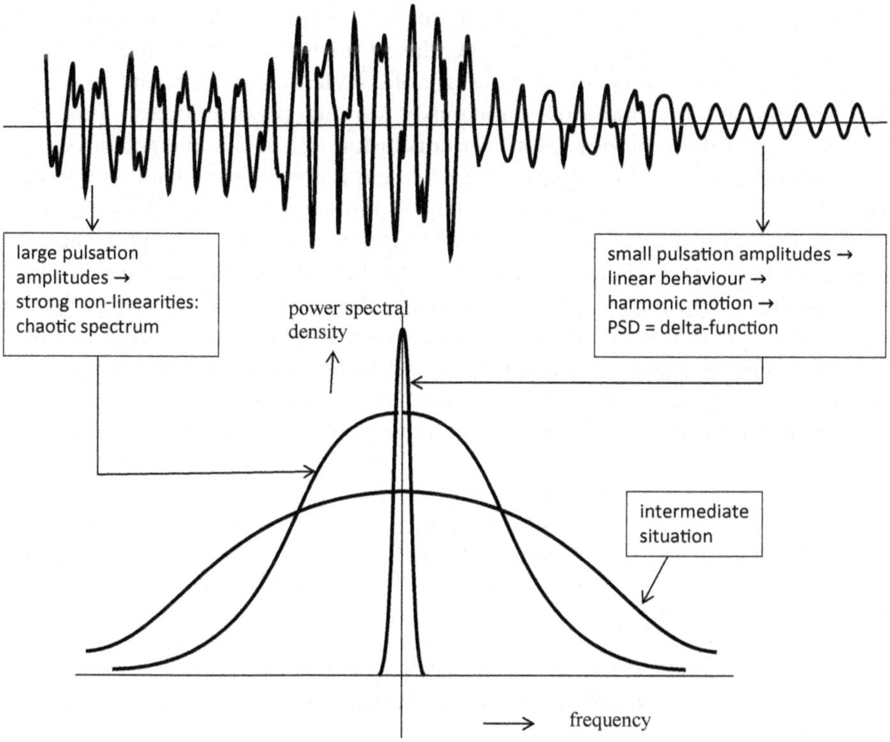

Fig. F.4.3 Schematic showing relation between pulsation amplitudes and the power spectral density

analysis of the linearised equations of motion has some relevance for the chaotic motion as well.

Now consider the autocorrelation of the function $\frac{\Delta s}{s_e}(t)$, expressing how much correlation exists between the value of the signal at some point in time and the value at the next instant. For large amplitudes of the pulsation and thus chaotic motion fully developed, the middle part of the PSD curve represents a broad band signal tending to a white noise signature. In the case of exact white noise the autocorrelation function would be a delta function, meaning that there is no correlation at all between the value of the signal at some point in time and the value at the next instant. Consequently, the large amplitude, chaotic pulsation of $\frac{\Delta s}{s_e}(t)$ will show a very small autocorrelation as well. In other words, the function $\frac{\Delta s}{s_e}(t)$ will jump up and down in a random and almost discontinuous way.

Now, looking back at Eq. (F.3.8), it has already be mentioned that one might consider the r.h.s. of the equation of motion as a pseudoforce, similar to d'Alembert's concept of inertia forces. The l.h.s. of the equation would, without these pseudo forces, represent harmonic motion at a single frequency. The pseudo forces represent the non-linear effects, giving an additional forced excitation of the motion. If the resulting motion is chaotic, this is due to the non-linear effects in the r.h.s. of the equation of motion, and the lack of correlation in the chaotic motion can thus be ascribed to almost discontinuously varying pseudo forces. This is an essential observation, in view of a later conclusion (when the interaction between the droplet's motion and the electromagnetic field is considered) that discontinuous forces on a droplet will cause a temporary suppression of the radiation, during a short period following the jump of the external force. It will be shown that there is a brief delay between the jump of the force and the ensuing start of a wave emission.

A chaotic motion may, according to the above, be viewed as an "ongoing sequence of transients ", i.e. at each point in time there is an attempt to start up radiation, but all the time this just leads to a "rejected take off " of the radiation, since before a radiating state can develop, there is already a next jump of the pseudo forces.

All this can be summarised as in the diagram given below.

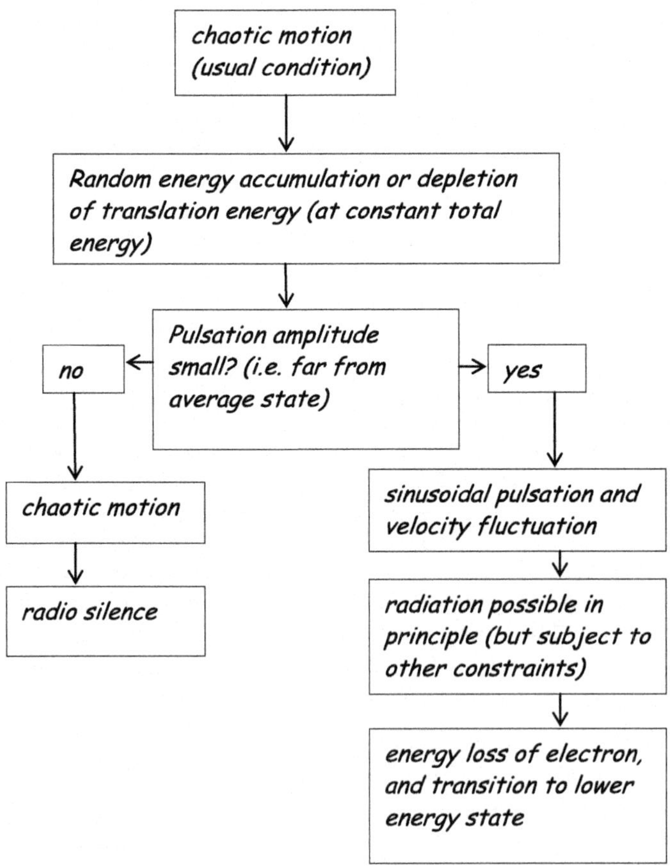

The remarkable conclusion which may be drawn from this summary of the present chapter is, that the transition from one energy level to a lower level (here it is meant the *total* energy of the droplet) is governed by the process of chaos. Chaos is not pure randomness, as earlier remarked. Chaos is just as much a solution of the equations of motion as the harmonic motion occurring during the extremes of very small pulsation amplitudes. The conclusion is therefore that the initiation of a transition between energy levels is not a matter of pure chance (i.e. just governed by laws of probability). A transition is a causal process, although its occurrence is unpredictable due to the chaotic character of the motion.

Finally a few words about possible simplifications of the equations of motion. Equation (F.4.3) and further were amongst others based on the scenario that there are no losses by radiation. This was until now considered to be an artificial simplification, which could be implemented by choosing the artificial force $F(t)$ such that it cancels the term expressing the radiation resistance. Now we have seen that during the chaotic motion the radiation will be suppressed anyway. In the chaotic state, corresponding to large amplitude variations of $\frac{\Delta s}{s_e}(t)$, the translation equation can therefore be modified compared with Eq. (F.1.4). The term $\frac{2}{3}\frac{q^2}{4\pi\epsilon_0}\frac{\ddot{v}_m}{c^3}$, representing the recoil force due to radiation, will drop from the equation. In the special case that there is no external potential gradient, i.e. $q\,E_{ext} = 0$ the translation equation becomes $\dot{p} = 0$ or $p = constant$, even without an intervention by an artificial external force $F(t)$.

In contrast, in the case of small amplitudes of $\frac{\Delta s}{s_e}(t)$ such a simplification is not possible, and the term $\frac{2}{3}\frac{q^2}{4\pi\epsilon_0}\frac{\ddot{v}_m}{c^3}$ must be retained in the translation equation. On the other hand, the pulsation equation may then be completely linearised in $\frac{\Delta s}{s_e}$.

F.5 Equations of Motion Linearised for Small Pulsations

As discussed, harmonic motion is associated with small amplitude pulsations. In order to perform a more detailed analysis of this state, it is now assumed that the pulsation amplitude is very small, with $\Delta s/s_e \ll 1$ so that linearisation in $\Delta s/s_e$ is allowed. In this case the r.h.s. of Eq. (F.3.8) is neglected, and the following equation of motion results:

$$\begin{aligned}\left(m^*_{em_0}\right)\frac{\ddot{\Delta s}}{s_e} + \frac{\left(m_{em_0}.c^2\right)}{s_e^2}\mu\tau\left(1-\frac{1}{\mu}-\frac{1}{\tau}+\frac{1}{\mu\tau}\right)\frac{\Delta s}{s_e}\left[1-\frac{2}{\mu\tau\left(1-\frac{1}{\mu}-\frac{1}{\tau}+\frac{1}{\mu\tau}\right)}\frac{v_m^2}{c^2}\right] \\ = -\frac{\left(m_{em_0}.c^2\right)}{s_e^2}\frac{v_m^2}{c^2}\end{aligned} \quad (F.5.1)$$

The translation equation still reads

$$\frac{dp}{dt} = \frac{2}{3}\frac{q^2}{4\pi\epsilon_0}\frac{\ddot{v}_m}{c^3} + q.E_{external} + F(t) \tag{F.5.2}$$

but the expressions for the momentum and velocity are now linearised:

$$p(t) = \frac{1}{2}m_{em_0}\, v_m(t).\left[1 - 2\frac{\Delta s}{s_e}\right] \tag{F.5.3}$$

or conversely

$$v_m(t) = \frac{p(t)}{\frac{1}{2}m_{em_0}}\left[1 + 2\frac{\Delta s}{s_e}\right] \tag{F.5.4}$$

F.6 Short Notations for the Linearised Equations of Motion

In the following the short notations m and m^* will be used to denote the constant equilibrium values of the droplet's mass and generalised mass, with

$$m = \frac{1}{2}m_{em_0} = \frac{1}{2}\frac{1}{c^2}\frac{q^2}{4\pi\varepsilon_0}\frac{\alpha}{s_e} \tag{F.6.1}$$

$$m^* = \frac{1}{2}m^*_{em_0} = \frac{1}{2}\frac{1}{c^2}\frac{q^2}{4\pi\varepsilon_0}\frac{\alpha^*}{s_e} \tag{F.6.2}$$

Furthermore, the index "m" for "mid point" will be dropped from the notation v_m, since in the remainder of the text no ambiguity can exist about the translation velocity, which always is the velocity of the mid-point of the droplet.

In the case of the pulsation equation we introduce the short notations

$$\omega_0^2 = \frac{m}{m^*}\frac{c^2}{s_e^2}\mu\tau\left(1 - \frac{1}{\mu} - \frac{1}{\tau} + \frac{1}{\mu\tau}\right) \tag{F.6.3}$$

$$k = \frac{1}{c^2}\frac{1}{\mu\tau}\frac{1}{\left(1 - \frac{1}{\mu} - \frac{1}{\tau} + \frac{1}{\mu\tau}\right)} \tag{F.6.4}$$

The often occurring product $k.\omega_0^2$ is thus simply

$$k.\omega_0^2 = \frac{m}{m^*}\frac{1}{s_e^2} = \frac{\alpha/\alpha^*}{s_e^2} \tag{F.6.5}$$

This consequently leads to the following form for the pulsation equation:

$$\frac{\ddot{\Delta s}}{s_e} + \omega_0^2(1 - 2\,k\,v^2)\frac{\Delta s}{s_e} = -k\,\omega_0^2\,v^2 \tag{F.6.6}$$

In the case of the translation equation we introduce the short notation

$$\gamma = \frac{2}{3}\frac{q^2}{4\pi\varepsilon_0}\frac{1}{c^3} \tag{F.6.7}$$

so that the translation equation in short notation reads

$$\frac{dp}{dt} = \gamma\,\ddot{v} + q.E_{external} + F(t) \tag{F.6.8}$$

together with

$$p = m\,v\left[1 - 2\frac{\Delta s}{s_e}\right] \tag{F.6.9}$$

or

$$v = \frac{p}{m}\left[1 + 2\frac{\Delta s}{s_e}\right] \tag{F.6.10}$$

F.7 Solution of the Linearised Equations of Motion (Undamped Conditions, No Potential Gradients)

In the case of an undamped motion the linearised equations of motion allow a closed form solution. Under "undamped motion" is understood a motion with constant total energy, obtained by taking the artificially added force $F(t)$ such that it cancels the radiation resistance. Under such circumstances the motion of the droplet will still cause radiation and loss of energy by radiation, but the force $F(t)$ supplies sufficient energy to keep the total energy of the droplet unchanged.

If furthermore the situation is considered where the droplet is not subjected to external potential gradients, i.e. if $q.E_{external} = 0$, the momentum equation reduces to

$$\dot{p} = 0 \tag{F.7.1}$$

or

$$p = p_0 \tag{F.7.2}$$

with p_0 a constant.

Combination of Eqs. (F.6.6) and (F.6.10) yields, strictly keeping only the linear terms:

$$\frac{\ddot{\Delta s}}{s_e} + \omega_0^2 \left[1 + 2\,k \left(\frac{p_0}{m}\right)^2\right] \frac{\Delta s}{s_e} = -k\,\omega_0^2 \left(\frac{p_0}{m}\right)^2 \tag{F.7.3}$$

Note the change of sign of the second term within the square brackets, compared with Eq. (F.6.6).

The solution $\frac{\Delta s}{s_e}(t)$ of this linear equation is the sum of the particular solution and the solution of the homogeneous equation. On account of the very small value of k the homogeneous equation is satisfied by a harmonic time function with angular frequency

$$\omega = \omega_0 \left[1 + k \left(\frac{p_0}{m}\right)^2\right] \tag{F.7.4}$$

so that

$$\frac{\Delta s}{s_e}(t) = S\,cos(\omega\,t + \varepsilon) - k \left(\frac{p_0}{m}\right)^2 \tag{F.7.5}$$

where non-linear terms in k have been omitted.

The corresponding velocity is according to (F.6.10)

$$v(t) = \left(\frac{p_0}{m}\right) \left[1 + 2\,S\,cos(\omega\,t + \varepsilon) - 2k \left(\frac{p_0}{m}\right)^2\right] \tag{F.7.6}$$

The time-averaged value of the velocity is therefore *un*equal to $\left(\frac{p_0}{m}\right)$, and is given by

$$v_{av} = \left(\frac{p_0}{m}\right)\left[1 - 2k\left(\frac{p_0}{m}\right)^2\right] = \left(\frac{p_0}{m}\right)\cdot\left(\frac{\omega_0}{\omega}\right)^2 \tag{F.7.7}$$

It should be noted that by v_{av} is *not* meant the velocity averaged over *all* time, which would include the reversions of the direction of the momentum caused by the encounters with the walls of the potential box. Here the notation v_{av} is reserved for the average taken over a single sweep of the droplet from left to right within the box (or a sweep in the other direction with negative momentum), or at least an average taken over a sufficiently large number of pulsation cycles.

To the present accuracy, where higher orders of k are neglected, we may instead of Eq. (F.7.4) also write

$$\omega = \omega_0\left[1 + k\,v_{av}^2\right] \tag{F.7.8}$$

In the following the energy of the motion will be considered. Here we meet a conceptual problem because the energy depends on quadratic terms of the pulsation, whereas in the above shown derivations the pulsation was derived from a linearised equation of motion.

In fact, this problem is not dissimilar to the usual theory of linear oscillators. The mathematical model of a linear oscillator is often from the physical point of view obtained by linearising the degrees of freedom, whilst nevertheless energy considerations are based on quadratic terms in these same degrees of freedom. What is then tacitly being assumed is, that the mathematically linear equations describe a real linear system, disregarding how it physically was derived. This same approach will be taken here.

The sum of the kinetic and potential energy in the pulsation is:

$$\left(T_{puls} + U_{puls}\right) = \frac{1}{2}m^*\left[\left(\dot{\Delta s}\right)^2 + \omega_0^2(\Delta s)^2\right] \tag{F.7.9}$$

On substituting Eq. (F.7.5) the expression becomes (also using (F.7.4) and retaining only terms linear in k):

$$\left(T_{puls} + U_{puls}\right) = \frac{1}{2}m^*\,s_e^2\omega_0^2\left[S^2 + 2\,S^2k\left(\frac{p_0}{m}\right)^2\sin^2\omega t - 2\,Sk\left(\frac{p_0}{m}\right)^2\cos\omega t\right] \tag{F.7.10}$$

The translation energy of a droplet of charge inside a square potential box consists of kinetic energy only, which using Eq. (F.3.2) for m_{em} and the instantaneous mass $-\frac{1}{2}m_{em_0} + m_{em} = m\left[1 - 2\frac{\Delta s}{s_e} + 2\left(\frac{\Delta s}{s_e}\right)^2\right]$ can be written as:

$$T_{trans} = \frac{1}{2}\frac{p_0^2}{m}\left[1 + 2\frac{\Delta s}{s_e} + 2\left(\frac{\Delta s}{s_e}\right)^2\right] \tag{F.7.11}$$

Using Eq. (F.7.5) and substituting (F.6.5) yields (again neglecting terms of order $O(k^2)$):

$$T_{trans} = \frac{1}{2} m^* s_e^2 . \omega_0^2 k \left(\frac{p_0}{m}\right)^2 \left[1 + 2 S \cos \omega t + 2 S^2 \cos^2 \omega t\right] \tag{F.7.12}$$

The instantaneous value of the total energy E is

$$E = \left(T_{puls} + U_{puls}\right) + T_{trans} = \frac{1}{2} m^* s_e^2 \omega_0^2 \left[S^2 + k\left(\frac{p_0}{m}\right)^2 + 2 S^2 k \left(\frac{p_0}{m}\right)^2\right] \tag{F.7.13}$$

It is seen that the total energy is constant, i.e. it does not fluctuate with the high frequency of the pulsation, which is in accordance with the cancellation of the radiation resistance by the assumed external force $F(t)$. It should be realised, however, that even under these steady conditions there is a fast fluctuation of the translation energy and the pulsation energy separately, as is seen from Eqs. (F.7.10) and (F.7.12). These equations describe the fast exchange of energy between the two components of energy, as was derived more generally by Eqs. (F.2.9) and (F.2.12).

The motion of the droplet with a constant energy and with constant amplitude harmonic velocity variations has artificially been achieved by adding the external force $F(t) = -\gamma \ddot{v}$ to the equations of motion. This force performs work, the work per second given by

$$F.v = -\gamma v \ddot{v} = -\frac{1}{2} \gamma \frac{d^2(v^2)}{dt^2} + \gamma \dot{v}^2 \tag{F.7.14}$$

The significance of the second form is, that the term $\gamma \dot{v}^2$ equals the energy flow in the radiation escaping to infinity (see later chapters on the radiation field). It thus represents the loss of energy from the system. The other term in the last equality of (F.7.14) is the socalled "Schott energy", which is often interpreted as energy periodically stored in the field and recovered again. It does not contribute to a permanent loss of energy from the system.

We see that the "ripples" in the velocity of the droplet cause energy radiation, which loss of energy is compensated by the work done by the external force.

From (F.7.6) it is derived

$$\frac{d^2(v^2)}{dt^2} = \left(\frac{p_0}{m}\right)^2 \left[-4S\omega^2 \cos(\omega t) - 8S^2 \omega^2 \cos(2\omega t) - 8kS\omega^2 \left(\frac{p_0}{m}\right)^2 \cos(\omega t)\right] \tag{F.7.15}$$

$$\dot{v}^2 = 4\left(\frac{p_0}{m}\right)^2 S^2 \omega^2 \sin^2(\omega t) = 2\left(\frac{p_0}{m}\right)^2 S^2 \omega^2 [1 - \cos(2\omega t)] \tag{F.7.16}$$

The work per second by the external force $F(t)$ is, averaged over one cycle of the pulsation, or averaged over a long time interval covering a sufficiently large number of cycles:

$$\overline{F.v} = 2\gamma\left(\frac{p_0}{m}\right)^2 S^2\omega^2 \tag{F.7.17}$$

The solution described by the Eqs. (F.7.4) through (F.7.6) contains two constants, viz. p_0 and S. These constants must be determined by additional boundary- or initial conditions. Because of the relation (F.7.13) the pair of constants p_0 and S can be exchanged for the pair p_0 and E.

An estimate of the order of magnitude of the pulsation amplitude S in Eqs. (F.7.5) and (F.7.6) is possible by assuming equipartition of energy. The following analysis has just the purpose to get some feeling for orders of magnitude, regardless whether such a condition of equipartition is actually ever realised. The state of equipartition entails that, averaged over a sufficiently large number of pulsation cycles, the average energy in the translation equals the average energy in the pulsation.

The time averaged value of the pulsation energy is (see (F.7.10)):

$$\left(\overline{T_{puls} + U_{puls}}\right) = \frac{1}{2}m^*\, s_e^2\omega_0^2\, S^2\left[1 + k\left(\frac{p_0}{m}\right)^2\right] \tag{F.7.18}$$

and the time averaged value of the translation energy is

$$\overline{T_{trans}} = \frac{1}{2}m^*\, s_e^2\omega_0^2\, k\left(\frac{p_0}{m}\right)^2\left[1 + S^2\right] \tag{F.7.19}$$

Applying the condition expressing equipartition

$$\left(\overline{T_{puls} + U_{puls}}\right) = \overline{T_{trans}} \tag{F.7.20}$$

yields

$$S^2 = k\left(\frac{p_0}{m}\right)^2 \tag{F.7.21}$$

Substituting this result into (F.7.13) and using (F.6.5) gives for the total energy

$$E = \left(T_{puls} + U_{puls}\right) + T_{trans} = \frac{p_0^2}{m}\left[1 + k\left(\frac{p_0}{m}\right)^2\right] = \frac{p_0^2}{m}\frac{\omega}{\omega_0} \quad (equipartition) \tag{F.7.22}$$

It is interesting to note that the sum of the average energies exactly equals the average of the total energy (the "average" of the total energy naturally equals the total energy since total energy is constant):

$$\left(\overline{T_{puls} + U_{puls}}\right) + \overline{T_{trans}} = \frac{1}{2} m^* s_e^2 \omega_0^2 \left[S^2 + k\left(\frac{p_0}{m}\right)^2 + 2\,S^2 k\left(\frac{p_0}{m}\right)^2 \right] \qquad \text{(F.7.23)}$$

which according to Eq. (F.6.11) equals the total energy E.

F.8 Asymptotic Approximation for the Damped Motion

Starting from the solution for the case of an undamped motion with constant energy, an approximate solution can be found for the case where the droplet gradually loses its energy by radiation. The constant energy solution was obtained by adding the artificial force $F(t)$ to the translation equation so that $\gamma\,\ddot{v} + F(t) = 0$. The loss of energy by radiation is then compensated by the work done by the artificial force. We can now gradually reduce the influence of such a force until it is zero, and determine an asymptotic solution for the resulting equations of motion. We take $\gamma\,\ddot{v} + F(t) = \varepsilon.\gamma\,\ddot{v}$ with ε an asymptotic expansion parameter. In the limit $\varepsilon \to 0$ one obtains the undamped solution, whereas for $\varepsilon = 1$ the artificial force would be completely zero. It is questionable whether the asymptotic expansion is still valid for the case $\varepsilon = 1$, but at least we can see what will be the tendency when shifting away from the undamped solution towards a solution with energy loss by radiation.

The equations of motion are now, going back to (F.6.6) and (F.6.8):

$$\left(\frac{\ddot{\Delta s}}{s_e}\right) + \omega_0^2\left[1 - 2kv^2\right]\left(\frac{\Delta s}{s_e}\right) = -k\omega_0^2 v^2 \qquad \text{(F.8.1)}$$

$$\dot{p} = \varepsilon\gamma\ddot{v} \qquad \text{(F.8.2)}$$

In order to reduce the number of variables, the momentum p can as follows be expressed in terms of the velocity v. A first integral of (F.8.2) is

$$p = p_0 + \varepsilon\gamma\dot{v} \qquad \text{(F.8.3)}$$

where the constant of integration p_0 is the same as the constant momentum of the undamped solution, as may be seen from the limit $\varepsilon \to 0$ (in principle there can also be a constant term of order $O(\varepsilon)$ but this does not have an influence on the solution, and will be omitted). The momentum can be rewritten in terms of the constant mass

m and the pulsation amplitude $(\Delta s/s_e)$, and reads under the assumption of small pulsation amplitude (see Eq. (F.6.9)):

$$m\left[1-2\left(\frac{\Delta s}{s_e}\right)\right]v = p_0 + \varepsilon\gamma\dot{v} \tag{F.8.4}$$

or

$$v = \left[\frac{p_0}{m} + \varepsilon\frac{\gamma}{m}\dot{v}\right].\left[1+2\left(\frac{\Delta s}{s_e}\right)\right] \tag{F.8.5}$$

so that we can make the following substitution into (F.8.1):

$$v^2 = \left(\frac{p_0}{m}\right)^2\left[1+4\left(\frac{\Delta s}{s_e}\right)\right] + 2\varepsilon\frac{p_0}{m}\frac{\gamma}{m}\dot{v}\left[1+4\left(\frac{\Delta s}{s_e}\right)\right] + O(\varepsilon^2) \tag{F.8.6}$$

Performing the substitution of (F.8.6) into (F.8.1), and taking into account the linearisation in $(\Delta s/s_e)$:

$$\begin{aligned}\left(\frac{\ddot{\Delta s}}{s_e}\right) &+ \omega_0^2\left[1-2k\left(\frac{p_0}{m}\right)^2 - 4\varepsilon.k\frac{p_0}{m}\frac{\gamma}{m}\dot{v}\right]\left(\frac{\Delta s}{s_e}\right)\\ &= -k\omega_0^2\left(\frac{p_0}{m}\right)^2\left[1+4\left(\frac{\Delta s}{s_e}\right)\right] - 2\varepsilon.k\omega_0^2\frac{p_0}{m}\frac{\gamma}{m}\dot{v}\left[1+4\left(\frac{\Delta s}{s_e}\right)\right] + O(\varepsilon^2)\end{aligned} \tag{F.8.7}$$

or

$$\begin{aligned}\left(\frac{\ddot{\Delta s}}{s_e}\right) &+ \omega_0^2\left[1+2k\left(\frac{p_0}{m}\right)^2\right]\left(\frac{\Delta s}{s_e}\right)\\ &= -k\omega_0^2\left(\frac{p_0}{m}\right)^2 - 2\varepsilon.k\omega_0^2\frac{p_0}{m}\frac{\gamma}{m}\dot{v}\left[1+2\left(\frac{\Delta s}{s_e}\right)\right] + O(\varepsilon^2)\end{aligned} \tag{F.8.8}$$

Note the change in sign inside the square brackets. This equation reduces to the equation for the undamped motion in the case $\varepsilon \to 0$. An asymptotic approximation can thus be made by assuming the solution to have the asymptotic form

$$\left(\frac{\Delta s}{s_e}\right) = \left(\frac{\Delta s}{s_e}\right)_0 + \varepsilon\left(\frac{\Delta s}{s_e}\right)_1 + O(\varepsilon^2) \tag{F.8.9}$$

$$v = v_0 + \varepsilon v_1 + O(\varepsilon^2) \tag{F.8.10}$$

$$p = p_0 + \varepsilon p_1 + O(\varepsilon^2) \tag{F.8.11}$$

Substituting this into (F.8.8) and subtracting the undamped solution, we find the equation which has to be satisfied by $(\Delta s/s_e)_1$:

$$\left(\frac{\ddot{\Delta s}}{s_e}\right)_1 + \omega_0^2\left[1+2k\left(\frac{p_0}{m}\right)^2\right]\left(\frac{\Delta s}{s_e}\right)_1 = -k\omega_0^2 2\frac{p_0}{m}\frac{\gamma}{m}\dot{v}_0\left[1+2\left(\frac{\Delta s}{s_e}\right)_0\right] \quad \text{(F.8.12)}$$

From the undamped solution determined in Chap. F.7 we know

$$v_0 = \frac{p_0}{m}\left[1+2\left(\frac{\Delta s}{s_e}\right)_0\right] \quad \text{(F.8.13)}$$

or

$$\dot{v}_0 = 2\frac{p_0}{m}\left(\frac{\dot{\Delta s}}{s_e}\right)_0 \quad \text{(F.8.14)}$$

which can be substituted into (F.8.12), so that after linearisation of the pulsation amplitude:

$$\left(\frac{\ddot{\Delta s}}{s_e}\right)_1 + \omega_0^2\left[1+2k\left(\frac{p_0}{m}\right)^2\right]\left(\frac{\Delta s}{s_e}\right)_1 = -k\omega_0^2 4\left(\frac{p_0}{m}\right)^2\frac{\gamma}{m}\left(\frac{\dot{\Delta s}}{s_e}\right)_0 \quad \text{(F.8.15)}$$

We sum this equation (after multiplication with ε) with the equation for the undamped motion:

$$\left(\frac{\ddot{\Delta s}}{s_e}\right)_0 + \omega_0^2\left[1+2k\left(\frac{p_0}{m}\right)^2\right]\left(\frac{\Delta s}{s_e}\right)_0 = -k\omega_0^2\left(\frac{p_0}{m}\right)^2 \quad \text{(F.8.16)}$$

which leads, after adding an asymptotically irrelevant term to the r.h.s., to:

$$\left(\frac{\ddot{\Delta s}}{s_e}\right) + 4\varepsilon.k\omega_0^2\left(\frac{p_0}{m}\right)^2\frac{\gamma}{m}\left(\frac{\dot{\Delta s}}{s_e}\right) + \omega_0^2\left[1+2k\left(\frac{p_0}{m}\right)^2\right]\left(\frac{\Delta s}{s_e}\right) = -k\omega_0^2\left(\frac{p_0}{m}\right)^2 \quad \text{(F.8.17)}$$

The resulting differential equation for the pulsation amplitude describes a damped harmonic pulsation. The damping is very small on account of the presence of the product $k\frac{\gamma}{m}$, so that the solution may be approximated by

$$\left(\frac{\Delta s}{s_e}\right) = Se^{-\eta t}\cos(\omega t+\delta) - k\left(\frac{p_0}{m}\right)^2 \quad \text{(F.8.18)}$$

with

$$\eta = 2\varepsilon.k\omega_0^2\left(\frac{p_0}{m}\right)^2\frac{\gamma}{m} \tag{F.8.19}$$

$$\omega = \omega_0\left[1 + k\left(\frac{p_0}{m}\right)^2\right] \tag{F.8.20}$$

Comparing with the undamped solution, it appears that the amplitude of the pulsation has now become $Se^{-\eta t}$ instead of the constant amplitude S. The pulsation still has the form of a constant value on which an oscillation is superimposed, although the oscillation is now slowly damping away.

Starting from (F.8.5) we can determine an asymptotic solution for the velocity:

$$\begin{aligned} v &= \left[\frac{p_0}{m} + \varepsilon\frac{\gamma}{m}\dot{v}\right].\left[1 + 2\left(\frac{\Delta s}{s_e}\right)\right] \\ &= \frac{p_0}{m}\left[1 + 2\left(\frac{\Delta s}{s_e}\right)\right] + \varepsilon\frac{\gamma}{m}[\dot{v}_0 + \varepsilon\dot{v}_1].\left[1 + 2\left(\frac{\Delta s}{s_e}\right)_0 + 2\varepsilon\left(\frac{\Delta s}{s_e}\right)_1\right] + O(\varepsilon^2) \\ &= \frac{p_0}{m}\left[1 + 2\left(\frac{\Delta s}{s_e}\right)\right] + \varepsilon\frac{\gamma}{m}\dot{v}_0\left[1 + 2\left(\frac{\Delta s}{s_e}\right)_0\right] + O(\varepsilon^2) \end{aligned} \tag{F.8.21}$$

Again using (F.8.14), linearising and adding an asymptotically irrelevant term to the r.h.s.:

$$\begin{aligned} v &= \frac{p_0}{m}\left[1 + 2\left(\frac{\Delta s}{s_e}\right)\right] + 2\varepsilon\frac{\gamma}{m}\frac{p_0}{m}\left(\frac{\dot{\Delta s}}{s_e}\right)_0 + O(\varepsilon^2) \\ &= \frac{p_0}{m}\left[1 + 2Se^{-\eta t}\cos(\omega t + \delta) - 2k\left(\frac{p_0}{m}\right)^2\right] - 2\varepsilon\frac{\gamma}{m}\frac{p_0}{m}Se^{-\eta t}\omega\sin(\omega t + \delta) + O(\varepsilon^2) \end{aligned} \tag{F.8.22}$$

The velocity also has the form of a constant on which a damped oscillation is superimposed. For $\varepsilon \neq 0$ it is seen that the oscillation amplitude is initially ($t = 0$) slightly larger than in the steady case, and comprises a part which is 90° out of phase compared with the undamped oscillation.

In order to determine the momentum from (F.8.3) we need the acceleration, obtained by differentiating (F.8.22), taking into account that $\eta = O(\varepsilon)$:

$$\dot{v} = -2\frac{p_0}{m}Se^{-\eta t}\left[\eta\cos(\omega t + \delta) + \omega\sin(\omega t + \delta) + \varepsilon\frac{\gamma}{m}\omega^2\cos(\omega t + \delta)\right] + O(\varepsilon^2) \tag{F.8.23}$$

The momentum thus becomes

$$\begin{aligned} p &= p_0 + \varepsilon\gamma\dot{v} \\ &= p_0 - 2\varepsilon.\gamma\frac{p_0}{m}Se^{-\eta t}\omega\,\sin(\omega t+\delta)+O(\varepsilon^2) \end{aligned} \tag{F.8.24}$$

The momentum is no longer constant as it was in the steady case, which is understandable because of the fluctuating character of the r.h.s. of the differential equation $\dot{p} = \varepsilon\gamma\ddot{v}$. The time-averaged momentum (i.e. the momentum averaged over a large number of pulsation cycles) is nevertheless constant. This is in accordance with the lack of any sustained external force acting on the droplet. The shedding of energy during a damped motion thus only affects the amplitudes of the pulsation and velocity, and does not go at the cost of the average momentum. This conclusion is a bit surprising, because it goes against expectations from the physical point of view.

The order of magnitude of the time to halve the oscillations can be estimated as follows:

$$T_{1/2} = \frac{ln2}{\eta} \tag{F.8.25}$$

Comparing this characteristic time with the period of a pulsation cycle (see (F.8.19) and (F.8.20)):

$$T_{1/2}.f_0 = \frac{ln2}{\eta}\frac{\omega_0}{2\pi} = \frac{ln2}{2\pi}\frac{1}{2\varepsilon k\,\omega_0\left(\frac{p_0}{m}\right)^2\frac{\gamma}{m}} \tag{F.8.26}$$

where the value of the constants is given by (see Eqs. (F.6.1), (F.6.7) and (F.6.3), (F.6.4)):

$$\frac{\gamma}{m} = \frac{4}{3}\frac{1}{c}\frac{s_e}{\alpha} \tag{F.8.27}$$

$$k\,\omega_0 = \sqrt{\frac{\alpha}{\alpha^*}\frac{1}{cs_e}\frac{1}{\sqrt{\mu\tau}}} \tag{F.8.28}$$

so that the denominator in (F.8.26) may be written as

$$2\varepsilon k\,\omega_0\left(\frac{p_0}{m}\right)^2\frac{\gamma}{m} = \varepsilon\frac{8\pi}{3\beta}\frac{(p_0/m)^2}{c^2}\frac{1}{\sqrt{\mu\tau}} \tag{F.8.29}$$

with

$$\beta = \sqrt{\pi^2 \alpha . \alpha^*} \tag{F.8.30}$$

which finally results in

$$T_{1/2}.f_0 = O\left(\frac{1}{\varepsilon}\sqrt{\mu\tau}\right) \tag{F.8.31}$$

showing that the typical time to damp out the oscillations is large compared with the period of the pulsation.

Finally, the energy during the damped motion could be considered. In view of the complicated expressions, a complete derivation will be omitted. A rather more qualitative inspection of the expressions can be given, by noting that in the expressions for the pulsation and velocity the damped case is found from the undamped case by a replacement of S (undamped case) by $(S\,e^{-\eta t})$ (damped case), apart from extra oscillations due to the Schott energy. Similarly, we might expect for the time averaged energy an expression obtained by a simple modification of Eq. (F.7.13):

$$\overline{E} = \frac{1}{2}\frac{m}{k}\left[(S\,e^{-\eta t})^2 \frac{\omega^2}{\omega_0^2} + k\left(\frac{p_0}{m}\right)^2\right] \tag{F.8.32}$$

from which, substituting for η:

$$\frac{d\overline{E}}{dt} = -2(S\,e^{-\eta t})^2\,\omega^2\left(\frac{p_0}{m}\right)^2\gamma \tag{F.8.33}$$

This expression is in agreement with Eq. (F.7.17), which gave the average radiated power in the case of undamped motion:

$$\overline{F.v} = 2\,\gamma\left(\frac{p_0}{m}\right)^2 S^2\omega^2 \tag{F.7.17}$$

Appendix G
De Broglie's Formula and Planck's Constant; the Concept of Photons

G.1 Assigning a Wavelength to the Velocity Dependent Part of the Pulsation Frequency

The pulsation frequency as well as the frequency of the velocity fluctuations, both in the damped and in the undamped case, were found to be (see Eq. (F.7.8))

$$\omega = \omega_0\left[1 + k\, v_{av}^2\right] \tag{G.1.1}$$

It can be written as

$$\omega = \omega_0 + \Delta\omega \tag{G.1.2}$$

where $\Delta\omega$ stands for the frequency *shift* due to velocity. The basic frequency ω_0 will be called the "zero-speed frequency". The frequency shift equals

$$\Delta\omega = k\,\omega_0\, v_{av}^2 \tag{G.1.3}$$

Physically the frequency shift $\Delta\omega$ might be viewed as the frequency of beats that would occur in the—for now hypothetical—case that there is an extra excitation of the equations of motion in the "zero speed" frequency. Later it will be seen that this "hypothetical" mixing of the frequencies ω and ω_0 comes about in a physically natural way when the equations of motion for the "free" droplet (as derived in the above) are combined with the boundary conditions imposed by the presence of a potential box. For now we will consider this mixing of frequencies as a thought experiment, which can be implemented in the equations for the undamped case by taking the fictive force in the momentum equation (F.6.8) as:

$$F(t) = -\gamma\,\ddot{v} + \widehat{F}\,cos(\omega_0 + \delta)t \tag{G.1.4}$$

T. van Holten, *The Atomic World Spooky? It Ain't Necessarily So!*
DOI 10.2991/978-94-6239-234-2

so that the momentum equation (with the assumption of zero external potential) becomes

$$\dot{p} = \widehat{F}\, cos(\omega_0 + \delta)t \tag{G.1.5}$$

or

$$p = p_0 + \frac{\widehat{F}}{\omega_0} sin(\omega_0 + \delta)t \tag{G.1.6}$$

If now the same procedure is followed as shown in Chap. F.7 for the undamped case, and assuming that $\frac{\widehat{F}}{\omega_0}$ is small, we find as equation of motion for the pulsation:

$$\frac{\ddot{\Delta s}}{s_e} + \omega_0^2 \left[1 + 2\,k\left(\frac{p_0}{m}\right)^2\right] \frac{\Delta s}{s_e} = -k\,\omega_0^2 \left(\frac{p_0}{m}\right)^2 - 2\,k\,\omega_0^2 \left(\frac{p_0}{m}\right) \frac{\widehat{F}}{\omega_0} sin(\omega_0 + \delta)t \tag{G.1.7}$$

This equation shows that there is indeed an additional forcing of the pulsation in the frequency ω_0. The homogeneous equation has as solution a harmonic oscillation with frequency $\omega = \omega_0 \left[1 + k\left(\frac{p_0}{m}\right)^2\right] = \omega_0 \left[1 + k\, v_{av}^2\right]$. We thus see that the complete solution will show up interference between the particular solution and the homogeneous solution, which as a consequence of the close proximity of the values of ω and ω_0 leads to a beat phenomenon.

The "beat frequency" (see the note below on the nomenclature and some other properties of beats) is $\Delta\omega = (\omega - \omega_0)$ and thus equals $\Delta\omega = k\,\omega_0\, v_{av}^2$. Within the framework of this physical interpretation of the frequency shift $\Delta\omega$, one may assign a wave length λ to $\Delta\omega$. Physically, the interpretation of a "wavelength assigned to a *shift* in frequency" is, that it is the distance travelled by the droplet between two consecutive beats. The wavelength is then defined by

$$\lambda = 2\pi \frac{v_{av}}{\Delta\omega} = \frac{2\pi}{k\,\omega_0} \frac{1}{v_{av}} \tag{G.1.8}$$

Note: nomenclature and properties of beats.

Caution is needed when speaking about "the beat frequency", a careful definition is needed. In general, if we have interference between two signals $cos(\omega t)$ and $cos(\omega_0 t)$ then the sum signal

$$\begin{aligned} f(t) &= \cos(\omega t) + \cos(\omega_0 t) \\ &= 2.\cos\left(\frac{\omega + \omega_0}{2}\, t\right).\cos\left(\frac{\omega - \omega_0}{2}\, t\right) \end{aligned} \tag{G.1.9}$$

has periodicity with a much lower frequency than the separate signals. It is even possible that there is no periodicity at all, i.e. that the repetition period T is infinite.

The repetition period of the sum signal is found by requiring that $f(t+T)=f(t)$ for arbitrary time t. Substituting $f(t)$ from (G.1.9), this requirement reads:

$$\cos\omega(t+T)+\cos\omega_0(t+T)=\cos(\omega t)+\cos(\omega_0 t) \tag{G.1.10}$$

or:

$$\begin{aligned} &cos(\omega t).cos(\omega T)-sin(\omega t).sin(\omega T)\\ &\quad+cos(\omega_0 t).cos(\omega_0 T)-sin(\omega_0 t).sin(\omega_0 T)=cos(\omega t)+cos(\omega_0 t) \end{aligned} \tag{G.1.11}$$

This equality is satisfied if <u>simultaneously</u>

$$\omega T=2\pi.n \quad (n=1,2,3,\ldots) \tag{G.1.12}$$

(where $n=0$ has been excluded because it would imply $T=0$) <u>and</u>

$$\omega_0 T=2\pi.k \quad (k=1,2,3,\ldots) \tag{G.1.13}$$

If $\omega\neq\omega_0$, say $\omega-\omega_0>0$, then $n-k>0$ and $(n-k)$ equals an integer larger than 0:

$$m=(n-k) \quad (m=1,2,3,\ldots.).$$

The combined conditions (H.2.12) and (H.2.13) may thus be replaced by the requirements

$$(\omega-\omega_0)T=2\pi.m \quad (m=1,2,3,\ldots.) \tag{G.1.14}$$

<u>and</u>

$$\omega_0 T=2\pi.k \quad (k=1,2,3,\ldots) \tag{G.1.13}$$

These conditions are graphically sketched in Fig. G.1.1, for $m=1$.

It is seen that ω_0 and $(\omega-\omega_0)$ must be commensurate in order to have a finite repetition time T, i.e. there must exist a time T where $\omega_0=(\omega-\omega_0)\frac{k}{m}$, so that $(\omega-\omega_0)/\omega_0$ is a rational fraction. For completely arbitrary values of ω and ω_0 this may not be the case, and the repetition time becomes $T\rightarrow\infty$. If ω_0 and $(\omega-\omega_0)$ are commensurate, even then the repetition time may be very large.

What is shown in Fig. G.1.1 is a rather special situation: at the first instant of time where $(\omega - \omega_0)t = 2\pi$ it so happens that also $\omega_0 t$ equals a -large—integer times 2π. Here we have assumed that the ratio $\omega_0/(\omega - \omega_0)$ equals an integer k.

Often one encounters the situation that ω and ω_0 are very large, but that their difference is much smaller. In such a case the time t_1 for which $(\omega - \omega_0)t_1 = 2\pi$ and the nearest time t_2 for which $\omega_0 t_2/(2\pi)$ is an integer are so close that one may, to an accuracy considered acceptable, ignore the condition (G.1.13). Both in the case that condition (G.1.13) is indeed ignored (or considered to be approximated well enough) as well as in the special case that $t_1 = t_2$ so that the condition (G.1.13) is exactly satisfied, the repetition period is

$$T = \frac{2\pi.m}{(\omega - \omega_0)} \quad (m = 1, 2, 3, \ldots) \tag{G.1.15}$$

Equation (G.1.15) expresses for $m = 1$ the potentially smallest repetition time of the signal. Whether in an actual case this smallest possible time is realised depends on the condition (G.1.13).

Often we will meet the situation that at the time $t = \frac{2\pi}{(\omega - \omega_0)}$ the argument $\omega_0 t$ of the cosine $\cos \omega_0 t$ equals only an integer number times π: $\omega_0 t = k.\pi$. In such a case the smallest repetition time strictly speaking corresponds to $m = 2$ so that $T = \frac{4\pi}{(\omega - \omega_0)}$ at which time the condition $\omega_0 T = k.2\pi$ is satisfied. However, if the physically relevant time scale is so large that an interval $\Delta t = \frac{2\pi}{\omega_0}$ may be considered to be negligibly small, then in practice the repetition time becomes again $T = \frac{2\pi}{(\omega - \omega_0)}$.

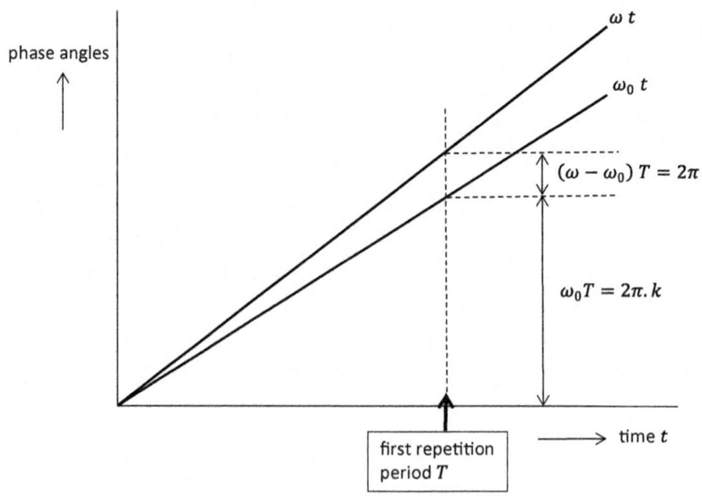

Fig. G.1.1 Graphic representation of the conditions (G.1.13) and (G.1.14) for $m = 1$ and if $\omega_0 = k.(\omega - \omega_0)$

Definition: *the difference frequency* $(\omega - \omega_0)$—or: the frequency shift due to velocity—*will be called the "beat frequency"*. It is the frequency corresponding to the potentially smallest repetition time ($m = 1$).

Why the nomenclature "beat" is used for this frequency is clear from the second equality of Eq. (G.1.9):

$$f(t) = 2.\cos\left(\frac{\omega + \omega_0}{2}t\right).\cos\left(\frac{\omega - \omega_0}{2}t\right) \tag{G.1.16}$$

which shows that $f(t)$ may be considered as a fast oscillation with a variable amplitude. At first sight it might seem from (G.1.16) that the period of the amplitude variation is not in agreement with (G.1.15) and that the frequency of the amplitude variations is half the value given by (G.1.15). However, it must be realised that $f(t)$ is here written as a product of two terms, and if both cosines become -1 then $f(t)$ has the same value again as at time $t = 0$. A more general proof that at all times the repetition period of (G.1.16) equals the period (G.1.15) can be given as follows. The proof given below is based on the assumption that there exists a finite repetition time T which satisfies both the conditions (G.1.13) and (G.1.14), implying the commensurate condition $\omega_0 = (\omega - \omega_0)\frac{k}{m}$.

The first cosine in (G.1.16) may be written at time $t + T$ as

$$\begin{aligned}
&\cos\left[\frac{\omega + \omega_0}{2}(t + T)\right] \\
&\quad = \cos\left[\frac{\omega + \omega_0}{2}t + \frac{\omega - \omega_0}{2}T + \omega_0 T\right] \\
&\quad = \cos\left[\frac{\omega + \omega_0}{2}t + \pi(2k + m)\right] \\
&\quad = \pm\cos\left[\frac{\omega + \omega_0}{2}t\right]
\end{aligned} \tag{G.1.17}$$

where the positive sign is associated with even values of the integer *m*, and the negative sign with odd values.

The second cosine in (G.1.16) may for $t + T$ be written as

$$\cos\left[\frac{\omega - \omega_0}{2}(t + T)\right] = \cos\left[\frac{\omega - \omega_0}{2}t + \frac{\omega - \omega_0}{2}T\right] \tag{G.1.18}$$

Again substituting (G.1.15) we find

$$\cos\left[\frac{\omega - \omega_0}{2}(t + T)\right] = \cos\left[\frac{\omega - \omega_0}{2}t + m.\pi\right] = \pm\cos\left[\frac{\omega + \omega_0}{2}t\right] \tag{G.1.19}$$

where the positive sign is associated with even values of the integer m, and the negative sign with odd values. The product of the cosines thus satisfies the equality

$$\begin{aligned} f(t+T) &= 2.\cos\left[\frac{\omega+\omega_0}{2}(t+T)\right].\cos\left[\frac{\omega-\omega_0}{2}(t+T)\right] \\ &= 2.\cos\left[\frac{\omega+\omega_0}{2}t\right].\cos\left[\frac{\omega-\omega_0}{2}t\right] = f(t) \end{aligned} \tag{G.1.20}$$

for all integer values of m, if T complies with the two conditions (G.1.13) and (G.1.14).

The function $f(t)$ has been sketched in Fig. G.1.2, which shows these conclusions graphically.

Beats in the case of unequal amplitude waves.

We now consider the function

$$f(t) = a.\cos\omega t + b.\cos\omega_0 t \tag{G.1.21}$$

where $a < b$

Schematically drawing the envelopes of the fast oscillation of $f(t)$ for decreasing values of a/b gives the picture of Fig. G.1.3. For small values of a/b the envelope approaches a cosine-shape. If the envelope is denoted $g(t)$ then

$$g(t) = a.cos(\omega - \omega_0)t + b \quad \textit{for small}\; a/b \tag{G.1.22}$$

as will be shown below. The curve $f(t)$ for small values of a/b then has the shape as schematically shown in Fig. G.1.4.

The proof that the envelope $g(t)$ of the function $f(t) = a.\cos\omega t + b.\cos\omega_0 t$ has a sinusoidal shape for small a/b goes as follows. If the frequencies ω and ω_0 are

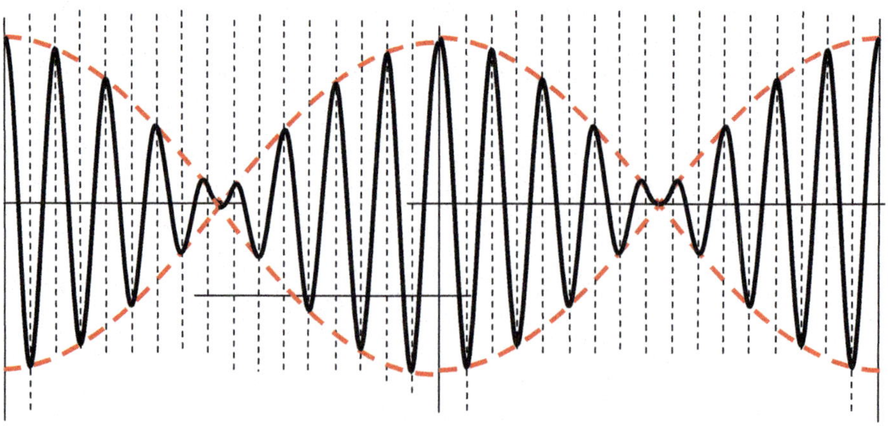

Fig. G.1.2 Beat resulting from superposition of equal amplitude vibrations with slightly different frequencies

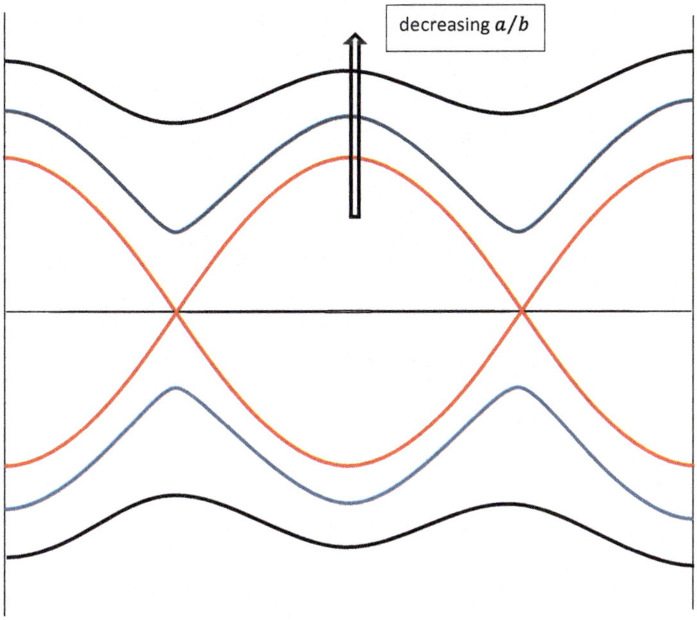

Fig. G.1.3 Beats resulting from superposition of *un*equal amplitude vibrations

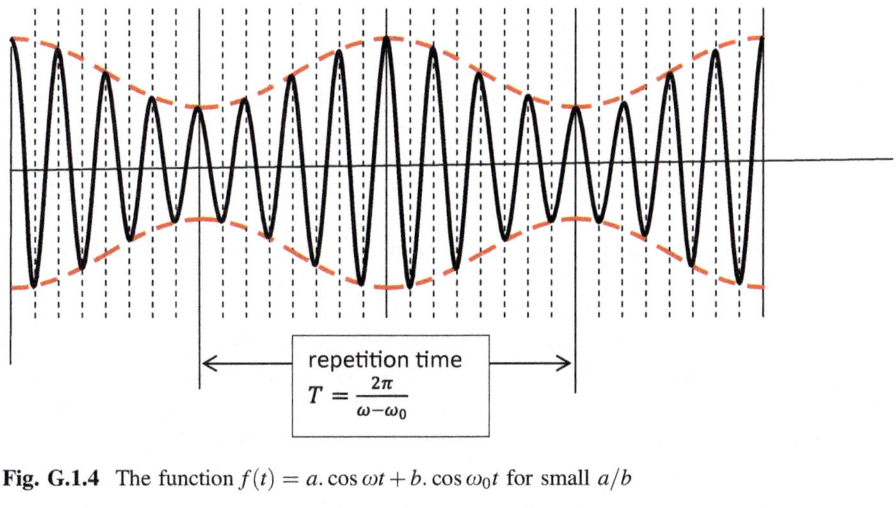

Fig. G.1.4 The function $f(t) = a.\cos\omega t + b.\cos\omega_0 t$ for small a/b

large and their difference $\omega - \omega_0$ is very small compared with these frequencies, the envelope $g(t)$ practically coincides with a curve through the maxima of $f(t)$. The times t_m where these maxima occur follow from the equation

$$-a\,\omega \sin \omega t_m - b\omega_0 \sin \omega_0 t_m = 0 \tag{G.1.23}$$

so that

$$\sin\omega_0 t_m = -\frac{a}{b}\frac{\omega}{\omega_0} \sin \omega t_m \tag{G.1.24}$$

$$\cos\omega_0 t_m = \sqrt{1 - \sin^2 \omega_0 t_m} \approx 1 - \frac{1}{2}\left(\frac{a}{b}\right)^2 \left(\frac{\omega}{\omega_0}\right)^2 \sin^2 \omega t_m \quad \textit{for small } a/b \tag{G.1.25}$$

The function $f(t)$ may be written in the form

$$\begin{aligned} f(t) &= a.\cos \omega t + b.\cos \omega_0 t \\ &= b\left[\frac{a}{b}\cos\{(\omega - \omega_0)t + \omega_0 t\} + \cos \omega_0 t\right] \\ &= b\left[\frac{a}{b}\cos(\omega - \omega_0)t.\cos \omega_0 t - \frac{a}{b}\sin(\omega - \omega_0)t.\sin \omega_0 t + \cos \omega_0 t\right] \end{aligned} \tag{G.1.26}$$

The value of this function at the times $t = t_m$ is found by substituting (G.1.24) and (G.1.25) into (G.1.26). For small a/b we neglect the squares $\left(\frac{a}{b}\right)^2$ whilst retaining only the terms linear in $\frac{a}{b}$, from which:

$$f(t_m) \approx a.\cos(\omega - \omega_0)t_m + b \tag{G.1.27}$$

Since it was assumed that the envelope of $f(t)$ practically coincides with a curve passing through the maxima of $f(t)$, i.e. a curve through the points $f(t_m)$, it is thus seen that the enveloping curve is given by

$$g(t) = a.\cos(\omega - \omega_0)t + b \quad \textit{for small } a/b \tag{G.1.28}$$

It is helpful to use another interpretation of this result, in terms of the *stroboscopic image* of $a.\cos \omega t$ when it is sampled with the frequency ω_0.

Definition of the stroboscopic image. Another way to represent the interference phenomenon between the two cosines with slightly different frequencies and a large difference of amplitude is to use the concept of "stroboscopic image". It is useful to determine what is the relation between the beat phenomenon and the stroboscopic image. Imagine that the function $\cos \omega t$ is sampled with sampling frequency ω_0, i.e. we determine the values $g(t_i) = \cos \omega t_i$ where the times t_i correspond with the maxima of the function $cos\, \omega_0 t$. Thus, $t_i = i\frac{2\pi}{\omega_0}$ $(i = 0, 1, 2, \ldots.)$.

If we would have the situation of Figs. G.1.1 and G.1.4 where

$$(\omega - \omega_0)T = 2\pi, \quad \omega_0 T = k.2\pi \tag{G.1.29}$$

then

$$\frac{t_i}{T} = \frac{i}{k} \tag{G.1.30}$$

We can write $g(t)$ in the form

$$\begin{aligned} g(t) &= \cos \omega t = \cos[(\omega - \omega_0)t + \omega_0 t] \\ &= \cos(\omega - \omega_0)t \,.\, \cos \omega_0 t - \sin(\omega - \omega_0)t \,.\, \sin \omega_0 t \end{aligned} \tag{G.1.31}$$

and according to (G.1.30):

$$g(t_i) = \cos\left(2\pi \frac{t_i}{T}\right).cos(2\pi\,.i) - \sin\left(2\pi \frac{t_i}{T}\right).sin(2\pi\,.i) = \cos\left(2\pi \frac{t_i}{T}\right) \tag{G.1.32}$$

Comparing with Fig. G.1.4, the times t_i correspond with the vertical dashed lines in this figure. The function $\cos\left(2\pi \frac{t}{T}\right)$ on which the points $g(t_i)$ according to (G.1.32) are situated corresponds with the upper envelope of the function $f(t) = a.cos(\omega t) + b.\ \cos(\omega_0 t),\ \ \frac{a}{b} \ll 1$, apart from a constant vertical shift. The *stroboscopic* viewing thus traces out the amplitude variations due to the *beat phenomenon*.

If one would double the frequency of the stroboscopic sampling, one would in addition get images of $g(t) = \cos \omega t$ picked out at the intermediate times

$$t_j = T\frac{i - \frac{1}{2}}{k} \quad (i = 1,\ 2,\ 3,\ \ldots.) \tag{G.1.33}$$

so that one gets the extra values

$$g(t_j) = \cos\left(2\pi \frac{t_j}{T}\right).cos\left[2\pi(i - \frac{1}{2})\right] - \sin\left(2\pi \frac{t_j}{T}\right).sin\left[2\pi\left(i - \frac{1}{2}\right)\right] = -\cos\left(2\pi \frac{t_i}{T}\right) \tag{G.1.34}$$

which traces out the lower envelope of the beat phenomenon shown in Fig. G.1.4, again apart from a constant.

The advantage of using the stroboscopic viewing as an intermediate step to determine the interference pattern is, that stroboscopic viewing complies with several simple and easy to understand rules:

(1) the stroboscopic image has the same form as the sampled time function, be it that it is stretched to a much larger repetition period.
(2) the number of "slow" stroboscopic cycles is equal to the difference in the number of "fast" cycles of the two interfering waves. See for an illustration Fig. G.1.5 where the interference is visualised by overlapping line patterns.

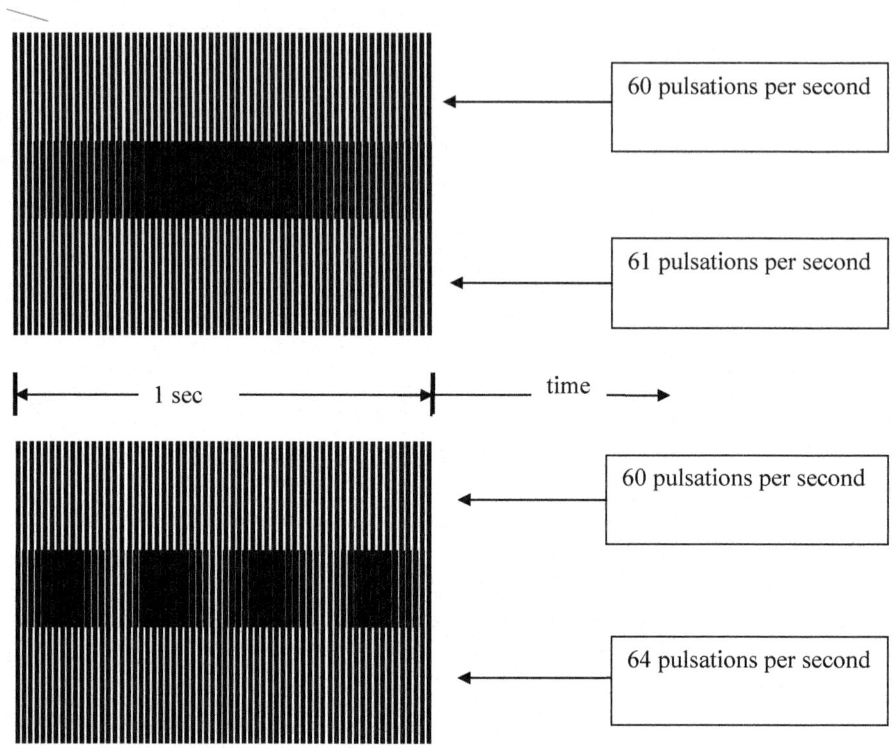

Fig. G.1.5 Some properties of beats, illustrated by overlapping *white-black line* patterns. *Top* the two interfering waves differ one cycle in the period shown. *Middle* four cycles difference

(3) the phase of the stroboscopic image, which in particular is important in the begin- and endpoints, is the same as the phase of the wave which is being "viewed" by the stroboscopic sampling.

The case of frequency-modulated (FM) signals. Loosely speaking, a frequency-modulated signal is—according to the definition that will be used in this text—a "near-harmonic function", closely resembling e.g. a cosine $\cos[\omega t]$, where the frequency $\omega(t)$ displays a slow variation in time; "slow" meaning a variation with a very much larger time scale than the period of the cosine. According to the definitions used here, any type of near-harmonic function $g(t)$, subject either to amplitude-, frequency- or phase modulation, will have the form

$$g(t) = A(t)\cos\left[\int_0^t \{\omega + \Delta\omega(\tau)\}d\tau + \varepsilon(t)\right] \tag{G.1.35}$$

where it is assumed that the amplitude, frequency and phase are "slow" functions, in the sense described above.

The use of the terminology "frequency" etc. at time t in this case requires some further consideration, because it implies an "instantaneous frequency" or "instantaneous period", which is a contradictio in terminis. The physical justification for using this terminology can be based on the concept of "the osculating cosine". In each point of the near-harmonic function one can define a cosine-function which has in this point a contact of second order, i.e. the values of the near-harmonic function and the osculating cosine are equal, as well as their first and second derivatives. If the near-harmonic function is denoted by $g(t)$, then the parameters of the cosine $A.\cos(\omega t+\varepsilon)$ which osculates in the point of time t_0 are given by the relations

$$\omega^2(t_0) = -\frac{\ddot{g}}{g}(t_0) \tag{G.1.36}$$

$$A^2(t_0) = \left[g^2 - \dot{g}^2\frac{g}{\ddot{g}}\right]_{t_0} \tag{G.1.37}$$

$$\omega(t_0).t_0 + \varepsilon(t_0) = \arccos\left(\frac{g}{A}\right)_{t_0} + n.2\pi \tag{G.1.38}$$

In this way one can give a physical interpretation to the instantaneous frequency, instantaneous amplitude and instantaneous phase in each point of the near-harmonic function. The definition of "near harmonic function" is then, that it is a function where the so defined instantaneous parameters ω, A and ε show a slow variation in time, i.e. the characteristic time of the variations is large compared with the time $\frac{2\pi}{\omega}$. If the instantaneous frequency ω remains constant in time, then Eq. (G.1.36) indicates that the function is completely harmonic, with a constant amplitude A and a constant phase angle ε.

If the near-harmonic (frequency- or otherwise modulated) signal is periodic with a repetition period T_{FM} (the index relating to the frequency modulation) then the instantaneous parameters repeat themselves. This situation is sketched in Fig. G.1.6.

The next Fig. G.1.7 shows the corresponding phase of the cosine $\cos\left[\int_0^t \{\omega + \Delta\omega(\tau)\}d\tau\right]$. Since T_{FM} is the repetition time of the near harmonic function, the phase at this time should be $n.2\pi$. This forms a restriction to the integrand $\{\omega+\Delta\omega(\tau)\}$, which thus has to satisfy the condition

$$\int_0^{T_{FM}} \{\omega + \Delta\omega(\tau)\}d\tau = n.2\pi \tag{G.1.39}$$

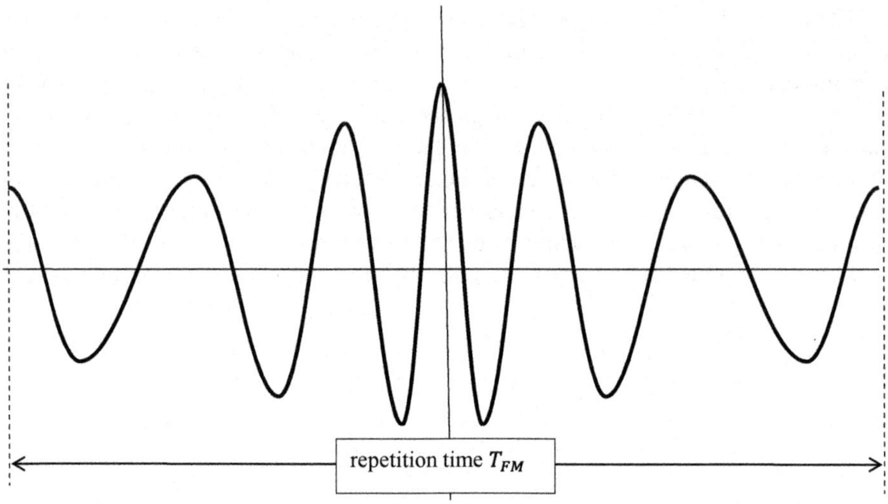

Fig. G.1.6 Periodic near-harmonic function with repetition time T_{FM}

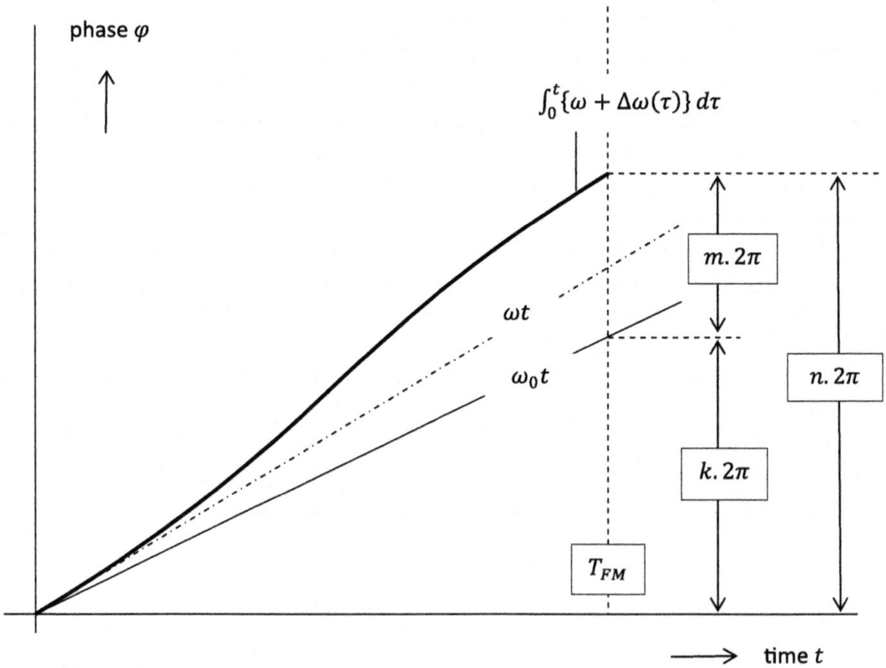

Fig. G.1.7 Smallest possible repetition period of signal which is the sum of a near harmonic function (Fig. G.1.6) and a pure harmonic function $\cos(\omega_0 t)$

If we now have interference of the signal of Fig. G.1.6 with another signal $\cos(\omega_0 t)$ having a constant frequency, in the most general case there need not exist a finite repetition time of the sum signal (see the text above, applicable to Fig. G.1.1).

However, under certain conditions there can be a finite repetition time of the sum signal, and the smallest possible repetition time may even be equal to T_{FM}, if the following additional condition is satisfied (see Fig. G.1.7):

$$\omega_0.T_{FM} = k.2\pi \tag{G.1.40}$$

Combining the latter condition with (G.1.39) yields:

$$\int_0^{T_{FM}} \{\omega + \Delta\omega(\tau)\} d\tau - \omega_0.T_{FM} = m.2\pi \tag{G.1.41}$$

Definition: the beat frequency in the case of interference between a near harmonic function and a pure harmonic function is defined as $\omega_{beat} = 2\pi/T_{FM}$, thus corresponding to the smallest possible repetion time subject to the conditions (G.1.40) and (G.1.41).

The shape of the interference signal. For simplicity we will take a near harmonic function with constant amplitude, having a local maximum at $t = 0$:

$$g(t) = \cos\left[\int_0^t \{\omega + \Delta\omega(\tau)\} d\tau\right] \tag{G.1.42}$$

The frequency modulated signal is sketched in a highly schematic form in Fig. G.1.8: the signal has been divided in three parts, all of them containing constant frequency signals. However, the middle part has a higher frequency than the outer parts. In other words: $\Delta\omega = 0$ in the outer parts, and $\Delta\omega$ is unequal to zero but constant in the middle part.

Now the interference is considered with a signal having the frequency ω everywhere. The result is shown by the black lines in the second diagram of Fig. G.1.8, in the form of the stroboscopic image. To start with, the FM-signal and the mixing signal coincide for some time, so that the stroboscopic image is just a straight line. Entering the second part of the signal, there is a constant frequency difference between the two signals. The figure has been drawn such that in the second time interval there is just one cycle difference between the two "fast" signals, so that the stroboscopic image shows one "slow" cycle of a harmonic function. The last part of the interference is shown up again as a straight line.

The red line now shows the strobe image in a more realistic case where the variation of frequency of the FM-signal is spread out over the entire repetition time, with a maximum in the middle of the interval. One then expects a smooth picture, if the "instantaneous period" (as defined above during the discussion of the osculating sine) varies smoothly from a relatively long period to a shorter period in the middle of the beat, and to a longer instantaneous period again at the end of the time interval of the beat.

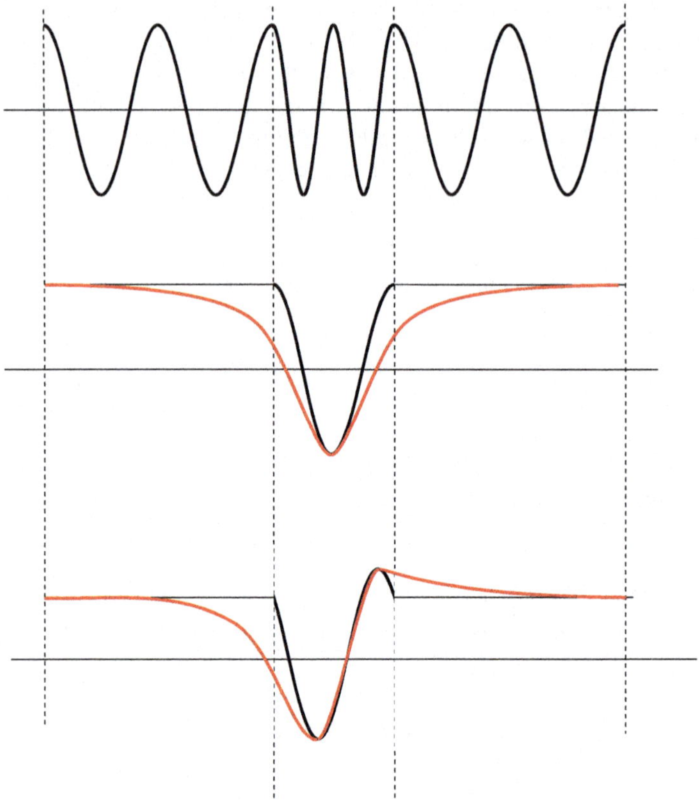

Fig. G.1.8 Simplified frequency modulated signal, and interference patterns after mixing with constant frequency signal (*black lines*). *Red lines* expected shape of stroboscopic image in the case of continuously varying frequency

Finally, in the lower part of Fig. G.1.8 the situation is sketched where the two interfering signals are phase-shifted at time $t = 0 : g(t) = \cos\left[\int_0^t \{\omega + \Delta\omega(\tau)\}d\tau\right]$ is mixed with the constant frequency but phase-shifted signal $\cos(\omega t + \varepsilon)$.

What is important to note for future reference is, that the interference pattern now no longer displays any symmetry or anti-symmetry with respect to the mid-time of the repetition period.

End of note on nomenclature and properties of beats.

G.2 De Broglie's Formula and Planck's Constant

Now, we can define a certain kind of "average momentum" of the droplet as

$$p_{av} = m\, v_{av} \tag{G.2.1}$$

Note that this is how the momentum would be defined in the usual quantum mechanics where the mass is taken to be a constant. But recall that the so defined p_{av} is unequal to p_0 which denotes the constant momentum in a potential box in the case of a variable mass. The relation between p_{av} and p_0 is given by Eq. (F.7.7):

$$p_{av} = p_0 \left(\frac{\omega_0}{\omega}\right)^2 \tag{G.2.2}$$

It may be verified, using Eq. (G.1.8) that we obtain from (G.2.1) the relation

$$p_{av} = m\, v_{av} = m \frac{2\pi}{k\, \omega_0} \frac{1}{\lambda} = \frac{K}{\lambda} \tag{G.2.3}$$

where the constant K is equal to

$$K = m \frac{2\pi}{k\, \omega_0} \tag{G.2.4}$$

Recalling the meaning of the short notations ω_0, k and m (see Eqs. (F.6.3), (F.6.4) and (F.6.1)):

$$\omega_0 = \sqrt{\frac{m}{m^*} \frac{c^2}{s_e^2} \mu\tau \left(1 - \frac{1}{\mu} - \frac{1}{\tau} + \frac{1}{\mu\tau}\right)} \approx \frac{c}{s_e} \sqrt{\frac{\alpha}{\alpha^*}} \sqrt{\mu\tau} \tag{G.2.5}$$

$$k = \frac{1}{\mu\tau\left(1 - \frac{1}{\mu} - \frac{1}{\tau} + \frac{1}{\mu\tau}\right)} \frac{1}{c^2} \approx \frac{1}{\mu\tau} \frac{1}{c^2} \tag{G.2.6}$$

$$m = \frac{1}{2} m_{em_0} = \frac{1}{2} \frac{1}{c^2} \frac{q^2}{4\pi\varepsilon_0} \frac{\alpha}{s_e} \tag{G.2.7}$$

we find

$$K = \frac{1}{c} \frac{q^2}{4\pi\epsilon_0} \sqrt{\mu\tau} \sqrt{\pi^2\, \alpha^* \alpha} = \frac{1}{c} \frac{q^2}{4\pi\epsilon_0} \sqrt{\mu\tau}\, .\beta \tag{G.2.8}$$

with the factor β:

$$\beta = \sqrt{\pi^2\, \alpha^* \alpha} \tag{G.2.9}$$

This factor depends on the charge distribution in the droplet, and has the order unity.

Apart from the factor β which depends on the model assumptions, all the other factors in the expression for K are constants of nature.

Inserting the following numerical values of the constants occurring in the expression for K:

$$c = 2.9979 \times 10^{8}\, m/s \quad (speed\, of light\, in\, vacuum)$$

$$q = 1.6021 \times 10^{-19}\, C \quad (unit\, charge)$$

$$\varepsilon_0 = 8.8544 \times 10^{-12} C^2/N.m^2 \quad (vacuum\, permittivity)$$

$$\mu = \frac{105.7}{0.511} = 206.85 \quad (ratio\, of\, muon\, and\, electron\, masses)$$

$$\tau = \frac{1776.8}{0.511} = 3477.1 \quad (ratio\, of\, tau\, and\, electron\, masses)$$

we find

$K = 6.52 \times 10^{-34}.\beta$ with the dimension $[J.s]$

Nothing can be said about the exact value of the constant β which depends on the non-dimensional shape of the droplet, except that it must be of order unity. If we put $\beta = 1$, the value of K is nearly equal to Planck's constant $h = 6.6256 \times 10^{-34}\, J.s$ (difference with K ca. 1.5 %).

This result suggests that the relation $p_{av} = \frac{K}{\lambda}$ corresponds with De Broglie's relation $p = \frac{h}{\lambda}$, and that the matter waves of quantum mechanics might be

compared with the train of beats in the dynamics of a pulsating unit charge, if the dynamics are mixed with the "zero-speed" frequency.

G.3 Time- and Length Scale of the Pulsation

It is useful to get some feeling for the orders of magnitude of on the one hand the pulsation frequency and the wavelength of the pulsation, and on the other hand the frequency and wavelength of the above defined beats.

The wavelength of the pulsation is defined as the distance travelled by the pulsating particle between two consecutive cycles of the pulsation. Starting from the relation (G.2.4) between the constants k (occurring in the formula $\omega = \omega_0(1 + k\, v^2)$) and $K(\approx h)$:

$$k = \frac{m}{K}\frac{2\pi}{\omega_0} = \frac{m}{K}\frac{1}{f_0} \tag{G.3.1}$$

one may write

$$f_0 = \frac{m}{K}\frac{1}{k} = \frac{m}{K}\mu\tau\, c^2 \tag{G.3.2}$$

Using the electron rest mass

$$m_{electron} = 9.1091 \times 10^{-31}\, kg$$

and the earlier given values of $c,\ \mu,\ \tau\, and\, h$ we find

$$f_0 = 0.888 \times 10^{26} \quad sec^{-1}$$

The characteristic time scale of the pulsation is therefore of the order of 10^{-26} sec.

The wavelength corresponding to the "zero speed" frequency $f_0 = \frac{m}{K}\mu\tau\, c^2$ is given by

$$\lambda_0 = \frac{v_{av}}{f_0} \tag{G.3.3}$$

It can be quantified by using Eqs. (G.2.7) and (G.2.8) to determine the ratio

$$\frac{K}{m} = \frac{\beta}{\alpha/2}\, c\, s_e\sqrt{\mu\tau} \tag{G.3.4}$$

and from (G.3.2)

$$\frac{1}{f_0} = \frac{K}{m}\frac{1}{\mu\tau\, c^2} \tag{G.3.5}$$

This gives, together with (G.3.3):

$$\frac{\lambda_0}{s_e} = \frac{v_{av}}{c}\frac{1}{\sqrt{\mu\tau}}\frac{\beta}{\alpha/2} \tag{G.3.6}$$

This expression shows that the wavelength of the pulsation is extremely small, in fact several orders smaller than the length of the droplet.

An estimate of the actual value of s_e may be obtained by assuming that it is of the same order of magnitude as the "classical electron radius" r_0 which is defined as

$$r_0 = \frac{1}{c^2}\frac{q^2}{4\pi\epsilon_0}\frac{1}{m_{electron}} \tag{G.3.7}$$

It has the value $r_0 = 2.82 \times 10^{-15}\, m$, a value obtained by assuming in Lorentz's electron model the electromagnetic mass to be equal to the measured mass of an electron (neglecting a factor 2/3 which depends on the assumed charge distribution). Therefore

$$\frac{\lambda_0}{r_0} = O\left(\frac{v_{av}}{c}\frac{1}{\sqrt{\mu\tau}}\right) \tag{G.3.8}$$

G.4 Time- and Length Scale of the "de Broglie Waves"

The frequency shift due to a translation velocity is according to Eqs. (G.1.3) and (G.2.6):

$$\Delta\omega = k\,\omega_0 v_{av}^2 = \omega_0 . \frac{1}{\mu\tau}\frac{v_{av}^2}{c^2} \tag{G.4.1}$$

With the factor $\mu\tau = 206.85 \times 3477.1 = 0.72 \times 10^6$ it is seen that this frequency is very much lower than de pulsation frequency itself.

As discussed, a possible physical interpretation of the frequency shift is, that it is the beat frequency due to interference between the field of a moving droplet and the field of a stationary charge. Using this interpretation of the frequency shift $\Delta\omega$, one may thus assign a wavelength λ to the frequency shift, and define this wavelength λ as the distance travelled by the droplet between two consecutive beats (but note carefully the definition of "beat" and the associated wavelength as discussed in Chap. G.1).

The wavelength thus defined, and shown to be equivalent to the wavelength of "De Broglie matter waves", has an order of magnitude given by (see Eqs. (G.3.6) and (G.4.1)):

$$\frac{\lambda}{s_e} = \frac{\lambda_0}{s_e} . \frac{\omega_0}{\Delta\omega} = O\left(\frac{\sqrt{\mu\tau}}{v_{av}/c}\right) \tag{G.4.2}$$

It is seen that for typical electron velocities v_{av}/c between 0.1 and 0.5 the wavelength of the beats is in the order of some 5000 times the length of the droplet. This is an atomic scale, in contrast to the wavelength of the pulsations themselves which is a small fraction of the size of sub-atomic particles.

Another fundamental difference between the wave lengths of the pulsations and the beats is, that the wavelength of the beats is inversely proportional to the translation speed, in contrast to the pulsation wavelength which is directly proportional.

G.5 Order of Magnitude of the Time Scale of Damping by Radiation

In case the droplet is left to itself without the intervention of an artificial external force $F(t)$, it was established in Chap. F.8 that the resulting damping by radiation is characterised by the time $T_{1/2}$ to halve the pulsation and velocity oscillations:

$$T_{1/2}.f_0 = O\left(\sqrt{\mu\tau}\right) \tag{G.5.1}$$

We now compare this characteristic time with the period of the beats in the field (or: the period of De Broglie waves), using Eq. (G.4.1):

$$T_{1/2}.\Delta f = \left(T_{1/2}.f_0\right).\frac{\Delta\omega}{\omega_0} = O\left(\frac{1}{\sqrt{\mu\tau}}\frac{v_{av}^2}{c^2}\right) \tag{G.5.2}$$

This result shows that the decay time during a damped motion is short compared with the period of a "De Broglie wave". Considering that the wavelength of a "De Broglie wave" has an atomic scale (or in the case of the one-dimensional oscillator: the scale of the potential well), it may be concluded that the path length covered during the process of shedding energy by radiation is small compared with the dimensions of the well.

G.6 On the Concept of Photons

As derived in Chap. G.3, the characteristic time scale of the pulsation is of the order of 10^{-26} s. Now, the modeling of an electron as a pulsating droplet of charge was shown to be compatible with usual quantum mechanics concepts, if the model is interpreted as a time-averaged representation of an electron subject to the "zitter" motion. The "zitter" is the result of the spontaneous appearance and annihilation of virtual electron-positron pairs in vacuum. The life-time of a virtual electron-positron pair is in the order of 10^{-21} s. This must also be the characteristic time-scale of the "zitter". If time-scales of the order of 10^{-21} s are thus considered to be unobservably small so that averaging is applied, it is consistent that phenomena of an even smaller time-scale, like the pulsation, are also treated by averaging them *when it is tried to observe them*. The latter must be stressed: the averaging process is applied as a simulation of the process of observation, it does *not* imply the non-existence of the fast phenomena, even if they fall outside our observation window.

Within the scope of the time-averaged electron model, i.e. when analysing the pulsating droplet of charge instead of the actual zittering electron, the characteristic time-scale of pulsations of 10^{-26} s is therefore <u>un</u>observable. This leads to some curious conclusions.

To anticipate the results from the later chapters on radiation, those particular components of the magnetic and electrical field ($\underline{B}$ and $\underline{E}$ respectively) that contribute to the loss of energy from the system by radiation are given by

$$B_\chi = -\frac{1}{c^3}\frac{q}{4\pi\varepsilon_0}\frac{\sin\varphi}{\rho}\dot{v}(t-\rho/c) \tag{G.6.1}$$

$$E_\varphi = -\frac{1}{c^2}\frac{q}{4\pi\varepsilon_0}\frac{\sin\varphi}{\rho}\dot{v}(t-\rho/c) \tag{G.6.2}$$

where ρ and φ are polar coordinates describing the far field (Fig. B.2.2) and the notation $\dot{v}(t-\rho/c)$ symbolyses that these field strengths are subject to time retardation.

In the undamped case it was found:

$$v(t) = \left(\frac{p_0}{m}\right)\left[1 + 2S\cos(\omega t + \varepsilon) - 2k\left(\frac{p_0}{m}\right)^2\right] \tag{G.6.3}$$

so that $\dot{v}$ in (G.6.1) and (G.6.2) equals:

$$\dot{v} = -2S\left(\frac{p_0}{m}\right)\omega\sin[\omega(t-\rho/c)+\varepsilon] \tag{G.6.4}$$

Both the components of the field strength (G.6.1) and (G.6.2) therefore have the form of a fast oscillation with frequency ω around a zero mean. The frequency ω is so high that it lies outside the observation window, and will be averaged out during an attempt to observe the field. Because of the zero mean value, observation of the electromagnetic field thus leads to a "null result". This is the logical consequence within the framework of the theory based on a pulsating droplet of charge.

The conclusion is different when we consider the energy flux in the field. Some other results from the theory of radiation were already mentioned in Chap. F.7. The energy in the radiation that will eventually escape to infinity was there stated to be $\gamma\dot{v}^2$, which according to Eq. (F.7.16) leads to the energy flow travelling to infinity:

$$\frac{dE_{rad}}{dt} = \gamma 2\left(\frac{p_0}{m}\right)^2 S^2\omega^2[1 - \cos 2[\omega(t - \rho/c) + \varepsilon]] \tag{G.6.5}$$

This expression comprises an—unobservable—high frequency component, but a nonzero mean value. The cause of this nonzero mean is related to the fact that the energy flow involves a product of the field strengths. What therefore can be observed is the average power which was also given in Eq. (F.7.17):

$$\frac{d\overline{E_{rad}}}{dt} = \gamma 2\left(\frac{p_0}{m}\right)^2 S^2\omega^2 \quad (undamped\ case) \tag{G.6.6}$$

In the case of the undamped motion this average power is constant in time. The artificially added force continues to replenish the radiated energy, so that the process of radiation does not stop.

In the case of the damped motion, where such an artificial energy source is absent, one would after the start of radiation see a decay of the radiated energy to ultimately zero:

$$\frac{d\overline{E_{rad}}}{dt} = \gamma 2\left(\frac{p_0}{m}\right)^2 \left(S\,e^{-\eta(t-\rho/c)}\right)^2 \omega^2 \quad (damped\ case) \tag{G.6.7}$$

Here the symbol $\frac{d\overline{E_{rad}}}{dt}$ means: averaged over a large number of pulsation cycles. The rate of decay is given by the value of η. It was already determined that the typical decaying time is large compared with the pulsation period (so that the averaging process as defined above is still meaningful), but small compared with the time needed by the droplet to traverse the width of the potential box. In the far field, on a time scale comparable with this "return period" of the droplet (which was the same as the characteristic time forming the basis for the asymptotic expansions of the first few chapters) the shedding of energy by the droplet will appear as a brief "flash" of energy travelling to infinity. However, no "supporting" field strengths can be observed, as would be the case in a normal radio transmission. Due to the limitations of observability inherent in the modeling of an electron as a pulsating droplet, we thus "see" a packet of electromagnetic energy travelling away from the droplet, without discernable electromagnetic fields.

There are no observable frequencies associated with the packet of energy sent out by the free droplet with damping. This is different in the case that there is an - hypothetical as yet—excitation of the equations of motion in the zero-speed frequency. In that case we had found that the pulsation—and therefore also the velocity—contains both the frequencies ω and ω_0, resulting in a beat phenomenon. Now consider the associated electromagnetic field. According to Eqs. (G.6.1) and (G.6.2) above, the field still consists of unobservable high frequency oscillations around a zero mean value. The field is still "invisible". The energy per second leaving the system in the form of radiation now has the form

$$\begin{aligned}\frac{dE_{rad}}{dt} &= \gamma \dot{v}^2 \div (\sin \omega t + \sin \omega_0 t)^2 \\ &= 1 + \cos(\omega - \omega_0)t + high\,frequency\,terms\end{aligned} \tag{G.6.8}$$

There is now a low frequency "visible" in the radiation, viz. the beat frequency or, which is the same: the frequency of the "De Broglie matter waves". In the case of the damped motion we again obtain an energy packet of short time- and spacewise dimensions, and we see that the energy packet does contain an observable frequency, viz. the beat frequency. Unfortunately, at this point we are not yet able to tell how much energy in total is present in the energy packet. Later it will appear that this total energy in the packet E_{rad} is a simple function of the beat frequency $\Delta\omega = (\omega - \omega_0)$. These variables are later found to be related to each other as:

$$E_{rad} = \frac{K}{2\pi}\Delta\omega \approx h\Delta f \tag{G.6.9}$$

where K is the constant defined in the theory of the pulsating droplet by Eq. (G.2.4) and $\Delta\omega$ is the angular frequency of the beats. It has already appeared that $K \approx h$ with h Planck's constant, whereas the beat frequency $\Delta f = (\omega - \omega_0)/(2\pi)$ was identified as the "De Broglie matter wave frequency". The relation (G.6.9) thus happens to be the same as applies to photons in the usual quantum theory. This conclusion is of course subject to the proviso's that the assumed extra excitation of the equations of motion in the zero-speed frequency indeed is a real physical phenomenon in a potential well, and that we can show that the total energy contained in a "packet of radiation" indeed has the value given by the first equality of (G.6.9). Both these matters are the subject of later chapters.

Summing up the "hard"conclusions of this chapter and ignoring for now Eq. (G.6.9): we find the curious phenomenon of "invisible" field strengths coupled with "observable" energy flux. This energy flux has a limited duration: it comes in "flashes". Furthermore, the flux may be expected to have a definite characteristic frequency.

All this strongly reminds one of the concept of energy packets (photons) of the usual quantum theory. The latter concept entails packets of electromagnetic energy travelling in isolation through space. Or, formulated in terms of quantum field theory: the ensemble of all these isolated packets constitute the field. At the very least it can now be said that the present theory of the pulsating droplet of charge is not contradictory to this concept of the usual quantum theory. This compatibility—or if one likes: these complementary concepts of how the electromagnetic field looks like—might provide a mental picture of the photon which is able to reconcile the "anti-intuitive" concept in quantum theory of a *particle* (the photon) endowed with a certain *frequency*.

Appendix H
The Droplet of Charge Within a Potential Well: Energy Quantisation

H.1 The Boundary Conditions Associated with a Potential Box; Solution of the Equation of Motion

The development of the theory up to this point concerned the freely moving droplet of charge. For the case where the droplet is moving back and forth within a potential box, we will have to apply additional boundary conditions.

For now, consider again the situation where an artificial force maintains an undamped motion of the droplet by suppressing the radiation resistance force. If we consider the time intervals *between* the bounces against the box walls, the linearised equations of motion are the same as in Chap. F.7:

$$\dot{p} = \gamma \ddot{v} + q.E_{ext} + F(t) = 0 \tag{H.1.1}$$

or

$$p = \pm p_0 \tag{H.1.2}$$

with p_0 a constant, and the sign depending on whether the droplet is moving from left to right in the box, or vice versa. The points in time when the droplet hits the walls (where $p = 0$) are not covered by (H.1.1). In these points the force due to the box potential is impulsive, and is thus described by a delta-function. The function $p(t)$ can on the interval $-\infty < t < \infty$ be described as an infinite "row of battlements" (Fig. H.1.1), if the hits against the potential walls are included.

The pulsation equation is

$$\frac{\ddot{\Delta s}}{s_e} + \omega_0^2 \left[1 + 2k\left(\frac{p}{m}\right)^2\right] \frac{\Delta s}{s_e} = -k\,\omega_0^2 \left(\frac{p}{m}\right)^2 \tag{H.1.3}$$

If the momentum squared were truly a constant in time: $p^2 = p_0^2$ (instead of the square of a "battlement function") this would result in a pulsation with the angular frequency

T. van Holten, *The Atomic World Spooky? It Ain't Necessarily So!*
DOI 10.2991/978-94-6239-234-2

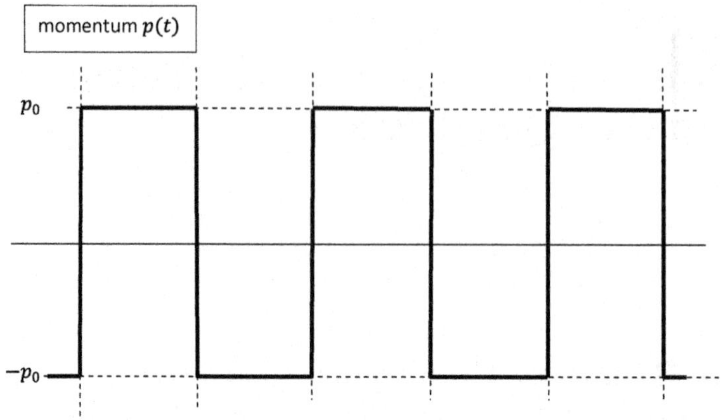

Fig. H.1.1 Momentum of the droplet in a potential box, as a function of time

$$\omega = \omega_0 \left[1 + k \left(\frac{p_0}{m} \right)^2 \right] \quad (p^2 \; constant) \tag{H.1.4}$$

However, because in reality the time function $p^2(t)$ has the constant value p_0^2 except in a finite number of points, the function $p^2(t)$ has periodicity. We must thus treat Eq. (H.1.3) as a differential equation with variable coefficients. According to the theory by Floquet-Ljapunov about this type of equations, the solution $\frac{\Delta s}{s_e}(t)$ will have the same periodicity as $p^2(t)$. The fundamental period of $\frac{\Delta s}{s_e}(t)$ therefore equals the time taken by the droplet to travel from wall to wall. The corresponding fundamental frequency which will occur in the solution of Eq. (H.1.3) is thus very much smaller than the frequency ω. The conclusion is, that it is essential to treat Eq. (H.1.3) as a linear differential equation with *variable* coefficients.

We start to consider more carefully the origins of the box model. Physically, an infinite potential wall at a sharply defined position is unrealistic. The model of the box must be considered as a limiting case of a potential well with rounded corners, as sketched in Fig. H.1.2.

It means that a droplet "bouncing back and forth" within the box in actual fact will gradually decelerate when it approaches a wall, comes to a stop, and then accelerates in the opposite direction so that it has obtained the full momentum again after having travelled a short distance.

The next Fig. H.1.3 shows the variation in time of the momentum:

The momentum squared, which occurs in the pulsation equation, would in a physically realistic situation show a variation in time as sketched in the next Fig. H.1.4:

In the limiting case of the potential box the momentum squared thus should be represented as sketched in Fig. H.1.5.

This representation of the momentum squared shows that there is clearly periodicity, even in the limiting case of the exact potential box.

The equation of motion to be solved is

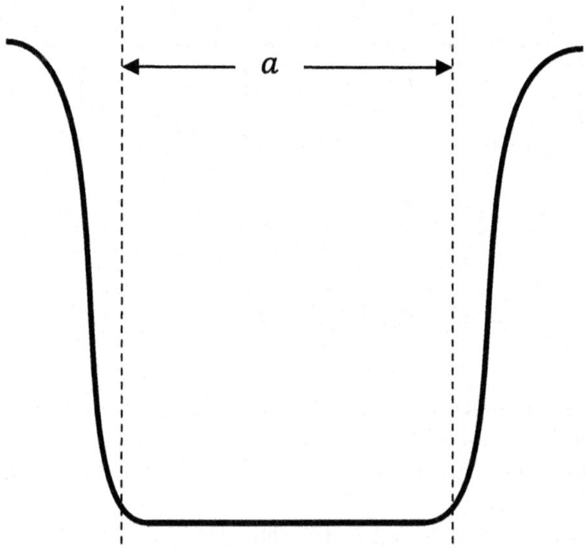

Fig. H.1.2 Physical potential well, from which the potential box is derived as a limiting case

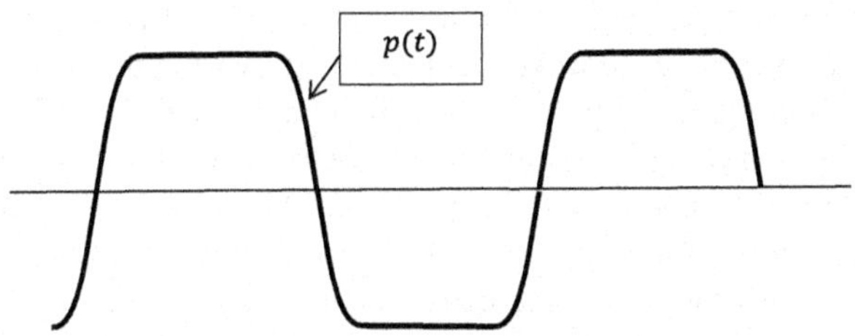

Fig. H.1.3 The variation of the momentum in a physical potential well

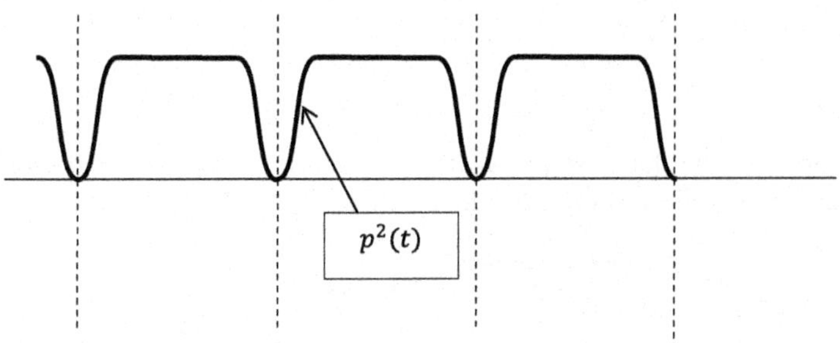

Fig. H.1.4 Momentum squared as a function of time in a physically realistic potential well

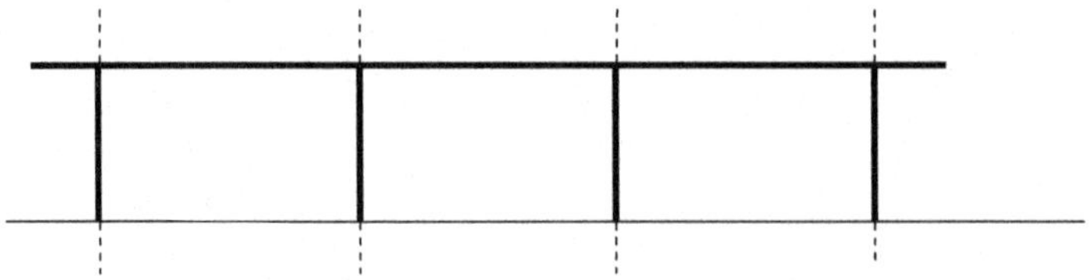

Fig. H.1.5 Momentum squared as a function of time in the limit of a box

$$\frac{\ddot{\Delta s}}{s_e} + \omega_0^2\left[1 + 2k\left(\frac{p(t)}{m}\right)^2\right]\frac{\Delta s}{s_e} = -k\,\omega_0^2\left(\frac{p(t)}{m}\right)^2 \tag{H.1.5}$$

An asymptotic approach may give insight in the character of the solution. Such an approximation is possible because the variation of the "stiffness" coefficient is small due to the small value of the constant k. An asymptotic expansion for $k \to 0$ may therefore be attempted.

The function $p^2(t)$ as sketched in Fig. H.1.6 will be considered. In contrast to Fig. H.1.4, the time variation of $p^2(t)$ is here drawn such that it is not symmetric w.r.t. the middle of a single pass of the droplet from left to right or v.v., although it is symmetric w.r.t. the "bouncing times" (these bouncing times indicated by the vertical dashed lines in Fig. H.1.6). The repetition period T is in accordance defined as the time the droplet needs to travel back and forth within the well. The reason for this choice will become apparent later, in Chap. H.3, where it is found that certain descriptors of the motion are not always symmetric w.r.t. the middle of the passes. To stress this, this lack of symmetry w.r.t. the middle of each sweep is symbolised by an asymmetry in the function $p^2(t)$, although it must be stressed that this is not necessarily the descriptor that is the cause of the asymmetry.

Denoting the repetition time by T, this type of function can be represented by a cosine-series:

$$\frac{p^2}{m}(t) = a_0 + \sum_{n=1}^{\infty} a_n.\,\cos\left(n.2\pi\frac{t}{T}\right) = a_0 + \sum_{n=1}^{\infty} a_n.\,\cos(n.\Omega t) \tag{H.1.6}$$

with $\Omega = \frac{2\pi}{T}$.

Since one of the later steps of the asymptotic analysis will consist of a term-by-term integration of (H.1.5), the constant a_0 has to be singled out because it

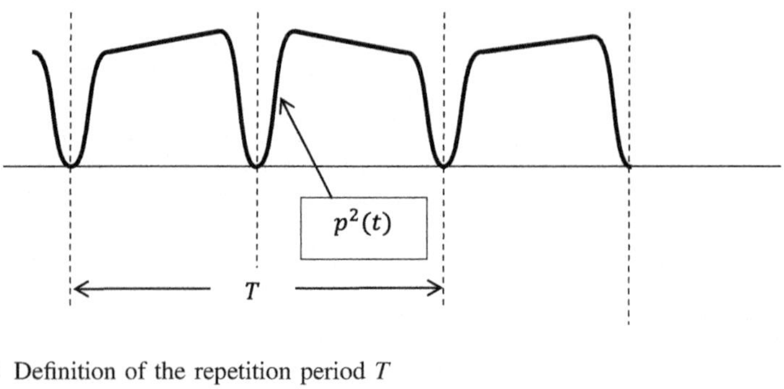

Fig. H.1.6 Definition of the repetition period T

does not have periodicity with the period T. It will be shown that a term-by-term integration, although it strictly is not allowed in the un-approximated equation (H.1.5) because the principle of superposition is not applicable, is nevertheless consistent with the asymptotic approximation. We then write

$$\frac{\ddot{\Delta s}}{s_e} + \omega_0^2\left[(1+2k\,a_0) + 2k\sum_{n=1}^{\infty} a_n.\ \cos(n.\Omega t)\right]\frac{\Delta s}{s_e} = -k\,\omega_0^2\left[a_0 + \sum_{n=1}^{\infty} a_n.\ \cos(n.\Omega t)\right] \tag{H.1.7}$$

Neglecting terms of $O(k^2)$ the differential equation can next be written in the form

$$\frac{\ddot{\Delta s}}{s_e} + \omega^2\left[1 + 2k\sum_{n=1}^{\infty} a_n.\ \cos(n.\Omega t)\right]\frac{\Delta s}{s_e} = -k\,\omega^2\left[a_0 + \sum_{n=1}^{\infty} a_n.\ \cos(n.\Omega t)\right] \tag{H.1.8}$$

where

$$\omega = \omega_0(1 + k\,a_0) \tag{H.1.9}$$

Now assuming an asymptotic representation of the time-function $\frac{\Delta s}{s_e}(t)$:

$$\frac{\Delta s}{s_e}(t) = \left(\frac{\Delta s}{s_e}\right)_0 + k.\left(\frac{\Delta s}{s_e}\right)_1 + \ldots. \quad (k \to 0) \tag{H.1.10}$$

the equation of motion (H.1.5) finally takes the form

$$\left[\left(\frac{\ddot{\Delta s}}{s_e}\right)_0 + \omega^2\left(\frac{\Delta s}{s_e}\right)_0\right] + k.\left[\left(\frac{\ddot{\Delta s}}{s_e}\right)_1 + \omega^2\left(\frac{\Delta s}{s_e}\right)_1\right] + k.\omega^2\left[2\left(\frac{\Delta s}{s_e}\right)_0\sum\ldots + \left\{a_0 + \sum\ldots\right\}\right] + O(k^2) = 0 \tag{H.1.11}$$

In the limit $k \to 0$ the differential equation to be satisfied by $\left(\frac{\Delta s}{s_e}\right)_0$ is

$$\left(\frac{\ddot{\Delta s}}{s_e}\right)_0 + \omega_0^2\left(\frac{\Delta s}{s_e}\right)_0 = 0 \tag{H.1.12}$$

The solution of $\left(\frac{\Delta s}{s_e}\right)_0$ is a harmonic function with angular frequency ω_0:

$$\left(\frac{\Delta s}{s_e}\right)_0 = a.\cos\omega_0 t + b.\sin\omega_0 t \tag{H.1.13}$$

From Eq. (H.1.11) it is seen that an approximation up to and including the $O(k)$ is determined by the differential equation

$$\left[\left(\frac{\ddot{\Delta s}}{s_e}\right) + \omega^2\left(\frac{\Delta s}{s_e}\right)\right] + k.\omega^2\left[2\left(\frac{\Delta s}{s_e}\right)_0 \sum \ldots + \left\{a_0 + \sum \ldots\right\}\right] = 0 \tag{H.1.14}$$

This equation includes the next higher order terms (of order $O(k)$), but strictly speaking it also includes a few terms of an irrelevant order (order $O(k^2)$). This has been done to avoid an unwanted—and physically unrealistic—singularity (a resonance condition between $\left(\frac{\Delta s}{s_e}\right)_0$ and $\left(\frac{\Delta s}{s_e}\right)_1$ only caused artificially by the asymptotic procedure).

Substituting (H.1.13) into (H.1.14) gives

$$\begin{aligned}\left(\frac{\ddot{\Delta s}}{s_e}\right) + \omega^2\left(\frac{\Delta s}{s_e}\right) &= -k\,\omega^2\left\{a_0 + \sum_{n=1}^{\infty} a_n.\ \cos(n.\Omega t)\right\}\\ &+ -2\,k\,\omega^2\sum_{n=1}^{\infty} a_n\left[\frac{a}{2}\{\cos(\omega_0 + n\Omega)t + \cos(\omega_0 - n\Omega)t\}\right.\\ &\left. + \frac{b}{2}\{\sin(\omega_0 + n\Omega)t + \sin(\omega_0 - n\Omega)t\}\right]\end{aligned} \tag{H.1.15}$$

The solution of the homogeneous equation is a harmonic function with the high frequency $\omega = \omega_0(1 + k\,a_0)$. Its repetition time is very much smaller than the repetition time T of the signal $p^2(t)$, although it may have the same periodicity if

$$\omega = \omega_0(1 + k\,a_0) = q\,\Omega \quad (q\ integer) \tag{H.1.16}$$

The particular solution associated with the first term in the r.h.s. of Eq. (H.1.15) has the form

$$\left(\frac{\Delta s}{s_e}\right)_{part\#1} = -k\,\omega^2\sum_0^{\infty}\frac{a_n\cos n\Omega t}{\omega^2 - (n\Omega)^2} \tag{H.1.17}$$

It automatically has the required periodicity with the period T.

For the purpose of inspecting the form of the second part of the particular solution, i.e. the part associated with the sum in the r.h.s. $-2\,k\,\omega^2\sum$, consider one of the terms separately as an example, for instance

$$\left(\frac{\ddot{\Delta s}}{s_e}\right) + \omega^2\left(\frac{\Delta s}{s_e}\right) = -2\,k\,\omega^2\,a_n\frac{a}{2}\cos(\omega_0 + n\Omega)t + \ldots. \tag{H.1.18}$$

The particular solution is

$$\left(\frac{\Delta s}{s_e}\right)_{part\#2} = -2\,k\,a_n \frac{a}{2} \frac{\omega^2}{\omega^2 - (\omega_0 + n\Omega)^2} \cos(\omega_0 + n\Omega)t + \ldots \quad \text{(H.1.19)}$$

This is again a high-frequency signal. How nevertheless periodicity with the low frequency Ω arises, is seen when taking together the high-frequency terms contributed by the homogeneous solution and the particular solution: $\left(\frac{\Delta s}{s_e}\right)_{homog} + \left(\frac{\Delta s}{s_e}\right)_{part\#2}$.

We then obtain a high-frequency oscillation with *variable* amplitude. Recall the general considerations about beat phenomena of Chap. G.1, where the case has been considered of the sum of two oscillations with a small difference of their frequency, and a large difference of their amplitude:

$$f(t) = a.\cos\omega_2 t + b.\cos\omega_1 t \quad (a/b \ll 1) \quad \text{(H.1.20)}$$

In Chap. G.1 it was derived that this sum function has an envelope

$$g(t) = a\cos(\omega_2 - \omega_1)t + b \quad (a/b \ll 1) \quad \text{(H.1.21)}$$

and Fig. G.1.4 gave a schematic picture of the high-frequency signal with a low-frequency amplitude modulation. Taking into account that k in (H.1.19) is small, in the case at hand where we are summing functions like $B.\cos\omega t$ (homogeneous solution) and $A.\cos(\omega_0 + n\Omega)t$ (particular solution, $A = O(k)$) we will find as the sum function a high-frequency oscillation with an envelope (i.e. an amplitude variation)

$$g(t) = A\cos(\omega - \omega_0 - n\Omega)t + B \quad (A/B \ll 1) \quad \text{(H.1.22)}$$

or, substituting $\omega = \omega_0(1 + k\,a_0)$:

$$g(t) = A\cos(k\,\omega_0 a_0 - n\Omega)t + B \quad (A/B \ll 1) \quad \text{(H.1.23)}$$

We must require that the complete solution of (H.1.5) has the same periodicity as $\left(\frac{p(t)}{m}\right)^2$. This is a general result from the theory by Floquet and Ljapunov about differential equations with variable coefficients. The correct periodicity is only obtained when the beat phenomenon expressed by Eq. (H.1.23) also has this same periodicity. It therefore follows that we must require

$$(k\,\omega_0 a_0 - n\Omega) = m\,\Omega \quad (\mathrm{m} = 1, 2, 3, \ldots.) \quad \text{(H.1.24)}$$

or

$$k\,\omega_0 a_0 = p.\Omega \quad (\mathrm{p} = 1,\, 2,\, 3,\, \ldots.) \tag{H.1.25}$$

Combining this with Eq. (H.1.16) shows that the periodicity with frequency $m\,\Omega$ also requires that

$$\omega_0 = r\,\Omega \quad (r\,integer) \tag{H.1.26}$$

The coefficient a_0 has the physical meaning of the average value of $\left(\frac{p(t)}{m}\right)^2$, and can be related to the average of the kinetic energy defined as $\overline{E_k} = \frac{1}{2}m\,\overline{\left(\frac{p(t)}{m}\right)^2}$, so that Eq. (H.1.25) may be written as

$$2\frac{k\,\omega_0}{m}\,\overline{E_k} = p.\Omega \tag{H.1.27}$$

According to Eq. (G.2.4): $K = m\frac{2\pi}{k\,\omega_0}$. Furthermore equating the constant K to Planck's constant h and using the notation $\hbar = h/2\pi$, Equation (H.1.27) transforms into

$$\overline{E_k} = p.\frac{1}{2}\hbar\,\Omega \tag{H.1.28}$$

Taking the limit of an exact potential box, i.e. proceeding from Fig. H.1.4 to Fig. H.1.5, and substituting $\overline{E_k} = \frac{1}{2}m\,v_{av}^2$ and $\Omega = \frac{2\pi}{T} = 2\pi\frac{v_{av}}{2a}$, it follows that the permitted values of v_{av} are:

$$v_{av} = p.\frac{h}{2\,m\,a} \quad (p = 1,\, 2,\, 3,\, \ldots) \quad (quantisation\ in\ potential\ box) \tag{H.1.29}$$

and the total energy—which in a box equals the kinetic energy—is

$$E = \frac{h^2}{8\,m.a^2}.p^2 \quad (p = 1,\, 2,\, 3,\, \ldots) \quad (quantisation\ in\ potential\ box) \tag{H.1.30}$$

This is exactly the same quantisation as found in the usual quantum mechanics by solving Schrödinger's equation in a potential box.

H.2 Alternative Solution of the Equation of Motion in a Potential Box

The present chapter gives an alternative procedure for solving the equation of motion in a potential box, now starting directly from the momentum variations as shown in Fig. H.1.5. It will give some additional insight, thanks to the different point of view taken.

In order to solve the differential equation (H.1.5) with variable coefficients, repeated here:

$$\frac{\ddot{\Delta s}}{s_e} + \omega_0^2 \left[1 + 2k \left(\frac{p(t)}{m} \right)^2 \right] \frac{\Delta s}{s_e} = -k\,\omega_0^2 \left(\frac{p(t)}{m} \right)^2 \tag{H.2.1}$$

and in order to explicitly preserve the time variations of the momentum squared, the graph of Fig. H.1.5 is considered as the sum of two separate block functions, as shown in Fig. H.2.1.

The first one, denoted as $\left(\frac{p(t)}{m}\right)_1^2$, is a regular block function with equal lengths of the stretches with $p(t) = p_0$ and the stretches with $p(t) = 0$. The second block function $\left(\frac{p(t)}{m}\right)_2^2$ in principle has unequal lengths of the two stretches, as shown by the dashed lines in the figure. It is only when we take the limit of an exact potential box that the second function becomes a time-shifted replica of the first one (shown by the drawn lines).

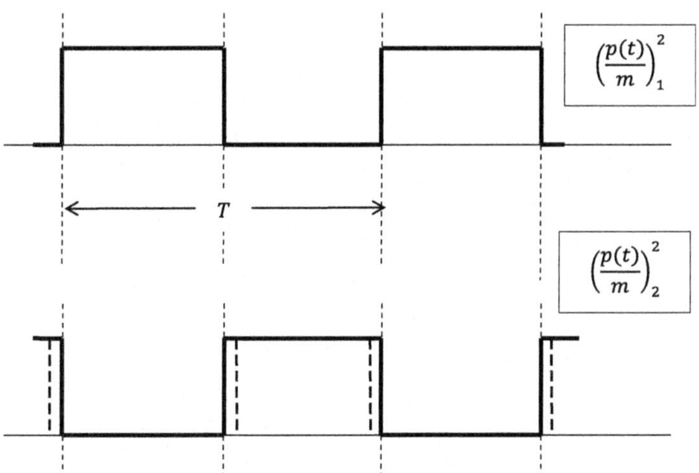

Fig. H.2.1 The function $p^2(t)$ of Fig. H.1.5 represented by the sum of two block functions

Note that the differential equation (H.2.1) does not allow superposition to find the solution associated with the sum of the two block functions.

If one would "feed" the equation of motion (H.2.1) with just the time function $\left(\frac{p(t)}{m}\right)^2_1$ the solution, denoted as $\left(\frac{\Delta s(t)}{s_e}\right)_1$, will show periodicity with the period T indicated in the figure. Schematically:

$$\left(\frac{p(t)}{m}\right)^2_1 \rightarrow \left(\frac{\Delta s(t)}{s_e}\right)_1$$

The response to this kind of block function is considered in detail in the book by J.P. den Hartog: Mechanical Vibrations, Chap. 8 on systems with variable elasticity. In this book special attention is given to the stability of the solution. The matter of stability is not (yet) considered here.

The way to solve the equation (H.2.1) is taken from Den Hartog. During the first half of the period in the interval $0 < t < T/2$, the solution of the pulsation equation is:

$$\frac{\Delta s}{s_e} = A\,cos(\omega t + \varepsilon) - k\left(\frac{p_0}{m}\right)^2 \tag{H.2.2}$$

with the frequency

$$\omega = \omega_0\left[1 + k\left(\frac{p_0}{m}\right)^2\right] \tag{H.2.3}$$

During the second half of the period, i.e. in the interval $T/2 < t < T$ where the momentum squared is zero, we find

$$\frac{\Delta s}{s_e} = B\,cos(\omega_0 t + \delta) \tag{H.2.4}$$

The constants A, B, ε *and* δ are determined by requiring that $\frac{\Delta s}{s_e}(t)$ is continuous and has continuous derivatives at the times $t = 0$ and $t = T/2$. This leads to four algebraic equations determining the four constants:

$$A\cos\varepsilon - k\left(\frac{p_0}{m}\right)^2 = B\cos\delta \tag{H.2.5}$$

$$A\,\omega\sin\varepsilon = B\,\omega_0\sin\delta \tag{H.2.6}$$

$$A\cos\left(\omega\frac{T}{2} + \varepsilon\right) - k\left(\frac{p_0}{m}\right)^2 = B\cos\left(\omega_0\frac{T}{2} + \delta\right) \tag{H.2.7}$$

$$A\,\omega \sin\left(\omega \frac{T}{2} + \varepsilon\right) = B\,\omega_0 \sin\left(\omega_0 \frac{T}{2} + \delta\right) \tag{H.2.8}$$

Without completely working out the solution of this set of algebraic equations, using Cramer's rule shows that both the coefficients A and B have the form:

$$A, B = -k\left(\frac{p_0}{m}\right)^2 .[\mathit{factor}\; O(1)] \tag{H.2.9}$$

This result will be needed later.

If on the other hand we would feed the pulsation equation by substituting the second block function of Fig. H.2.1 (drawn lines; limit of the exact potential box) the solution will be denoted by $\left(\frac{\Delta s(t)}{s_e}\right)_2$:

$$\left(\frac{p(t)}{m}\right)^2_2 \rightarrow \left(\frac{\Delta s(t)}{s_e}\right)_2$$

This solution can be obtained from $\left(\frac{\Delta s(t)}{s_e}\right)_1$ by a shift of the time scale over half a period:

$$\left(\frac{p(t)}{m}\right)^2_2 = \left(\frac{p(t + T/2)}{m}\right)^2_1 \tag{H.2.10}$$

and

$$\left(\frac{\Delta s(t)}{s_e}\right)_2 = \left(\frac{\Delta s(t + T/2)}{s_e}\right)_1 \tag{H.2.11}$$

The complete solution for $\left(\frac{\Delta s(t)}{s_e}\right)$ must be obtained by solving the equation of motion

$$\begin{aligned} &\frac{\ddot{\Delta s}}{s_e} + \omega_0^2\left[1 + 2\,k\left(\frac{p(t)}{m}\right)^2_1 + 2k\left(\frac{p(t)}{m}\right)^2_2\right]\frac{\Delta s}{s_e} \\ &= -k\,\omega_0^2\left[\left(\frac{p(t)}{m}\right)^2_1 + \left(\frac{p(t)}{m}\right)^2_2\right] \end{aligned} \tag{H.2.12}$$

where the sum of the two block functions of Fig. H.2.1, i.e. the function shown in Fig. H.1.5 has been used to "feed" the pulsation equation. As stated before, the complete solution is *not* the sum of $\left(\frac{\Delta s(t)}{s_e}\right)_1$ and $\left(\frac{\Delta s(t)}{s_e}\right)_2$ because the principle of superposition cannot be applied. However, it will be shown now that only an error of the order $O(k^2)$ results if we approximate the solution $\left(\frac{\Delta s(t)}{s_e}\right)$ by the sum of $\left(\frac{\Delta s(t)}{s_e}\right)_1$ and $\left(\frac{\Delta s(t)}{s_e}\right)_2$. Substituting $\left(\frac{\Delta s(t)}{s_e}\right)_1 + \left(\frac{\Delta s(t)}{s_e}\right)_2$ for $\frac{\Delta s}{s_e}$ into Eq. (H.2.12) yields

$$2k\omega_0^2\left[\left(\frac{\Delta s(t)}{s_e}\right)_1 \cdot \left(\frac{p(t)}{m}\right)_2^2 + \left(\frac{\Delta s(t)}{s_e}\right)_2 \cdot \left(\frac{p(t)}{m}\right)_1^2\right] = 0 \tag{H.2.13}$$

which cannot be true exactly and indeed shows that the principle of superposition is not applicable. However, in view of the result (H.2.9) it is seen that the l.h.s of Eq. (H.2.13) is of the order $O(k^2)$. Terms of this small order have everywhere been neglected in the theory developed so far, so that it can be concluded that the approximate solution

$$\left(\frac{\Delta s(t)}{s_e}\right) \approx \left(\frac{\Delta s(t)}{s_e}\right)_1 + \left(\frac{\Delta s(t)}{s_e}\right)_2 \tag{H.2.14}$$

satisfies the differential Eq. (H.2.12) with sufficient accuracy. Using Eqs. (H.2.2) and (H.2.4) we thus find in the entire time interval $0<t<T$

$$\frac{\Delta s}{s_e} = A\,cos(\omega t + \varepsilon) - k\left(\frac{p_0}{m}\right)^2 + B\,cos(\omega_0 t + \delta) \tag{H.2.15}$$

which clearly shows a beat phenomenon in the pulsation, caused by the special boundary conditions imposed by the fact that the droplet is bouncing back and forth within the potential box. We have found that the boundary conditions of the potential box lead to a pulsation containing beats with the low frequency $\Delta\omega = \omega - \omega_0$ (recall the definition of "beat frequency" as discussed in Chap. G.1). This explains the earlier mentioned result that the periodicity of the pulsation must have a very much larger time scale than the period associated with just the frequency ω. The beat frequency is the same as the De Broglie frequency of matter waves.

The function $\frac{\Delta s}{s_e}(t)$ as given by Eq. (H.2.15) must be periodic with period T (see Fig. H.2.1), i.e. the total time needed by the droplet to move back and forth within the box. According to Chap. G.1 the repetition period T is related to ω and ω_0 by the combined requirement

$$\Delta\omega.T = (\omega - \omega_0)T = 2\pi.m\ (m = 1, 2, 3, \ldots)\ and\ \omega_0 T = 2\pi.k\ (k = 1, 2, 3, \ldots) \tag{H.2.16}$$

Since $\Delta\omega$ is related to the average velocity of the droplet by (G.1.3):

$$\Delta\omega = k\,\omega_0\,v_{av}^2 \tag{H.2.17}$$

we find

$$T = \frac{2\pi}{k\,\omega_0\,v_{av}^2}.m \quad (m = 1, 2, 3, \ldots) \tag{H.2.18}$$

The relation between T and v_{av} according to (H.2.18) is required by the fact that the motion of the droplet should be periodic with the period T. Essentially, this is a *mathematical* requirement following from the theory by Floquet and Liapunov concerning linear differential equation with periodic coefficients.

There is a second relation between T and v_{av} which follows from the *physical* consideration that in the time T the droplet must have traversed twice (back *and* forth, see Fig. H.2.1) the width a of the potential box:

$$T = \frac{2\,a}{v_{av}} \tag{H.2.19}$$

Combining (H.2.18) and (H.2.19) we find that v_{av} is restricted to a series of discrete values:

$$v_{av} = \frac{\pi}{k\,\omega_0\,a}.m \quad (m = 1, 2, 3, \ldots.) \tag{H.2.20}$$

This result can also be stated in a different way, by considering the wavelength of the beat phenomenon in the pulsation:

$$\lambda = v_{av}\frac{2\pi}{\Delta\omega} = \frac{2\pi}{k\,\omega_0\,v_{av}} \tag{H.2.21}$$

or, combining with Eq. (H.2.20):

$$a = (\lambda/2).m \quad (m = 1, 2, 3, \ldots.) \tag{H.2.22}$$

In words: the box can contain only integer multiples of half the wavelength of the beats (or, which is the same: half the De Broglie wavelength).

This is the same result as follows from the usual quantum theory. However, there is also a difference. In the usual quantum theory, where there is no pulsation and therefore no fluctuation of the mass and velocity, the expression for the energy of the droplet would be

$$E = \frac{1}{2}m\,v_{av}^2 \quad (\textit{usual quantum theory}) \tag{H.2.23}$$

Using Eq. (G.2.4) we can write v_{av}, Eq. (H.2.20), in terms of the constant $K(\approx h)$:

$$v_{av} = \frac{\pi}{k\,\omega_0\,a}.n = \frac{1}{2}\frac{K}{m\,a}.n \quad (n = 1, 2, 3, \ldots.) \tag{H.2.24}$$

where the integer counter m has been replaced by n in order to avoid confusion with the mass m. Eq. (H.2.24) is a relation from the droplet theory. According to it, it should be found that in the usual quantum theory (equating K and Planck's constant h):

$$E = \frac{1}{2}m\,v_{av}^2 = \frac{h^2}{8\,m\,a^2}.n^2 \quad (n = 1, 2, 3, \ldots.) \tag{H.2.25}$$

This is indeed the expression for the allowed energy levels in a potential box as found in the literature. The difference with the droplet theory is, that a pulsating droplet has more energy than just the translation energy associated with a constant velocity. Actually, the total energy, including the energy associated with the pulsation and fluctuating velocity, is in droplet theory approximately twice as much as in the usual quantum theory. This situation is somewhat similar to the derivation of De Broglie's momentum relation using the droplet theory. Here we had to define a certain kind of momentum p_{av} which is a natural quantity in the usual quantum theory, and it could then be shown that De Broglie's relation is valid for this quantity. At the same time, the so defined p_{av} is neither in meaning nor in value equal to the quantity p_0 which is the "natural" quantity occurring in the droplet theory.

Likewise, within the framework of the droplet theory the more "natural" and more relevant quantity is the average speed v_{av}, instead of the "kinetic energy" $\frac{1}{2}m\,v_{av}^2$ which involves the square and products of averages.

Note on the applicability of the results

In the previous two chapters a rather artificial situation was considered, where an external force was inserted into the equations of motion in order to suppress the radiation resistance. However, the results obtained have a wider applicability, viz. to the middle frequency found in the power spectral density curve of the chaotic state.

In Chap. F.4 a qualitative discussion was given of the complete non-linear equations of motion. It was argued that the most generally and most often occurring motion of the droplet would be one where the pulsation and velocity fluctuations would have a chaotic character, varying around a constant average velocity. This chaotic motion would automatically block any radiation (as will be shown in a later chapter), so that the radiation resistance term would drop from the equations of motion, and no damping of the motion would occur. The chaotic motion is in the frequency domain described by a power spectral density (PSD) curve characterised by a spread of frequencies around a certain middle frequency. The expectation was put forward that this middle frequency of the PSD curve would correspond with the frequency of the delta function into which the PSD curve degenerates in the case of small amplitude fluctuations (corresponding to harmonic velocity oscillations). Looking back at Fig. F.4.3, a definite "rhythm" can be discerned in the chaotic time signal. The frequency associated with this "rhythm" is thus assumed to be the same as the middle frequency of the spectrum sketched in the lower part of the figure.

The conclusions we have arrived at in the last two chapters, where the interaction with the potential box has been considered for the case without radiation resistance, therefore lead to the expectation that the middle frequency of the PSD curve of the chaotic state is in the same way as in the previous chapter subject to an interference phenomenon giving rise to the "De Broglie" (beat) frequency. One may expect that in the real non-radiating chaotic state the same quantisation will be found, although the quantisation will be somewhat more "fuzzy" because of the frequency spread around the middle frequency.

H.3 The Phase of the Beat Phenomenon in a Potential Box

The quantisation of energy levels was derived by considering solely the repetition times of the pulsation signal and the wavelengths of the beats present in this signal. The way these beats are spatially distributed in the potential well did not matter yet. The spatial distribution of the beats (or, what amounts to the same: the phase of the pulsation signal in the time domain) is the subject of the present chapter. For reasons of simplicity the discussion will remain restricted to the special case of the potential box.

We consider a physically possible potential well as shown in Fig. H.1.2, which has a form which in the limit can transform into a box. If the force exerted by the well-potential is $F(z)$, given as a function of the position of the droplet, the momentum equation is

$$\frac{d}{dt}(mv) = F(z) \tag{H.3.1}$$

or, completely expressed in terms of the position coordinate:

$$v.\frac{d}{dz}(mv) = F(z) \tag{H.3.2}$$

or

$$\frac{dp}{dz} = \frac{F(z)}{v(z)} \tag{H.3.3}$$

The increase of the momentum when the droplet moves from point z_1 to z_2 is:

$$p(z_2) - p(z_1) = \int_1^2 \frac{F}{v}\, dz \tag{H.3.4}$$

If we trace the droplet whilst it is moving in the direction of the wall, from a point on the horizontal plateau of the potential up to the point where the droplet reverses its direction of motion:

$$\int_1^{\textit{return point}} dp = -p_{to\,wall}(z_1) = \int_1^{\textit{return point}} \frac{F(z)}{v_{to\,wall}(z)}\, dz \tag{H.3.5}$$

After the bounce:

$$\int_{\textit{return point}}^1 dp = p_{from\,wall}(z_1) = \int_{\textit{return point}}^1 \frac{F(z)}{v_{from\,wall}(z)}\, dz \tag{H.3.6}$$

Comparing (H.3.5) and (H.3.6) it is concluded that the absolute value of the momentum $|p(z_1)|$, both when approaching the wall and after the bounce back, is the same *on condition* that the function $v_{from\,wall}(z) = -v_{to\,wall}(z)$, i.e. on condition that at any point the velocity $v(z)$ (including its fluctuations around the average value and the slow amplitude beats) is the same—apart from its sign—irrespective of the direction of motion of the droplet.

A alternative reasoning in the time domain yields the same result, as will now be shown.

If the time of a particular bounce against the potential wall is called t_0, we see from Fig. H.1.3 that the variation of the momentum around these points is anti-symmetric (in a stationary state), i.e.

$$p(t_0 + \Delta t) = -p(t_0 - \Delta t) \tag{H.3.7}$$

According to Chap. F.6 the instantaneous velocity depends both on p and $\frac{\Delta s}{s_e}$:

$$v(t) = \frac{p(t)}{m}\left[1 + 2\frac{\Delta s}{s_e}(t)\right] \tag{F.6.10}$$

or

$$v(t_0 + \Delta t) = \frac{p(t_0 + \Delta t)}{m}\left[1 + 2\frac{\Delta s}{s_e}(t_0 + \Delta t)\right] \tag{H.3.8}$$

If we would assume that the velocity is symmetric w.r.t. the times of momentum reversal, i.e. if it is assumed $v(t_0 + \Delta t) = v(t_0 - \Delta t)$ then the consequence would be (combining Eqs. (H.3.7) and (H.3.8)):

$$\frac{\Delta s}{s_e}(t_0 + \Delta t) + \frac{\Delta s}{s_e}(t_0 - \Delta t) = -1 \tag{H.3.9}$$

which would be incompatible with the linearisation condition $\left|\frac{\Delta s}{s_e}\right| \ll 1$ on which (F.6.10) is based.

We therefore must conclude that the velocity, like the momentum, is antisymmetric (in a stationary state) w.r.t. the reversal times, despite the mass fluctuations:

$$v(t_0 + \Delta t) = -v(t_0 - \Delta t) \tag{H.3.10}$$

As a further consequence it may be concluded that the pulsation, in contrast to this, is symmetric in the reversal points (in a stationary state):

$$\frac{\Delta s}{s_e}(t_0 + \Delta t) = \frac{\Delta s}{s_e}(t_0 - \Delta t) \tag{H.3.11}$$

In the main text the symmetry of the pulsation w.r.t. the wall is called "path retraceability". This term refers to the mechanical analogue that can be made of the linearised equations of motion, where the pulsation is analoguous to the nodding motion in the mechanical analogue. The translating and "nodding" mass in the analogue then traces out the same path before and after the bounce against the wall of the simulated potential box.

From (H.3.5) and (H.3.6) it is seen that the absolute value of the momentum of the droplet will be altered by a bounce against a potential wall, if the condition of path retraceability is not satisfied.

Let us first consider the *stationary* state, where the absolute value of the momentum is not changed by the bounces against the potential walls. The requirement that in this case the pulsation should be symmetric w.r.t. the walls may then pictorially be presented as in the following Fig. H.3.1.

We see that only a multiple of *half* the beat-wavelengths can satisfy the stationary state. This is the same conclusion as was reached in Chap. H.2. In one of the following chapters the quantisation of energy in a potential box will be revisited, based on this geometric criterion.

Now, the visualisation of the stationary state as given in Fig. H.3.1 is somewhat too simplified. It is based on harmonic pulsation, which cannot be stationary since it would lead to radiation. An actual stationary state without radiation is only possible if the pulsation is chaotic. A more realistic—although still very schematic—picture of the possible stationary states is given in the next picture Fig. H.3.2.

Interpreting this picture it should be realised that in actual fact the frequency of the pulsations is very much higher than sketched. The picture drawn in Fig. H.3.3 therefore gives a better impression of the actual situation. It is seen that, despite the

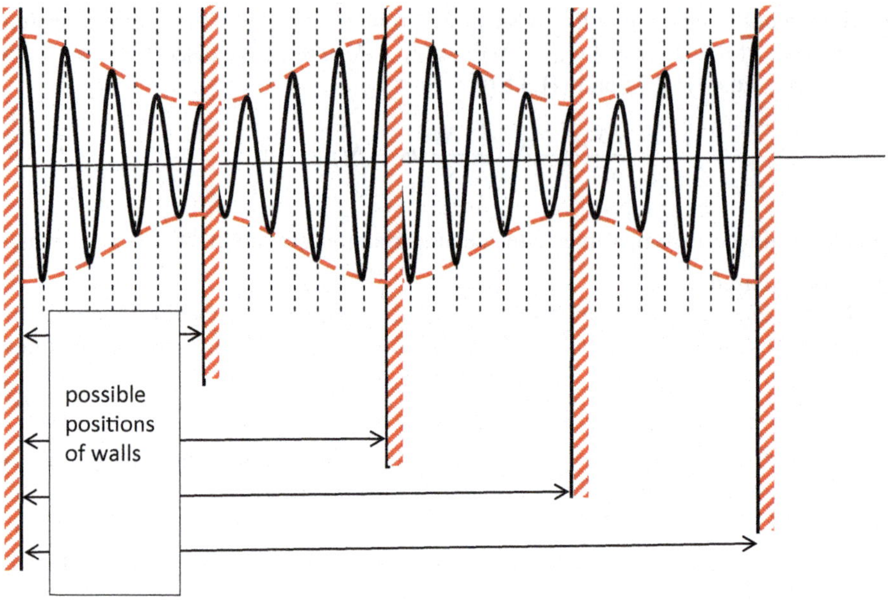

Fig. H.3.1 The possible positions of the potential walls w.r.t. the pulsation, in a stationary state

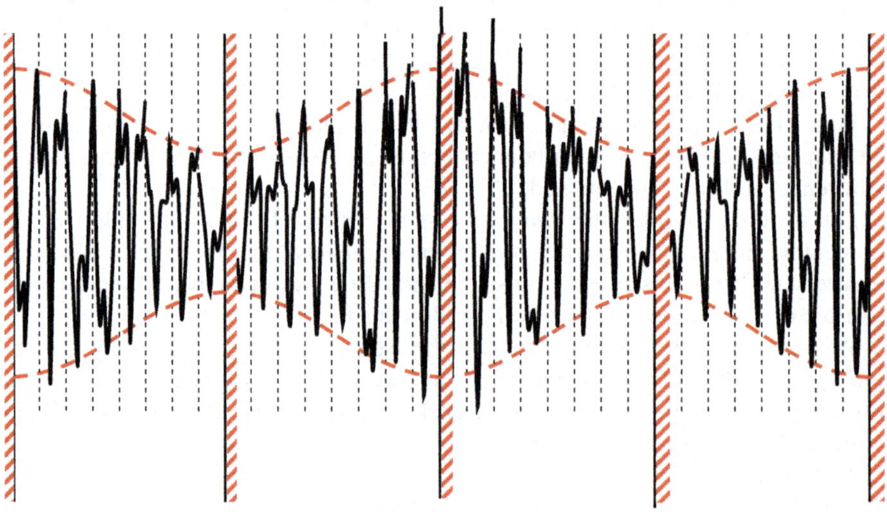

Fig. H.3.2 The chaotic pulsation subject to beats, and the position of the potential walls in a stationary state

chaotic state, the beat pattern is still very prominent. The sketch in Fig. H.3.3 also elucidates why the linearised analysis remains to have relevance even in the case of chaotic motion (the linearised result is shown by the red dashed lines). The linearised analyses are strictly valid only for the middle frequency of the chaotic spectrum, but they nevertheless give relevant information about the beat phenomena in the actual chaotic situation.

If the spurious long-term changes in the energy partition are also superimposed on the pattern shown in Fig. H.3.3, one may have a degeneration from the chaotic state into the state of harmonic motion. The vague grey contours in Fig. H.3.3 then shrink into the dashed red curve. Compare Fig. F.4.2, where a free-flying droplet was considered. In the latter case the shady contours belonging to the chaotic motion may shrink into the constant amplitude dashed red lines. The droplet inside a box may experience something similar when its motion degenerates into harmonic motion. Immediately after the degeneration radiation will start up. What the dashed red curve in Fig. H.3.3 shows are just the *initial* conditions of a damped motion. If the droplet is in a certain position within the box, and if at this position a degeneration happens to take place, the red dashed line gives the amplitude of the pulsation at that time (as well as the instantaneous rate of change of the amplitude). After radiation has started up, the pulsation amplitude will be damped and is no longer given by the red curve. The red curve just gives the initial conditions for the damped motion occurring directly after the droplet has left the chaotic state.

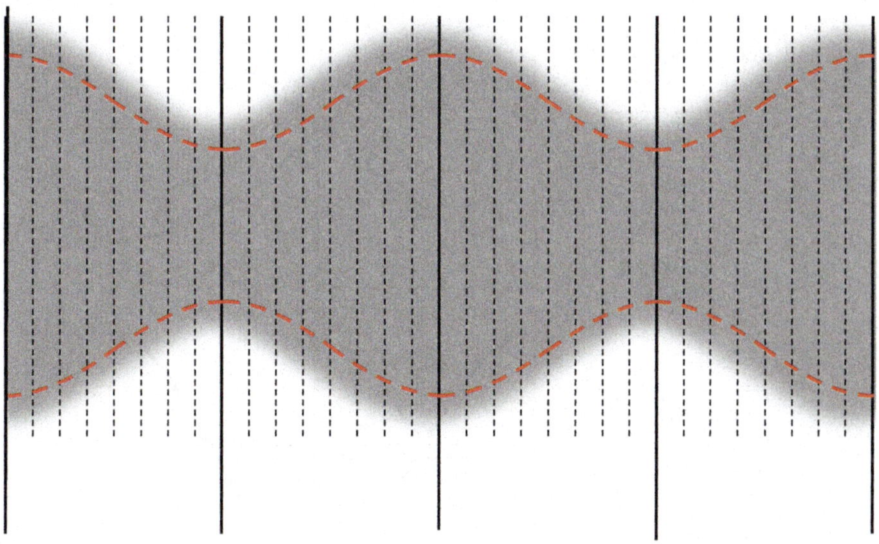

Fig. H.3.3 The beat pattern in the chaotic state, taking into account the large difference between the beat frequency and the pulsation frequency

H.4 Alternative Formulation of the Quantisation Condition Within a Potential Box

In Chap. G.1 it was mentioned that the shape of the beat patterns is the same (apart from a constant shift) as the stroboscopic image obtained by sampling the pulsation having the velocity dependent frequency ω, with the zero-speed frequency ω_0.

The conclusion from Chap. H.3 concerning the phase of the beats w.r.t. the walls of the potential box (i.e. the conclusion that in a stationary state the walls should coincide with maxima or minima of the beats) leads to a simple quantisation criterion. This requirement amounts to the requirement that in a stationary state the slope of the strobe image in the reversal points should be zero. This condition is sketched below in Fig. H.4.1, for the lowest two energy levels.

At each energy level, there are two different modes of pulsation corresponding to two different possibilities of the strobe image, in Fig. H.8.1 indicated by a drawn line and a dashed line. Both pulsation modes represent half a beat period.

It is clear that the other possible stroboscopic images, associated with higher energy levels, represent a series of pulsation modes containing a full beat, $1\frac{1}{2}$ beat, etc. This is in agreement with the analytic results of Chap. H.2, where it was concluded that in a stationary state a box can only contain integer multiples of half the wavelength of a De Broglie wave.

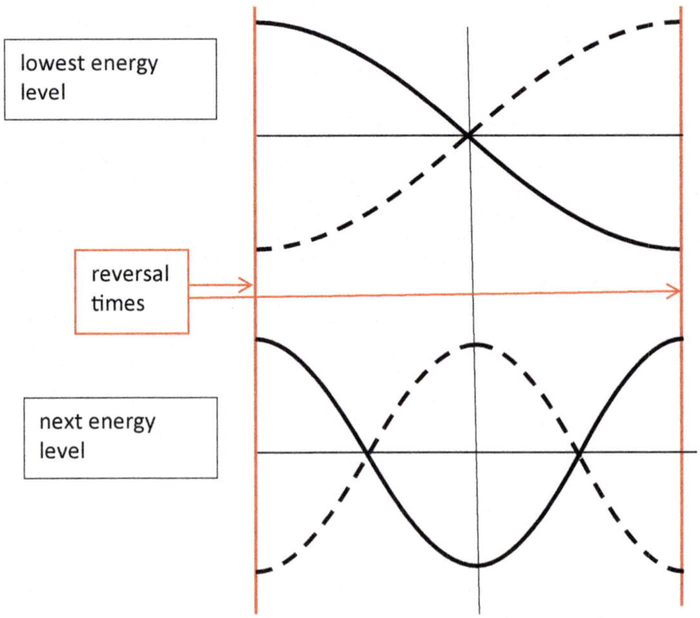

Fig. H.4.1 Stroboscopic images of the pulsation within a potential box (stroboscope fed by the zero-speed frequency ω_0)

H.5 Energy Width and Time-Energy Uncertainty (Qualitative)

Looking back at Fig. H.3.2, it is clear that in the case of the actual chaotic pulsation motion one can never *exactly* satisfy the requirement for a stationary state, which according to Eq. (H.3.11) said that the pulsation should be symmetric w.r.t. the reversal times. A chaotic signal never repeats itself, so that the requirement of symmetry w.r.t. the reversal time cannot be met. During an actual bounce against the wall of the box, according to Eqs. (H.3.5) and (H.3.6) a change of the momentum of the droplet (the absolute value) will occur, comparing its value before and after the bounce. Remember that in a box the momentum p_0 is constant during the time intervals in between the consecutive bounces, due to the absence of a force on the droplet. The encounters with the walls cause that $|p_0|$ jumps up and down a bit, comparing the different passes of the droplet back and forth within the box.

Equation (F.4.8) gives the relation between the momentum and translational energy:

$$E_{transl} = \frac{p_0^2}{2.m(s)} = \frac{p_0^2}{2m}\left[1 + 2\frac{\Delta s}{s_e} + 2\left(\frac{\Delta s}{s_e}\right)^2 + \ldots\right] \tag{H.5.1}$$

The jumps of $|p_0|$ are thus associated with jumps of the average value of the translational energy. There is an amount of *uncertainty* in the translational energy, so to speak. It should be realised that the static walls of the box can never change the *total* energy of the droplet. The uncertainty in the translational energy is only possible because of the two communicating energy reservoirs we have in the case of the droplet (translational energy and energy in the elongation). The quantisation of the translation energy in a box therefore is not sharp, but rather a bit fuzzy.

In the usual quantum theory one finds that the quantisation of energy is not sharp either: a small spread of the allowed energy level occurs around the energy levels indicated by Eq. (H.2.25), except in the lowest energy level (the socalled "zero-point energy") which is sharply defined. In quantum theory the spread of energy is associated with the uncertainty principle, in particular with the time-energy uncertainty expressed by

$$\Delta t.\Delta E \geq h \tag{H.5.2}$$

The uncertainty in time is, to explain the "energy width", related to the fact that the quantised state can undergo a transition at some unpredictable time to a lower energy state, by emitting a photon. To quote Alonso and Finn (Fundamental University Physics, vol.III, Quantum and Statistical physics) for a further explanation: "Let us consider an electron in an excited stationary state in an atom. The electron after a certain time will suffer a radiative transition into another stationary state of less energy. However, we have no means of predicting with certainty how long the electron will remain in the stationary state before making the transition. The most we can talk about is the probability per unit time that the electron will jump into a lower energy state. Therefore the average length of time the electron is in the stationary state, also called the lifetime of the state, and which is inversely proportional to the transition probability, is known with an uncertainty Δt. Hence the energy of the stationary state of the electron is not known precisely but has an uncertainty ΔE, such that relation (H.5.2) holds. Often ΔE is designated as the energy width of the state whose energy is most probably between $E - \frac{1}{2}\Delta E$ and $E + \frac{1}{2}\Delta E$ (Fig. H.5.1). We may assume that Δt is of the order of magnitude of the lifetime of the excited state. Thus, the shorter the lifetime of an excited state, the larger the uncertainty in the energy of the state. For the ground state, whose lifetime is infinite because a system which is at its ground state cannot suffer a transition to a stationary state of lower energy, we have $\Delta t \sim \infty$. This yields $\Delta E = 0$ and the energy of the ground state can be determined accurately".

A similar relation between energy width and the life-time of a particular stationary state may be expected in the case of the droplet theory. Equation (H.5.1) shows that the translational energy depends not only on the momentum p_0, but also on the chaotic fluctuations of $\frac{\Delta s}{s_e}$. In the case of chaotic pulsation, the higher order terms in the expansion between the square brackets in (H.5.1) may cause, what in Chap. F.4 was called, "long-term" variations of the translational energy. The indication "long-term" refers to the fact that these fluctuations can accumulate during intervals of time spanning many pulsation cycles. However, considered on the time-scale of the back and forth movement within the well, these so-called "long-term" variations of the energy are very short. They are spurious energy

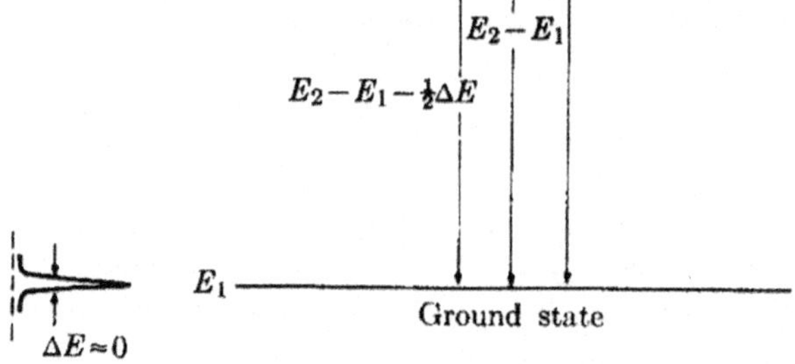

Fig. H.5.1 Energy width and transition energy. From: Alonso and Finn, Fundamental University Physics

fluctuations superimposed on the spread of energy levels associated with the phenomenon of energy width. These spurious energy fluctuations have a statistical character: their duration as well as their amplitude may vary randomly.

Now, we have seen that during *extremes* of the energy fluctuations there can be a "degeneration" of the motion into harmonic motion, with ensuing radiation and transition towards a lower energy state. Because these spurious energy fluctuations are superimposed on the energy spread due to energy width, a large energy width leads to a larger probability per unit of time that the "degeneration threshold" can be reached and a transition occurs, than when the energy width is small. If the energy width is small, the probability of a transition is reduced, and we will have to wait a longer time before such a transition will occur.

The conclusion is, that in the droplet theory a similar relation exists between energy width and transition probability as in the usual quantum mechanics. In usual quantum mechanics this relation is attributed to the energy-time uncertainty. In the droplet theory the relation between energy width and transition probability is basically caused by the fact that here there are two communicating energy reservoirs in the system, that can "borrow" energy from each other. In the droplet theory there is no violation of the law of conservation of energy, and nevertheless there is an amount of "uncertainty" in the *translational* energy.

All this may explain why there is any relation at all between energy-width and transition probability, but it does not explain why under certain circumstances the energy width is larger than in other situations. It does not explain why each of the allowed energy levels in a potential box will possess its own characteristic life-time. In particular, it neither gives an explanation why the lowest energy level (the zero-point energy) apparently does not have any energy width at all, and why the spurious energy fluctuations coming on top of the energy width in the case of the lowest level never lead to transition. This matter has to be taken up in a later chapter.

H.6 Stability of the Stationary States (Qualitative)

The question addressed here is the following. The bounces of the droplet against the potential walls always cause a change of the absolute value of the momentum $|p_0|$, so that the momentum to the right before the bounce is slightly different from the momentum to the left after the bounce. As pointed out, this leads to differences in the partition of the total energy, and causes a spread of translational energy (the energy width). The question is: could these changes of momentum and translational energy accumulate after many passes of the droplet between the walls, so that the average translational energy drifts away from the stationary value? This question essentially concerns the stability of the stationary states.

The spread of momentum values between the different passes of the droplet is according to De Broglie's formula accompanied by a spread of the wavelengths of the beats. In general, these wavelengths then never fit exactly within the width of the potential box. In other words: the energy width is accompanied by a certain amount of "fuzziness" of the "De Broglie waves".

What happens if there is a large discrepancy between the length of the "De Broglie waves" and the width of the box is sketched in Fig. H.6.1.

Here a pass of the droplet from left to right is considered. It has been assumed that the momentum is momentarily smaller than corresponds with the exact stationary state. A smaller momentum implies a larger wavelength so that there is no exact fit between the walls of the box. We see that the symmetry of the pulsation w.r.t. the wall, which is a requirement for the stationary state, is lost. The droplet will now be bounced back with a rather different momentum (in absolute terms) than it possessed when approaching the wall. The momentum jump during the collision with the wall will be larger than just the difference caused by the normal "chaotic asymmetry", because it is enhanced by the non-zero gradient of the amplitude of the pulsations. The result may be assumed to be a change of the momentum in such a sense that there is a tendency to restore the conditions of a stationary state. The larger the momentum is off the stationary value, the larger is the tendency to change the momentum by the collisions with the walls. This is imagined to be the mechanism ensuring the stability of the stationary state.

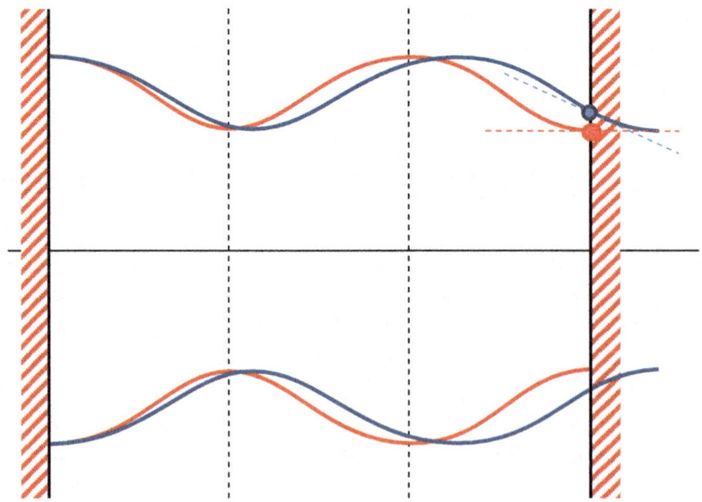

Fig. H.6.1 Stability of the stationary state. *Red line* is pulsation pattern in exact stationary state. *Blue line* droplet has smaller momentum, and has a longer "De Broglie wavelength

H.7 The Relation Between the Pulsation Beats and "Position Probability Density"

In usual quantum mechanics the matter waves give information about the probability to observe an electron at a particular position. For instance, the first three wave functions in a potential box, as determined by solving the Schrödinger-equation, are shown in Fig. H.7.1 on the left side. They are given by

$$\Psi_n \div \sin\left(n\,.\pi\frac{z}{a}\right) \quad (n = 1,\, 2,\, 3.\ldots.) \tag{H.7.1}$$

meaning that, in a stationary state, the width of the box can contain an integer number of half " De Broglie " wavelengths.

On the right hand side of Fig. H.7.1 the corresponding PPD-curves (Position Probability Density) are shown, giving information about the probability to observe the electron within an interval Δz of the box width. The Position Probability Density is given by the proportionality

$$PPD \div (\Psi_n)^2 \tag{H.7.2}$$

In particular the zero-points of the PPD-curves are noteworthy. In these positions the probability to observe the particle is zero. Some books on quantum mechanics give an even stricter interpretation, by stating that the particle never passes these points.

In the droplet theory it will be found, as shown below, that the less strict first interpretation is applicable here. The droplet will pass these zero-points unhindered on its way back and forth between the walls of the potential box. However, these positions are characterised by the fact that no transition towards another energy state can be initiated there. This means that no radiation will be emitted or absorbed whenever the droplet is in or near one of the zero-points. In other words, no interaction with the outside worlds (= observation) is possible.

Preceding a more careful analysis of this phenomenon, first a heuristic explanation will be given. In Fig. H.7.2 the first three energy states in a potential box are once again shown, where the results from usual quantum mechanics are compared with those obtained by the droplet theory. It is seen that the zero-points of the PPD-curve (corresponding with the zero-points of the wave functions) have the same position as the "zero-slope positions" of the stroboscopic images of the pulsation. This is not surprising, because in these "zero-slope positions" the amplitude of the pulsations is constant. This is physically the same situation as encountered in the case of a free-flying droplet which is not constrained by potential walls. Such a free-flying droplet everywhere has the same amplitude of its pulsation. From physics it is known that a free-flying droplet, like a free-flying electron, does not have the possibility to spontaneously undergo a transition towards another energy level. Therefore, in the points of zero-slope of the strobe images there is neither a possibility for a transition, and no observation of the droplet is possible in

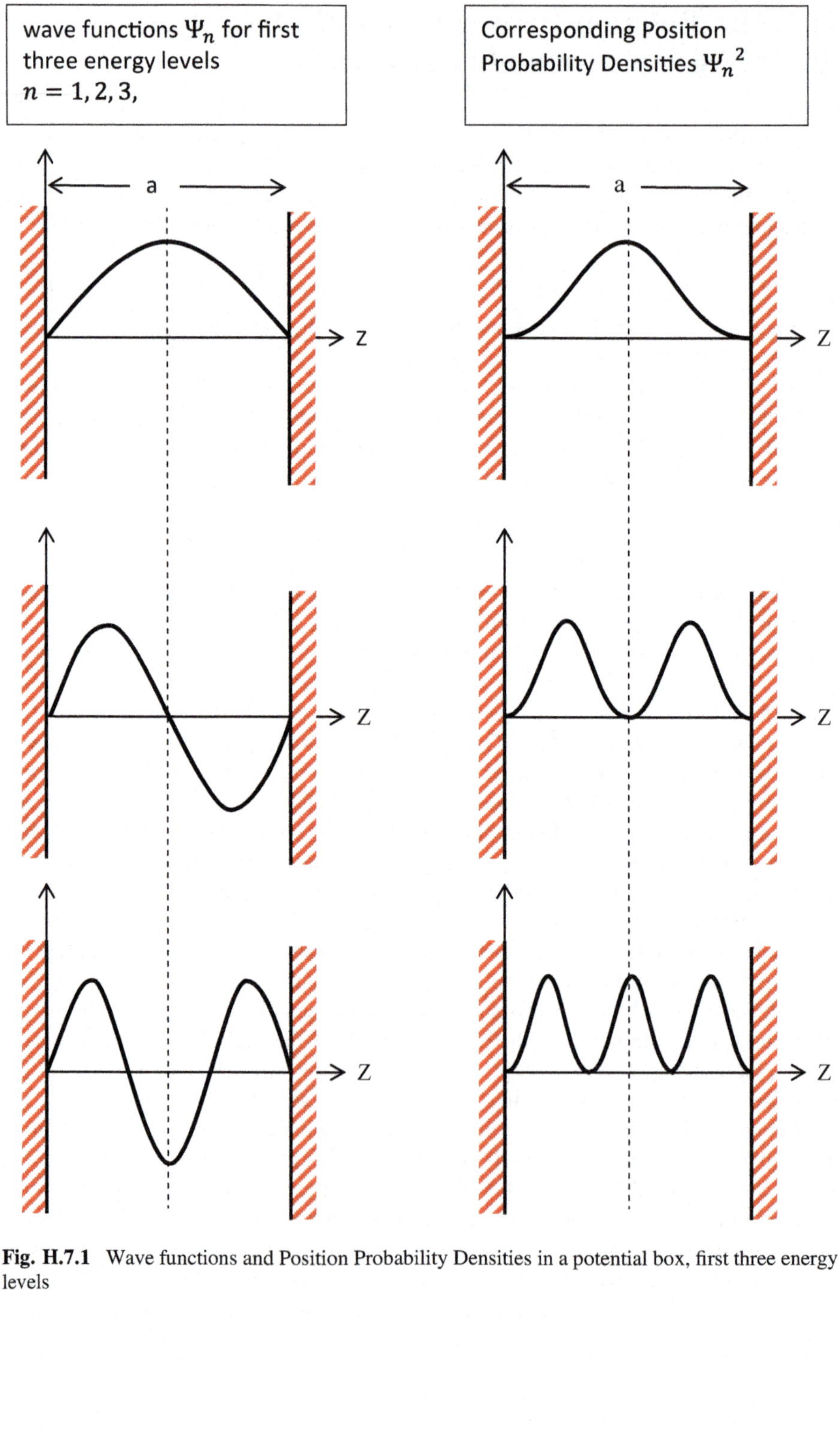

Fig. H.7.1 Wave functions and Position Probability Densities in a potential box, first three energy levels

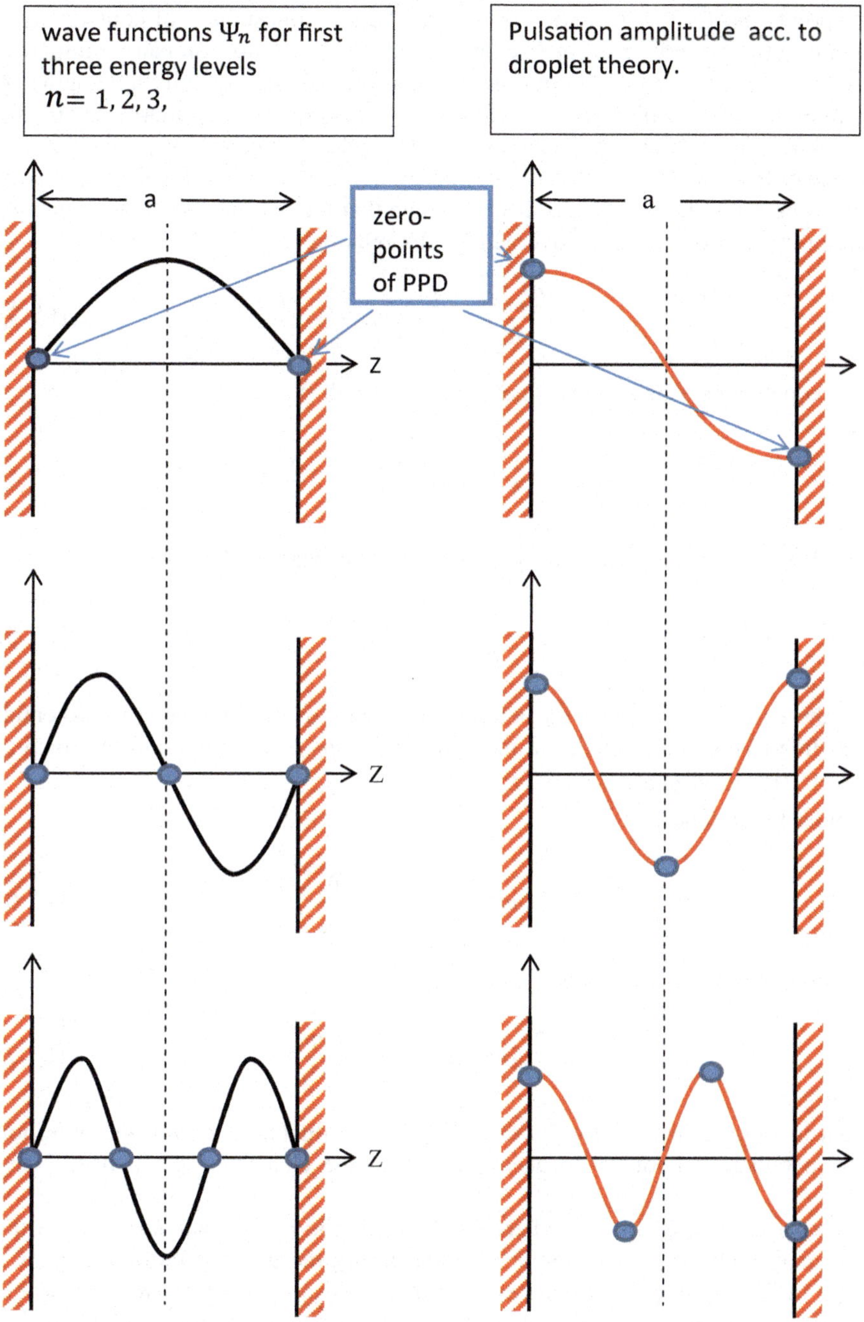

Fig. H.7.2 Wave functions and pulsation amplitudes (in terms of stroboscopic images) in a potential box, first three energy levels

these points. So far the heuristic reasoning. The next question is of course why a constant pulsation amplitude physically implies that no transition can be initiated, whereas a variable amplitude does allow radiation and consequently a damping of the motion. The explanation in found by considering in more detail the phase-difference between the pulsation and the velocity fluctuations.

Chapter F.2 considered the possible energy transfers between translation energy and the energy in the elongation. It was shown that the partition of the total energy between these two modes is governed by the relation

$$\frac{dE_{transl}}{dt} = \frac{1}{2} v^2 \, m_{em} \frac{\dot{s}}{s} \tag{H.7.3}$$

which equation becomes, when linearising for small $\frac{\Delta s}{s_e}$:

$$\frac{dE_{transl}}{dt} = \frac{1}{2} v^2 \, m_{em_0} \frac{\dot{\Delta s}}{s_e} \left(1 - 2\frac{\Delta s}{s_e}\right) \tag{H.7.4}$$

Furthermore, when linearising we also have the relation (F.6.10):

$$v = \frac{p}{m} \left(1 + 2\frac{\Delta s}{s_e}\right) \tag{H.7.5}$$

In the case that there are no potential gradients or any influence by potential walls (free-flying conditions), the momentum $p = p_0$ is constant, and the fluctuations of $v(t)$ are in phase with $\frac{\Delta s}{s_e}(t)$.

Under these conditions

$$\frac{dE_{transl}}{dt} = m\left(\frac{p_0}{m}\right)^2 . \left(1 + 2\frac{\Delta s}{s_e}\right) \frac{\dot{\Delta s}}{s_e} \tag{H.7.6}$$

From Chap. F.7, Eq. (F.7.5):

$$\frac{\Delta s}{s_e}(t) = S\, cos(\omega t + \varepsilon) - k\left(\frac{p_0}{m}\right)^2 \tag{H.7.7}$$

so that it follows that the translation energy E_{transl} fluctuates *within* one pulsation cycle, but there cannot be an accumulation of the change of E_{transl} over many cycles.

Comparing with Chap. F.8, where the damped motion was analysed (still for the free-flying condition), the difference is that during the damped motion there is a change of E_{transl} which accumulates over many pulsation cycles. This is possible only, because in this case there is a phase difference between $v(t)$ and $\frac{\Delta s}{s_e}(t)$ (see Eq. (F.8.22)). For instance, write

$$\frac{\Delta s}{s_e}(t) = A + S.\cos\omega t \tag{H.7.8}$$

and

$$v(t) = v_{av} + V.\cos(\omega t + \varepsilon) \tag{H.7.9}$$

then

$$\begin{aligned}\frac{dE_{transl}}{dt} &= \frac{1}{2}v^2\, m_{em_0}\frac{\dot{\Delta s}}{s_e}\left(1 - 2\frac{\Delta s}{s_e}\right)\\ &= m.\left[v_{av}^2 + 2\,V\,v_{av}\cos(\omega t + \varepsilon)\right].[1 - 2A - 2S\cos\omega t].[-\omega S\,sin\,\omega t]\end{aligned} \tag{H.7.10}$$

which contains a term

$$m\,\omega\,S\,V\,v_{av}(1 - 2A)\sin\varepsilon \tag{H.7.11}$$

Thanks to the phase difference ε the term can contribute to a change of E_{transl} over a longer time period than just one pulsation cycle. This is of course as it should be during the damped motion.

Now, the problem is in starting up the radiation (or: initialising the damped motion) from initial conditions associated with the free-flying, undamped conditions. The latter entails that there is no phase-difference between $v(t)$ and $\frac{\Delta s}{s_e}(t)$, whereas as soon as damping starts a phase-difference is essential: the phase difference is a "conditio sine qua non". The conclusion must be, that the free-flying situation *cannot provide the proper initial conditions* for a transition to the damped motion. This explains why a droplet of charge (or for that matter: an electron) cannot enter a transition towards another energy level (and thus cannot be observed) if it is outside a potential well. The conditions inside a potential well will next be considered.

Inside a well a slow variation of the pulsation amplitude will take place ("slow" compared with the pulsation period), as well as a slow variation of the amplitude of the velocity fluctuations. According to the table in Chap. F.4 on page 431 increasing amplitudes are associated with increasing E_{transl}, at the cost of the potential energy in the elongation, even if the momentum is constant. According to the above given heuristic reasoning such a slow increase, over a time interval spanning many pulsation cycles, is possible only if there is a phase difference between $v(t)$ and $\frac{\Delta s}{s_e}(t)$. This phase difference becomes zero in the maxima of minima of the envelope of the pulsation oscillations, because in these points there is no variation of the translation energy. The phase difference reaches the largest value somewhere in between the maxima or minima of the envelope of the pulsation oscillations. Where the phase difference is practically zero, these conditions prevent an initialisation of damping by radiation. However, at other points such a damped motion can be started up, the easier the larger the phase difference. The boundary conditions inside a well therefore make at some places in the well a transition possible, in contrast to the free-flying situation.

In Chaps. H.1 and H.2 the velocity $v(t)$ was not worked out, so that the phase difference with $\frac{\Delta s}{s_e}(t)$ was not made explicit. later, in Chap. H.10 regarding the general well, it will be shown (see Eq. (H.10.4)) that the velocity and pulsation in a potential well are related to each other by a first-order differential equation of the form

$$\dot{v} = f(t).\frac{\dot{\Delta s}}{s_e} + \dot{f}\,\frac{\Delta s}{s_e} \tag{H.7.12}$$

where $f(t)$ is a slow time-function compared with the fast oscillating pulsation $\frac{\Delta s}{s_e}$. From this equation it may be concluded, even without explicitly working out the solution of (H.7.12), that in general there can be a phase difference between v and $\frac{\Delta s}{s_e}$ when the droplet obeys the equations of motion associated with a potential well.

H.8 The Rectangular Potential Well (Qualitative)

The rectangular potential well differs from the potential box by the fact that the "walls" are not infinitely high. It is sketched in Fig. H.8.1. We now have two different cases.

The first possibility is that the (mean) translation energy of the droplet is less than the height of the potential walls, so that the droplet is constrained to remain within the well, moving back and forth between the two walls. This case is characterised in the figure as $E_{transl} < E_0$. The variation of v^2 (mean value, for now neglecting the random variations associated with energy width) in time is

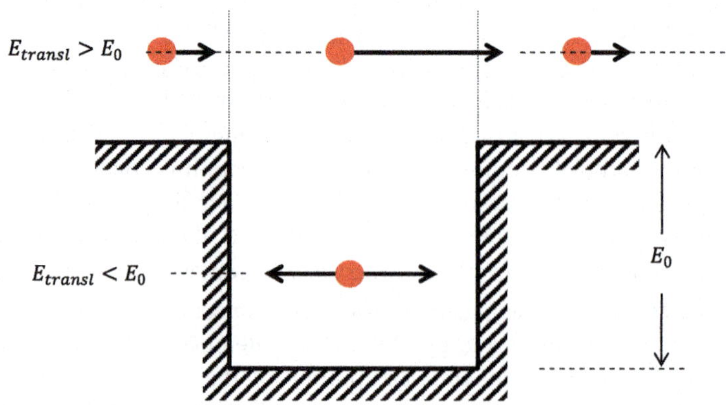

Fig. H.8.1 Rectangular potential well

schematically shown in the lower part of Fig. H.8.2, which corresponds with Fig. H.1.5 for the potential box.

The second case is indicated as $E_{transl} > E_0$. The droplet then possesses a positive velocity everywhere (assuming that the droplet flies from left to right), although it shows a temporary increase of velocity as long as it is in the region of the smaller potential energy. The variation of v^2 with time (again: the mean value) is schematically shown in the upper part of Fig. H.8.2.

It should be emphasised that the first part of the following discussion is restricted to the behaviour of droplets having the "middle frequency" of the PSD curve (Fig. F.4.3) of the actually chaotically moving droplet, as if there existed a harmonic pulsation without amplitude variations. Later it will be considered what the effect is of the actual spread of translational energy around this average situation, associated with chaotic motion.

The discussion is qualitative only, without a quantification of the allowed energy levels. This is not really a handicap, since in the books on the usual quantum mechanics a solution is neither readily found (the solution can be determined only numerically, so that the rectangular well is didactically not very useful). Therefore, no easy comparison with usual quantum mechanics results would be possible to assess the validity of any quantified expressions. However, the present qualitative discussion of the rectangular well leads to some new concepts, which can be used to advantage during a later discussion of the parabolic potential well. The parabolic well is, in contrast, amenable to closed form quantitative solutions, both in droplet theory as well as in usual quantum theory, so that some form of validation of the concepts developed in the present chapter will become available later.

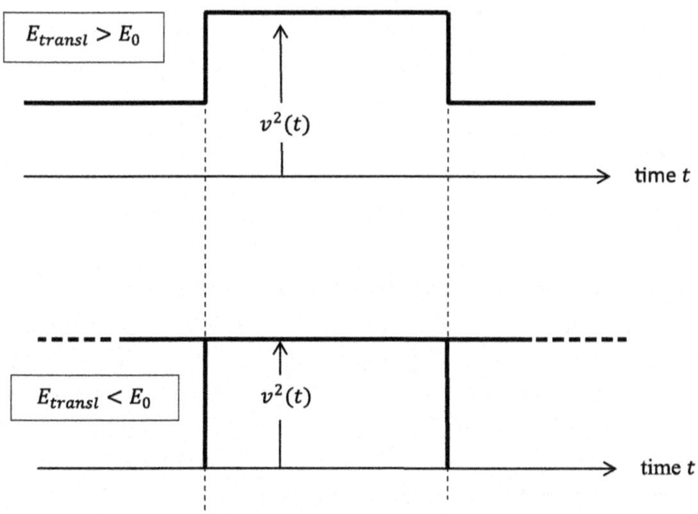

Fig. H.8.2 Velocity squared as function of time. *Top* "free" electron passing a rectangular dip in the potential. *Below* "bound" electron moving back and forth within a rectangular potential well

The functions $v^2(t)$ in Fig. H.8.2 are drastically different between the two cases, and can explain why no energy quantisation occurs in the case of a free electron, whereas it does occur in the case of a bound electron. The variation of $v^2(t)$ for $E_{transl} > E_0$ is not a periodic function in time. In the frequency domain it could only be described by a Fourier-*integral*, implying that the fundamental "repetition" frequency is zero. In contrast, the case $E_{transl} < E_0$ is periodic, it can be described by a Fourier-*series* and has a frequency content with a lower bound on the frequency. The function $v^2(t)$ in this case was earlier discussed more extensively.

Formally, the solution of the pulsation equation

$$\frac{\ddot{\Delta s}}{s_e} + \omega_0^2\left[1 + 2\,k\,v^2(t)\right]\frac{\Delta s}{s_e} = -k\,\omega_0^2 v^2(t) \tag{H.8.1}$$

can be determined in the case $E_{transl} > E_0$ by the same procedure (taken from Den Hartogh) as was shown in Chap. H.1 for the potential box. Let us denote the lower level of $v^2(t)$ in the upper part of Fig. H.8.2 by v^2_{low} and similarly the higher level by v^2_{high}. In the region $v^2(t) = v^2_{low}$ the solution of the pulsation equation is

$$\frac{\Delta s}{s_e}(t) = A\,\cos(\omega_{low}t + \varepsilon) - k\,v^2_{low} \tag{H.8.2}$$

with the frequency

$$\omega_{low} = \omega_0\left[1 + k\,v^2_{low}\right] \tag{H.8.3}$$

Similarly, in the region $v^2(t) = v^2_{high}$ we find

$$\frac{\Delta s}{s_e}(t) = B\,cos\left(\omega_{high}t + \delta\right) - k\,v^2_{high} \tag{H.8.4}$$

with the frequency

$$\omega_{high} = \omega_0\left[1 + k\,v^2_{high}\right] \tag{H.8.5}$$

The constants A, B, ε *and* δ are determined by requiring that $\frac{\Delta s}{s_e}(t)$ is continuous and has continuous derivatives at all times. This leads to four algebraic equations determining the four constants. The equations (H.8.2) and (H.8.4) together constitute the complete solution of $\frac{\Delta s}{s_e}(t)$. At no instant of time the solution shows a frequency content consisting of more than one frequency, and there is no question of any interference effects. Both frequencies ω_{high} and ω_{low} are very high frequencies, and there is no requirement of periodicity which would form any restriction for the value of these two frequencies. Consequently, there is neither any restriction imposed on the translation energy of the "free flying" droplet.

In contrast, the droplet having a smaller energy $E_{transl} < E_0$ (i.e. the droplet bound inside the well) is subject to restrictions, because the mathematical solution for $\frac{\Delta s}{s_e}(t)$ has to fulfill the requirement of periodicity. The relevant frequencies in this case are $\omega = \omega_0[1 + k\,v^2]$ and the zero-speed frequency ω_0 itself. The periodicity required has a so much lower frequency than either ω or ω_0, that this fact alone points at the conclusion that there must exist a beat phenomenon with a frequency depending in some way on the difference frequency $\Delta\omega = \omega - \omega_0$. This heuristic reasoning will be used later, when we come to the analysis of the parabolic potential well.

In the case of the rectangular potential well, the heuristic reasoning can easily be corroborated by an actual mathematical analysis, practically the same as the analysis shown in Chap. H.1, which therefore will be recalled just scantily. Here we showed that it is allowed, within the accuracy pursued, to write at any arbitrary point in time t the complete solution $\frac{\Delta s}{s_e}(t)$ as the sum of the contributions by two rows of "battlements", shifted in time with respect to each other. Using this conclusion again in the case of the rectangular well, the superposition then leads to the conclusion that at any time the function $\frac{\Delta s}{s_e}(t)$ is the sum of two oscillations, so that both the frequencies ω and ω_0 are mixed, and indeed lead to a low frequency beat phenomenon. The requirement of periodicity then subsequently leads to the requirement that the fundamental frequency ω_0, the higher frequency ω and the beat phenomenon must be *commensurate*. In other words:

$$\omega = k.\frac{2\pi}{T} \quad (k = 1,\ 2,\ 3,\ \ldots) \tag{H.8.6}$$

as well as

$$\omega_0 = n.\frac{2\pi}{T} \quad (n = 1,\ 2,\ 3,\ \ldots) \tag{H.8.7}$$

and therefore also

$$\omega - \omega_0 = m.\frac{2\pi}{T} \quad (m = 1,\ 2,\ 3,\ \ldots) \tag{H.8.8}$$

where T is the required repetition time of $\frac{\Delta s}{s_e}(t)$.

As was seen in Chap. H.2 this requirement of commensurability leads to a restriction of the possible values of the translation energy, in other words: it leads to quantisation of the energy levels.

All this was discussed extensively in the case of the potential box. We can see that the same type of reasoning must lead to quantisation of the energy levels in a rectangular potential well.

At first sight one might even think that the allowed energy levels in a rectangular well are the same as those in a potential box, because the frequency of the back and forth motion of the droplet would be the same, if the width of the rectangular well is equal to the width of the potential box so that the repetition period is the same in both cases. However, this would not be a correct conclusion. This matter is taken up in the next chapter.

H.9 Tunneling Effect (Qualitative)

To explain why the energy quantisation in a rectangular well is different from that in a box, we must recall that the considerations in the above are strictly valid only for the mean value of the velocity squared.

We have seen in Chap. H.3 that around this mean value fluctuations of v^2 will occur, due to the random energy exchange between the translational energy and the energy in the elongation, even under conditions of constant total energy. The situation was summarised in the upper part of Fig. F.4.3.

In the case of the bound droplet these random fluctuations of the translation energy were called the "energy width" of the quantisation.

Looking at Fig. H.8.1 one can easily imagine a droplet with a *mean* kinetic energy $E_{mean} < E_0$ which has temporarily a higher energy than E_0: $E_{tempo} > E_0$. If such an upward fluctuation of the energy coincides with the encounter of the droplet with the wall of the potential well, it will allow the droplet to escape from the well.

Similarly, a free droplet may suffer a "down fluctuation" of its translational energy just at the moment it passes an edge of the potential well, so that it will be reflected by the wall. Therefore, there is a chance that even the free flying droplets are affected by the presence of the potential well and that their direction of flight is reversed.

In usual quantum mechanics these effects are well known. The possible escape of an electron from a potential well in which it in principle is locked up according to its energy, is called the "tunneling effect". The partial reflection of a stream of particles in the region of abrupt variations of the potential is also well known in quantum theory, and is here called the "wave-like property" of particles, because such a reflection is a typical property of waves meeting discontinuities in e.g. a propagation channel.

Once again, in usual quantum theory these effects are attributed to the uncertainty relations. The uncertainty permits fluctuations of the energy to take place even under non-radiating conditions, when in principle the energy should be constant.

In the droplet theory the random fluctuations of the translational energy, permitted because there exists a communicating energy reservoir in the form of elongation energy, explain this "tunneling" and "wave-like" behaviour. This is a second example where the existence of the second degree of freedom (i.e. the elongation), "invisible" in quantum mechanics, is equivalent to "uncertainty".

The tunneling effect has some consequences for the quantisation values of the energy. Consider a droplet with a mean energy E_{mean} not far below the depth E_0 of the potential well. Large *positive* excursions of the translation energy will often lead to an escape of the droplet from the well (taking a statistical point of a view: many observations of many droplets, all starting their life in the well with the same E_{mean}). Very large positive fluctuations are thus cut off from the statistics: the droplets undergoing large "up-excursions" of the translational energy are lost. On the other hand, *negative* excursions do not have this effect, no matter how large they are. The distribution of energies becomes skew, and strictly speaking it is no longer correct to call the *starting* energy to be the same as the *mean* energy. Looking at a large ensemble of observations, the actual mean energy will be lower. To use a metaphore: if we have a large collection of observations, a least squares approximating function can be affected rather drastically by eliminating the outliers from the observations.

For this reason one cannot expect that the rectangular potential well, with its tunneling effects, will have the same quantisation levels as the potential box.

H.10 Energy Quantisation in General Potential Wells, Incl. Application to Parabolic Well

We start again from the general, linearised equations of motion for the droplet:

$$\frac{\ddot{\Delta s}}{s_e} + \omega_0^2 (1 - 2\,k\,v^2)\frac{\Delta s}{s_e} = -k\,\omega_0^2\,v^2 \tag{F.6.6}$$

$$\frac{dp}{dt} = \gamma\,\ddot{v} + q.E_{external} + F(t) \tag{F.6.8}$$

where

$$p = m\,v\left[1 - 2\frac{\Delta s}{s_e}\right] \tag{F.6.9}$$

Once again we assume a constant energy motion, which is implemented in the equations by choosing the external force such that it cancels the radiation resistance.

In contrast to the earlier developed theory, we assume that there is an external potential, so that $q.E_{external} \neq 0$. Accordingly, we split the variations of the velocity in two parts:

$$v(t) = v_{marble}(t) + \Delta v(t) \tag{H.10.1}$$

where the symbol $v_{marble}(t)$ stands for the velocity variations a rigid marble (or: a droplet without pulsation) would experience under the influence of the external potential. This definition has as a consequence:

$$m\,\dot{v}_{marble} - q.E_{external} = 0 \tag{H.10.2}$$

The translation equation of motion (F.6.8), taking into account (F.6.9), then reads

$$m\left(\dot{v}_{marble} + \dot{\Delta v}\right).\left(1 - 2\frac{\Delta s}{s_e}\right) - 2m(v_{marble} + \Delta v)\frac{\dot{\Delta s}}{s_e} - q.E_{external} = 0 \tag{H.10.3}$$

which after substitution of Eq. (H.10.2) gives

$$\dot{\Delta v} = 2\frac{d}{dt}\left(v_{marble}.\frac{\Delta s}{s_e}\right) \tag{H.10.4}$$

or, integrated

$$\Delta v = 2\,v_{marble}.\frac{\Delta s}{s_e} \tag{H.10.5}$$

so that

$$v = v_{marble} + \Delta v = v_{marble}\left(1 + 2\frac{\Delta s}{s_e}\right) \tag{H.10.6}$$

Substitution of this result into the pulsation equation and properly linearising:

$$\frac{\ddot{\Delta s}}{s_e} + \omega_0^2\left[1 + 2\,k\,v_{marble}^2(t)\right]\frac{\Delta s}{s_e} = -k\,\omega_0^2 v_{marble}^2(t) \tag{H.10.7}$$

The r.h.s. of this equation has a slow variation in time compared with the time scale of the pulsation, so that the particular solution may be approximated by the quasi steady solution

$$\left(\frac{\Delta s}{s_e}\right)_{partic} = -k\,v_{marble}^2(t) \tag{H.10.8}$$

The homogeneous solution is determined by the equation

$$\frac{\ddot{\Delta s}}{s_e} + \omega_0^2\left[1 + 2\,k\,v_{marble}^2(t)\right]\frac{\Delta s}{s_e} = 0 \tag{H.10.9}$$

We assume that the potential well has such a shape that the droplet is forced to move back and forth within the well and does not have the possibility to escape. Accordingly we can define a repetition time T and a repetition frequency $\Omega = \frac{2\pi}{T}$ as defined in Fig. H.10.1.

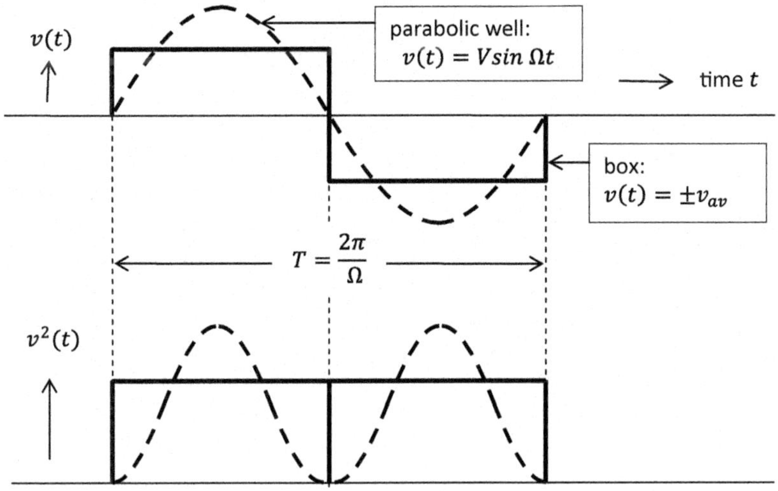

Fig. H.10.1 Definition of repetition time and -frequency. Examples: box and parabolic well

The examples given in the figure are the velocity in the case of a potential box and a parabolic potential well.

Both the particular solution determined by (H.10.8) as well as the homogeneous solution from (H.10.9) have the same repetition time. In Chap. H.1 it was shown that the complete—although asymptotically approximated—solution will have the form of a superposition of two oscillations, i.e. a near-harmonic oscillation having the variable frequency $\omega(t) = \omega_0\left[1 + k\,v^2_{marble}(t)\right]$ and a constant frequency oscillation with the "zero-speed frequency" ω_0.

In a potential *box* the solution of the pulsation $\frac{\Delta s}{S_e}(t)$ has the special characteristic that $\omega(t)$ is constant, and in that particular—but exceptional—case the corresponding wave form $\cos\omega t$ is not near-harmonic but is pure harmonic.

In order to keep the discussion general, we use the theory of Chap. G.1 for the interference effects of a near-harmonic oscillation which is mixed with a pure harmonic oscillation. In Chap. G.1 it was derived that for a given repetition time T_{FM} two conditions must be satisfied:

$$\omega_0.T_{FM} = k.2\pi \tag{G.1.40}$$

and

$$\int_0^{T_{FM}} \{\omega + \Delta\omega(\tau)\}d\tau - \omega_0.T_{FM} = n.2\pi \quad (n = 1, 2, 3, \ldots) \tag{G.1.41}$$

In the case at hand we must substitute into these conditions $\omega = \omega_0$, $T_{FM} = T$ and $\Delta\omega(\tau) = k\,\omega_0 v^2_{marble}(t)$ so that it is required that

$$k\,\omega_0 \int_0^{2\pi/\Omega} v^2_{marble}(\tau)d\tau = n.2\pi \quad (n = 1, 2, 3, \ldots) \tag{H.10.10}$$

The non-dimensional form of $v^2_{marble}(\tau)$ is completely determined by the potential variation within the well. The amplitude, however, is determined by the initial conditions. From (H.10.10) it is seen that this amplitude is quantised, since there are only discrete values which can satisfy the condition. Let us assume that a function $(v^2_{marble})_n$ in some way has been determined which satisfies condition (H.10.10) for the value of the counter n. The next possibility to satisfy (H.10.10) is denoted $(v^2_{marble})_{n+1}$. From (H.10.10) we then find:

$$k\,\omega_0 \int_0^{2\pi/\Omega} \left[\left(v^2_{marble}\right)_{n+1} - \left(v^2_{marble}\right)_n\right] d\tau = 2\pi \quad (n = 1, 2, 3, \ldots) \tag{H.10.11}$$

In the case of a potential box this would lead to the condition

$$k\,\omega_0 \left[\left(v^2_{av}\right)_{n+1} - \left(v^2_{av}\right)_n\right] \frac{2\pi}{\Omega} = 2\pi \tag{H.10.12}$$

which can be rearranged to obtain the form

$$\frac{k\,\omega_0}{m}\left[\frac{1}{2}m\left(v^2_{av}\right)_{n+1} - \frac{1}{2}m\left(v^2_{av}\right)_n\right] = \frac{1}{2}\Omega \tag{H.10.13}$$

where the notation m stands for the average mass of the droplet.

In Chap. G.2, Eq. (G.2.4), it was determined that the combination $\left(\frac{k\,\omega_0}{m}\right)$ equals $\frac{2\pi}{K}$, whilst it was found that $K \approx h$, with h Planck's constant. Very often in quantum mechanics the notation $\hbar$ is used for $h/(2\pi) = \hbar$.

Using this, Eq. (H.10.13) takes the form (because in a box the energy of the droplet is entirely made up of kinetic energy):

$$E_{n+1} - E_n = \frac{1}{2}\hbar\Omega \quad (potential\,box) \tag{H.10.14}$$

This gives the difference between the consecutive discrete energy levels in a potential box. Note that we cannot yet compare this expression with the results found in Chap. H.2, because the value of Ω depends on the actual energy.

It is at this stage of the discussion neither possible to derive the actual values of the energy, because we cannot derive by the above shown method the lowest energy level (the socalled zero-point energy). The reason is that we cannot determine the distance between E_1 and $E_0 = 0$, since the case $v_{av} = 0$ (or $\Omega = \infty$) falls outside the scope of the theory, which is based on interference between a wave with $\omega \neq \omega_0$ and another wave with $\omega = \omega_0$. In the case $v_{av} = 0$ these two waves become identical and there is no interference pattern (leaving aside that in usual quantum theory the state $E_0 = 0$ is impossible).

Before embarking on a separate determination of the zero-point energy, it is useful to mention the equivalent result for the parabolic potential well. Here we must substitute into (H.10.11):

$$v(t) = V \sin \Omega t \tag{H.10.15}$$

which leads to

$$E_{n+1} - E_n = \hbar\Omega \quad (parabolic\, well) \tag{H.10.16}$$

In this case the repetition frequency does not depend on the actual value of the energy, and is thus a constant (compare a pendulum, where the repetition time is not influenced by the amplitude of the swings). The distance between the consecutive energy levels is therefore constant in a parabolic well, and does not depend on which level is considered. Again, the present approach does not give information about the actual values of the allowed energy levels.

In the case of the parabolic potential well it will be shown that the equations of motion for the droplet can be transformed such that a Mathieu equation is obtained. Solutions for this type of equation are known, and they therefore can provide the necessary information to determine the zero-point energy. Having obtained this additional information, combining with Eq. (H.10.16) yields the complete sequence of allowed energy levels within a parabolic well. The reason why this additional piece of information can be generalised to a procedure which is applicable to wells of a different shape than parabolic is the following.

In the book by Den Hartogh: Mechanical Vibrations, equations of motion of the type (H.10.9) for mechanical systems are analysed. It concerns mechanical mass-spring systems with a forced variation of the spring stiffness. It is there remarked that "feeding" the equation of motion with $v^2(t)$ either in the form of a block function or a harmonic function (see lower part of Fig. H.10.1) does not make an essential difference for the general character of the solution. Actually, if the "harmonic feed" is approximated by a series of blocks, the solutions are good approximations of each other. In the present case, we are just interested in the general character of the solution, which thus can be obtained by studying the exact Mathieu solutions.

For our purpose Eq. (H.10.9) is, for $v(t) = V sin\, \Omega t$, rewritten like

$$\frac{\ddot{\Delta s}}{s_e} + \omega_n^2 \left[1 - 2\,k\,V_{marble}^2 \cos^2(\Omega\, t)\right] \frac{\Delta s}{s_e} = 0 \tag{H.10.17}$$

where

$$\omega_n^2 = \omega_0^2 \left(1 + 2\,kV_{marble}^2\right) \tag{H.10.18}$$

Equation (H.10.17) has the form of the Mathieu equation. In Morse and Feshbach: Methods of Theoretical Physics, vol.I, pp. 556 and following, this equation is analysed in the form

$$\frac{d^2\psi}{d\phi^2} + (b - h^2 \cos^2 \phi)\psi = 0 \tag{H.10.19}$$

which corresponds with our equation of motion when choosing

$$\psi = \frac{\Delta s}{s_e} \tag{H.10.20}$$

$$\phi = \Omega t \tag{H.10.21}$$

$$b = \left(\frac{\omega_n}{\Omega}\right)^2 \tag{H.10.22}$$

$$h^2 = 2\,k\,V_{marble}^2 \left(\frac{\omega_n}{\Omega}\right)^2 \tag{H.10.23}$$

In Fig. H.10.2, copied from the book by Morse and Feshbach, some of the characteristics of the solution of the Mathieu equation are shown. The drawn lines in the figure correspond with solutions where after each fundamental period T the value of $\frac{\Delta s}{s_e}(t)$ is the same, i.e. the drawn lines correspond to $\frac{\Delta s}{s_e}(t+T) = \frac{\Delta s}{s_e}(t)$. The regions of the diagram indicated as "stable" indicate that here an exponential decrease of $\frac{\Delta s}{s_e}(t)$ has occurred after each repetition period: $\frac{\Delta s}{s_e}(t+T) = e^{-\alpha T}\frac{\Delta s}{s_e}(t)$. Conversely, in the "unstable" regions the solution increases exponentially.

In our case, we are considering the constant energy motion of a droplet within a parabolic potential well, without energy loss by radiation (undamped motion). We are thus dealing with a steady situation, where consequently those particular

Fig. H.10.2 Stability diagram of the solutions of the Mathieu equation, taken from Morse and Feshbach, Methods of Theoretical Physics

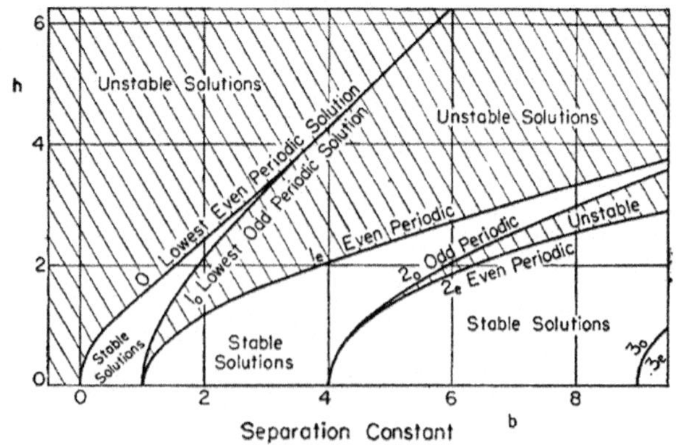

Values of separation constants for periodic solutions of the Mathieu equation.

combinations of b and h are applicable which correspond with the drawn lines in the diagram. In the case at hand the value of the parameter h is very small, due to the multiplication in Eq. (H.10.23) by the small factor k. The points in the diagram representing the motion of our droplet are thus, with good approximation, the points where the drawn lines intersect the abscissa. We see that the possibility to have an undamped motion $\frac{\Delta s}{s_e}(t)$ corresponds with

$$b = \left(\frac{\omega_n}{\Omega}\right)^2 = 1,\ 4,\ 9,\ \ldots \tag{H.10.24}$$

or

$$\frac{\omega_n}{\Omega} = p \quad (p = 1,\ 2,\ 3,\ \ldots) \tag{H.10.25}$$

Using Eq. (H.10.18) it is found

$$\omega_n = \omega_0\left(1 + kV^2_{marble}\right) \tag{H.10.26}$$

so that the condition $\frac{\omega_n}{\Omega} = p$ may be written as

$$\frac{\omega_0}{\Omega}\left(1 + kV^2_{marble}\right) = p \quad (p = 1,\ 2,\ 3,\ \ldots.) \tag{H.10.27}$$

or:

$$\frac{\omega_0}{\Omega} + k\,\omega_0\,.V^2_{marble}.\frac{1}{\Omega} = p \quad (p = 1,\ 2,\ 3,\ \ldots.) \tag{H.10.28}$$

It should be realised that the counter p is an extremely large integer, in view of $\frac{\omega_0}{\Omega}$ being in the l.h.s. of Eq. (H.10.28). This inconvenience can be repaired by subtracting from (H.10.28) the earlier result $\omega_0/\Omega = m \quad (m = 1,\ 2,\ 3,\ \ldots)$ obtained from (G.1.40), which contains a similarly large integer. We then obtain

$$k\,\omega_0\,.V^2_{marble}.\frac{1}{\Omega} = n \quad (n = 1,\ 2,\ 3,\ \ldots.) \tag{H.10.29}$$

What we are interested in is the lowest possible value of the energy which is, using the same steps as in (H.10.13) and (H.10.14) and taking $n = 1$:

$$E_1 = \frac{1}{2}\hbar\,\Omega \tag{H.10.30}$$

Note that the solution of the Mathieu equation gives for the higher energy levels more possibilities than the earlier result (H.10.16). One should interpret this such that the Mathieu equation yields potential solutions, from which only certain solutions can satisfy the actual physical boundary conditions. What can be concluded from the Mathieu equation is, that the lowest possible value of the energy is given by (H.10.30). No matter the physical boundary conditions, there is no smaller value of the energy than (H.10.30) able to comply with the equation of motion.

Combining Eqs. (H.10.16) and (H.10.30) then yields

$$E = \hbar\Omega\frac{2n-1}{2} \quad (n = 1, 2, 3, \ldots.) \quad (\textit{parabolic well}) \tag{H.10.31}$$

which is exactly the expression given in the literature for the energy quantisation in a parabolic potential well.

It is shown by Den Hartogh that feeding the equation of motion with a block function instead of a harmonic function leads to the same series (H.10.25). In the case of the potential box we thus can combine the lowest possible energy according to (H.10.30) with the earlier result (H.10.14) to find for the energy levels within a box:

$$E = \frac{1}{2}\hbar\Omega.n \quad (n = 1, 2, 3, \ldots) \quad (\textit{potential box}) \tag{H.10.32}$$

In this expression the repetition frequency Ω and the energy E are interrelated by

$$E = \frac{1}{2}m\,v_{av}^2 = \frac{1}{2}m\left(\frac{2a}{T}\right)^2 = \frac{1}{2}m\left(a\frac{\Omega}{\pi}\right)^2 \tag{H.10.33}$$

where a is the width of the box. Combining (H.10.32) and (H.10.33) then after some further manipulation leads to:

$$E = \frac{\hbar^2\pi^2 n^2}{2ma^2} \quad (n = 1, 2, 3, \ldots.) \quad (\textit{potential box}) \tag{H.10.34}$$

which is in agreement with the results from usual quantum mechanics, as well as in accordance with (H.2.11).

A last question is, why these quantisation levels are not influenced by the spurious energy fluctuations associated with the socalled "energy width". As we have seen, in the case of the rectangular potential well these energy fluctuations caused that the quantisation of the mean energy was somewhat modified w.r.t. the box because, when performing a great many experiments, there is a non-zero probability to lose by the tunneling effect the electrons with the highest energy level.

The situation in the case of the parabolic potential well is sketched in the following Fig. H.10.3. It is seen that excursions of the droplet's motion outside the classical limits are possible due to the energy width, as confirmed by the usual quantum theory. However, no electrons will become "lost", the wall of the parabolic well will always reflect the droplets back. There will not be a cut-off of the maximum energy due to vanishing of electrons from the statistics, and the mean quantised energy level will not be affected by the energy width.

The same applies to the quantisation levels in a potential box: the electrons are always reflected by the walls. In this case, there is not even the possibility that electrons exceed the classical limits of motion, since the slope of the walls is infinite. Compare with Fig. H.10.3 where the finite slope enabled electrons with a higher energy than the nominal quantisation to exceed these classical limits.

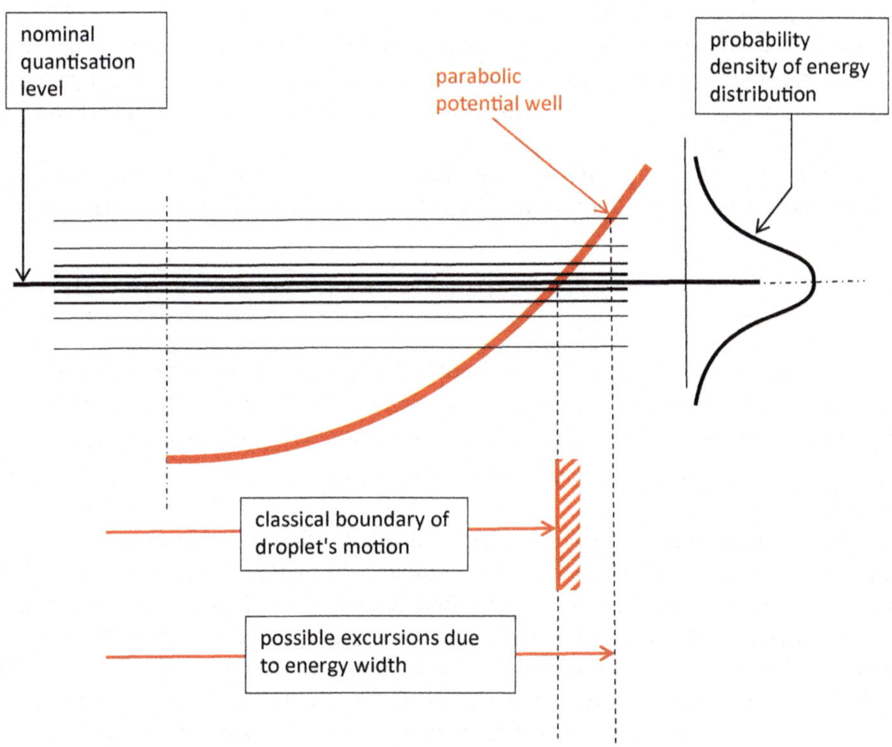

Fig. H.10.3 The possibility to exceed the classical limits of motion

H.11 Bohr's Radiation Formula; Superposition of States; Planck's Energy-Frequency Relation

In this chapter the radiation will be considered emitted during a transition from one stationary state to another. First, this discussion will be based on purely mathematical arguments, without bothering whether the solution which is allowed from a purely mathematical viewpoint will or can actually be realised in the physical situation. This will be followed by a more physical discussion, which unfortunately is not complete yet. The physical discussion points out some fundamental problems inherent in the purely mathematical results, and it tentatively proposes in which direction to search for a solution of the noted problems.

The stationary states determined for the potential box and the parabolic well essentially are eigenfunctions associated with series of eigenvalues. Both the box and the parabolic well constituted of eigenvalue problems, because of the requirement that the beat patterns (the eigenfunctions) would "fit" into the available width of the box or well, if a stationary state is to be realised.

In the case of the box an exact fit was only possible for a set of discrete values of v_{av}^2 (the set of eigenvalues). The eigenfunctions were approximated by beat patterns due to two harmonic waves, one with the zero-speed frequency and the other with the frequency belonging to the speed v_{av}.

In the case of the parabolic well likewise we have been dealing with an eigenvalue problem. In Chap. H.10 it was found that the differential equation

$$\frac{\ddot{\Delta s}}{s_e} + \omega_0^2\left[1 + 2\,k\,v_{marble}^2\,(t)\right]\frac{\Delta s}{s_e} = -k\,\omega_0^2 v_{marble}^2(t) \qquad \text{(H.10.7)}$$

where $v_{marble}^2\,(t)$ has a repetition frequency Ω, only has stationary solutions for certain discrete amplitudes of $v_{marble}^2\,(t)$ (these amplitudes being the eigenvalues). For each eigenvalue we find a specific time function $\frac{\Delta s}{s_e}(t)$ satisfying the differential equation: the eigenfunction. The eigenfunctions in this case were Mathieu-functions.

The eigenfunctions of Eq. (H.10.7) form a set of orthogonal functions, as seen in the above given examples. Such a set of orthogonal functions represents a so-called complete set, meaning that any arbitrary function can be built up by summing them, each function multiplied by a suitable factor. Arbitrary functions of both position and time can thus be represented by a sum of eigenfunctions with time-variable coefficients. The beat patterns *during* a transition, changing in shape as time proceeds, therefore can be described mathematically by a sum of the appropriate eigenfunctions with coefficients that are functions of time.

Let us assume that the time-dependent beat pattern during a transition can be expressed as the sum of two stationary states, viz. the states describing the beginning and end of the transition.

A transition from one state to another is then described by the gradual reduction to zero of one of the coefficients, and a simultaneous increase from zero to a non-zero value of the other coefficient. The transition is thus described (still purely mathematically) by the sum of two eigenfunctions, both with time-varying coefficients. Physically it means that the original state damps out, whereas the new state gradually rises in importance until it has taken over, and the coefficient of the original state has become zero. This view implies that during the transition process the two states are both non-zero at the same time, in other words they overlap in time.

Let us take the potential box as an example to see the consequences, this being the simplest case. Let us call the beat frequency of the original state ω_1. It is given by

$$\omega_1 = k\,\omega_0\,v_{av_1}^2 \tag{H.11.1}$$

whereas likewise the beat frequency of the final state is given by

$$\omega_2 = k\,\omega_0\,v_{av_2}^2 \tag{H.11.2}$$

When during the transition both these frequencies are found together in the radiation flux, the flux will show up an observable frequency $\omega_1 - \omega_2$ (see Chap. G.6 concerning the special meaning attached here to "observability").

The frequency found in the energy flux during the transition is:

$$\omega_1 - \omega_2 = k\,\omega_0\left[v_{av_1}^2 - v_{av_2}^2\right] \tag{H.11.3}$$

which may be written as

$$\omega_1 - \omega_2 = 2\frac{k\,\omega_0}{m}\left[\frac{1}{2}m\,v_{av_1}^2 - \frac{1}{2}m\,v_{av_2}^2\right] \tag{H.11.4}$$

or, using $\frac{k\,\omega_0}{m} = \frac{1}{\hbar}$ (see Eq. (G.2.4) where $K \rightarrow h$ and $\hbar = \frac{h}{2\pi}$):

$$f_{flux}.h = 2(E_1 - E_2) \tag{H.11.5}$$

Usually, we interpret the "colour of light" as the frequency of the electric and magnetic field components, both having *half* the frequency of the associated energy flux. Even though in the present case these field components are not observable, the observable frequency f_{flux} will still be associated with the "light frequency" $f_{light} = \frac{1}{2}f_{flux}$ according to the usual interpretation. As a result we find from (H.11.5):

$$f_{light}.h = (E_1 - E_2) \tag{H.11.6}$$

which corresponds exactly with Bohr's theory in usual quantum mechanics stating that the frequency of light emitted during a transition is proportional to the difference in energy levels of the two states involved, with Planck's constant h as a proportionality factor.

Interpreting the brief energy flux during the transition as a photon (see Chap. G.6), and considering that the total energy contained in such an "energy packet" must be $(E_1 - E_2)$, we find Planck's expression applicable to photons:

$$E_{photon} = h.f_{light} \tag{H.11.7}$$

Fundamental critique on the purely mathematical procedure to describe transition.

From the physical point of view the above given description of the transition process is not satisfactory. It is based on the mathematical view that the event of a transition can always be described as a superposition of stationary states. This would include the start and end of the transition. Conceptually, it would mean that some information about all the other possible stationary states is already contained in any one state. In other words: any stationary state is considered as a *superposition* of all the possible states, although only one of them is actually realised, and the other states are just potentially there, they are dormant so to speak, until a transition occurs.

Such a view of superposition is essentially non-causal. How would the droplet, being in a particular mode of motion, have any "knowledge" about other mathematically possible, but physically non-existing modes? Within the framework of a strictly causal view, one should be able to point out a definite physical mechanism causing that there is already "something" of the other modes present. This book searches for causal interpretations of quantum mechanics, and we therefore have the task to try to find a causal sequence of events during a transition.

The answer is possibly found in the earlier Fig. G.1.3, repeated here as Fig. H.11.1, and in this case to be interpreted in the space domain instead of the time domain (as was done in Chap. G.1). Such an interchange of the time- and space domains is easily imagined in the case of a potential *box*.

What is shown are the envelopes $g(z)$ of the high frequency function

$$f(z) = a.\cos\left(n.2\pi\frac{z}{w}\right) + b.\cos\left(m.2\pi\frac{z}{w}\right) \tag{H.11.8}$$

for different values of the ratio a/b. Consider first the black curve in Fig. H.11.1. This curve is applicable for small values of a/b. As was shown in Chap. G.1 the envelope $g(z)$ of the high frequency curve $f(z)$ approaches the limit

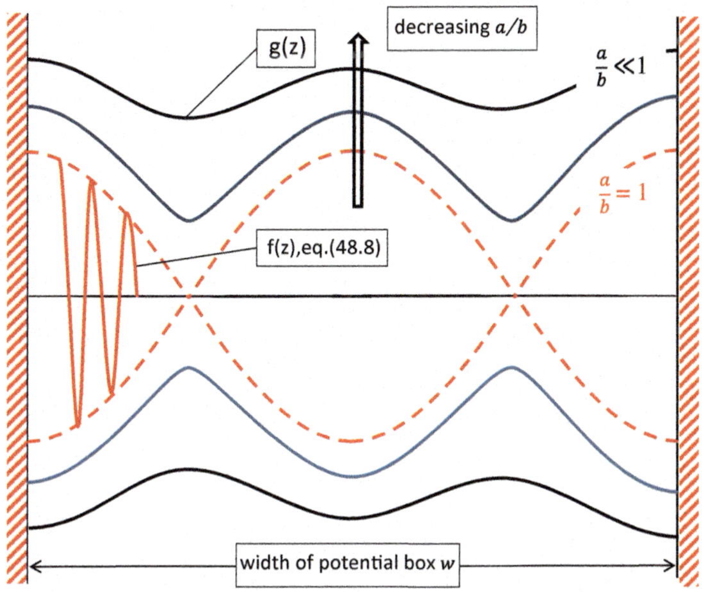

Fig. H.11.1 Beats resulting from superposition of *un*equal amplitude vibrations

$$g(z) \xrightarrow{small\, a/b} a.\cos\left[(n-m).2\pi\frac{z}{w}\right] + b \quad (envelope\ of\ f(z)) \tag{H.11.9}$$

In the figure, the example shown is $(n - m) = 2$, so that the envelope for small a/b goes through two harmonic cycles within the width of the box.

What we now see is, that the envelope may have a completely different frequency content for other values of the ratio a/b. The envelope shapes drawn in the figure for values between 0 and 1 suggests that there may be a whole range of wavelengths present in the envelope, even though the high frequency signal $f(z)$ itself contains only two wavelengths, independent from the value of a/b.

The envelopes for arbitrary a/b between 0 and 1 will contain wavelengths both larger as well as smaller than the wavelength in Eq. (H.11.9). Nevertheless, any envelope as sketched in Fig. H.11.1 satisfies the requirement for a stationary state, i.e. the requirement that the slope must be zero at both ends. For any arbitrary value of a/b the associated envelope (and therefore: the beat pattern) is apparently the sum of an infinite series of the eigenfunctions, all of them multiplied by non-zero coefficients except in the limit $a/b \to 0$ in which limit all the coefficients approach zero except one.

In actual fact, the envelope of the pulsation oscillations inside a potential box is the black curve in Fig. H.11.1, valid for *small* values of a/b, but *not* representing the limit $a/b \to 0$. The stationary state where the transition begins therefore already comprises the frequency of the state where the transition ends. The superposition of states is thus a real physical phenomenon instead of just a mathematical artifice.

Appendix I
Miscellaneous Subjects

I.1 A Mechanical Analogon of the Linearised Undamped Equations of Motion

The coupling between two oscillators, as in the equation of motion for the droplet of charge, occurs too in the case of the double pendulum (see the notations in the schematic).

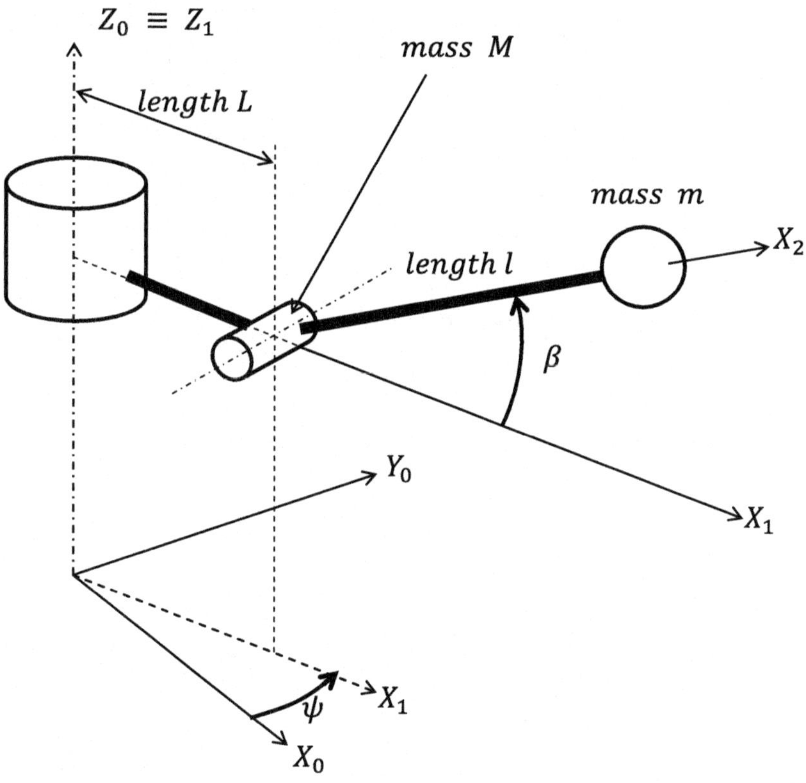

T. van Holten, *The Atomic World Spooky? It Ain't Necessarily So!*
DOI 10.2991/978-94-6239-234-2

The position vector of the mass m can be written as

$$\underline{r} = L(1 \quad 0 \quad 0)\begin{Bmatrix} \underline{i_1} \\ \underline{j_1} \\ \underline{k_1} \end{Bmatrix} + l(1 \quad 0 \quad 0)\begin{Bmatrix} \underline{i_2} \\ \underline{j_2} \\ \underline{k_2} \end{Bmatrix} \tag{I.1.1}$$

where the product of a row and a column is defined according to the rules of linear algebra. The symbols $\underline{i_1}$, $\underline{j_1}$ and $\underline{k_1}$ denote the unit-vectors of the coordinate system $(X_1 \quad Y_1 \quad Z_1)$, which system is obtained by a rotation over the angle ψ with respect to the inertial system $(X_0 \quad Y_0 \quad Z_0)$. The system $(X_2 \quad Y_2 \quad Z_2)$ with unit-vectors $\underline{i_2}$, $\underline{j_2}$ and $\underline{k_2}$ is obtained by a further rotation over the angle β.

A short notation for (I.1.1) is

$$\underline{r} = L(1 \quad 0 \quad 0)\{\underline{E}_1\} + l(1 \quad 0 \quad 0)\{\underline{E}_2\} \tag{I.1.2}$$

The transformation matrix $[\psi]$ relating the systems 0 and 1 is given by

$$\{\underline{E}_1\} = [\psi]\{\underline{E}_0\} \quad [\psi] = \begin{bmatrix} \cos\psi & \sin\psi & 0 \\ -\sin\psi & \cos\psi & 0 \\ 0 & 0 & 1 \end{bmatrix} \tag{I.1.3}$$

and the transformation matrix $[\beta]$ relating the systems 1 and 2 is given by

$$\{\underline{E}_2\} = [\beta]\{\underline{E}_1\} \quad [\beta] = \begin{bmatrix} \cos\beta & 0 & \sin\beta \\ 0 & 1 & 0 \\ -\sin\beta & 0 & \cos\beta \end{bmatrix} \tag{I.1.4}$$

The angular velocity $\underline{\omega}_1$ of system 1 is

$$\underline{\omega}_1 = \dot{\psi}(0 \quad 0 \quad 1)\{\underline{E}_1\} \tag{I.1.5}$$

Using the following property of transformation matrices

$$[\beta]^{-1} = [\beta]^T \tag{I.1.6}$$

the angular velocity of system 2 is

$$\begin{aligned} \underline{\omega}_2 &= \underline{\omega}_1 + \dot{\beta}(0 \quad -1 \quad 0)\{\underline{E}_2\} \\ &= \dot{\psi}(0 \quad 0 \quad 1)[\beta]^T\{\underline{E}_2\} + \dot{\beta}(0 \quad -1 \quad 0)\{\underline{E}_2\} \\ &= (\dot{\psi}\sin\beta \quad -\dot{\beta} \quad \dot{\psi}\cos\beta)\{\underline{E}_2\} \end{aligned} \tag{I.1.7}$$

Finally, use will be made of the "rotation operator" $[\omega_\times]$, relating the time derivatives of the unit vectors to the unit vectors themselves:

$$\{\underline{\dot{E}}\} = [\omega_\times]\{\underline{E}\} \quad [\omega_\times] = \begin{bmatrix} 0 & r & -q \\ -r & 0 & p \\ q & -p & 0 \end{bmatrix} \tag{I.1.8}$$

where p, q and r are defined as the components of the angular velocity of the system with unit-vectors $\{\underline{E}\}$:

$$\underline{\omega} = (p \quad q \quad r)\{\underline{E}\} \tag{I.1.9}$$

The index " × " in the notation of the rotation operator is associated with the "cross-product" of vectors. In general:

$$\underline{a} \times \underline{b} = \begin{bmatrix} 0 & b_z & -b_y \\ -b_z & 0 & b_x \\ b_y & -b_x & 0 \end{bmatrix} (a_x \quad a_y \quad a_z)\{\underline{E}\} \tag{I.1.10}$$

The time derivative of the position vector $\underline{r}$ as given by (I.1.2) is thus written as:

$$\begin{aligned} \underline{\dot{r}} &= L(1 \quad 0 \quad 0)\{\underline{\dot{E}}_1\} + l(1 \quad 0 \quad 0)\{\underline{\dot{E}}_2\} \\ &= L(1 \quad 0 \quad 0)[\omega_{1_\times}][\beta]^T\{\underline{E}_2\} + l(1 \quad 0 \quad 0)[\omega_{2_\times}]\{\underline{E}_2\} \end{aligned} \tag{I.1.11}$$

A vector is unchanged when written in its transposed form:

$$\underline{\dot{r}} = -L\{\underline{E}_2\}^T[\beta][\omega_{1_\times}]\begin{Bmatrix} 1 \\ 0 \\ 0 \end{Bmatrix} - l\{\underline{E}_2\}^T[\omega_{2_\times}]\begin{Bmatrix} 1 \\ 0 \\ 0 \end{Bmatrix} \tag{I.1.12}$$

The product $\underline{\dot{r}}.\underline{\dot{r}}$ follows by multiplying (I.1.11) and (I.1.12). The total kinetic energy of the system follows by writing out the matrix multiplications:

$$T = \frac{1}{2}M.L^2.\dot{\psi}^2 + \frac{1}{2}m\left[\dot{\psi}^2.\{L + l\cos\beta\}^2 + l^2.\dot{\beta}^2\right] \tag{I.1.13}$$

If springs are assumed in both the degrees of freedom, the equations of motion are:

$$\begin{aligned} &\frac{d}{dt}\left(\frac{\partial T}{\partial \dot{\psi}}\right) - \frac{\partial T}{\partial \psi} + \frac{\partial U}{\partial \psi} = M.L^2.\ddot{\psi} + m\{L + l\cos\beta\}^2.\ddot{\psi} - 2.m.\dot{\psi}.\dot{\beta}.\{L + l\cos\beta\}.l\sin\beta + K_\psi.\psi = 0 \\ &\frac{d}{dt}\left(\frac{\partial T}{\partial \dot{\beta}}\right) - \frac{\partial T}{\partial \beta} + \frac{\partial U}{\partial \beta} = m.l^2\ddot{\beta} + m.\dot{\psi}^2\{L + l\cos\beta\}.l\sin\beta + K_\beta(\beta - \beta_0) = 0 \end{aligned} \tag{I.1.14}$$

Linearising β around β_0 gives:

$$
\begin{aligned}
&M.L^2.\ddot{\psi}+m.\ddot{\psi}.(L+l.\cos\beta_0)^2-2m.l.\ddot{\psi}.\Delta\beta.\sin\beta_0.(L+l.\cos\beta_0)\\
&\quad+-2m.l.\dot{\psi}.\Delta\dot{\beta}.\sin\beta_0.(L+l.\cos\beta_0)+K_\psi.\psi=0\\
&m.l^2.\Delta\ddot{\beta}+m.l.\dot{\psi}^2.\sin\beta_0.(L+l.\cos\beta_0)+m.l.\dot{\psi}^2.\Delta\beta.\cos\beta_0.(L+l.\cos\beta_0)\\
&\quad+-m.l^2.\dot{\psi}^2.\Delta\beta.\sin^2\beta_0+K_\beta.\Delta\beta=0
\end{aligned}
\tag{I.1.15}
$$

or alternatively, in order to let the dimensions of these equations of motion correspond with those of the electromagnetic equations:

$$
\begin{aligned}
&M.(\ddot{\psi}.L)+m.(\ddot{\psi}.L).(1+l/L.\cos\beta_0)^2-2m.(l/L).(\ddot{\psi}.L).\Delta\beta.\sin\beta_0.(1+l/L.\cos\beta_0)\\
&\quad+-2m.(l/L).(\dot{\psi}.L).\Delta\dot{\beta}.\sin\beta_0.(1+l/L.\cos\beta_0)+(K_\psi/L^2).(\psi.L)=0\\
&m.l^2.\Delta\ddot{\beta}+m.(l/L).\dot{\psi}^2.\sin\beta_0.(1+l/L.\cos\beta_0)+m.(l/L).(\dot{\psi}.L)^2.\Delta\beta.\cos\beta_0.(1+l/L.\cos\beta_0)+\\
&\quad-m.(l/L)^2.(\dot{\psi}.L)^2.\Delta\beta.\sin^2\beta_0+K_\beta.\Delta\beta=0
\end{aligned}
\tag{I.1.16}
$$

These equations, by an appropriate choice of coefficients, can be made to be the analogon of the equations of motion for a droplet of charge:

$$
\begin{aligned}
&m_{bare}.\ddot{z}_m+m_{em_0}\left(1-\frac{\Delta s}{s_e}\right).\ddot{z}_m+q.k.z_m=m_{em_0}.\dot{z}_m\frac{\Delta\dot{s}}{s_e}.\\
&\left(m^*_{em_0}.s_e^2+m^*_{bare}.s_e^2\right)\frac{\Delta\ddot{s}}{s_e}+\frac{1}{2}m_{em_0}.c^2\mu\tau\frac{\Delta s}{s_e}-m_{em_0}.\frac{\Delta s}{s_e}.\dot{z}_m^2+\frac{1}{2}m_{em_0}.\dot{z}_m^2=0
\end{aligned}
\tag{I.1.17}
$$

The following correspondences must then exist:

$$
\begin{aligned}
\psi.L &\quad\leftrightarrow\quad z_m\\
\Delta\beta &\quad\leftrightarrow\quad \frac{\Delta s}{s_e}
\end{aligned}
\tag{I.1.18}
$$

The requirements for the analogy of the translation equations are furthermore:

$$
\begin{aligned}
&M+m(1+l/L.\cos\beta_0)^2 \quad\leftrightarrow\quad m_{bare}+m_{em_0} \quad (coeff.van\ \ \ddot{\psi}L\ \ en\ \ \ddot{z}_m)\\
&-2m.(l/L).\sin\beta_0.(1+l/L.\cos\beta_0) \quad\leftrightarrow\quad -m_{em_0} \quad (coeff.van\ \ \ddot{\psi}L.\Delta\beta\ \ en\ \ \ddot{z}_m.\Delta s/s_e)\\
&-2m.(l/L).\sin\beta_0.(1+l/L.\cos\beta_0) \quad\leftrightarrow\quad -m_{em_0} \quad (coeff.van\ \ \dot{\psi}L.\Delta\dot{\beta}\ \ en\ \ \dot{z}_m.\Delta\dot{s}/s_e)\\
&\frac{K_\psi}{L^2} \quad\leftrightarrow\quad q.k
\end{aligned}
\tag{I.1.19}
$$

whereas the pulsation equation requires:

$$m.l^2 \quad \leftrightarrow \quad \left(m^*_{emo_0}.s_e^2 + m^*_{bare}.s_e^2\right) \quad (coeff.van \quad \Delta\ddot{\beta} \quad en \quad \Delta\ddot{s}/s_e)$$

$$m.(l/L).\sin\beta_0.(1 + l/L.\cos\beta_0) \quad \leftrightarrow \quad \frac{1}{2}m_{em_0} \quad (coeff.van \quad (\dot{\psi}L)^2 \quad en \quad \dot{z}_m^2)$$

$$m.(l/L).\cos\beta_0.(1 + l/L.\cos\beta_0) - m.(l/L)^2.\sin^2\beta_0 \quad \leftrightarrow \quad -m_{em_0} \quad (coeff.van \quad \Delta\beta.(\dot{\psi}L)^2 \quad en \quad \Delta s/s_e.\dot{z}_m^2)$$

$$K_\beta \quad \leftrightarrow \quad \frac{1}{2}m_{em_0}c^2\mu\tau \quad (coeff.van \quad \Delta\beta \quad en \quad \Delta s/s_e)$$

(I.1.20)

Analogy is possible if the following two equations are satisfied:

$$m(1 + l/L.\cos\beta_0)^2 = 2m.(l/L).\sin\beta_0.(1 + l/L.\cos\beta_0)$$

$$\begin{aligned} &m.(l/L).\cos\beta_0.(1 + l/L.\cos\beta_0) - m.(l/L)^2.\sin^2\beta_0 \\ &\quad + 2m.(l/L).\sin\beta_0.(1 + l/L.\cos\beta_0) \\ &\quad = 0 \end{aligned} \tag{I.1.21}$$

The first of these conditions reads, written in the alternative form:

$$1 + l/L.\cos\beta_0 = 2.(l/L).\sin\beta_0 \tag{I.1.22}$$

This may be substituted into the second condition:

$$2.\cos\beta_0.\sin\beta_0 + 3\sin^2\beta_0 = 0 \tag{I.1.23}$$

or

$$\begin{aligned} \tan\beta_0 &= -\frac{2}{3} \\ \beta_0 &= -33,7^0 \end{aligned} \tag{I.1.24}$$

I.2 A Heuristic Derivation of the Schrödinger's Equation for the Steady State

As explained in Chap. G.1 in the "Note on the nomenclature and the properties of beats", a close relationship exists between on the one hand the beat pattern of the pulsation (i.e. the envelope of the pulsations or the velocity fluctuations) and on the other hand the stroboscopic image obtained by sampling the pulsation, having the frequency $\omega = \omega_0(1 + k\,{v_{av}}^2)$, with the zero-speed frequency ω_0. The stroboscopic image has—apart from a vertical shift—the same shape as the beats, i.e. the strobe

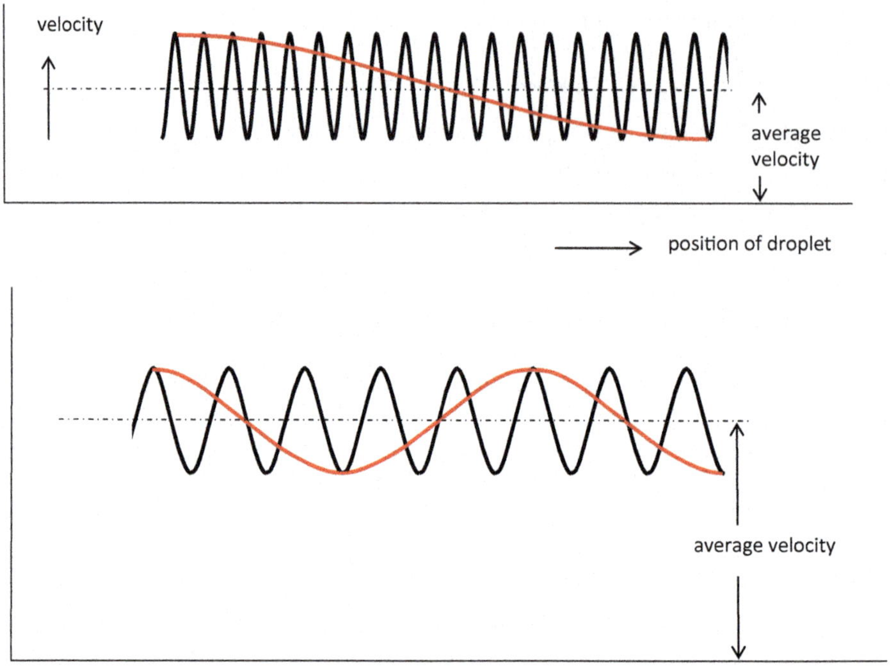

Fig. I.2.1 Comparison of velocity fluctuations and strobe images at different average velocities

image is similar to the shape of the envelope of the pulsations. In the following the stroboscopic image will be used as a tool for the explanation why the beat patterns comply with Schrödinger's equation.

In Fig. I.2.1a comparison is shown between the velocity fluctuations and the strobe images at two different average velocities of the droplet. It is seen that the wavelength of the velocity fluctuations themselves increases with the average velocity, whereas the wavelength of the strobe image becomes smaller when the average velocity increases. The latter is in agreement with De Broglie's relation $p = \frac{h}{\lambda}$.

In Fig. I.2.1 we have the situation of constant average velocity. What one would expect to see in the case of a variable average velocity is sketched in the next Fig. I.2.2.

The upper part of Fig. I.2.2 shows the expected pattern of the velocity fluctuations themselves, if on the left end and on the right end the average velocity decreases to ultimately zero, as would be the case in a potential well.

In the lower part of Fig. I.2.2 the expected stroboscopic pattern is shown. In the middle of the well the "local wavelength" will be smaller than near the boundaries of the well. In fact, according to De Broglie's relation exactly in the reversal points on the walls of the well the "local wavelength" would become infinite. The terminology "local wavelength" was defined in Chap. G.1.

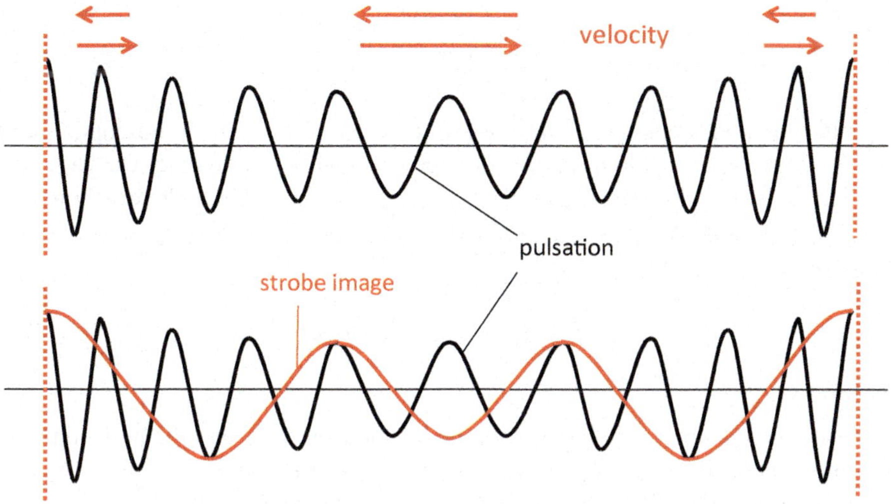

Fig. I.2.2 Expected patterns of velocity fluctuations and stroboscopic image in case the average velocity is continuously variable, e.g. in a parabolic well

In the "Note on the nomenclature and the properties of beats" of Chap. G.1 the "local wavelength" of a curve was defined by using the osculating sine. An osculating sine is the harmonic function that in the point considered has a second order contact with a "near-harmonic" curve.

Let in a point z_1 of the near-harmonic curve $f(z)$ the function value be $f(z_1)$, and the first and second derivatives $f'(z_1)$ resp. $f''(z_1)$.The requirements on the osculating sine, given by $g(z) = g_0 \cos\left(2\pi\frac{z}{\lambda} + \varepsilon\right)$ are:

$$g(z_1) = g_0 \cos\left(2\pi\frac{z_1}{\lambda} + \varepsilon\right) = f(z_1) \tag{I.2.1}$$

$$g'(z_1) = -g_0 \cdot \left(\frac{2\pi}{\lambda}\right) \sin\left(2\pi\frac{z_1}{\lambda} + \varepsilon\right) = f'(z_1) \tag{I.2.2}$$

$$g''(z_1) = -g_0 \left(\frac{2\pi}{\lambda}\right)^2 \cos\left(2\pi\frac{z_1}{\lambda} + \varepsilon\right) = f''(z_1) \tag{I.2.3}$$

from which the wavelength of the osculating sine follows:

$$\left(\frac{2\pi}{\lambda}\right)^2 = -\frac{f''(z_1)}{f(z_1)} \tag{I.2.4}$$

By definition the local wavelength $\lambda(z_1)$ in the point z_1 of the curve $f(z)$ is equal to the wavelength of the osculating sine, so that:

$$\left(\frac{2\pi}{\lambda(z_1)}\right)^2 = -\frac{f''(z_1)}{f(z_1)} \tag{I.2.5}$$

If the curve $f(z)$ is the red curve of Fig. I.2.2, i.e. if it is the stroboscopic image of the velocity fluctuations of the droplet one can, by assuming De Broglie's relation to be locally valid, equate $\lambda(z_1)$ to $\frac{h}{p(z_1)}$. Because this relation should be valid in each point of the curve $f(z)$, the following differential equation will be satisfied by $f(z)$:

$$\hbar^2 f''(z) + p^2(z).f(z) = 0 \tag{I.2.6}$$

where $\hbar = \frac{h}{2\pi}$.

Using the relation $E_k = \frac{p^2}{2m}$ between the kinetic energy E_k and the momentum p, and the relation $E = E_p + E_k$ between kinetic energy E_k, potential energy E_p and total energy E, this differential equation may be brought in the form

$$-\frac{\hbar^2}{2m} f''(z) + E_p(z).f(z) = E.f(z) \tag{I.2.7}$$

which is the same as the 1-D time-independent Schrödinger equation. "Time-independent" is perhaps an inadequate term, because it in reality is the equation for the spatial part of the wave function, obtained from the time-dependent Schrodinger equation by the process of separation of variables.

Appendix J
The Radiation Field

J.1 The Far Field of the Scalar Potential of the Entire Droplet

The far field of the scalar potential $\phi_{far}(x, y, z, t)$ induced by a point charge dq which at the *retarded* time $t - \rho/c$ (with $\rho = \sqrt{x^2 + y^2 + z^2}$) is in the position (x_0, y_o, z_o) and has a velocity in z-direction $v = \dot{z}_0$, was determined in section C.3:

$$\begin{aligned}\phi_{far}\cdot\frac{4\pi\varepsilon_0}{dq} &= \frac{1}{\rho} + \frac{1}{2}\frac{\partial^2}{\partial x^2}\left(\frac{x_0^2}{\rho}\right) + \frac{1}{2}\frac{\partial^2}{\partial y^2}\left(\frac{y_0^2}{\rho}\right) + \frac{1}{2}\frac{\partial^2}{\partial z^2}\left(\frac{z_0^2(t-\rho/c)}{\rho}\right)\\ &+ -\frac{\partial}{\partial x}\left(\frac{x_0}{\rho}\right) - \frac{\partial}{\partial y}\left(\frac{y_0}{\rho}\right) - \frac{\partial}{\partial z}\left(\frac{z_0(t-\rho/c)}{\rho}\right) + \frac{\partial^2}{\partial x.\partial y}\left(\frac{x_0.y_0}{\rho}\right) + \frac{\partial^2}{\partial x.\partial z}\left(\frac{x_0.z_0(t-\rho/c)}{\rho}\right) + \frac{\partial^2}{\partial y.\partial z}\left(\frac{y_0.z_0(t-\rho/c)}{\rho}\right)\\ &\text{up to } O(a/\lambda)^3\end{aligned} \tag{J.1.1}$$

What we are especially interested in, is the part of the outward travelling energy flux which falls off with distance like $1/\rho^2$. Terms in the energy flux that diminish faster with distance cannot contribute to the energy loss from the system by radiation. Later it will appear that for this purpose we are only interested in terms of order $O(1/\rho)$ in the far field ϕ_{far}. Such terms are contributed only by the retardation effects, as may be seen by considering for instance the term

$$-\frac{\partial}{\partial z}\left(\frac{z_0(t-\rho/c)}{\rho}\right) = \left[\frac{z_0(t-\rho/c)}{\rho^2} + \frac{\dot{z}_0(t-\rho/c)}{c}\frac{1}{\rho}\right]\frac{z}{\rho} \tag{J.1.2}$$

where subsequently may be substituted $\dot{z}_0(t - \rho/c) = v(t - \rho/c)$ and $z/\rho = \cos\varphi$.

The latter change from a Cartesian system of coordinates to a spherical system shows more clearly that only the second term between the square brackets, originating from the differentiation of the retardation, contributes a term of order $O(1/\rho)$.

For the purpose of an examination of all the poles in expression (J.1.1) the definitions used in the spherical system of coordinates are repeated here.

T. van Holten, *The Atomic World Spooky? It Ain't Necessarily So!*
DOI 10.2991/978-94-6239-234-2

The transformation between the Cartesian system and the spherical system is

$$x = \rho.\sin\varphi.\cos\chi$$
$$y = \rho.\sin\varphi.\sin\chi$$
$$z = \rho.\cos\varphi$$

as shown in Fig. J.1.1:

We then find:

$$\phi_{far}\frac{4\pi\varepsilon_0}{dq} = \frac{1}{\rho} + \frac{v^2 + z_0.\dot{v}}{c^2}.\frac{cos^2\varphi}{\rho} + \frac{v}{c}\frac{cos\varphi}{\rho} + x_0\frac{\dot{v}}{c^2}.\frac{sin\varphi\cos\varphi\cos\chi}{\rho} + y_0\frac{\dot{v}}{c^2}.\frac{sin\varphi\cos\varphi\sin\chi}{\rho} + O\left(\frac{1}{\rho^2}\right) \quad (J.1.3)$$

where $\phi_{far}(\rho, \varphi, t)$ is determined at time t, and those variables in the right hand side that depend on time are to be taken at the retarded time: $z_0(t-\rho/c)$, $v(t-\rho/c)\, and\, \dot{v}(t-\rho/c)$.

This is the contribution of one charge element of the complete droplet. To find the entire far field an integration is performed over all the charge elements. The last two terms in Eq. (J.1.3) depending on x_0 and y_0 can immediately be neglected, because of the rotational symmetry of the droplet. For the purpose of an integration of the other terms the expression is changed from the Lagrangian to the Eulerian description, using

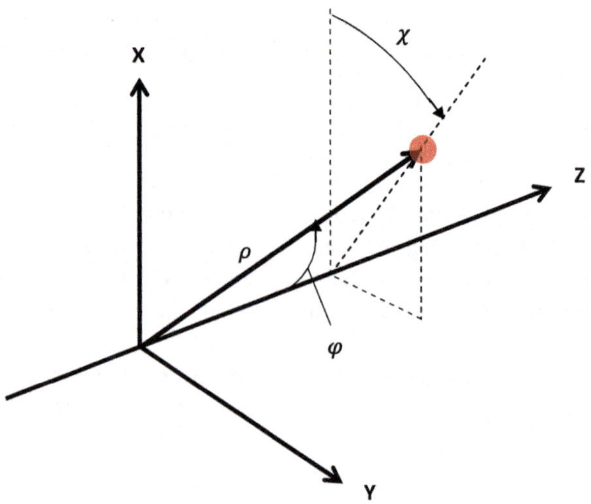

Fig. J.1.1 Spherical coordinate system used in far field

$$dq \xrightarrow{\text{Eulerian description}} \sigma.dx_0.dy_0.dz_0 \tag{J.1.4}$$

$$\begin{aligned} v(t-\rho/c) &\xrightarrow{\text{Eulerian description}} v(z_0, t-\rho/c) \\ &= v_m(t-\rho/c) + \frac{\dot{s}}{s}(t-\rho/c).[z_0 - z_m(t-\rho/c)] \end{aligned} \tag{J.1.5}$$

Using some further simplifications explained in section D.4, the other time dependent terms are replaced by (see Eq. (D.4.4))

$$\begin{aligned} v^2(t-\rho/c) &\xrightarrow{\text{Eulerian description}} v_m^2(t-\rho/c) \\ &+ 2.v_m(t-\rho/c).\frac{\dot{s}}{s}(t-\rho/c).[z_0 - z_m(t-\rho/c)] \end{aligned} \tag{J.1.6}$$

$$\begin{aligned} \dot{v}(t-\rho/c) &\xrightarrow{\text{Eulerian description}} \frac{Dv(z_0, t-\rho/c)}{Dt} \\ &= \dot{v}_m(t-\rho/c) + \frac{\ddot{s}}{s}(t-\rho/c).[z_0 - z_m(t-\rho/c)] \end{aligned} \tag{J.1.7}$$

Taking into account the symmetry of the droplet, as well as the assumption $s/a \ll 1$ we finally obtain (only retaining terms of order $O(1/\rho)$:

$$\Phi_{far}(\rho, \varphi, t) = \frac{q}{4\pi\varepsilon_0}\left[\frac{1}{\rho} + \frac{v_m}{c}\frac{\cos\varphi}{\rho} + \frac{v_m^2 + z_m.\dot{v}_m}{c^2}\frac{\cos^2\varphi}{\rho}\right]_{t-\rho/c} \tag{J.1.8}$$

J.2 The Far Field of the Vector Potential of the Entire Droplet

According to Chap. C.5, Eq. (C.5.5) the far field of the vector potential as induced by a single point charge is:

$$\begin{aligned} \frac{a_{z_{far}}(x,y,z,t)}{dq/(c^2.4\pi\varepsilon_0)} &= \frac{v(t-\rho/c)}{\rho} \\ &+ -\frac{\partial}{\partial x}\left(\frac{x_0.v(t-\rho/c)}{\rho}\right) - \frac{\partial}{\partial y}\left(\frac{y_0.v(t-\rho/c)}{\rho}\right) - \frac{\partial}{\partial z}\left(\frac{z_0(t-\rho/c).v(t-\rho/c)}{\rho}\right) \end{aligned} \tag{J.2.1}$$

Working out the differentiations and transforming to a spherical coordinate system:

$$\frac{a_{z_{far}}}{dq/(c^2.4\pi\varepsilon_0)} = \frac{v}{\rho} + \frac{x_0.\dot{v}\, sin\varphi.cos\chi}{c\qquad\rho} + \frac{y_0.\dot{v}\, sin\varphi.sin\chi}{c\qquad\rho} + \frac{v^2 + z_0\dot{v}\, cos\varphi}{c\qquad\rho} + O\left(\frac{1}{\rho^2}\right) \tag{J.2.2}$$

Integrating over all the elements of the droplet, and neglecting irrelevant terms as has been done in the case of the scalar potential gives the following final result:

$$A_{z_{far}}(\rho, \varphi.t) = \frac{q}{4\pi\varepsilon_0}\frac{1}{c^2}\left[\frac{v_m}{\rho} + \frac{v_m^2 + z_m.\dot{v}_m\, cos\varphi}{c\qquad\rho}\right]_{t-\rho/c} \tag{J.2.3}$$

J.3 Field Strengths in the Far Field

The vectorpotential $\underline{A}$ has no other components than the Z-component, and is thus written in terms of spherical coordinates as

$$\underline{A} = A_z\,\underline{k} = A_z\left(\underline{a}_\rho \cos\varphi - \underline{a}_\varphi \sin\varphi\right) \tag{J.3.1}$$

where $\underline{k}$, $\underline{a}_\rho, \underline{a}_\varphi$ are unit vectors in Z-direction, resp. in the direction of increasing ρ or φ.

Taking into account that the potentials are rotationally symmetric around the Z-axis, the magnetic field strength $\underline{B}$ is given by

$$\underline{B} = rot\underline{A} = \underline{a}_\chi \frac{1}{\rho}\left[\frac{\partial(\rho.A_\varphi)}{\partial\rho} - \frac{\partial A_\rho}{\partial\varphi}\right] \tag{J.3.2}$$

Substituting the following relations, according to Eq. (J.3.1)

$$A_\varphi = -A_z \sin\varphi \tag{J.3.3}$$

$$A_\rho = A_z \cos\varphi \tag{J.3.4}$$

we find

$$\underline{B} = -\underline{a}_\chi\left[\frac{\partial A_z}{\partial\rho} sin\varphi + \frac{\partial A_z}{\partial\varphi}\frac{cos\varphi}{\rho}\right] \tag{J.3.5}$$

where Eq. (J.2.3) must be substituted for A_z.

When determining the derivative w.r.t. ρ in this expression one should be careful, in view of the fact that the terms in the r.h.s. of Eq. (J.2.3) are retarded, which gives an additional term upon differentiation.

The electric field strength is

$$\underline{E} = -grad\,\Phi - \frac{\partial \underline{A}}{\partial t} = -\underline{a}_\rho \frac{\partial \Phi}{\partial \rho} - \underline{a}_\varphi \frac{1}{\rho}\frac{\partial \Phi}{\partial \varphi} - \frac{\partial \underline{A}}{\partial t} \tag{J.3.6}$$

Substituting $\underline{A}$ according to Eq. (J.3.1):

$$\underline{E} = -\underline{a}_\rho \left[\frac{\partial \Phi}{\partial \rho} + \frac{\partial A_z}{\partial t}\cos\varphi\right] - \underline{a}_\varphi \left[\frac{1}{\rho}\frac{\partial \Phi}{\partial \varphi} - \frac{\partial A_z}{\partial t} sin\varphi\right] \tag{J.3.7}$$

J.4 Poynting Vector

The Poynting vector $\underline{S}$ gives the flow of energy density in the field. It is defined as

$$\underline{S} = \varepsilon_0 c^2.\underline{E} \times \underline{B} \tag{J.4.1}$$

Due to the rotational symmetry of the fields we have in our case:

$$\underline{S} = \varepsilon_0 c^2.\begin{vmatrix} \underline{a}_\rho & \underline{a}_\varphi & \underline{a}_\chi \\ E_\varrho & E_\varphi & 0 \\ 0 & 0 & B_\chi \end{vmatrix} = \varepsilon_0 c^2.B_\chi\left[\underline{a}_\varrho\, E_\varphi - \underline{a}_\varphi E_\varrho\right] \tag{J.4.2}$$

The energy radiated outwards, away from the charge configuration is then given by the radial component of the Poynting vector:

$$S_\varrho = \varepsilon_0 c^2.B_\chi.E_\varphi = \varepsilon_0 c^2.\left[\frac{\partial A_z}{\partial \rho} sin\varphi + \frac{\partial A_z}{\partial \varphi}\frac{cos\varphi}{\rho}\right].\left[\frac{1}{\rho}\frac{\partial \Phi}{\partial \varphi} - \frac{\partial A_z}{\partial t} sin\varphi\right] \tag{J.4.3}$$

If we want to calculate the total energy which is radiated to infinity, only the terms contribute that fall off with distance as $1/\varrho^2$. Terms falling off faster with distance do not contribute to the flow of energy density integrated over a spherical surface surrounding the charge configuration.

Because S_ϱ consists of the product of two terms, in each bracketed term separately we have to search for terms falling off with distance as $1/\varrho$.

Working out the individual terms in Eq. (J.4.3):

$$\frac{\partial \Phi}{\partial \varphi}\frac{4\pi\varepsilon_0}{q} = -\frac{v_m}{c}\frac{sin\varphi}{\rho} - \frac{v_m^2 + z_m.\dot{v}_m}{c^2}\frac{2\cos\varphi\,\sin\varphi}{\rho} + O\left(\frac{1}{\rho^2}\right) \tag{J.4.4}$$

$$\frac{\partial A_z}{\partial \rho}\frac{(c^2.4\pi\varepsilon_0)}{q} = -\frac{\dot{v}_m}{c}\frac{1}{\rho} - \frac{3v_m\,\dot{v}_m + z_m\ddot{v}_m}{c^2}\frac{\cos\varphi}{\rho} + O\left(\frac{1}{\rho^2}\right) = -\frac{1}{c}\frac{\partial A_z}{\partial t} + O\left(\frac{1}{\rho^2}\right) \tag{J.4.5}$$

$$\frac{\partial A_z}{\partial \varphi}\frac{(c^2.4\pi\varepsilon_0)}{q} = -\frac{v_m^2 + z_m.\dot{v}_m}{c}\frac{\cos\varphi}{\rho} + O\left(\frac{1}{\rho^2}\right) \tag{J.4.6}$$

$$\frac{\partial A_z}{\partial t}\frac{(c^2.4\pi\varepsilon_0)}{q} = \frac{\dot{v}_m}{\rho} + \frac{3v_m\,\dot{v}_m + z_m\,\ddot{v}_m}{c}\frac{\cos\varphi}{\rho} + O\left(\frac{1}{\rho^3}\right) \tag{J.4.7}$$

and collecting:

$$S_\rho = \frac{1}{c^3}\frac{q^2}{4\pi\varepsilon_0}\frac{\sin^2\varphi}{4\pi\rho^2}(\dot{v}_m)^2\left[1 + 3\frac{v_m + z_m\ddot{v}_m/\dot{v}_m}{c}\cos\varphi\right]^2 + O\left(\frac{1}{\rho^3}\right) \tag{J.4.8}$$

The second term between the square brackets is strictly speaking asymptotically of an irrelevant order of magnitude. It is shown, because it does agree with exact theories showing that the radiation pattern with increasing velocity tilts forward in the direction of motion.

Integrating over the entire surface of a sphere concentric with the origin and neglecting the additional term between the square brackets, the total energy radiated per unit of time is:

$$\frac{dE_{rad}}{dt} = \frac{2}{3}(\dot{v}_m)^2\frac{1}{c^3}\frac{q^2}{4\pi\varepsilon_0}\bigg|_{t-\rho/c} \tag{J.4.9}$$

This result is in agreement with the formula by Larmor for the radiated energy by a harmonically oscillating dipole. Note, however, that in the above given derivation *no* assumptions have been made about the type of motion (whether the time history of the acceleration is harmonic or not). The expression (J.4.9) is according to the present theory *instantaneously* valid (as opposed to the classical result by Larmor) with the only restriction that the theory assumed a droplet of charge which is restrained to move in a small region around the origin. No further restrictions apply.

Another interesting result is, that the next term in S_ϱ, of order $O\left(\frac{1}{\rho^3}\right)$, has a much larger coefficient (the term $\frac{1}{c^2}$ is lacking in the product of the derivative of the scalar potential). It means that somewhat closer to the origin this term will dominate the term of $O\left(\frac{1}{\rho^2}\right)$, although it falls off faster with distance and is thus unable to contribute to the total loss of energy by radiation.

Finally note that the pulsation of the droplet does not contribute at all to the energy loss, at least not in the present asymptotic approximation.

Appendix K
Combining the Droplet's Motion and the Radiation Field: Bohr's Radio Silence

K.1 Balance of Energy in the Field (Outgoing Waves)

The conclusion that Eq. (J.4.9) is instantaneously valid and that its applicability is not restricted to harmonic oscillations of the charge (this in contrast to Larmor's theory), allows an analysis of the energy balance in the field which is more detailed than usually found in the literature. Let us, nevertheless, start a discussion of the energy balance by considering harmonic motion first.

The equation of motion of the droplet is written as (leaving out for convenience the index m of "mid point")

$$m.\dot{v}(t) - \gamma.\ddot{v}(t) = F(t) \tag{K.1.1}$$

with

$$\gamma = \frac{2}{3}\frac{1}{c^3}\frac{q^2}{4\pi\varepsilon_0} \tag{K.1.2}$$

It has been assumed that the mass is constant, implying that the effects due to the mass variations are absorbed in the definition of the external force F.

The work per second performed by the external force, denoted as the power P, is:

$$P = F.v = (m.\dot{v} - \gamma.\ddot{v}).v = d/dt\left(\frac{1}{2}m\,v^2\right) - \gamma\,v\,\ddot{v} \tag{K.1.3}$$

The first term in the r.h.s. equals the time rate of change of the kinetic energy of the droplet. The second term will be called the power $P_{rad.resistance}$ associated with the radiation resistance $-\gamma\,\ddot{v}$. This power can be written as the sum of two contributions:

$$P_{rad.resistance} = -\gamma\,v\,\ddot{v} = -\frac{d^2}{dt^2}\left(\frac{1}{2}\gamma\,v^2\right) + \gamma.\,\dot{v}^2 \tag{K.1.4}$$

The last term $\gamma.\,\dot{v}^2$ also has a clear physical meaning. We take a spherical control volume centered in the origin, and sufficiently large so that the outer surface with

T. van Holten, *The Atomic World Spooky? It Ain't Necessarily So!*
DOI 10.2991/978-94-6239-234-2

the radius ρ is in the far field. The energy lost from this system by radiation, at time t passing the surface of the control volume, is related to the acceleration of the droplet at some earlier time:

$$\frac{dE_{rad}}{dt}(t) = \gamma . \dot{v}^2 (t - \rho/c) \tag{K.1.5}$$

A causal interpretation of the term $\gamma . \dot{v}^2$ in the expression for $P_{rad.resistance}$ would be, to consider the droplet as the source of the radiated energy. At time t, this source injects per second an amount of energy $\gamma . \dot{v}^2$ into the control volume, which will subsequently leave the volume at a later time $t + \rho/c$ on its way to infinity. The power $\gamma . \dot{v}^2$ in $P_{rad.resistance}$ will accordingly be described as the "radiation power" P_{rad}, not to be confused with $\frac{dE_{rad}}{dt}$: P_{rad} denotes a part of $P_{rad.resistance}$ at a certain time, whilst $\frac{dE_{rad}}{dt}$ stands for the non-returning energy escaping through a surface $\rho = constant$ at a later time.

The expression for the power $P_{rad.resistance}$ contains another term $-\frac{d^2}{dt^2}\left(\frac{1}{2}\gamma v^2\right)$. A physical interpretation is possible by assuming that the electromagnetic field, apart from transmitting energy to infinity, also acts as an accumulator of energy, which is able to absorb and release energy. The part $-\frac{d^2}{dt^2}\left(\frac{1}{2}\gamma v^2\right)$ of the power $P_{rad.resistance}$ can thus be considered as an external powersource which feeds the storage of energy in the field or taps from it. Accordingly, this part will be called the storage power $P_{storage} = -\frac{d^2}{dt^2}\left(\frac{1}{2}\gamma v^2\right)$.

Equation (K.1.4) thus reads, in more descriptive notations:

$$P_{rad.resistance} = P_{rad} + P_{storage} \tag{K.1.6}$$

The three quantities occurring in this equation are depicted in Fig. K.1.1 as time functions, for the special case of harmonic motion of the droplet:

$$v(t) = V \cos \omega t \tag{K.1.7}$$

$$\dot{v}(t) = -V\omega \sin \omega t \tag{K.1.8}$$

$$\ddot{v}(t) = -V\omega^2 \cos \omega t \tag{K.1.9}$$

from which:

$$P_{rad.resistance} = -\gamma v \ddot{v} = \gamma V^2 \omega^2 \cos^2 \omega t \tag{K.1.10}$$

$$P_{rad} = \gamma . \dot{v}^2 = \gamma V^2 \omega^2 \sin^2 \omega t \tag{K.1.11}$$

$$P_{storage} = -\frac{d^2}{dt^2}\left(\frac{1}{2}\gamma v^2\right) = \gamma V^2 \omega^2 \cos(2\omega t) \tag{K.1.12}$$

In the Fig. K.1.1 the three powers are sketched.

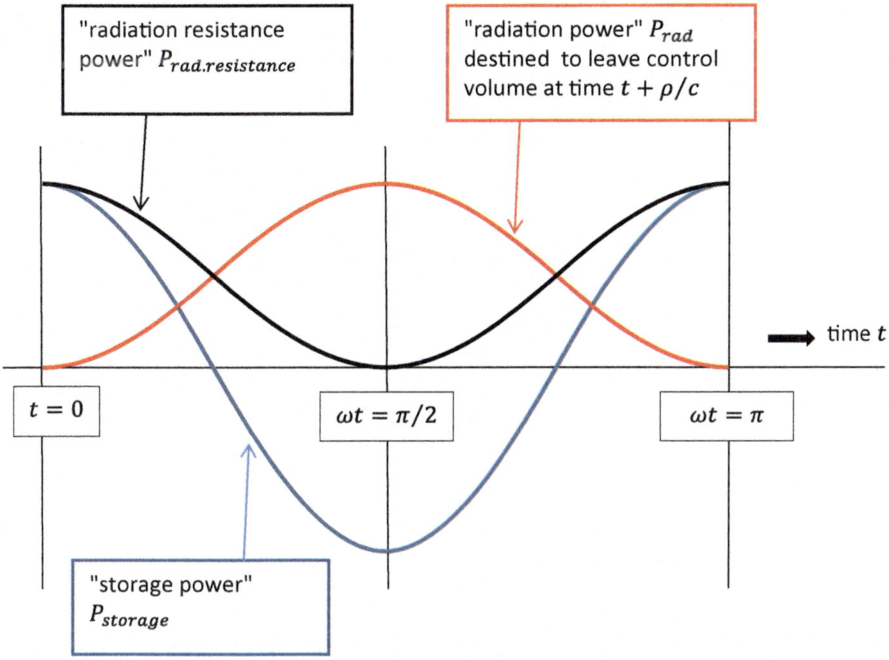

Fig. K.1.1 Break-down of the power, sinusoidal motion

If we now consider the energy balance of a control volume within a sphere $\rho = constant$, we get to see a different picture. Per second an amount of energy is injected into it equal to the radiation resistance power $P_{rad.resistance}$ of the droplet. An amount of energy (per second) $\frac{dE_{rad}}{dt}$ is lost through its outer surface. This amount equals the value of P_{rad} of the droplet *at an earlier time*. The concept of energy stored in the field implies that through the surface of a controlvolume also a fraction of the stored power is transferred to the next larger controlvolume. According to the analysis of the Poynting vector, which gives the flux of energy in the field, the energy transfer associated with storing energy drops off with distance like $1/\rho^3$, i.e. faster than the flux associated with radiation. We assume that the transfer of storage energy becomes relatively negligible if we take the controlvolume under consideration to be sufficiently large.

The energy balance of the control volume thus becomes as sketched below in Fig. K.1.2.

It is seen that the radiation loss through the outer surface to infinity does not equal the radiation power by the droplet at all times, in fact in general it does not. Most of the time the power radiated to infinity is partially "driven" by the energy stored in the field. This is especially clearly seen at $t = \frac{\pi}{2}\frac{1}{\omega}$, where the power input is zero, whilst the rate of energy storage is negative in combination with an equal positive radiation loss.

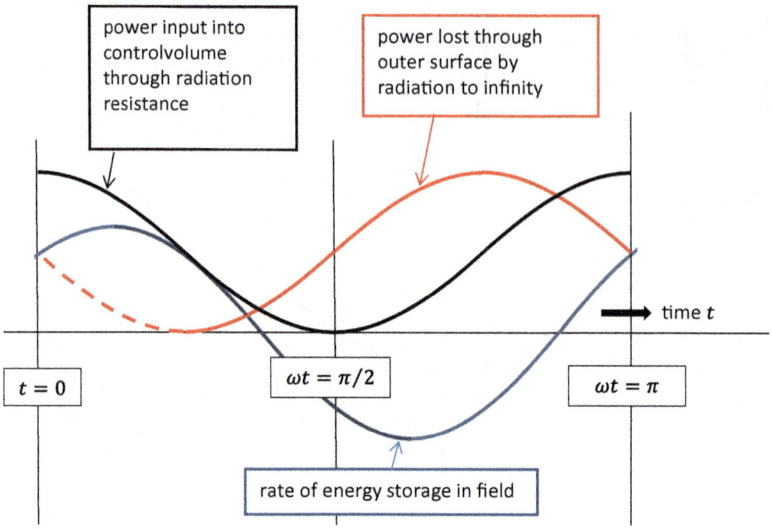

Fig. K.1.2 Alternative break-down of power, sinusoidal motion

K.2 Transient Phenomena and Non-causality

What should be emphasised is, that the above given results are valid only for a harmonic motion which continues unchanged between $t = -\infty$ and $t = +\infty$. If the motion would start from rest at some time later than $t = -\infty$, the results become quite different from those shown.

This is shown in the Fig. K.2.1. It has here been assumed that the motion of the droplet starts at time $t = \frac{\pi}{2}\frac{1}{\omega}$. The energy input into a spherical control volume then has a time history as shown by the black line. The power going out (red line) remains zero as long as the radiation wave has not reached a distance ρ from the origin. At the moment the wave has reached the surface of the control volume, the power lost by radiation jumps to the same value as in the earlier figure concerning the continuous motion.

The blue line, representing the rate of energy accumulation in the field (equal to the difference between energy in and out) then also has a shape quite different from the one in the earlier Fig. K.1.2, until the time $t = \frac{3}{4}\pi\frac{1}{\omega}$. After this point in time the situation *seems* to have returned to the same situation as shown for the continuous motion. In fact, it has not. The rate of accumulation may be the same, the amount of accumulated energy itself will still have a different value, and this stays so when the time proceeds, i.e. for $t > \frac{3}{4}\pi\frac{1}{\omega}$ up to infinity.

The latter means that Fig. K.2.1 cannot be the complete story. On physical grounds one would expect that the situation should gradually return to that sketched in Fig. K.1.2 if the transient start up process is sufficiently far in the past. One would expect that the transient process would involve additional phenomena,

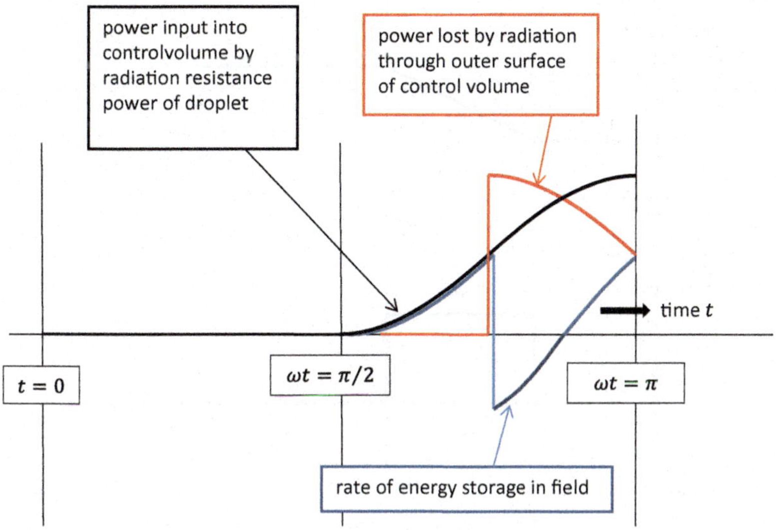

Fig. K.2.1 Break-down of power, start-up of sinusoidal motion

coming on top of the quasi-steady processes, and damping out gradually in such a way that, after a sufficiently long time interval, one comes back *exactly* to the steady situation. According to the present theory this is not the case, contrary to physical expectations. The conclusion is, that the present theory is probably not complete yet. A more detailed investigation of transient phenomena is clearly warranted.

As a simplified model of a transient motion we take a motion with constant velocity (v_1) until t_1 (Fig. K.2.2). From the time t_1 until t_2 there is a constant acceleration up to the velocity v_2, after which point in time the velocity is kept constant again.

The interesting aspect of this kind of motion is, that during the constant acceleration phase the jerk is zero, so that there is no radiation resistance. Seemingly in contradiction with this is that the radiation itself, which is proportional to $(\text{acceleration})^2$, is not zero during this time interval.

Nevertheless, we can show that energy is conserved, by looking more carefully at the phenomena at exactly the points in time t_1 and t_2. The magnitude of the acceleration is $a = (v_2 - v_1)/(t_2 - t_1)$, and can be represented as a time function by $a[H(t - t_1) - H(t - t_2)]$, where H denotes Heaviside's unit step function. The jerk equals $da/dt = a.\delta(t - t_1) - a.\delta(t - t_2)$, where δ denotes the delta-function.

The force on the droplet, effecting this motion, is as before:

$$F = m\dot{v} - \gamma\ddot{v} \tag{K.2.1}$$

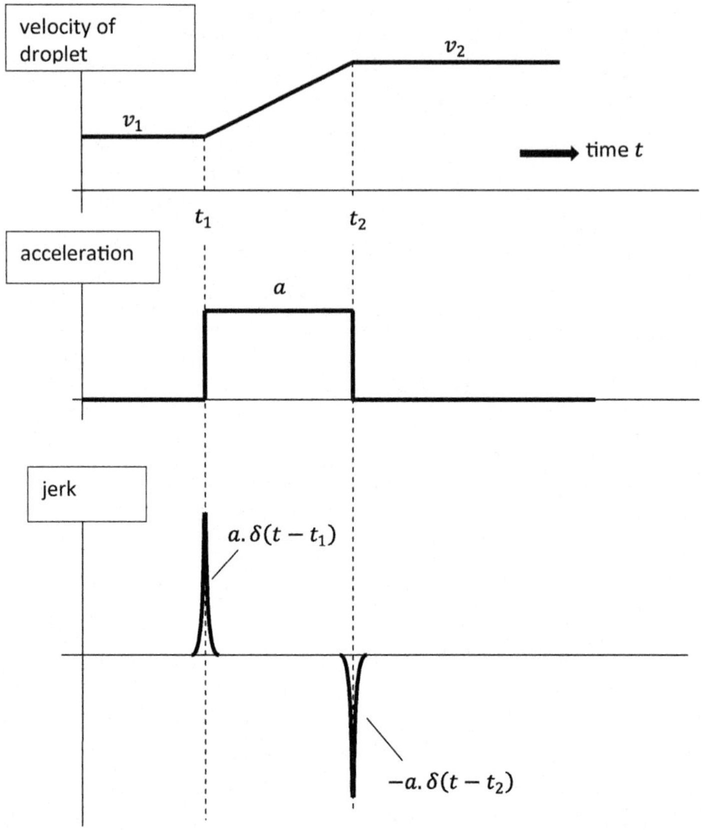

Fig. K.2.2 Simple transient motion

The work done by the external force during $t = -\infty$ to $t = +\infty$ is

$$\begin{aligned} W &= \int_{-\infty}^{+\infty} F.v.dt = \int_{-\infty}^{+\infty} (m\dot{v} - \gamma\ddot{v})v.dt \\ &= \frac{1}{2}m\left(v_2^2 - v_1^2\right) - \gamma.a \int_{-\infty}^{+\infty} v[\delta(t-t_1) - \delta(t-t_2)]dt \\ &= \frac{1}{2}m\left(v_2^2 - v_1^2\right) - \gamma.a(v_1 - v_2) = \frac{1}{2}m\left(v_2^2 - v_1^2\right) + \gamma.a^2(t_2 - t_1) \end{aligned} \tag{K.2.2}$$

Now, consider a spherical control surface in the far field with radius ρ. The energy per second flowing outwards through this surface is $\frac{dE_{rad}}{dt} = \gamma\dot{v}^2(t - \rho/c)$, so that the total energy escaping to infinity is

$$E_{rad} = \gamma \int_{-\infty}^{+\infty} \dot{v}^2(t - \rho/c).dt = \gamma.a^2(t_2 - t_1) \tag{K.2.3}$$

We see that the energy balance is correct: the work by the external force is found back as the sum of the increase of kinetic energy of the droplet and the energy lost by radiation.

Nevertheless, there is something wrong with causality. If we consider the external force needed to realise this kind of motion, the variation of it in time is:

$$F(t) = m\,\dot{v}(t) - \gamma\,\ddot{v}(t) = m\,a[H(t - t_1) - H(t - t_2)] - \gamma a.[\delta(t - t_1) - \delta(t - t_2)] \tag{K.2.4}$$

At time $t = t_1$ this is a *negative* impulsive function. At the very moment the acceleration is started, the force has a direction opposite to the acceleration.

During the acceleration phase itself, the force has a normal behaviour again: it has a magnitude just capable to increase the kinetic energy.

At the end of the accelerating phase, when the motion returns to a constant velocity, the force injects a positive impulse into the system in the direction of the velocity, which again goes against physical expectation.

The cause of the trouble is the peculiar form of the "damping" term in the equation of motion:

$$F = m\,\dot{v} - \gamma\,\ddot{v} \tag{K.2.1}$$

We tend to think about the term $-\gamma\,\ddot{v}$ as "just another form of damping" comparable with the more usual term $+D\,v$ for damping. This idea finds its origin in most textbooks where just a continuous harmonic motion is considered without starting transients. In the case of a harmonic motion without a starting transient we assume that the force $F(t)$ is varying in time in such a way, that harmonic motion results: $v = V\,\cos\omega t$. In that case $\ddot{v} = -V\,\omega^2\,\cos\omega t = -\omega^2\,v$, and the equation of motion becomes:

$$m\,\dot{v} + \gamma\,\omega^2\,v = F \quad (harmonic\,motion) \tag{K.2.5}$$

so that the radiation resistance term indeed acts like a normal damping/dissipation term. The time variation of the force $F(t)$ is also harmonic, as we find in the usual dynamics of bodies, so that no suspicions are raised.

However, if we now consider the transient case where $F = 0$ for $t \leq 0$ after which $F(t)$ jumps to a harmonic variation, the term $-\gamma\,\ddot{v}$ in the equation of motion $F = m\,\dot{v} - \gamma\,\ddot{v}$ plays havoc.

The usual procedure to calculate the transient motion would be to determine the particular solution and add the solution of the homogeneous equation $m\,\dot{v} - \gamma\,\ddot{v} = 0$ multiplied by a suitable constant, which is chosen such that the initial conditions during the starting-up process are satisfied. For a usual dynamical system the contribution by the homogeneous solution damps out in time, so that we after some time come back to the same harmonic motion as found for the motion without transients.

If one would adhere to this usual procedure in the case of the equation of motion $F = m\,\dot{v} - \gamma\,\ddot{v}$, we find as solution of the homogeneous equation

$$v(t) = v(0).\,e^{(m/\gamma).t} \tag{K.2.6}$$

which does not damp out, but on the contrary grows infinitely large for $t \to +\infty$.

In the older literature about Lorentz's electron model (by e.g. Abraham, see for a full review Yaghjian's more recent book "Relativistic Dynamics of a Charged Sphere") this behaviour was called the "runaway motion" of an electron, which is clearly not in agreement with actual observations.

An attempt to better conform to physical reality would be, to consider both the time domains $t<0$ and $t\geq 0$, and to require that $F = 0\;(t<0)$ whereas the force at time $t = 0$ may jump to a harmonic variation in the domain $t\geq 0$. To include the jump itself, we have to add the time domain $t<0$ in the analysis. In the domain $t\geq 0$ we now assume that the solution of the homogeneous equation is multiplied by an integration constant equal to zero, to avoid the runaway motion. In the domain $t<0$ we also determine the solution, and require it to "fit" with the solution for $t\geq 0$.

First consider the solution for $t\geq 0$. We here have a harmonically forced motion, and the particular solution of the equation of motion will also be harmonic. If the position variation $z(t)$ of the droplet is taken such that it initially is $z(0) = 0$, then:

$$z(t) = Z\sin\omega t \tag{K.2.7}$$

$$v(t) = Z\,\omega\cos\omega t \tag{K.2.8}$$

$$\dot{v}(t) = -Z\,\omega^2\sin\omega t \tag{K.2.9}$$

$$\ddot{v}(t) = -Z\,\omega^3\cos\omega t \tag{K.2.10}$$

The force $F(t)$ causing this motion is thus given by

$$F(t) = m\,\dot{v} - \gamma\,\ddot{v} = m\left(-Z\,\omega^2\sin\omega t\right) - \gamma\left(-Z\,\omega^3\cos\omega t\right) \quad (t\geq 0) \tag{K.2.11}$$

The force is indeed harmonic, and it jumps in $t = 0$ from $F\uparrow 0 = 0$ to the value $F\downarrow 0 = \gamma\,Z\,\omega^3$.

In the interval $-\infty<t<0$ the force is required to be zero at all times. The equation of motion in this interval is then

$$m\,\dot{v} - \gamma\,\ddot{v} = 0 \quad (t<0) \tag{K.2.12}$$

with the solution

$$\dot{v}(t) = a.e^{(m/\gamma)t} \tag{K.2.13}$$

Integrating w.r.t. to time gives:

$$v(t) = \frac{a}{(m/\gamma)} e^{(m/\gamma)t} + b \tag{K.2.14}$$

$$z(t) = \frac{a}{(m/\gamma)^2} e^{(m/\gamma)t} + b.t + c \tag{K.2.15}$$

In these expressions the integration constants a, b and c must be chosen such that the physical boundary conditions are satisfied.

It follows immediately that $b = 0$ to avoid that the position of the droplet is $z = -\infty$ for $t \to -\infty$.

We must furthermore require that the position of the droplet is not discontinuous at $t = 0$, from which we obtain

$$c = -\frac{a}{(m/\gamma)^2} \tag{K.2.16}$$

Finally, we can require that the velocity is neither discontinous at $t = 0$ (although its graph will show a "kink" at this point), which determines the constant a:

$$a = Z\,\omega(m/\gamma) \tag{K.2.17}$$

We thus find—alas—that $a \neq 0$ and that the acceleration in the interval $t < 0$ is:

$$\dot{v}(t) = Z\,\omega(m/\gamma).e^{(m/\gamma)t} \tag{K.2.18}$$

In the older literature about Lorentz's electron model this behaviour is called "pre-acceleration": there is an acceleration of the droplet *before* there is any force acting on it.

This clearly violates causality. If we require that causality is preserved, then the conclusion must be that something is wrong with the equation of motion.

The characteristic duration of this pre-acceleration is extremely small, it is of the order of the length of time a ray of light needs to travel from one side of Lorentz's atom model to the other side. Even so, the anti-causality of the pre-acceleration cannot be tolerated, obviously.

All these problems mentioned in this section concerning transient motions are caused by the fact that the term $-\gamma\,\ddot{v}$ does *not* have the same character as a simple damping/dissipation term. Yaghjian in his earlier mentioned book proposes to repair the situation and to remove the anti-causality by adding a time function which multiplies the radiation resistance, so that the equation of motion becomes

$$F(t) = m\,\dot{v} - \eta(t).\gamma\,\ddot{v} \tag{K.2.19}$$

The function $\eta(t)$ must have the effect to suppress the radiation resistance during a brief period of time, just long enough to make transient motions satisfy causality. However, Yaghjian does not discuss what could be the physical origin of such a suppression of the radiation resistance force. Adding the extra function $\eta(t)$ is mainly a mathematical trick, and is from the physical point a "deus ex machina" device.

K.3 A Natural Way to Recover Causality During Transients

In the following it will be argued that the mentioned problems can easily be solved, both in a mathematically valid as well as in a physically plausible and entirely acceptable way. The riddle can be solved by realising that we did *not* make use of *all* the possible mathematical solutions of Maxwell's equations. If we do use the neglected solutions that are mathematically possible, this gives additional freedom to satisfy extra boundary conditions. The extra freedom can be used to satisfy the requirement of causality.

As mentioned, until now we did not use the complete solution of the Maxwell equations. The far field of the droplet was tacitly assumed to consist only of outgoing waves. It is usual to discard the mathematically equally valid solution of incoming waves, on the basis of a causality argument. This is called Sommerfeld's radiation condition, based on the argument that incoming waves must have been sent by a source outside the system considered. These incoming waves must be such that they exactly fit the motions of the droplet near the centre of the system, and they have to do so at an instant of time long after the waves have been emitted by the external radiation source. This situation seems to violate causality. For this reason incoming waves are usually not accepted as feasible components of the complete solution, except in the case of hard boundaries around the system in which case incoming waves are the result of reflections.

However, this causality argument is *not* applicable to transient motions, as will be shown now.

In the following Fig. K.3.1 two wave "blocks" are shown travelling in opposite directions. One may imagine these blocks to schematically depict—idealised—deformations in a long rope under tension. Alternatively, one could imagine them to symbolise the vertical displacement of the water surface in a large reservoir.

When these blocks meet each other we get superposition, which leads to the stepped contour shown by the dashed line.

At a certain moment the two blocks completely overlap, and their sum will be the rectangular block with the same length as the individual blocks, but with a double height.

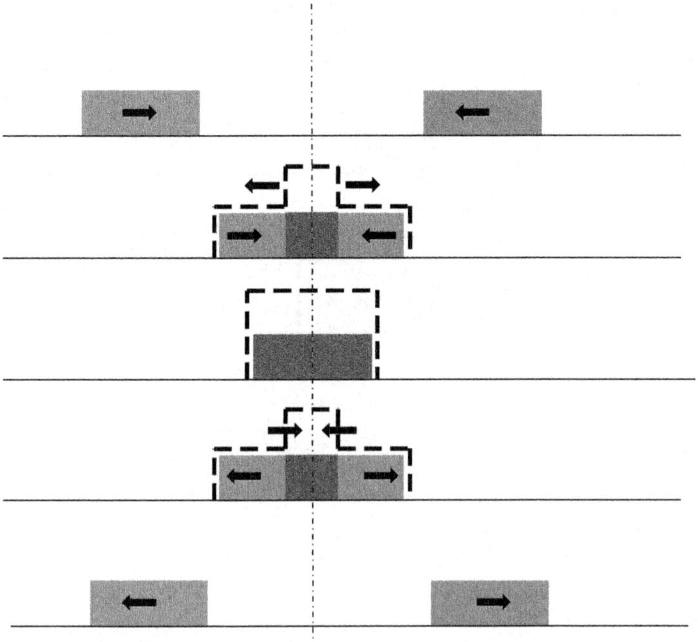

Fig. K.3.1 Two wave blocks moving in opposite directions

Whilst passing each other, the vertical displacement in the centre again deforms into a stepped contour, and we finally end up with two separate wave blocks moving away from each other.

The situation is symmetric w.r.t. the vertical dash-dot line. We might therefore consider only the plane to the right of the symmetry line, and consider the line of symmetry as a reflecting wall. Furthermore, consider only what happens from the time onwards when the two waves completely overlap. This is sketched in a separate Fig. (K.3.2).

What we see here is initially a displacement in the centre which could for instance be caused by filling a tub, and then suddenly removing the walls of the tub. The bulge of water existing at the very first instant after the removal of the tub walls, will subsequently deform to send out a wave to the right. Here we see a typical transient process which shows how the wave travelling to the right comes into existence and starts to move away, originating from a static displacement.

We also see here that mathematically the transient process can—for a brief moment—be described by two waves, travelling in opposite direction. At the very first moment, when the contour of the displacement is still given by a block (with double the height of the finally resulting wave"block", and half the length), the process is decribed by an "outgoing" and an "incoming" wave of equal strength.

As time proceeds, the transient process is still decribed by two waves moving in opposite direction, although the "incoming" wave block gradually becomes shorter:

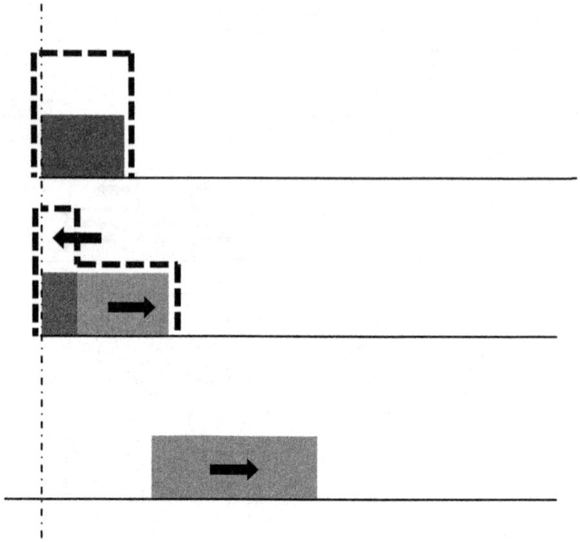

Fig. K.3.2 Starting-up of wave

more and more of the incoming wave has been reflected, to form a part of the outgoing wave.

The final situation is, that the "incoming" wave completely disappears, and only an "outgoing" wave is left: *after* the transient the situation is described solely by an outward moving wave.

The conclusion is that, in order to describe the transient process itself, we need waves in both directions. Now, in the next section a matched asymptotic analysis has been made where the field is represented by an incoming wave (in contrast to the earlier analysis which considered an outgoing wave). The incoming wave is a perfectly valid solution of the Maxwell equations, and the derivations follow the same procedure as earlier shown in the case of an outgoing wave.

What is found from this analysis is, that the self forces on a droplet are indeed affected by the propagation direction in the far field. The most important conclusion for the present discussion is, that the radiation resistance changes sign. Physically this is to be expected, because there is now energy supply from outside the system, instead of dissipation by the loss of energy escaping to infinity.

Remarkably, the electromagnetic mass is not affected at all by the direction of the waves in the far field.

If we have a field consisting of both an incoming as well as an outgoing wave, and if these waves have equal strength, then the equation of motion for the droplet contains the same mass effects as shown above in earlier sections of this text. These mass effects include the influence of pulsation of the droplet, which gives a variable mass in the equation for the translation. However, in the case of an equal strength of incoming and outgoing waves the radiation resistance term drops from the equation of motion. The combination of equally strong in- and outward moving waves is precisely the situation encountered during the very first instant of time of a transient process.

After this very first instant of time, the influence of the incoming wave gradually subsides, as is shown by the simplified sketch in Fig. K.3.2. We then see that the radiation resistance term in the equation of motion is gradually increasing from zero, and will have obtained its final value when the transient process has ended.

This is exactly the kind of behaviour which is expressed by Yaghjian's $\eta(t)$-multiplier. The $\eta(t)$ multiplier function is needed to solve the causality problem, and—as is shown by Yaghjian—is indeed successful in restoring causality. We now have found the physical basis for the occurrence of the multiplier function, and have shown that it directly follows from a mathematically sound solution of Maxwell's equations.

K.4 The "Radio Silence" Usually Maintained by Electrons in Motion Proposed Explanation

It was one of the postulates in Bohr's atom model that electrons "circling" the nucleus of an atom do not radiate away their energy, despite the fact that they constantly experience an acceleration whilst orbiting. This could not be explained by classical electrodynamics, which predicts that an accelerated charge always radiates and thus continuously loses its energy. To "explain" why electrons stay in their constant-energy orbit (until the sparse moments when the electron jumps to another orbit), Bohr's postulate was needed as a sort of additional law of physics, although it seems to infringe upon the laws of electrodynamics. The later quantum mechanics theory based on Schrödinger's equation neither gave a real "explanation" for this conflict with Maxwell's laws, it just gave a mathematically more refined and more generally valid description of the observed phenomena.

Now, what is being done in the present book is, to analyse strictly one-dimensional motion in an atomic sized oscillator. The present theory is thus unable to predict what happens in three-dimensional circumstances like an electron orbiting a nucleus. Nevertheless, in one-dimensional oscillators the same phenomenon of "radio silence" occurs, and one can suspect that the underlying reasons for this phenomenon will be the same as in the three-dimensional case.

In fact, in the case of one-dimensional motion we can now put together the pieces of the jig-saw puzzle, and we can propose an explanation for the phenomenon of radio silence. An explanation which does *not* imply a conflict with Maxwell's laws.

First we derived the equations of motion for an electron model which not only translates but also pulsates. The two equations of motion are mutually coupled due to the effect of pulsation. The coupling in fact gives rise to a set of non-linear differential equations, describing chaotic motion and a process of random energy exchanges between the two degrees of freedom.

It is only if we assume that the fluctuations of the translation velocity are small, that the equations of motion can be approximated by linear equations, having a solution of harmonic movement. This can happen, when during the chaotic motion there is momentarily much of the total energy accumulated in the elongation "spring", so that the ripples in the translation and pulsation become small. We then find that the motion almost attains the character of a well ordered harmonic motion.

During the greater part of the time we are dealing with highly non-linear equations of motion, which consequently predict a chaotic translation motion. One of the essences of chaotic motion is, that the time history of particular combinations of velocity, acceleration etc. is never repeated. At every instant of time the motion is different from any other instant.

We could therefore interpret the chaotic motion as *a continuous transient process*. As has been shown, during transients the radiation resistance is temporarily suppressed, and the loss of energy by radiation is disrupted. A *continuous* transient process may thus be expected to be described by equations of motion where the radiation resistance term is absent all the time (at least as long as the chaotic motion persists), with an attendant absence of radiation.

This is the process that is here proposed to explain why Bohr's "radio silence" had to be postulated. In fact, no separate postulate appears to be needed. The temporary absence of radiation and radiation resistance indeed follows as a consequence of Maxwell's laws themselves instead of being in conflict with them.

During the phases where a droplet displays chaotic motion it cannot be observed by any radiation signals sent by it. On the other hand, the chaotic motion and the accompanying energy fluctuations occurring in both the translation and pulsation may give rise to the tunneling effect. Observation of the electron is therefore not impossible, be it by means of the tunneling effect and not by radiation.

During the phases where a large part of the energy resides in the elongation "spring", so that the translation motion is going to resemble that of a linear oscillator with the possibility of harmonic motion, the suppression of radiation resistance is lost. During such a phase the electron may shed its energy in the form of radiation, so that it can be observed in this way.

K.5 The Radiation Field Associated with Incoming Waves

In the case of a field that is a combination of outgoing and incoming waves of equal strength, one may expect that no energy is lost from the system. Usually, this situation is encountered if there are hard barriers surrounding the system, preventing any energy loss, so that a field of standing waves is built up. In the absence of such barriers, it is usual to discard the occurrence of incoming waves (Sommerfeld's radiation condition) because it would then require an external source sending waves that exactly fit the motion of the charge in the centre at a future time. This would seem to contradict causality. However, a field of standing waves is *mathematically* an allowed solution of Maxwell's equations, even if there are no barriers. It depends on the specific circumstances of the case considered whether such a field is *physically* acceptable.

A physical discussion was already given in an earlier chapter. We restrict ourselves here to the mathematical form of a field of incoming waves.

The matter of incoming or outgoing waves did not enter the analysis of the near field. Retardation effects were taken into account by solving a Poisson equation instead of the quasi-stationary Laplace equation, no matter whether these retardations were positive or negative. In the matching condition, requiring that the outer expansion of the near field should equal the inner expansion of the far field, as a consequence the outer expansion of the near field is not affected by any assumptions about the in- or outgoing character of the waves. The outer expansion of the near field of the scalar potential was (Eq. (C.4.1)):

$$\begin{aligned}\left(\textit{outer expansion of } \phi_{near}.\frac{4\pi\varepsilon_0}{dq}\right) &= (\textit{quasi}-\textit{static poles}).\frac{4\pi\varepsilon_0}{dq}\\ &+ -\frac{1}{2}\frac{\dot{v}}{c^2}z\left[\frac{1}{\rho}-\frac{\partial}{\partial x}\left(\frac{x_0}{\rho}\right)-\frac{\partial}{\partial y}\left(\frac{y_0}{\rho}\right)-\frac{\partial}{\partial z}\left(\frac{z_0}{\rho}\right)\right]+\frac{1}{2}\frac{v^2}{c^2}z\frac{\partial}{\partial z}\left(\frac{1}{\rho}\right)\\ &+\left(\textit{outer expansion of } \Delta\phi.\frac{4\pi\varepsilon_0}{dq}\right)\end{aligned} \tag{K.5.1}$$

where the quasi-static poles, i.e. the outer expansion of the solutions of the Laplace equation, were given by (C.2.4) and (C.2.5):

$$\begin{aligned}(\textit{quasi}-\textit{static poles}).\frac{4\pi\varepsilon_0}{dq} &= \frac{1}{\rho}+\frac{1}{2}\frac{\partial^2}{\partial x^2}\left(\frac{x_0^2}{\rho}\right)+\frac{1}{2}\frac{\partial^2}{\partial y^2}\left(\frac{y_0^2}{\rho}\right)+\frac{1}{2}\frac{\partial^2}{\partial z^2}\left(\frac{z_0^2}{\rho}\right)\\ &+ -\frac{\partial}{\partial x}\left(\frac{x_0}{\rho}\right)-\frac{\partial}{\partial y}\left(\frac{y_0}{\rho}\right)-\frac{\partial}{\partial z}\left(\frac{z_0}{\rho}\right)+\frac{\partial^2}{\partial x.\partial y}\left(\frac{x_0.y_0}{\rho}\right)+\frac{\partial^2}{\partial x.\partial z}\left(\frac{x_0.z_0}{\rho}\right)+\frac{\partial^2}{\partial y.\partial z}\left(\frac{y_0.z_0}{\rho}\right)\end{aligned} \tag{K.5.2}$$

This expansion of the near field must be made to match with an inner expansion of the far field. Matching is achieved by choosing the far field such that it has the same type of poles. However, the far field explicitly contains retarded potentials. Now, if we consider *incoming* waves the far field will have the form

$$\begin{aligned}\phi_{far}.\frac{4\pi\varepsilon_0}{dq} &= \frac{1}{\rho}+\frac{1}{2}\frac{\partial^2}{\partial x^2}\left(\frac{x_0^2}{\rho}\right)+\frac{1}{2}\frac{\partial^2}{\partial y^2}\left(\frac{y_0^2}{\rho}\right)+\frac{1}{2}\frac{\partial^2}{\partial z^2}\left(\frac{z_0^2(t+\rho/c)}{\rho}\right)\\ &+ -\frac{\partial}{\partial x}\left(\frac{x_0}{\rho}\right)-\frac{\partial}{\partial y}\left(\frac{y_0}{\rho}\right)-\frac{\partial}{\partial z}\left(\frac{z_0(t+\rho/c)}{\rho}\right)+\frac{\partial^2}{\partial x.\partial y}\left(\frac{x_0.y_0}{\rho}\right)+\frac{\partial^2}{\partial x.\partial z}\left(\frac{x_0.z_0(t+\rho/c)}{\rho}\right)+\frac{\partial^2}{\partial y.\partial z}\left(\frac{y_0.z_0(t+\rho/c)}{\rho}\right)\\ &\text{up to } O(a/\lambda)^3\end{aligned} \tag{K.5.3}$$

Note that the retardation is now negative, for which reason the factors $(t - \rho/c)$ in Eq. (C.3.1) have been replaced by $(t + \rho/c)$ in the above Eq. (K.5.3).

The procedure to determine the inner expansion of the far field is the same as in section C.3. It will here be repeated, with due attention to the sign of the terms.

$$\begin{aligned}(\phi_{far})_{z-dipole}\frac{4\pi\varepsilon_0}{dq} &= -\frac{\partial}{\partial z}\left[\frac{z_0(t+\rho/c)}{\rho}\right]\\ &= \left[-\frac{1}{c}\frac{v(t+\rho/c)}{\rho} + \frac{z_0(t+\rho/c)}{\rho^2}\right].cos\varphi \xrightarrow{inner\ expansion}\\ &\xrightarrow{inner\ expansion}\left[\frac{z_0(t)}{\rho^2} - \frac{1}{2}\frac{\dot{v}(t)}{c^2} - \frac{1}{3}\rho\frac{\ddot{v}(t)}{c^3}\right]cos\varphi =\\ &= -\frac{\partial}{\partial z}\left[\frac{z_0(t)}{\rho}\right] - \frac{1}{2}\frac{z}{\rho}\frac{\dot{v}(t)}{c^2} - \frac{1}{3}z\frac{\ddot{v}(t)}{c^3}\end{aligned} \tag{K.5.4}$$

Comparing with Eq. (C.3.2) the only difference appears in the sign of the term with $\ddot{v}$.

Again the time functions $x_0.z_0(t-\rho/c)$, $y_0.z_0(t-\rho/c)\, and\, z_0^2(t-\rho/c)$ will be symbolized by the general notation $g(t-\rho/c)$:

$$\begin{aligned}(\phi_{far})_{xz-quadrupole}.\frac{4\pi\varepsilon_0}{dq} &= \frac{\partial^2}{\partial x.\partial z}\left[\frac{g(t+\rho/c)}{\rho}\right] = -z.\frac{\partial}{\partial x}\left[\frac{g}{\rho^3} - \frac{1}{\rho^2}\frac{g'}{c}\right]\\ &= \frac{xz}{\rho^5}\left[3.g(t+\rho/c) - 3.\left(\frac{\rho}{c}\right).g'(t+\rho/c) + \left(\frac{\rho}{c}\right)^2.g''(t+\rho/c)\right] \rightarrow\\ &\xrightarrow{inner\ expansion}\\ &\rightarrow 3\frac{xz}{\rho^5}\left[g(t) + \left(\frac{\rho}{c}\right).g'(t) + \frac{1}{2}\left(\frac{\rho}{c}\right)^2.g''(t) + \frac{1}{6}\left(\frac{\rho}{c}\right)^3.g'''(t)\right]\\ &+ -3\frac{xz}{\rho^5}\left(\frac{\rho}{c}\right).\left[g'(t) + \left(\frac{\rho}{c}\right).g''(t) + \frac{1}{2}\left(\frac{\rho}{c}\right)^2.g'''(t)\right]\\ &+ \frac{xz}{\rho^5}\left(\frac{\rho}{c}\right)^2.\left[g''(t) + \left(\frac{\rho}{c}\right).g'''(t)\right]\\ &= \frac{xz}{\rho^5}\left[3.g(t) - \frac{1}{2}\left(\frac{\rho}{c}\right)^2.g''(t)\right]\\ &= 3.g(t)\frac{xz}{\rho^5} - \frac{1}{2}\frac{g''(t)}{c^2}\frac{xz}{\rho^3}\\ &= \frac{\partial^2}{\partial x.\partial z}\left(\frac{g(t)}{\rho}\right) + \frac{1}{2}.z.\frac{\partial}{\partial x}\left(\frac{g''(t)/c^2}{\rho}\right)\end{aligned} \tag{K.5.5}$$

Substituting $g = x_0.z_0$:

$$(\phi_{far})_{xz-quadrupole}.\frac{4\pi\varepsilon_0}{dq} = \frac{\partial^2}{\partial x.\partial z}\left[\frac{x_0.z_0(t+\rho/c)}{\rho}\right] \xrightarrow{innerexpansion} \frac{\partial^2}{\partial x.\partial z}\left[\frac{x_0.z_0(t)}{\rho}\right] + \frac{z}{2}\frac{\dot{v}}{c^2}\frac{\partial}{\partial x}\left[\frac{x_0}{\rho}\right] \tag{K.5.6}$$

A cyclic change of coordinates shows:

$$(\phi_{far})_{yz-quadrupole}\cdot\frac{4\pi\varepsilon_0}{dq}=\frac{\partial^2}{\partial y.\partial z}\left[\frac{y_0.z_0(t+\rho/c)}{\rho}\right]\xrightarrow{inner\,expansion}\frac{\partial^2}{\partial y.\partial z}\left[\frac{y_0.z_0(t)}{\rho}\right]+\frac{z}{2}\frac{\dot{v}}{c^2}\frac{\partial}{\partial y}\left[\frac{y_0}{\rho}\right] \tag{K.5.7}$$

In the case of the term in Eq. (K.5.3) with the ZZ-quadrupole: $(\phi_{far})_{zz-quadrupole}\frac{4\pi\varepsilon_0}{dq}=\frac{1}{2}\frac{\partial^2}{\partial z^2}\left[\frac{z_0^2(t-\rho/c)}{\rho}\right]$ there is an additional term compared with the other quadrupoles (also note the factor $1/2$ multiplying this term in the complete far field (K.5.3)):

$$\begin{aligned}2.(\phi_{far})_{zz-quadrupole}\cdot\frac{4\pi\varepsilon_0}{dq}&=z.\frac{\partial}{\partial z}\left[-\frac{g}{\rho^3}+\frac{1}{\rho^2}\frac{g'}{c}\right]+\left[-\frac{g}{\rho^3}+\frac{1}{\rho^2}\frac{g'}{c}\right]\rightarrow\\ &\xrightarrow{inner\,expansion}\\ &\rightarrow\frac{z^2}{\rho^5}\left[3.g(t)-\frac{1}{2}\left(\frac{\rho}{c}\right)^2.g''(t)\right]+\frac{1}{\rho^3}\left[-g(t)+\frac{1}{2}\left(\frac{\rho}{c}\right)^2.g''(t)+\frac{1}{3}\left(\frac{\rho}{c}\right)^3.g'''(t)\right]\\ &=\frac{\partial^2}{\partial z^2}\left[\frac{g(t)}{\rho}\right]+\frac{1}{2}.z.\frac{\partial}{\partial z}\left[\frac{g''(t)/c^2}{\rho}\right]+\frac{1}{2}\left[\frac{g''(t)/c^2}{\rho}\right]+\frac{1}{3}\frac{g'''(t)}{c^3}\end{aligned} \tag{K.5.8}$$

After substitution of $g=z_0^2$:

$$(\phi_{far})_{zz-quadrupole}\cdot\frac{4\pi\varepsilon_0}{dq}\rightarrow\frac{1}{2}\frac{\partial^2}{\partial z^2}\left[\frac{z_0^2(t)}{\rho}\right]+\frac{z}{2}\frac{\partial}{\partial z}\left[\frac{(v^2/c^2+z_0.\dot{v}/c^2)}{\rho}\right]+\frac{1}{2}\left[\frac{v^2/c^2+z_0.\dot{v}/c^2}{\rho}\right]+\frac{1}{3}\frac{3v.\dot{v}+z_0.\ddot{v}}{c^3} \tag{K.5.9}$$

The complete inner expansion of the far field is, collecting all the terms:

$$\begin{aligned}&\left(inner\,expansion\,of\,\phi_{far.}\cdot\frac{4\pi\varepsilon_0}{dq}\right)=(quasi-static\,poles)\\ &+-\frac{1}{2}\frac{z}{\rho}\frac{\dot{v}(t)}{c^2}-\frac{1}{3}z\frac{\ddot{v}(t)}{c^3}+\frac{z}{2}\frac{\dot{v}}{c^2}\frac{\partial}{\partial x}\left[\frac{x_0}{\rho}\right]+\frac{z}{2}\frac{\dot{v}}{c^2}\frac{\partial}{\partial y}\left[\frac{y_0}{\rho}\right]\\ &+\frac{z}{2}\frac{\partial}{\partial z}\left[\frac{(v^2/c^2+z_0.\dot{v}/c^2)}{\rho}\right]+\frac{1}{2}\left[\frac{v^2/c^2+z_0.\dot{v}/c^2}{\rho}\right]+\frac{1}{3}\frac{3v.\dot{v}+z_0.\ddot{v}}{c^3}\end{aligned} \tag{K.5.10}$$

The matching condition yields the result

$$(\Delta\phi)\frac{4\pi\varepsilon_0}{dq}\xrightarrow{outer\,expansion}-\frac{1}{3}\frac{\ddot{v}(t)}{c^3}.z+\frac{1}{2}\frac{v^2/c^2+z_0.\dot{v}/c^2}{\rho}+\frac{1}{3}\frac{3v.\dot{v}+z_0.\ddot{v}}{c^3} \tag{K.5.11}$$

so that, in order to achieve the matching, in the near field a solution of the Laplace equation must be added of the form

$$(\Delta\phi)_{near}\frac{4\pi\varepsilon_0}{dq} = -\frac{1}{3}\frac{\ddot{v}(t)}{c^3}.z + \frac{1}{2}\frac{v^2/c^2 + z_0.\dot{v}/c^2}{r} + \frac{1}{3}\frac{3v.\dot{v} + z_0.\ddot{v}}{c^3} \tag{K.5.12}$$

We may furthermore add, to enhance the symmetry of the final expression, another term of irrelevant order

$$-\frac{1}{2}\frac{v^2}{c^2}.z_0.\frac{\partial}{\partial z}\left(\frac{1}{r}\right) \xrightarrow{outer\ expansion} O(a/\lambda)^4 \tag{K.5.13}$$

The near field of the scalar potential is given by

$$\phi_{near}\frac{4\pi\varepsilon_0}{dq} = \frac{1}{r} - \frac{z}{2}\left[\frac{\dot{v}}{c^2}\frac{1}{r} - \frac{v^2}{c^2}\frac{\partial}{\partial z}\left(\frac{1}{r}\right)\right] + \Delta\phi \tag{K.5.14}$$

which finally results in the following expression for the near field of the scalar potential:

$$(\phi_{near}) = \frac{dq}{4\pi\varepsilon_0}\left[\left(1 + \frac{1}{2}\frac{v^2}{c^2}\right).\frac{1}{r} + \frac{1}{2}\frac{v^2}{c^2}(z - z_0)\frac{\partial}{\partial z}\left(\frac{1}{r}\right) - \frac{1}{2}\frac{\dot{v}}{c^2}\frac{z - z_0}{r} - \frac{1}{3}\frac{\ddot{v}}{c^3}(z - z_0) + \frac{v.\dot{v}}{c^3}\right] \tag{K.5.15}$$

The component a_z of the vectorpotential is similarly derived. Its near field is given by Eq. (C.5.10):

$$(a_z)_{near}\frac{4\pi\varepsilon_0}{dq}c^2 = v\frac{1}{r} + \frac{\ddot{v}}{c^2}\frac{r}{2} - 3\frac{v.\dot{v}}{c^2}\frac{z}{2}\frac{1}{r} + \frac{v^3}{c^2}\frac{z}{2}\frac{\partial}{\partial z}\left(\frac{1}{r}\right) + \Delta(a_z)_{near} \tag{K.5.16}$$

The outer expansion of the near field is

$$\begin{aligned}
&(a_z)_{near}\frac{4\pi\varepsilon_0}{dq}c^2 \xrightarrow{outer\ expansion} \\
&\xrightarrow{outer\ expansion} v\left[\frac{1}{\rho} - \frac{\partial}{\partial x}\left(\frac{x_0}{\rho}\right) - \frac{\partial}{\partial y}\left(\frac{y_0}{\rho}\right) - \frac{\partial}{\partial z}\left(\frac{z_0}{\rho}\right)\right] \\
&+ \frac{1}{2}\frac{\ddot{v}}{c^2}\left[\rho - \frac{x.x_0}{\rho} - \frac{y.y_0}{\rho} - \frac{z.z_0}{\rho}\right] - \frac{3}{2}\frac{v.\dot{v}}{c^2}\frac{z}{\rho} \\
&+ (outer\ expansion\ of\ \Delta(a_z)_{near}\frac{4\pi\varepsilon_0}{dq}c^2)
\end{aligned} \tag{K.5.17}$$

The far field consists of the same poles as occur in the near field expansion, in the case of the incoming wave with a negative retardation:

$$
\begin{aligned}
(a_z)_{far}\frac{4\pi\varepsilon_0}{dq}c^2 &= \frac{v(t+\rho/c)}{\rho} \\
&+ -\frac{\partial}{\partial x}\left(\frac{x_0.v(t+\rho/c)}{\rho}\right) - \frac{\partial}{\partial y}\left(\frac{y_0.v(t+\rho/c)}{\rho}\right) - \frac{\partial}{\partial z}\left(\frac{z_0.v(t+\rho/c)}{\rho}\right) \\
&\xrightarrow{\text{inner expansion}} \frac{v(t)}{\rho} + \frac{\dot v}{c} + \frac{1}{2}\frac{\ddot v}{c^2}\rho - \frac{\partial}{\partial x}\left(\frac{x_0.v(t)}{\rho}\right) - \frac{1}{2}\frac{x_0.\ddot v(t)}{c^2}\frac{x}{\rho} \\
&+ -\frac{\partial}{\partial y}\left(\frac{y_0.v(t)}{\rho}\right) - \frac{1}{2}\frac{y_0.\ddot v(t)}{c^2}\frac{y}{\rho} - \frac{\partial}{\partial z}\left(\frac{z_0.v(t)}{\rho}\right) - \frac{1}{2}\frac{3v\dot v(t) + z_0.\ddot v(t)}{c^2}\frac{z}{\rho}
\end{aligned}
\tag{K.5.18}
$$

Matching gives:

$$
(outer\ expansion\ of\ \Delta(a_z)_{near}\frac{4\pi\varepsilon_0}{dq}c^2) = \frac{\dot v}{c} \tag{K.5.19}
$$

and

$$
\Delta(a_z)_{near}\frac{4\pi\varepsilon_0}{dq}c^2 = \frac{\dot v}{c} \tag{K.5.20}
$$

The complete near field is finally, after adding some terms of irrelevant order to enhance the symmetry:

$$
(a_z)_{near} = \frac{dq}{4\pi\varepsilon_0}\left[\frac{v}{c^2}\frac{1}{r} + \frac{\ddot v}{c^4}\frac{r}{2} - 3\frac{v.\dot v}{c^4}\frac{z-z_0}{2}\frac{1}{r} + \frac{v^3}{c^4}\frac{z-z_0}{2}\frac{\partial}{\partial z}\left(\frac{1}{r}\right) + \frac{\dot v}{c^3}\right] \tag{K.5.21}
$$

K.6 Self Force on a Droplet in the Case of an Incoming Wave

The Z-force on a unit charge situated in the near field is:

$$
\begin{aligned}
f_z &= -\frac{\partial \phi_{near}}{\partial z} - \frac{\partial (a_z)_{near}}{\partial t} \\
&= -\frac{dq}{4\pi\varepsilon_0}\left[\frac{\partial}{\partial z}\left(\frac{1}{r}\right) - \frac{1}{3}\frac{\ddot v}{c^3}\right] - \frac{dq}{4\pi\varepsilon_0}\left[\frac{\dot v}{c^2}\frac{1}{r} - \frac{v^2}{c^2}\frac{\partial}{\partial z}\left(\frac{1}{r}\right) + \frac{\ddot v}{c^3}\right] \\
&= -\frac{dq}{4\pi\varepsilon_0}\left[\left(1-\frac{v^2}{c^2}\right)\frac{\partial}{\partial z}\left(\frac{1}{r}\right) + \frac{\dot v}{c^2}\frac{1}{r} + \frac{2}{3}\frac{\ddot v}{c^3}\right]
\end{aligned}
\tag{K.6.1}
$$

where the slender body approximations (D.2.4) and (D.2.5) have been used, and terms of an irrelevant order have been omitted.

Comparing Eq. (D.2.6) for the force on a unit charge in the case of an outgoing wave with Eq. (K.6.1) derived for the incoming wave, it appears that the sign of the term $\frac{2}{3}\frac{\ddot{v}}{c^3}$ is the only difference.

A double integration of the force f_z over all the charge elements is performed in exactly the same way as shown before, and need not be repeated. Taking into account the change of sign in one of the terms leads immediately to the self force:

$$F_z(t) = -2\frac{U_{es}}{c^2}\dot{v}_m + 2\,v_m(\dot{s}/s)\frac{U_{es}}{c^2} - \frac{2}{3}\frac{q^2}{4\pi\varepsilon_0}\frac{\ddot{v}_m}{c^3} \quad (incoming\,wave) \qquad \text{(K.6.2)}$$

to be compared with

$$F_z(t) = -2\frac{U_{es}}{c^2}\dot{v}_m + 2\,v_m(\dot{s}/s)\frac{U_{es}}{c^2} + \frac{2}{3}\frac{q^2}{4\pi\varepsilon_0}\frac{\ddot{v}_m}{c^3} \quad (outgoing\,wave) \qquad \text{(K.6.3)}$$

From this result it appears that taking the sum of half the far field of an outgoing wave and half that of an incoming wave leads to a self force without radiation resistance:

$$F_z(t) = -2\frac{U_{es}}{c^2}\dot{v}_m + 2\,v_m(\dot{s}/s)\frac{U_{es}}{c^2} \quad (1/2\,incoming + 1/2\,outgoing\,wave) \qquad \text{(K.6.4)}$$

This is of course to be expected since we expect on physical grounds that the resulting field of standing waves does not transport any energy to infinity.

K.7 The Form of the Radiation Field When Radiation Resistance Is Absent

It is instructive to inspect in more detail the far radiation field in the case of absence of radiation resistance on the droplet, i.e. when there is a combination of incoming and outgoing waves of equal strength.

Using Eq. (J.1.8) the scalar potential of such a field is, retaining only the terms of order $O(1/\rho)$:

$$\begin{aligned}\Phi_{far}(\rho,\varphi,t) = {} & \frac{1}{2}\frac{q}{4\pi\varepsilon_0}\left[\frac{1}{\rho} + \frac{v_m^2 + z_m\dot{v}_m}{c^2}\frac{cos^2\varphi}{\rho}\right]_{t-\rho/c} \\ & + \frac{1}{2}\frac{q}{4\pi\varepsilon_0}\left[\frac{1}{\rho} + \frac{v_m^2 + z_m\dot{v}_m}{c^2}\frac{cos^2\varphi}{\rho}\right]_{t+\rho/c}\end{aligned} \qquad \text{(K.7.1)}$$

and according to Eq. (J.2.3):

$$A_{z_{far}}(\rho,\varphi,t)=\frac{1}{2}\frac{q}{4\pi\varepsilon_0}\left[\frac{v_m}{c^2}\frac{1}{\rho}\right]_{t-\rho/c}+\frac{1}{2}\frac{q}{4\pi\varepsilon_0}\left[\frac{v_m}{c^2}\frac{1}{\rho}\right]_{t+\rho/c} \tag{K.7.2}$$

The outgoing energy flux is given by the radial component of the Poynting vector (J.4.3):

$$S_\rho=\varepsilon_0 c^2.B_\chi.E_\varphi=\varepsilon_0 c^2.\left[\frac{\partial A_z}{\partial\rho}sin\varphi+\frac{\partial A_z}{\partial\varphi}\frac{cos\varphi}{\rho}\right].\left[\frac{1}{\rho}\frac{\partial\Phi}{\partial\varphi}-\frac{\partial A_z}{\partial t}sin\varphi\right] \tag{K.7.3}$$

Substituting the scalar and vector potential, and retaining only the components able to contribute to the energy flux at large distances, i.e. omitting the terms with a contribution to S_ϱ of order $O(1/\rho^3)$ and smaller:

$$\begin{aligned}S_\rho&=\frac{q^2}{4\pi\varepsilon_0}\frac{1}{4\pi}c^2\left[\frac{\partial A_z}{\partial\rho}sin\varphi\right].\left[-\frac{\partial A_z}{\partial t}sin\varphi\right]\\&=\frac{q^2}{4\pi\varepsilon_0}\frac{1}{4\pi}c^2\left[-\frac{1}{2}\frac{\dot v_m(t-\rho/c)}{c^3}\frac{sin\varphi}{\rho}+\frac{1}{2}\frac{\dot v_m(t+\rho/c)}{c^3}\frac{sin\varphi}{\rho}\right]\\&\quad.\left[-\frac{1}{2}\frac{\dot v_m(t-\rho/c)}{c^2}\frac{sin\varphi}{\rho}-\frac{1}{2}\frac{\dot v_m(t+\rho/c)}{c^2}\frac{sin\varphi}{\rho}\right]\end{aligned} \tag{K.7.4}$$

or

$$S_\rho=\frac{1}{4}\frac{q^2}{4\pi\varepsilon_0}\frac{1}{c^3}\frac{sin^2\varphi}{4\pi\rho^2}\left[\dot v_m^2(t-\rho/c)-\dot v_m^2(t+\rho/c)\right] \tag{K.7.5}$$

The radial radiation vanishes if the acceleration of the emitting droplet is constant in time. This is a situation that cannot exist indefinitely, and is therefore not interesting for a practical case.

In the case of a steady harmonic motion of the droplet, where

$$\dot v_m=K.cos(\omega.t) \tag{K.7.6}$$

the term between the square brackets in Eq. (K.7.5) becomes

$$\begin{aligned}\dot v_m^2(t-\rho/c)-\dot v_m^2(t+\rho/c)&=K^2\cos^2\omega(t-\rho/c)-K^2\cos^2\omega(t+\rho/c)\\&=K^2.\ \sin(2\,\omega\,t).sin\left(2\,\omega\frac{\rho}{c}\right)\end{aligned} \tag{K.7.7}$$

Spheres around the droplet with the radii $\rho=\frac{1}{2}\pi\frac{c}{\omega},\ \rho=\pi\frac{c}{\omega},\ \rho=\frac{3}{2}\pi\frac{c}{\omega},\ \ldots\ldots$ then act as "barriers" through which the net energy flux is zero. Again, the total energy lost by radiation is thus zero. The field has the character of standing waves with radial nodes, and with non-zero flux of energy in both directions in between these nodes.

Finally, there is the possibility that the accelerations are stochastic. In that case there is instantaneously a difference between $\dot{v}_m^2(t-\rho/c)$ and $\dot{v}_m^2(t+\rho/c)$, but averaged over a long time the outflow of energy will be zero (on condition that the stochastic motion is steady).

References

Alonso M., Finn E.J.: Fundamental University Physics, 1972, Addison-Wesley

Anderson R., Brady R.: Why bouncing droplets are a pretty good model of quantum mechanics, arXiv:1401.4356v1 [quant-ph] 16 Jan 2014

Brady R., Anderson R.: Maxwell's fluid model of magnetism, arXiv:1502.05926v1 feb 2015

Couder Y., Protière S., Fort E., Badouad A.: Dynamical phenomena: walking and orbiting droplets, Nature 437 (7056): 208–208, 2005

Den Hartog J.P.: Mechanical Vibrations, 1956, Dover Publications

Feynman R. P., Leighton R.B., Sands M.: The Feynman Lectures on Physics, 1963, Addison-Wesley

Feynman R. P.: QED, The strange theory of light and matter, 1985, Princeton University Press

Kelley D.H., Triana S.A., Zimmerman D.S., Tilgner A., Lathrop D.P.: Inertial waves driven by differential rotation in a planetary geometry. Geophysical and Astrophysical Fluid Dynamics, vol. 101, nos. 5–6, Oct–Nov 2007, 460–487

Kellogg O. D.: Foundations of Potential Theory, Frederick Ungar Publishing Company, 1926

Moláček J., Bush J.W.M.: Drops bouncing on a vibrating bath, J.Fluid Mechanics vol. 727, July 2013, pp. 582–611

Morse P.M., Feshbach H.: Methods of theoretical physics, 1953, McGraw-Hill

Pais A.: Inward Bound, of matter and forces in the physical world, Oxford University Press 1986

Schiff L. I.: Quantum Mechanics, McGraw-Hill Book Company, 3rd edition 1968

Thompson J.M.T., Stewart H.B.: Nonlinear dynamics and chaos, geometrical methods for engineers and scientists, 1986, Wiley

Vervoort L.: Comment on "Loophole-free Bell inequality violation using electron spins separated by 1.3 km", in arXiv 1602. 06870, Feb. 2016

Wolchover N.: Have we been interpreting quantum mechanics wrong this whole time? Quanta Magazine 06.30.14 (SimonsFoundation.org)

Yaghjian A. D.: Relativistic Dynamics of a Charged Sphere, updating the Lorentz-Abraham model, Lecture Notes in Physics Monograph 686. Springer, New York 2006.

T. van Holten, *The Atomic World Spooky? It Ain't Necessarily So!*
DOI 10.2991/978-94-6239-234-2

Index

T. van Holten, *The Atomic World Spooky? It Ain't Necessarily So!*
DOI 10.2991/978-94-6239-234-2

CPSIA information can be obtained
at www.ICGtesting.com
Printed in the USA
LVOW02*1617250617
539319LV00001B/73/P

9 789462 392335